Metal Nanocomposites in Nanotherapeutics for Oxidative Stress-Induced Metabolic Disorders

This book highlights the role and mechanism of different metal nanocomposites toward oxidative stress-induced metabolic disorders including metabolic pathways affected by oxidative stress and related pathophysiology. The book includes an illustrative discussion about the methods of synthesis, characterization, and biomedical applications of metal nanocomposites. It focuses on the therapeutic approaches for metabolic disorders due to oxidative stress by nano delivery systems. Moreover, the book includes chapters on nanotherapeutic approaches toward different diseases, including diabetes mellitus, obesity, cardiovascular disorders, cancers, and neurodegenerative diseases such as Alzheimer's disease and Parkinson's disease. This book is aimed at researchers and graduate students in nanocomposites, nano delivery systems, and bioengineering.

FEATURES

- Discusses nanocomposites in the field of therapy for diabetes, obesity, cardiovascular disorders, neurodegenerative diseases, and cancers
- Details the pathophysiology of oxidative stress-induced metabolic disorder
- Explains mechanisms of the antioxidant potential of metal nanocomposites
- Discusses pathways to elucidate the therapeutic activity
- Reviews specific and precise applications of metal nanocomposites against lifestyle-induced disorders

Emerging Materials and Technologies

Series Editor: Boris I. Kharissov

The *Emerging Materials and Technologies* series is devoted to highlighting publications centered on emerging advanced materials and novel technologies. Attention is paid to those newly discovered or applied materials with potential to solve pressing societal problems and improve quality of life, corresponding to environmental protection, medicine, communications, energy, transportation, advanced manufacturing, and related areas.

The series takes into account that, under present strong demands for energy, material, and cost savings, as well as heavy contamination problems and worldwide pandemic conditions, the area of emerging materials and related scalable technologies is a highly interdisciplinary field, with the need for researchers, professionals, and academics across the spectrum of engineering and technological disciplines. The main objective of this book series is to attract more attention to these materials and technologies and invite conversation among the international R&D community.

Nanomaterials for Energy Applications
Edited by L. Syam Sundar, Shaik Feroz, and Faramarz Djavanroodi

Wastewater Treatment with the Fenton Process: Principles and Applications
Dominika Bury, Piotr Marcinowski, Jan Bogacki, Michal Jakubczak, and Agnieszka Jastrzebska

Mechanical Behavior of Advanced Materials: Modeling and Simulation
Edited by Jia Li and Qihong Fang

Shape Memory Polymer Composites: Characterization and Modeling
Nilesh Tiwari and Kanif M. Markad

Impedance Spectroscopy and its Application in Biological Detection
Edited by Geeta Bhatt, Manoj Bhatt and Shantanu Bhattacharya

Nanofillers for Sustainable Applications
Edited by N.M Nurazzi, E. Bayraktar, M.N.F. Norrrahim, H.A. Aisyah, N. Abdullah, and M.R.M. Asyraf

Metal Nanocomposites in Nanotherapeutics for Oxidative Stress-Induced Metabolic Disorders
Edited by Anindita Behera

For more information about this series, please visit: www.routledge.com/Emerging-Materials-and-Technologies/book-series/CRCEMT

Metal Nanocomposites in Nanotherapeutics for Oxidative Stress-Induced Metabolic Disorders

Edited by
Anindita Behera

CRC Press
Taylor & Francis Group
Boca Raton London New York

CRC Press is an imprint of the
Taylor & Francis Group, an **informa** business

Designed cover image: © Shutterstock

First edition published 2024
by CRC Press
2385 NW Executive Center Drive, Suite 320, Boca Raton FL 33431

and by CRC Press
4 Park Square, Milton Park, Abingdon, Oxon, OX14 4RN

CRC Press is an imprint of Taylor & Francis Group, LLC

ISBN: 978-1-032-53259-2 (hbk)
ISBN: 978-1-032-62111-1 (pbk)
ISBN: 978-1-032-62113-5 (ebk)

DOI: 10.1201/9781032621135

Typeset in Sabon
by Apex CoVantage, LLC

Contents

3   The general concept of metal nanocomposites in nanotherapeutics for oxidative stress-induced metabolic disorders

57

JIGAR VYAS AND NENSI RAYTTHATHA

4   Nanotherapeutics for leptin resistance and obesity using metal nanocomposites

82

IPSA PADHY, BISWAJIT BANERJEE, AND TRIPTI SHARMA

7 Nanotherapeutics for diabetic nephropathy using metal nanocomposites

154

JHUMA SAMANTA, ASHOK BEHERA, AND AMAN CHAUDHARY

8 Nanotherapeutics for diabetic cardiomyopathy using metal nanocomposites

168

JAYARAMAN RAJANGAM, NARAHARI N. PALEI, G.S.N. KOTESWARA RAO,
R. PRAKASH, SAGILI VARALAKSHMI, HAJA BAVA BAKRUDEEN,
AND SHRUTI SRIVASTAVA

SOPAN NANGARE, JIDNYASA PANTWALAWALKAR, NAMDEO JADHAV,
ZAMIR KHAN, GANESH PATIL, MAHENDRA MAHAJAN, RUTUJA CHOUGALE,
AND PRAVIN PATIL

P. K. SAHU, P. S. KUMAR, AND S. K. PRUSTY

11 Nanotherapeutics for endometrial cancer using metal nanocomposites   228

SHRADDHA M. GUPTA, DINESH D. RISHIPATHAK, AND SIDDHARTH SINGH

12 Nanotherapeutics for colorectal cancer using metal nanocomposites   249

YUVRAJ PATIL AND SHVETANK BHATT

16 Nanotherapeutics for reversal of multidrug resistance in chemotherapy with metal nanocomposites

DEVESH U. KAPOOR, DIMPY RANI, MADAN MOHAN GUPTA, AND DEEPAK SHARMA

17 Nanotherapeutics for Alzheimer's disease using metal nanocomposites

NITIN VERMA, SHIWALI SHARMA, NIKITA THAKUR, NARINDERPAL KAUR, AND KAMAL DUA

18   Nanotherapeutics for Parkinson's disease using metal nanocomposites   392

RISHIKA DHAPOLA, PRAJJWAL SHARMA, SNEHA KUMARI, PUSHANK NAGAR,
BIKASH MEDHI, AND DIBBANTI HARIKRISHNA REDDY

Preface

An imbalance in the production and degradation of free radicals or superoxides causes oxidative stress. Nowadays, lifestyle, lack of exercise, and junk food are the main reasons behind oxidative stress. This excess of reactive oxygen species in our body is responsible for various metabolic disorders, such as cardiovascular diseases, diabetes mellitus, neurodegenerative diseases, cancer, and so on. The oxidative stress disturbs the different metabolic processes in our body; alters the genes; causes DNA damage or mutation; and affects different signaling pathways which are the primary pathophysiology behind oxidative stress-induced metabolic disorders. Hence, it is one of the approaches for the treatment of these metabolic disorders with antioxidants. Metal nanoparticles possess superior optical, magnetic, and electrical properties, and their small size, shape, high surface area, and tunable physicochemical properties make the metal nanoparticles or nanocomposites superior to their bulk metals. They can easily donate electrons or hydrogen to the ROS and are reported as efficient antioxidants. Nanocomposites may have a metal matrix, polymeric matrix, or ceramic matrix. In this book, the metal matrix nanocomposites and their applicability in the therapy of many metabolic disorders are discussed.

This book is published in the book series Emerging Materials and Technologies. The first chapter entails a close understanding of oxidative stress and the pathology of metabolic disorders due to oxidative stress. The second chapter briefly discusses the synthesis, characterization, and biomedical applications of metal nanocomposites. The third chapter discusses the application of metal nanocomposites in the therapy of oxidative stress-induced metabolic disorders. The fourth chapter reports the leptin resistance and occurrence of obesity and how the metal nanocomposites can elicit therapeutic efficacy against obesity. Chapters 5–9 deal with the applicability of metal nanocomposites against diabetes mellitus and the associated microvascular and macrovascular complications. The tenth chapter delves into the role of metal nanocomposites in amelioration of the cardiovascular complications like dyslipidemia.

Cancer is the most prevalent disease which affects the major mortality rate worldwide, and oxidative stress can disturb the signaling pathways, cell cycle and the overexpression of receptors on the cancer cells for uncontrolled growth, and proliferation of cancer cells. Antioxidants like metal nanocomposites can exhibit anticancer activity by active or passive targeting of the cancer cells. Chapters 11–15 discuss the different types of cancers treated by metal nanocomposites. The sixteenth chapter explores the application of metal nanocomposites in the reversal of multidrug resistance in cancer chemotherapy. Neurodegeneration is another metabolic disorder which involves a cascade of neuroinflammation, loss of plasticity, and finally neurodegeneration. Alzheimer's disease and Parkinson's disease are the most prevalent neurodegenerative disorders due to oxidative stress. Hence, chapters 17 and 18 discuss the antioxidant and therapeutic activity of metal nanocomposites against Alzheimer's disease and Parkinson's disease.

The editor is thankful to all the contributors for their contribution to drafting the chapters. The book would not have been possible without the assistance of friends and colleagues. Finally, I would like to express my gratitude to the CRC Press editors for their meticulous work on the publication.

Anindita Behera
Siksha 'O' Anusandhan Deemed to be University

About the editor

Anindita Behera is Associate Professor, Department of Pharmaceutical Analysis, Siksha 'O' Anusandhan Deemed to be University. She did her master's from Pune University, India, and PhD from Siksha 'O' Anusandhan Deemed to be University. She has developed new analytical methods for drug formulations used in the treatment of AIDS. Currently, she is involved in the green synthesis of nanomaterials targeting different life-risk diseases. She has more than 15 years of experience in research and academics. She has published several research and review articles in peer-reviewed journals. She has also contributed many chapters to books published in Elsevier, Springer, and Taylor & Francis Group. She has guided two PhD scholars; 11 postgraduate scholars, and 30 undergraduate students.

Contributors

Haja Bava Bakrudeen
AIMST University
Jalan Bedong—Semeling, Bedong, Kedah,
  Malaysia

Biswajit Banerjee
School of Pharmaceutical Sciences
Siksha 'O'Anusandhan (Deemed to be
  University)
Bhubaneswar, Odisha, India

Ashok Behera
DIT University
Makkawala, Dehradun, Uttarakhand, India

Asif Ahmad Bhat
Suresh Gyan Vihar University
Jagatpura, Jaipur, India

Shvetank Bhatt
Dr. Vishwanath Karad MITWPU
Kothrud, Pune, India

Aman Chaudhary
DIT University
Dehradun, Uttarakhand, India

Rutuja Chougale
Bharati Vidyapeeth College of Pharmacy
Kolhapur, Maharashtra, India

Piyush Dave
Suresh Gyan Vihar University
Jagatpura, Jaipur, India

Ditixa Desai
Maliba Pharmacy College
Uka Tarsadia University
Gujarat, India

Rishika Dhapola
Central University of Punjab
Bathinda, India

Kamal Dua
University of Technology Sydney
Australia

Rupesh K. Gautam
Indore Institute of Pharmacy, IIST Campus
Indore, Madhya Pradesh, India

Gaurav Gupta
Suresh Gyan Vihar University
Jagatpura, Jaipur, India

Manish Gupta
Suresh Gyan Vihar University
Jagatpura, Jaipur, India

Shraddha M. Gupta
University of Petroleum and Energy Studies
  (UPES)
Dehradun, India

Madan Mohan Gupta
University of the West Indies,
  St. Augustine
Trinidad and Tobago, West Indies

Namdeo Jadhav
Bharati Vidyapeeth College of Pharmacy
Kolhapur, Maharashtra, India

Shrikant Joshi
Uka Tarsadia University
Surat, Gujarat, India

Gajanan Kalyankar
Uka Tarsadia University
Surat, Gujarat, India

Anjoo Kamboj
Chandigarh College of Pharmacy
Mohali, Punjab, India

Devesh U. Kapoor
Dr. Dayaram Patel Pharmacy College
Bardoli, Gujarat, India

Rachel Kaul
Translational Health Science and
    Technology Institute
Faridabad, India

Narinderpal Kaur
Chitkara University School of Pharmacy
Chitkara University
Himachal Pradesh, India

Zamir Khan
H. R. Patel Institute of Pharmaceutical
    Education and Research
Dhule, Maharashtra, India

Nitesh Kumar
Amity University
Noida, Uttar Pradesh

P. S. Kumar
Siksha 'O' Anusandhan (Deemed to be
    University)
Bhubaneswar, Odisha, India

Varun Kumar
Translational Health Science and
    Technology Institute
Faridabad, India

Sneha Kumari
Department of Pharmacology
Central University of Punjab
Bathinda, India

Sandesh Lodha
Maliba Pharmacy College
Uka Tarsadia University
Surat, Gujarat, India,

Mahendra Mahajan
H. R. Patel Institute of Pharmaceutical
    Education and Research
Maharashtra, India

Hitesh Malhotra
Guru Gobind Singh College of
    Pharmacy
Haryana, India

Bikash Medhi
Department of Pharmacology,
    PGIMER
Chandigarh, India

Surbhi Mishra
Translational Health Science and
    Technology Institute
Faridabad, India

Pushank Nagar
Central University of Punjab
Bathinda, India

Sopan Nangare
H. R. Patel Institute of Pharmaceutical
    Education and Research
Maharashtra, INDIA

Ipsa Padhy
Siksha 'O'Anusandhan Deemed to be
    University
Bhubaneswar, Odisha, India

Narahari N. Palei
AMITY Institute of Pharmacy
AMITY University
Lucknow, Uttar Pradesh, India

Jidnyasa Pantwalawalkar
Department of Pharmaceutics
Bharati Vidyapeeth College of Pharmacy
Maharashtra, India

Hetal Patel
Maliba Pharmacy College
Uka Tarsadia University
Surat, Gujarat, India

Mitali Patel
Maliba Pharmacy College
Uka Tarsadia University
Bardoli, Gujarat

Ganesh Patil
H. R. Patel Institute of Pharmaceutical
   Education and Research
Maharashtra, INDIA

Pravin Patil
H. R. Patel Institute of Pharmaceutical
   Education and Research
Maharashtra, India

Yuvraj Patil
Dr. Vishwanath Karad MITWPU
Kothrud, Pune, India

R. Prakash
Crescent School of Pharmacy
BSACIST University
Vandalur, Chennai

S. K. Prusty
Siksha 'O' Anusandhan Deemed to be
   University
Bhubaneswar, Odisha, India

Jayaraman Rajangam
Shri Venkateshwara College of Pharmacy
Puducherry, India

Dimpy Rani
School of Medical and Allied Sciences
G D Goenka University
Gurugram, Haryana, India

G.S.N. Koteswara Rao
NMIMS
Mumbai, India

Nensi Raytthatha
Sigma Institute of Pharmacy
Vadodara, Gujarat

Dibbanti Harikrishna Reddy
Central University of Punjab
Bathinda, India

Dinesh D. Rishipathak
Mumbai Educational Trust's Institute of
   Pharmacy
Nashik, Maharashtra, India

P. K. Sahu
Siksha 'O' Anusandhan Deemed to be
   University
Bhubaneswar, Odisha, India

Mahendra Saini
Suresh Gyan Vihar University
Jagatpura, Jaipur, India

Jhuma Samanta
Kingstone Imperial Institute of Technology
   and Sciences
Dehradun, Uttarakhand, India

Priya Sen
School of Pharmacy
Suresh Gyan Vihar University
Jagatpura, Jaipur, India

Isha Shah
Sigma Institute of Pharmacy
Vadodara, Gujarat

Deepak Sharma
Institute of Pharmacy, Assam Don Bosco
   University
Tapesia, Assam, India

Prajjwal Sharma
Central University of Punjab
Bathinda, India

Pratishtha Sharma
LJ University
Ahmedabad Gujarat, India

Shiwali Sharma
Chitkara University
Himachal Pradesh, India

Tripti Sharma
Siksha 'O' Anusandhan Deemed to be
    University
Bhubaneswar, Odisha, India

Siddharth Singh
University of Petroleum and Energy Studies
    (UPES)
Dehradun, India

Shruti Srivastava
AMITY University
Lucknow, Uttar Pradesh, India

Nikita Thakur
Chitkara University
Himachal Pradesh, India

Naitik D. Trivedi
AR College of Pharmacy & GH Patel
    Institute of Pharmacy
Anand, Gujarat, India

Rutvi Vaidya
Maliba Pharmacy College
Uka Tarsadia University
Bardoli, Gujarat

Sagili Varalakshmi
Sree Vidyanikethan College of
    Pharmacy
Tirupati, Andhra Pradesh, India

Nitin Verma
Chitkara University School of
    Pharmacy
Chitkara University
Himachal Pradesh, India

Bhavin Vyas
Maliba Pharmacy College
Uka Tarsadia University
Surat, Gujarat, India

Jigar Vyas
Sigma Institute of Pharmacy
Vadodara, Gujarat

Oxidative stress-induced metabolic disorders

Mechanism and pathogenesis

Priya Sen, Manish Gupta, Mahendra Saini, Piyush Dave, Asif Ahmad Bhat, and Gaurav Gupta

LIST OF ABBREVIATIONS

8-oxoG	8-Oxo guanine
AGEs	Advanced glycation end products
ANF	Atrial natriuretic factor
ANT	Adenosine nucleotide translocator
ATP	Adenosine triphosphate
BMI	Body mass index
CAD	Coronary artery disease
CHF	Congestive heart failure
CRP	C-reactive protein
CVD	Cardiovascular disease
ERK	Extracellular signal-regulated kinase
FABP4	Fatty acid binding protein 4
FFA	Free fatty acids
FoxO	Forkhead box protein O1
GATA1	GATA binding protein 1
GSH	Glutathione
HDL	High-density lipoprotein
hMSCs	Human mesenchymal stem cells
IGF-1	Insulin growth factor-1
IL2RA	Interleukin-2 receptor subunit alpha
IL-6	Interleukin-6
LDL	Low-density lipoprotein
MAPK	Mitogen-activated protein kinase
NADPH	Nicotinamide adenine dinucleotide phosphate
NO_x	Nitrogen oxides
OXPHOS	Oxidative phosphorylation
P16[INK4a]	Cyclin-dependent kinase inhibitor 2A
PI3K	Phosphoinositide 3-kinase
PKC	Protein kinase C
PPARGC1A	PPARG coactivator 1 alpha
pRb	Retinoblastoma protein
ROS	Reactive oxygen species
TNF-α	Tumor necrosis factor-α

DOI: 10.1201/9781032621135-1

VEGF   Vascular endothelial growth factor
β-MHC   Beta-myosin heavy chain

1.1 INTRODUCTION

The body develops a variety of pathophysiological conditions known as metabolic diseases when regular metabolic functions are disrupted, resulting in redox and energy disequilibrium (Barber, Mead, & Shaw, 2006). According to the National Heart, Lung, and Blood Institute, a person must have at least three risk factors to be labeled with metabolic syndromes (Bar-Sela, Cohen, Ben-Arye, & Epelbaum, 2015). These risk factors cause cellular malfunction and redox disequilibrium, which advance a pro-oxidative surrounding and result in defective biomolecules (Bergeron, 1995). These biomolecules are highly adaptive and can encourage cell and tissue dysfunction, leading to metabolic illnesses. To create new indicators, molecular targets, and potent medications for preventing and treating these illnesses, it is now evident that there is a link between oxidative stress and metabolic diseases (Bosoi & Rose, 2013; Carlström, 2021).

The multifactorial disease known as metabolic syndrome raises the chance of diabetes and speeds up atherosclerosis (Carpita, Muti, & Dell'Osso, 2018). It has a high fatality rate and is linked to significant cardiovascular problems. According to the International Diabetes Federation, the most recent criteria for metabolic syndrome are defined by various groups of three or more of the following characteristics: high blood pressure and abdominal obesity, high blood glucose, and loss of physiological vasomotor activity (Chen, Bassot, Giuliani, & Simmen, 2021).

The interplay of genetic variations and environmental variables, which add to the deteriorating state of metabolic syndrome, has been shown in earlier research. Several research sources suggest that mitochondrial dysfunction and oxidative stress are significant causes of age-related neurological and metabolic disorders (Cheng et al., 2019). But there is still a lack of fundamental understanding of the etiology of the metabolic syndrome (Cho, Lee, Choi, Lee, & Lee, 2022) (Figure 1.1).

The Adult Treatment Panel III guidelines of the National Cholesterol (triglyceride) Education Program (NCEP) also established standards for the detection of metabolic syndrome, just like those of the WHO and other organizations (Chung & Kennedy, 2020; Courties, Sellam, & Berenbaum, 2017; Dewanjee et al., 2018). Westernized cultures have a very high incidence of metabolic syndrome compared to non-Westernized civilizations (Dinić et al., 2022). The metabolic syndrome was reviewed in an inter-representative US sample of approximately 12,363 women and men of about 20 years and older from the third National Health and Nutrition Examination Survey (El Hayek, Ernande, Benitah, Gomez, & Pereira, 2021). The metabolic syndrome is characterized by the adenosine triphosphate 3 diagnostic manual (elevated levels of triglycerides, abdominal overweight, lowering of good cholesterol, high blood pressure, and increased fasting blood glucose concentration) (Feriani et al., 2021). The disorder was present in 22.6% and 22.8% of women and men, respectively (Folli et al., 2011). Men who were average weight, overweight, or obese had a metabolic syndrome prevalence of 4.6%, 22.4%, and 59.6%, respectively. Physical inactivity was linked to a higher chance of getting the syndrome (Homma & Fujii, 2020). Additionally, it has been calculated that 80% of people with type II diabetes (Type II) and 43%

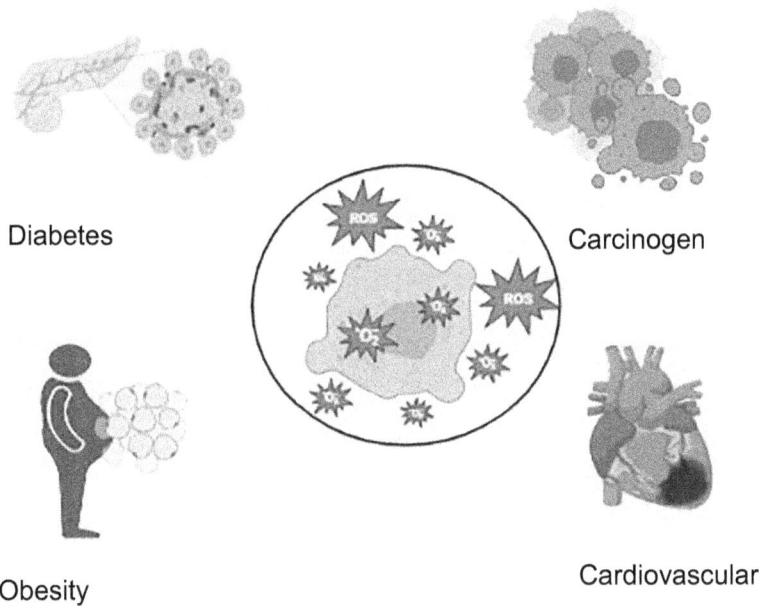

Figure 1.1 Different oxidative stress-induced diseases.

of people over 60 have metabolic syndrome. The current rate is probably greater because the projected incidence was founded on the findings of NHANES III (1988–1994) (Hopps, Noto, Caimi, & Averna, 2010).

Although it isn't the focus of this chapter, it should be noted that the word "metabolic syndrome" has been criticized for various reasons. Among them are, but not restricted to:

1. This same ATP-3 criterion does not explicitly assess insulin resistance, which is the only population in which it happens (James, Collins, Logan, & Murphy, 2012).
2. Many people could be significantly more at risk for cardiovascular disease (CVD) and not meet the somewhat artificial diagnostic cutoffs because they may be adequately resistant to insulin and have increased coronary artery disease (CAD) risk factors (Jiang, Ding, Huang, & Wang, 2022); and
3. Treating the individual variables may not be as effective as tackling the fundamental issue, which is typically insulin sensitivity brought on by lifestyle in people with hereditary susceptibility to the condition (Kang & Yang, 2020).

The word oxidative stress was first stated as "an imbalance in the prooxidant-antioxidant equilibrium in favor of the former" in the first part of the book on "Oxidative Stress" (Khalid, Alkaabi, Khan, & Adem, 2021). Prooxidants, antioxidants, and their internal and exogenous sources and biochemical sinks were discussed in the study titled "Biochemistry of Oxidative Stress" (Kondoh, Lleonart, Bernard, & Gil, 2007). The idea needed to be updated with the development of the redox signaling function, which led to the description given here (Table 1.1).

Table 1.1 Reactive oxygen species (ROS) and their corresponding antioxidants

Sl. No.	ROS	Antioxidant
1.	Hydrogen oxide radicals	Ascorbic acid, glutamine, polyphenols, α-lipoic acid
2.	Superoxide anion	Ascorbic acid, glutamine, polyphenols
3.	Dihydrogen dioxide	Ascorbic acid, glutamine, carotenoid, polyphenols, α-lipoic acid, tocopherol, ubidecarenone
4.	Lipid peroxidation	Carotenoid, tocopherol, polyphenols, benzoquinone

1.2 METABOLIC DISEASES AND OXIDATIVE STRESS

The primary definition of oxidative stress is an imbalance between the generation and degradation of ROS (Kushwaha, Kabra, Dubey, & Gupta, 2022). The data suggests that metabolic syndrome and increased systemic oxidative stress are closely related. In the animal models, there is a direct link between the existence of oxidative stress and elevated low-density lipoprotein (LDL) values and lower levels of high-density lipoprotein (HDL) (Kuzmenko, Udintsev, Klimentyeva, & Serebrov, 2016). Several models suggested an increase in oxidative stress in metabolic disorders. One of these processes, which may be caused by reduced HDL amounts in metabolic disorders, is the impaired good cholesterol (HDL)-enabled antioxidant mechanism (Lefranc, Friederich-Persson, Palacios-Ramirez, & Nguyen Dinh Cat, 2018).

When a person has metabolic syndrome, their antioxidant mechanisms are reported to be impaired and linked to increased oxidative stress and insulin intolerance (Li et al., 2019). Low HDL levels are negatively associated with oxidative stress indicators in the plasma. In contrast, low HDL levels in metabolic syndrome correlate with lipid peroxidation products (Lin, Xu, & Zhang, 2020). The concentration of denser and smaller LDL particles shifts, increasing the extent of LDL oxidation (Liu et al., 2019). According to studies, individuals with metabolic syndrome have higher amounts of oxidized LDL in their blood circulation, which raises their risk for myocardial infarction, atherosclerosis, and oxidative stress (Lu et al., 2021).

Additionally, metabolic syndrome can be accompanied by increased carbonylation, free radical attack of cellular proteins, NADPH oxidase activity, low GSH levels, and other conditions that boost ROS production. In reality, elevated oxidized LDL levels are associated with oxidative stress and low HDL in metabolic syndrome, which increases the chances of getting pathological conditions (Folli et al., 2011; Kuzmenko et al., 2016; Luc, Schramm-Luc, Guzik, & Mikolajczyk, 2019). Intracellular organelles like mitochondria are the storehouse of "energy currency" adenosine triphosphate (ATP) produced by the metabolism of nutrients that are essential for the health of cells (Manolagas, 2010). They are also in charge of numerous processes, including energy metabolism, the production of calcium homeostasis and free radicals, the cell cycle, and death (Maritim, Sanders, & Watkins, 2003). Their main job is producing ATP through oxidative phosphorylation (OXPHOS) in Krebs cycle-mediated substrate and fatty acid oxidation (Matsuda & Shimomura, 2013). Nowadays, it is known that pathological mitochondrial alterations are linked to altered mitochondrial functions, such as decreased oxidative capacity and antioxidant defense by the increased generation of diminished OXPHOS reactive oxygen species (ROS) and decreased ATP production (Matsuda & Shimomura, 2014). The change in mitochondrial fission and fusion processes as well as the suppression of mitophagy, a process that removes dysfunctional mitochondria, may be the cause of decreased mitochondrial biogenesis with age (Motataianu, Serban, Barcutean, & Balasa, 2022). Superoxide anions, hydroxyl, peroxyl, and other non-radicals with the ability to produce free radicals are all members of the family of free radicals known as ROS (Norenberg, Jayakumar, & Rama Rao, 2004). Although the production of ROS within the cells is a natural process, cells have a variety of protection mechanisms to prevent it. Lipids, DNA, and protein oxidative injury have all been linked to excessive ROS generation (Onyango, 2018).

The goal of this chapter is to provide a summary of the fundamental relationship between different symptoms of metabolic syndrome and oxidative stress (Parmeggiani & Vargas, 2018). We focus on these diseases because they are strongly associated with the oxidative damage carried on by elevated ROS generation, which leads to mitochondrial dysfunction (Pennathur & Heinecke, 2004). These disorders include overweight, CVD, carcinoma, and diabetes mellitus. Then, pharmacologic approaches taken from the lab to the patient's bedside will be offered to address mitochondrial malfunction and reduce the chance of metabolic syndrome (Pinazo-Durán, Zanón-Moreno, Gallego-Pinazo, & García-Medina, 2015) (Figure 1.2).

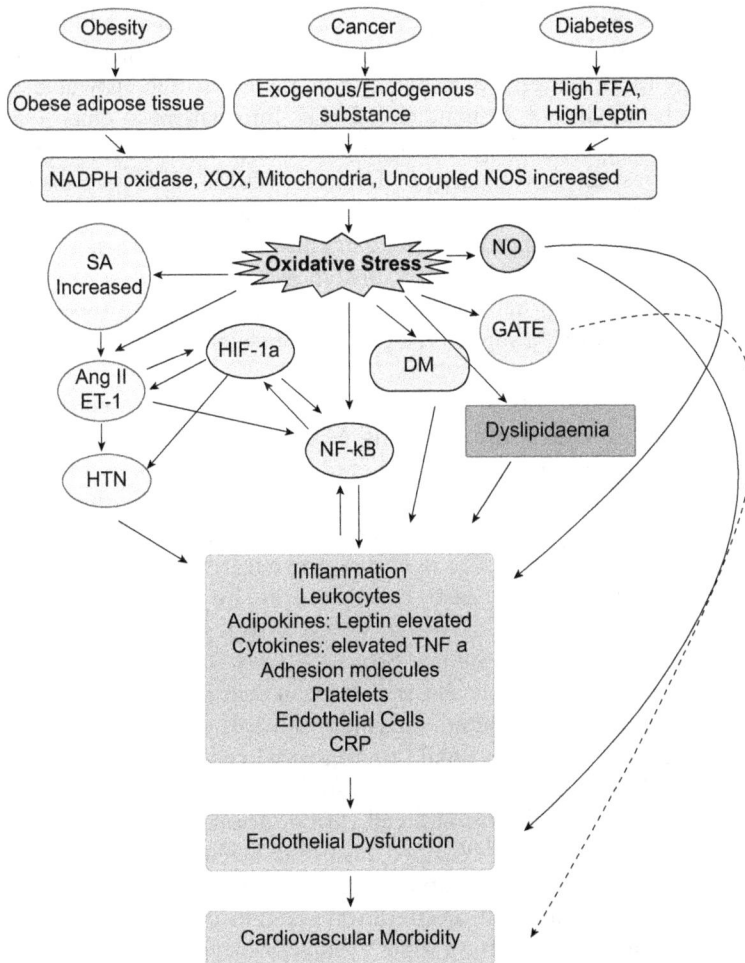

Figure 1.2 Different mechanisms involved in metabolic disorders associated with oxidative stress.

1.3 OXIDATIVE STRESS AND METABOLIC DISORDER LEADING TO DIABETES

Oxidative stress in diabetes causes metabolic disorders, raising blood sugar levels and reducing insulin sensitivity. Chronic diabetes can cause micro- or macrovascular damage leading to nephropathy, retinopathy, and neuropathy (Pokusa & Kráľová Trančíková, 2017). The majority of obese individuals acquire diabetes as a result of insulin resistance and subsequent increase in blood sugar (Rani, Deep, Singh, Palle, & Yadav, 2016). Oxidative stress has been associated with developing insulin resistance and changes in insulin signaling pathways and adipocytokines (Reis, Veloso, Mattos, Purish, & Nogueira-Machado, 2008). Elevated ROS levels in mice's liver and fat tissues treated with a high-fat diet have been linked to insulin resistance and altered by antioxidants. The major link between insulin sensitivity and obesity is adipocyte-derived oxidative stress, which includes leptin and free fatty acids (FFA) (Saeed, Kausar, Singh, Siddiqui, & Akhter, 2020). By turning on the mitochondria's oxidative phosphorylation uncouplers, the increased amounts of FFA can lead to mitochondrial malfunction (Sas, Robotka, Toldi, & Vécsei, 2007). There is an additional increase in ROS generation through defective mitochondria in the disruptive metabolic state brought on by a high-energy meal that results in higher glucose levels, free fatty acids, and insulin, as stated earlier (Scheen, Giraud, & Bendjelid, 2021). Impairment in glucose tolerance can result from insulin resistance, leading to continuous changes in the compensatory responses of insulin release (Sevastianos, Voulgaris, & Dourakis, 2020). This may disturb the establishment of type II diabetes by inhibiting insulin action and secretion. Since pancreatic cells do not produce enough antioxidants to defend against oxidative stress, a redox imbalance worsens β-cell mortality in pre-diabetic circumstances (Silveira Rossi et al., 2022). ROS-induced cell dysfunction, abnormal proliferation, and growth cause type II diabetes. The altered adipokine production brought on by obesity results in β-cell depletion (Simpson, Yen, & Appel, 2003).

Overproduction of ROS in obese individuals leads to diabetes mellitus, which causes the proliferation arrest of pancreatic beta cells. Many cells cannot re-enter cellular proliferation in addition to low cell cycle life (Song et al., 2022). ROS plays a significant part in the deregulation of pancreatic cell growth by changing the cell growth controllers, which contributes to the onset and sequence of diabetes (Sozen & Ozer, 2017). Increased ROS promotes insulin resistance, but ROS scavengers prevent it, according to genomic studies of insulin-resistant cellular models. Also, it demonstrated a two- to fivefold reduction in the proteins in charge of the G0/G1 switch, which is thought to control how quiescent cells enter the proliferation cycle (Spahis, Borys, & Levy, 2017). Reduced cyclin-dependent kinase 1 and cyclin B1 mRNA levels, which are in charge of the G1/S and G2/M transitions in these cells, are directly linked. CDK4, cyclin D1, and cyclin D2, crucial for the G1/S phase, are additional proteins linked to abnormal cell proliferation (Spoto, Pisano, & Zoccali, 2016). Adenovirus-mediated CDK4 and cyclin D1 expression promoted retinoblastoma protein phosphorylation and cell proliferation. Conversely, the smaller and less fertile CDK4 knockout mice developed insulin-deficient diabetes due to the loss of cell mass (Tabaei & Tabaee, 2019). Animals expressing mutant CDK4 exhibited pancreatic hyperplasia due to aberrant cell proliferation brought on by improper G1 arrest and non-binding of the cell cycle inhibitor P16[INK4a] (cyclin-dependent kinase inhibitor 2A). These findings suggest that changes in cell cycle element levels may have an impact on both the preservation of dry mass under basal conditions and their ability to respond to pathological conditions leading to diabetes (Sas et al., 2007; Scheen et al., 2021; Tappy & Lê, 2010; Tarnacka, Jopowicz, & Maślińska, 2021). Consequently, ROS controls the cell growth machinery to cause the pathophysiology of metabolic diseases (Tosti, Bertozzi, & Fontana, 2018).

Diabetes also affects how the cell cycle is regulated when the redox status changes. Hyperglycemia and glucotoxicity during diabetes are the main causes of oxidative stress (Turkmen, 2017). Diabetes-induced ROS reduces BCL-1 and cyclin-dependent protein kinase D2 but overexpresses cyclin-dependent kinases inhibitor protein (Umbayev et al., 2020). Furthermore, it is known that diabetes-induced inflammation and increased ROS cause forkhead box protein O1 (FoxO) transcription factors, which change the expression of several crucial proteins for cell cycle control, particularly those implicated in the G1/S transition (Un Nisa & Reza, 2020).

1.4 OXIDATIVE STRESS LEADING TO CVD

As a common side effect of dyslipidemia and diabetes, CVD is a major health concern throughout the globe as another metabolic disorder. Metabolic diseases play a major role in forming a pro-oxidative environment (Vega et al., 2016). Elevated myocardial lipid buildup and modification of substrate metabolic reactions in obesity are known to change blood flow dynamics and induce heart problems. For instance, it has been demonstrated that reduced systolic function is linked to increased myocardial neutral fats deposition and concentric left ventricular hypertrophy (LVH) (Vona, Gambardella, Cittadini, Straface, & Pietraforte, 2019).

Other factors linked to cardiovascular problems include estrogen, insulin growth factor-1 (IGF-1), C3a desArg protein (the protein that transports cholesteryl esters), retinal binding protein, fatty acid binding protein 4 (FABP4), leptin, and adiponectin. These are examples of adipocytokines and substances produced from adipose tissue that can directly impact cardiac shape and function (Wan et al., 2021). Adiponectin, a cytokine generated from white and brown adipose tissue, is crucial in metabolic conditions that result in heart failure. The levels of adiponectin are negatively connected with body mass index (BMI) in adults, thus those who are overweight or have diabetes have low amounts of the hormone, which raises LDL (bad cholesterol) and lowers HDL (good cholesterol) levels (Wang, Jia, & Rodrigues, 2017). Similar relationships exist between inflammatory cytokines and adiponectin like TNF-α and IL-6 in obesity, which likewise help to raise LDL and reduce HDL levels. Inflammation, a well-known reaction to cardiovascular risk factors, also persists with insulin resistance and adult-onset diabetes, raising ROS and C-reactive protein (CRP) levels and inducing endothelial dysfunction (Wang, Li, Li, & Lv, 2022).

These modifications raise purified intercellular adhesion molecule-1 and endothelial cell surface glycoprotein levels, which further attach LDL molecules to artery walls and enhance monocyte chemoattraction and CVD risk (Weiss & Hennet, 2017). Moreover, those with diabetes and obesity are likelier to have left ventricular hypertrophy when their adiponectin levels are low. Cardiac myocytes grow to enhance their work output as a result of cardiac hypertrophy, which happens as a coping mechanism to stress

(Umbayev et al., 2020; Wan et al., 2021; Whaley-Connell, McCullough & Sowers, 2011). Enhanced protein synthesis, the development of sarcomeres, the stimulation of rapid response genes (c-jun, c-fos, and c-myc), as well as the re-expression of embryonic genes like atrial natriuretic factor (ANF), beta-myosin heavy chain (β-MHC), skeletal alpha-actin, and GATA binding protein 1 (GATA-1) are the outcomes of this (Wu, Wu, & Wei, 2014). Several signaling pathways, including tyrosine kinase Src, GTP-binding protein Ras, protein kinase C (PKC), mitogen-activated protein kinase (MAPK), extracellular signal-regulated kinase (ERK), and phosphoinositide 3-kinase (PI3K), are active in hypertrophy. These changes at first lessen the increased strain, but prolonged hypertrophy ultimately results in cardiac necrosis and congestive heart failure (CHF) (Xu, Li, Adams, Kubena, & Guo, 2018).

It has been proposed that the common factor connecting many molecular illnesses is glycoxidative stress resulting in CVD (L. J. Yan, 2014). Diabetes and obesity cause more glycoxidation, which changes enzymatic processes, the way ligands bind to receptors, and the functioning and immunology of proteins. Advanced glycation end products (AGEs) accumulate due to oxidative stress brought on by hyperglycemia, which causes cell damage (Un Nisa & Reza, 2020; M. Yan et al., 2022). Non-enzymatic interactions between polypeptides, lipids, sugars, amino groups (reduction), and nucleic acids result in AGEs, which hasten the aging of proteins and lead to degenerative illnesses. AGEs can also directly affect the extracellular matrix and capillary structure and function by inducing the cross-linking of proteins like collagen, which promotes vascular elasticity (Xu et al., 2018; Yao, Li, & Zeng, 2020).

1.5 OXIDATIVE STRESS RESULTING IN CARCINOGENESIS

Obesity and metabolic disorders are significant risk factors for cancer. People with high BMI are more likely to have various malignancies, such as endometrial, colon, ovary, and breast cancers. Carcinoma due to obesity is thought to account for 20% of all cancer-causing factors. The redox change brought on by adipokines such as leptins, adiponectin, vascular endothelial growth factor (VEGF), tumor necrosis factor-α (TNF-α), and interleukin-6 (IL-6) is linked to an increased risk of development of cancer in obese people (Motataianu et al., 2022; Norenberg et al., 2004; Silveira Rossi et al., 2022; Yara, Lavoie, & Levy, 2015). Many studies have demonstrated that obesity increases oxidative stress by raising the quantity of ROS, which is a key factor in the development of cancers.

Some factors contributing to the ongoing oxidative stress in tumor cells include oncogenes (Ras2) activation, antioxidant deactivation, inflammation, activation of the NO_X pathway, and accumulation of metabolic waste products (Simpson et al., 2003). Interrupted redox balance, which affects multiple signal transduction linked to the proliferation of cells, induced cell death, resistance due to medications, and energy consumption, has been linked to the growth and spread of cancer. The inhibition of adenosine nucleotide translocator (ANT) causes a buildup of ATP inside the mitochondria, which decreases the exchange of free ATP with free ADP across the inner membrane of mitochondria (Song et al., 2022). The uncoupling effect causes molecular oxygen to reduce, releasing superoxide ions in the process partially. ROS buildup may aid in tumor growth by functioning as a signaling molecule or by encouraging genomic DNA mutation (Sozen & Ozer, 2017). ROS can also increase the growth of tumors by phosphorylating redox-sensitive kinases such as extracellular-signal-regulated kinases (ERKs), MAPK, and cyclins, which are necessary for the development and survival of cancerous cells (Spoto et al., 2016).

Genetic changes are one of the primary ways that oxidative stress exhibits its harmful consequences. ROS is known to harm bases and nucleotides and disrupts DNA strands, resulting in DNA damage (Yaribeygi, Farrokhi, Butler, & Sahebkar, 2019). The types of 8-hydroxylated guanine which are generated because of oxidized guanine lesions, like 8-oxo Guanine (8-oxoG), are primarily driven by ROS, which is thought to play a major role in the formation of tumors (Yazıcı & Sezer, 2017). These altered guanines can link with adenine and cytosine bases, leading to transversion mutations like G: C to T: A. The rise in the mutagenesis base 8-oxo-dG may accelerate cell mutation rates and/or disrupt DNA repair processes, ultimately defining tumor growth (Yun et al., 2020). Interestingly, 8-oxoG levels in oxidative stress-related tumors have been found to increase up to tenfold when compared to nearby normal tissues (Umbayev et al., 2020).

Epigenetic modifications can potentially affect the integrity of the genome. DNA methylation at cytosine residues is the most significant and extensively investigated epigenetic alteration. DNA methyltransferases initiate the process, which results in the synthesis of $5-CH_3$ cytosine by using S-adenosyl methionine (SAM) as the methyl group donor (Yun et al., 2020). DNA methylation controls the expression of many genes linked to diabetes, including interleukin-2 receptor subunit alpha (IL2RA) and PPARG coactivator 1 alpha (PPARGC1A). Recent research studies have demonstrated that obesity can change DNA methylation (Zhou et al., 2020).

1.6 OXIDATIVE STRESS LEADING TO OBESITY

Oxidative stress is a double-edged weapon since it can cause and result in obesity. Many investigations, including epidemiologic, animal, and clinical ones, have shown evidence that oxidation–reduction alteration and obesity are connected (Ziolkowska, Binienda, Jabłkowski, Szemraj, & Czarny, 2021). Higher levels of oxidative stress can result from several circumstances, such as a high fatty acid, simple sugars diet, and constant malnutrition, through promoting the action of PKC, the polyol pathway, cellular energy phosphorylation, glycoxidation, and nitrogen oxides (NO_x), among other intracellular processes (Bagul & Banerjee, 2013). Moreover, *in vitro* and *in vivo* research data indicates that oxidative stress may contribute to obesity by increasing the size of differentiated adipose tissues and promoting the proliferation of pre-adipocytes. Lipid metabolism is the process by which terminally differentiated pre-adipocytes rejoin the cell cycle and undergo proliferation, resulting in increased adipose mass (Bora & Shankarrao Adole, 2021). Pre-adipocytes multiply and differentiate into mature adipocytes as part of the two-step adipogenesis process. Both of these occurrences have been proven to involve ROS. Without insulin, H_2O_2 treatment of human and murine 3T3-L1 cells led to adipogenesis.

Additionally, it has been demonstrated that the ROS and NOX4 generated by mitochondria promote adipocyte growth in adipose-derived stem cells (Ceriello & Motz, 2004). The high levels of cellular damage, chronic adipocyte inflammation, fatty acid oxidation, excessive oxygen consumption, diet, and mitochondrial activity are just a few of the mechanisms that increase the production of ROS, which in turn plays a significant role in the development of metabolic disorders (Corrado, Cici, Rotondo, Maruotti, & Cantatore, 2020). Abnormal ROS generation promotes obesity and leads to cellular dysregulation in various

Table 1.2 Oxidative Stress and their biomarker and the related disorder

S. No.	Biomarkers	Oxidative target	Oxidative stress-induced diseases
1.	8-Oxo-7,8-dihydro-2′-deoxyguanosine	DNA	Cancer, diabetes mellitus, atherosclerosis
2.	8-Oxo-7,8-dihydrodeoxyguanine	DNA	Cancer, neurogenerative disorders
3.	Malondialdehyde (Propanedial)	Polyunsaturated fatty acids	atherosclerosis
4.	POE (20) sorbitan monooleate	Lysine, proline	Myocardial infarction

other organs. Adipogenesis has been found to boost cell cycle indicators for cell division like BCL-1, and cyclin E. Retinoblastoma protein (pRb) undergoes rapid dephosphorylation when glutathione reduction alters the cellular redox conditions *in vitro* (Donida et al., 2017). The E2F (transcriptional activation factor), a key regulator of the expression of genes involved in cell growth, especially those involved in the G1 and S phases of the cell cycle causes growth-arrest and post-confluent pre-adipocytes re-entrance of the cell cycle before proliferation and differentiation. E2F also controls 3T3-L1 adipocyte differentiation (Du, Shi, & Le, 2010). After the clonal development of adipocytes, the cells overexpress the cyclin-dependent kinase inhibitors p21 and p27, which stops the proliferation and promotes differentiation. ROS triggers PPAR, a direct target of E2F, to regulate adipocyte formation within human mesenchymal stem cells (hMSCs) (Farnaghi, Crawford, Xiao, & Prasadam, 2017). A well-known ROS quencher and antioxidant N-acetyl-l-cysteine (NAC) greatly slows down adipocyte development. These data support the notion that oxidative stress signals caused by ROS play a role in triggering lipid metabolism by controlling the cell cycle that encourages obesity (Feng et al., 2016) (Table 1.2).

1.7 CONCLUSION AND FUTURE PROSPECTS

Chronic non-communicable illnesses linked to lifestyle and nutrition are already significantly impacting global healthcare. To stop the spread of disorders associated with metabolic syndrome, a multifaceted plan for combating this epidemic must be swiftly developed and put into place. Together with encouraging people to adopt healthy lifestyles, a significant increase in funding for research into metabolic syndrome is required.

It is crucial to investigate the systems that upset the natural balance of adding oxygen and antioxidative processes since oxidative stress has become a key factor in chronic metabolic illnesses such as diabetes mellitus, overweight, carcinoma, and heart-related diseases. As previously mentioned, excess ROS and RNS production (from both endogenous and exogenous sources) cause the oxidation of all vital macromolecules necessary for life, such as lipids, proteins, and nucleic acids. The ongoing DNA damage brought on by oxidative stress may not only stimulate transcription factors but also cause genetic instability and cause proto-oncogene expression. It has also been demonstrated that insulin has a pro-tumorigenic potential that manifests in the form of excessive ROS production, which causes DNA damage, genomic instability, and carcinogenesis. Also, the harmed macro-biomolecules interfere with healthy cellular function, which causes illnesses linked to metabolic problems. New treatment approaches, such as naturally derived antioxidants derived from plants, are being researched to either stop the development of these health abnormalities or therapeutically intervene in them.

The impact of long-term inflammation and oxidative stress on stem cells is one of the newer fields of study. The impact of ROS on stem cells is particularly important because it may impair their capacity to refresh and revitalize different kinds of body cells throughout the organism's lifetime. Although cancer stem cells are the hardest to treat in the current cancer therapy regimens and frequently cause recurrence, it has been shown that ROS generation in dysregulation may provide healing potential. It is crucial to find biological targets that might aid in re-establishing the oxidative balance for improved health, given the catastrophic effects of oxidative stress and long-term infection in metabolic syndrome. Future research should concentrate on understanding the disease pathways and identifying shared targets to prevent or cure oxidative stress-induced diseases in patients with metabolic disorders.

REFERENCES

Bagul, P. K., & Banerjee, S. K. (2013). Insulin resistance, oxidative stress and cardiovascular complications: Role of sirtuins. *Curr Pharm Des, 19*(32), 5663–5677. doi:10.2174/13816128113199990372

Barber, S. C., Mead, R. J., & Shaw, P. J. (2006). Oxidative stress in ALS: A mechanism of neurodegeneration and a therapeutic target. *Biochim Biophys Acta, 1762*(11–12), 1051–1067. doi:10.1016/j.bbadis.2006.03.008

Bar-Sela, G., Cohen, M., Ben-Arye, E., & Epelbaum, R. (2015). The medical use of wheatgrass: Review of the gap between basic and clinical applications. *Mini Rev Med Chem, 15*(12), 1002–1010. doi:10.2174/1389557515121250731112836

Bergeron, C. (1995). Oxidative stress: Its role in the pathogenesis of amyotrophic lateral sclerosis. *J Neurol Sci, 129*(Suppl), 81–84. doi:10.1016/0022–510x(95)00071–9

Bora, S., & Shankarrao Adole, P. (2021). Carbonyl stress in diabetics with acute coronary syndrome. *Clin Chim Acta, 520*, 78–86. doi:10.1016/j.cca.2021.06.002

Bosoi, C. R., & Rose, C. F. (2013). Oxidative stress: A systemic factor implicated in the pathogenesis of hepatic encephalopathy. *Metab Brain Dis, 28*(2), 175–178. doi:10.1007/s11011-012-9351-5

Carlström, M. (2021). Nitric oxide signalling in kidney regulation and cardiometabolic health. *Nat Rev Nephrol, 17*(9), 575–590. doi:10.1038/s41581–021–00429-z

Carpita, B., Muti, D., & Dell'Osso, L. (2018). Oxidative stress, maternal diabetes, and autism spectrum disorders. *Oxid Med Cell Longev, 2018*, 3717215. doi:10.1155/2018/3717215

Ceriello, A., & Motz, E. (2004). Is oxidative stress the pathogenic mechanism underlying insulin resistance, diabetes, and cardiovascular disease? The common soil hypothesis revisited. *Arterioscler Thromb Vasc Biol, 24*(5), 816–823. doi:10.1161/01.atv.0000122852.22604.78

Chen, J., Bassot, A., Giuliani, F., & Simmen, T. (2021). Amyotrophic Lateral Sclerosis (ALS): Stressed by Dysfunctional Mitochondria-Endoplasmic Reticulum Contacts (MERCs). *Cells, 10*(7). doi:10.3390/cells10071789

Cheng, Y., Yu, X., Zhang, J., Chang, Y., Xue, M., Li, X., . . . Chen, L. (2019). Pancreatic kallikrein protects against diabetic retinopathy in KK Cg-A(y)/J and high-fat diet/streptozotocin-induced mouse models of type 2 diabetes. *Diabetologia, 62*(6), 1074–1086. doi:10.1007/s00125-019-4838-9

Cho, Y. H., Lee, Y., Choi, J. I., Lee, S. R., & Lee, S. Y. (2022). Biomarkers in metabolic syndrome. *Adv Clin Chem, 111*, 101–156. doi:10.1016/bs.acc.2022.07.003

Chung, M. C. M., & Kennedy, B. K. (2020). Aging: Mechanisms, measures, and interventions. *Proteomics, 20*(5–6), e1800336. doi:10.1002/pmic.201800336

Corrado, A., Cici, D., Rotondo, C., Maruotti, N., & Cantatore, F. P. (2020). Molecular basis of bone aging. *Int J Mol Sci, 21*(10). doi:10.3390/ijms21103679

Courties, A., Sellam, J., & Berenbaum, F. (2017). Metabolic syndrome-associated osteoarthritis. *Curr Opin Rheumatol, 29*(2), 214–222. doi:10.1097/bor.0000000000000373

Dewanjee, S., Das, S., Das, A. K., Bhattacharjee, N., Dihingia, A., Dua, T. K., . . . Manna, P. (2018). Molecular mechanism of diabetic neuropathy and its pharmacotherapeutic targets. *Eur J Pharmacol, 833*, 472–523. doi:10.1016/j.ejphar.2018.06.034

Dinić, S., Arambašić Jovanović, J., Uskoković, A., Mihailović, M., Grdović, N., Tolić, A., . . . Vidaković, M. (2022). Oxidative stress-mediated beta cell death and dysfunction as a target for diabetes management. *Front Endocrinol (Lausanne), 13*, 1006376. doi:10.3389/fendo.2022.1006376

Donida, B., Jacques, C. E. D., Mescka, C. P., Rodrigues, D. G. B., Marchetti, D. P., Ribas, G., . . . Vargas, C. R. (2017). Oxidative damage and redox in lysosomal storage disorders: Biochemical markers. *Clin Chim Acta, 466*, 46–53. doi:10.1016/j.cca.2017.01.007

Du, D., Shi, Y. H., & Le, G. W. (2010). Oxidative stress induced by high-glucose diet in liver of C57BL/6J mice and its underlying mechanism. *Mol Biol Rep, 37*(8), 3833–3839. doi:10.1007/s11033-010-0039-9

El Hayek, M. S., Ernande, L., Benitah, J. P., Gomez, A. M., & Pereira, L. (2021). The role of hyperglycaemia in the development of diabetic cardiomyopathy. *Arch Cardiovasc Dis, 114*(11), 748–760. doi:10.1016/j.acvd.2021.08.004

Farnaghi, S., Crawford, R., Xiao, Y., & Prasadam, I. (2017). Cholesterol metabolism in pathogenesis of osteoarthritis disease. *Int J Rheum Dis, 20*(2), 131–140. doi:10.1111/1756–185x.13061

Feng, B., Meng, R., Huang, B., Shen, S., Bi, Y., & Zhu, D. (2016). Silymarin alleviates hepatic oxidative stress and protects against metabolic disorders in high-fat diet-fed mice. *Free Radic Res, 50*(3), 314–327. doi:10.3109/10715762.2015.1116689

Feriani, A., Tir, M., Arafah, M., Gómez-Caravaca, A. M., Contreras, M. D. M., Nahdi, S., . . . Tlili, N. (2021). Schinus terebinthifolius fruits intake ameliorates metabolic disorders, inflammation, oxidative stress, and related vascular dysfunction, in atherogenic diet-induced obese rats: Insight of their chemical characterization using HPLC-ESI-QTOF-MS/MS. *J Ethnopharmacol, 269*, 113701. doi:10.1016/j.jep.2020.113701

Folli, F., Corradi, D., Fanti, P., Davalli, A., Paez, A., Giaccari, A., . . . Muscogiuri, G. (2011). The role of oxidative stress in the pathogenesis of type 2 diabetes mellitus micro- and macrovascular complications: Avenues for a mechanistic-based therapeutic approach. *Curr Diabetes Rev, 7*(5), 313–324. doi:10.2174/157339911797415585

Homma, T., & Fujii, J. (2020). Emerging connections between oxidative stress, defective proteolysis, and metabolic diseases. *Free Radic Res, 54*(11–12), 931–946. doi:10.1080/10715762.2020.1734588

Hopps, E., Noto, D., Caimi, G., & Averna, M. R. (2010). A novel component of the metabolic syndrome: The oxidative stress. *Nutr Metab Cardiovasc Dis, 20*(1), 72–77. doi:10.1016/j.numecd.2009.06.002

James, A. M., Collins, Y., Logan, A., & Murphy, M. P. (2012). Mitochondrial oxidative stress and the metabolic syndrome. *Trends Endocrinol Metab, 23*(9), 429–434. doi:10.1016/j.tem.2012.06.008

Jiang, M., Ding, H., Huang, Y., & Wang, L. (2022). Shear stress and metabolic disorders-two sides of the same plaque. *Antioxid Redox Signal, 37*(10–12), 820–841. doi:10.1089/ars.2021.0126

Kang, Q., & Yang, C. (2020). Oxidative stress and diabetic retinopathy: Molecular mechanisms, pathogenetic role and therapeutic implications. *Redox Biol, 37*, 101799. doi:10.1016/j.redox.2020.101799

Khalid, M., Alkaabi, J., Khan, M. A. B., & Adem, A. (2021). Insulin signal transduction perturbations in insulin resistance. *Int J Mol Sci, 22*(16). doi:10.3390/ijms22168590

Kondoh, H., Lleonart, M. E., Bernard, D., & Gil, J. (2007). Protection from oxidative stress by enhanced glycolysis: A possible mechanism of cellular immortalization. *Histol Histopathol, 22*(1), 85–90. doi:10.14670/hh-22.85

Kushwaha, K., Kabra, U., Dubey, R., & Gupta, J. (2022). Diabetic nephropathy: Pathogenesis to cure. *Curr Drug Targets, 23*(15), 1418–1429. doi:10.2174/1389450123666220820110801

Kuzmenko, D. I., Udintsev, S. N., Klimentyeva, T. K., & Serebrov, V. Y. (2016). Oxidative stress in adipose tissue as a primary link in pathogenesis of insulin resistance. *Biomed Khim, 62*(1), 14–21. doi:10.18097/pbmc20166201014

Lefranc, C., Friederich-Persson, M., Palacios-Ramirez, R., & Nguyen Dinh Cat, A. (2018). Mitochondrial oxidative stress in obesity: Role of the mineralocorticoid receptor. *J Endocrinol, 238*(3), R143–R159. doi:10.1530/joe-18-0163

Li, C., Zhang, J., Xue, M., Li, X., Han, F., Liu, X., . . . Chen, L. (2019). SGLT2 inhibition with empagliflozin attenuates myocardial oxidative stress and fibrosis in diabetic mice heart. *Cardiovasc Diabetol, 18*(1), 15. doi:10.1186/s12933-019-0816-2

Lin, Y., Xu, Y., & Zhang, Z. (2020). Sepsis-Induced Myocardial Dysfunction (SIMD): The pathophysiological mechanisms and therapeutic strategies targeting mitochondria. *Inflammation, 43*(4), 1184–1200. doi:10.1007/s10753–020–01233-w

Liu, F., Zhang, X., Zhao, B., Tan, X., Wang, L., & Liu, X. (2019). Role of food phytochemicals in the modulation of circadian clocks. *J Agric Food Chem, 67*(32), 8735–8739. doi:10.1021/acs.jafc.9b02263

Lu, J., Zhang, Y., Liang, J., Diao, J., Liu, P., & Zhao, H. (2021). Role of exosomal microRNAs and their crosstalk with oxidative stress in the pathogenesis of osteoporosis. *Oxid Med Cell Longev, 2021*, 6301433. doi:10.1155/2021/6301433

Luc, K., Schramm-Luc, A., Guzik, T. J., & Mikolajczyk, T. P. (2019). Oxidative stress and inflammatory markers in prediabetes and diabetes. *J Physiol Pharmacol, 70*(6). doi:10.26402/jpp.2019.6.01

Manolagas, S. C. (2010). From estrogen-centric to aging and oxidative stress: A revised perspective of the pathogenesis of osteoporosis. *Endocr Rev, 31*(3), 266–300. doi:10.1210/er.2009–0024

Maritim, A. C., Sanders, R. A., & Watkins, J. B., 3rd. (2003). Diabetes, oxidative stress, and antioxidants: A review. *J Biochem Mol Toxicol, 17*(1), 24–38. doi:10.1002/jbt.10058

Matsuda, M., & Shimomura, I. (2013). Increased oxidative stress in obesity: Implications for metabolic syndrome, diabetes, hypertension, dyslipidemia, atherosclerosis, and cancer. *Obes Res Clin Pract, 7*(5), e330–341. doi:10.1016/j.orcp.2013.05.004

Matsuda, M., & Shimomura, I. (2014). Roles of adiponectin and oxidative stress in obesity-associated metabolic and cardiovascular diseases. *Rev Endocr Metab Disord, 15*(1), 1–10. doi:10.1007/s11154-013-9271-7

Motataianu, A., Serban, G., Barcutean, L., & Balasa, R. (2022). Oxidative stress in amyotrophic lateral sclerosis: Synergy of genetic and environmental factors. *Int J Mol Sci, 23*(16). doi:10.3390/ijms23169339

Norenberg, M. D., Jayakumar, A. R., & Rama Rao, K. V. (2004). Oxidative stress in the pathogenesis of hepatic encephalopathy. *Metab Brain Dis, 19*(3–4), 313–329. doi:10.1023/b:mebr.0000043978.91675.79

Onyango, A. N. (2018). Cellular stresses and stress responses in the pathogenesis of insulin resistance. *Oxid Med Cell Longev, 2018*, 4321714. doi:10.1155/2018/4321714

Parmeggiani, B., & Vargas, C. R. (2018). Oxidative stress in urea cycle disorders: Findings from clinical and basic research. *Clin Chim Acta, 477*, 121–126. doi:10.1016/j.cca.2017.11.041

Pennathur, S., & Heinecke, J. W. (2004). Mechanisms of oxidative stress in diabetes: Implications for the pathogenesis of vascular disease and antioxidant therapy. *Front Biosci, 9*, 565–574. doi:10.2741/1257

Pinazo-Durán, M. D., Zanón-Moreno, V., Gallego-Pinazo, R., & García-Medina, J. J. (2015). Oxidative stress and mitochondrial failure in the pathogenesis of glaucoma neurodegeneration. *Prog Brain Res, 220*, 127–153. doi:10.1016/bs.pbr.2015.06.001

Pokusa, M., & Kráľová Trančíková, A. (2017). The central role of biometals maintains oxidative balance in the context of metabolic and neurodegenerative disorders. *Oxid Med Cell Longev, 2017*, 8210734. doi:10.1155/2017/8210734

Rani, V., Deep, G., Singh, R. K., Palle, K., & Yadav, U. C. (2016). Oxidative stress and metabolic disorders: Pathogenesis and therapeutic strategies. *Life Sci, 148*, 183–193. doi:10.1016/j.lfs.2016.02.002

Reis, J. S., Veloso, C. A., Mattos, R. T., Purish, S., & Nogueira-Machado, J. A. (2008). Oxidative stress: A review on metabolic signaling in type 1 diabetes. *Arq Bras Endocrinol Metabol, 52*(7), 1096–1105. doi:10.1590/s0004–27302008000700005

Saeed, M., Kausar, M. A., Singh, R., Siddiqui, A. J., & Akhter, A. (2020). The role of glyoxalase in glycation and carbonyl stress induced metabolic disorders. *Curr Protein Pept Sci, 21*(9), 846–859. doi:10.2174/1389203721666200505101734

Sas, K., Robotka, H., Toldi, J., & Vécsei, L. (2007). Mitochondria, metabolic disturbances, oxidative stress and the kynurenine system, with focus on neurodegenerative disorders. *J Neurol Sci, 257*(1–2), 221–239. doi:10.1016/j.jns.2007.01.033

Scheen, M., Giraud, R., & Bendjelid, K. (2021). Stress hyperglycemia, cardiac glucotoxicity, and critically ill patient outcomes current clinical and pathophysiological evidence. *Physiol Rep, 9*(2), e14713. doi:10.14814/phy2.14713

Sevastianos, V. A., Voulgaris, T. A., & Dourakis, S. P. (2020). Hepatitis C, systemic inflammation and oxidative stress: Correlations with metabolic diseases. *Expert Rev Gastroenterol Hepatol, 14*(1), 27–37. doi:10.1080/17474124.2020.1708191

Silveira Rossi, J. L., Barbalho, S. M., Reverete de Araujo, R., Bechara, M. D., Sloan, K. P., & Sloan, L. A. (2022). Metabolic syndrome and cardiovascular diseases: Going beyond traditional risk factors. *Diabetes Metab Res Rev, 38*(3), e3502. doi:10.1002/dmrr.3502

Simpson, E. P., Yen, A. A., & Appel, S. H. (2003). Oxidative Stress: A common denominator in the pathogenesis of amyotrophic lateral sclerosis. *Curr Opin Rheumatol, 15*(6), 730–736. doi:10.1097/00002281–200311000–00008

Song, Q. X., Sun, Y., Deng, K., Mei, J. Y., Chermansky, C. J., & Damaser, M. S. (2022). Potential role of oxidative stress in the pathogenesis of diabetic bladder dysfunction. *Nat Rev Urol, 19*(10), 581–596. doi:10.1038/s41585-022-00621-1

Sozen, E., & Ozer, N. K. (2017). Impact of high cholesterol and endoplasmic reticulum stress on metabolic diseases: An updated mini-review. *Redox Biol, 12*, 456–461. doi:10.1016/j.redox.2017.02.025

Spahis, S., Borys, J. M., & Levy, E. (2017). Metabolic syndrome as a multifaceted risk factor for oxidative stress. *Antioxid Redox Signal, 26*(9), 445–461. doi:10.1089/ars.2016.6756

Spoto, B., Pisano, A., & Zoccali, C. (2016). Insulin resistance in chronic kidney disease: A systematic review. *Am J Physiol Renal Physiol, 311*(6), F1087-F1108. doi:10.1152/ajprenal.00340.2016

Tabaei, S., & Tabaee, S. S. (2019). DNA methylation abnormalities in atherosclerosis. *Artif Cells Nanomed Biotechnol, 47*(1), 2031–2041. doi:10.1080/21691401.2019.1617724

Tappy, L., & Lê, K. A. (2010). Metabolic effects of fructose and the worldwide increase in obesity. *Physiol Rev, 90*(1), 23–46. doi:10.1152/physrev.00019.2009

Tarnacka, B., Jopowicz, A., & Maślińska, M. (2021). Copper, iron, and manganese toxicity in neuropsychiatric conditions. *Int J Mol Sci, 22*(15). doi:10.3390/ijms22157820

Tosti, V., Bertozzi, B., & Fontana, L. (2018). Health benefits of the Mediterranean diet: Metabolic and molecular mechanisms. *J Gerontol A Biol Sci Med Sci, 73*(3), 318–326. doi:10.1093/gerona/glx227

Turkmen, K. (2017). Inflammation, oxidative stress, apoptosis, and autophagy in diabetes mellitus and diabetic kidney disease: The four horsemen of the apocalypse. *Int Urol Nephrol, 49*(5), 837–844. doi:10.1007/s11255-016-1488-4

Umbayev, B., Askarova, S., Almabayeva, A., Saliev, T., Masoud, A. R., & Bulanin, D. (2020). Galactose-induced skin aging: The role of oxidative stress. *Oxid Med Cell Longev, 2020*, 7145656. doi:10.1155/2020/7145656

Un Nisa, K., & Reza, M. I. (2020). Key relevance of epigenetic programming of adiponectin gene in pathogenesis of metabolic disorders. *Endocr Metab Immune Disord Drug Targets, 20*(4), 506–517. doi:10.2174/1871530319666190801142637

Vega, C. C., Reyes-Castro, L. A., Rodríguez-González, G. L., Bautista, C. J., Vázquez-Martínez, M., Larrea, F., . . . Zambrano, E. (2016). Resveratrol partially prevents oxidative stress and metabolic dysfunction in pregnant rats fed a low protein diet and their offspring. *J Physiol, 594*(5), 1483–1499. doi:10.1113/jp271543

Vona, R., Gambardella, L., Cittadini, C., Straface, E., & Pietraforte, D. (2019). Biomarkers of oxidative stress in metabolic syndrome and associated diseases. *Oxid Med Cell Longev, 2019*, 8267234. doi:10.1155/2019/8267234

Wan, H., Zhao, S., Zeng, Q., Tan, Y., Zhang, C., Liu, L., & Qu, S. (2021). CircRNAs in diabetic cardiomyopathy. *Clin Chim Acta, 517*, 127–132. doi:10.1016/j.cca.2021.03.001

Wang, F., Jia, J., & Rodrigues, B. (2017). Autophagy, metabolic disease, and pathogenesis of heart dysfunction. *Can J Cardiol, 33*(7), 850–859. doi:10.1016/j.cjca.2017.01.002

Wang, M., Li, Y., Li, S., & Lv, J. (2022). Endothelial dysfunction and diabetic cardiomyopathy. *Front Endocrinol (Lausanne), 13*, 851941. doi:10.3389/fendo.2022.851941

Weiss, G. A., & Hennet, T. (2017). Mechanisms and consequences of intestinal dysbiosis. *Cell Mol Life Sci, 74*(16), 2959–2977. doi:10.1007/s00018-017-2509-x

Whaley-Connell, A., McCullough, P. A., & Sowers, J. R. (2011). The role of oxidative stress in the metabolic syndrome. *Rev Cardiovasc Med, 12*(1), 21–29. doi:10.3909/ricm0555

Wu, Y. T., Wu, S. B., & Wei, Y. H. (2014). Metabolic reprogramming of human cells in response to oxidative stress: Implications in the pathophysiology and therapy of mitochondrial diseases. *Curr Pharm Des, 20*(35), 5510–5526. doi:10.2174/1381612820666140306103401

Xu, H., Li, X., Adams, H., Kubena, K., & Guo, S. (2018). Etiology of metabolic syndrome and dietary intervention. *Int J Mol Sci, 20*(1). doi:10.3390/ijms20010128

Yan, L. J. (2014). Pathogenesis of chronic hyperglycemia: From reductive stress to oxidative stress. *J Diabetes Res, 2014*, 137919. doi:10.1155/2014/137919

Yan, M., Li, L., Wang, Q., Shao, X., Luo, Q., Liu, S., . . . Guo, J. (2022). The Chinese herbal medicine Fufang Zhenzhu Tiaozhi protects against diabetic cardiomyopathy by alleviating cardiac lipotoxicity-induced oxidative stress and NLRP3-dependent inflammasome activation. *Biomed Pharmacother, 148*, 112709. doi:10.1016/j.biopha.2022.112709

Yao, Y. S., Li, T. D., & Zeng, Z. H. (2020). Mechanisms underlying direct actions of hyperlipidemia on myocardium: An updated review. *Lipids Health Dis, 19*(1), 23. doi:10.1186/s12944-019-1171-8

Yara, S., Lavoie, J. C., & Levy, E. (2015). Oxidative stress and DNA methylation regulation in the metabolic syndrome. *Epigenomics, 7*(2), 283–300. doi:10.2217/epi.14.84

Yaribeygi, H., Farrokhi, F. R., Butler, A. E., & Sahebkar, A. (2019). Insulin resistance: Review of the underlying molecular mechanisms. *J Cell Physiol, 234*(6), 8152–8161. doi:10.1002/jcp.27603

Yazıcı, D., & Sezer, H. (2017). Insulin resistance, obesity and lipotoxicity. *Adv Exp Med Biol, 960*, 277–304. doi:10.1007/978-3-319-48382-5_12

Yun, H. R., Jo, Y. H., Kim, J., Shin, Y., Kim, S. S., & Choi, T. G. (2020). Roles of autophagy in oxidative stress. *Int J Mol Sci, 21*(9). doi:10.3390/ijms21093289

Zhou, R. P., Chen, Y., Wei, X., Yu, B., Xiong, Z. G., Lu, C., & Hu, W. (2020). Novel insights into ferroptosis: Implications for age-related diseases. *Theranostics, 10*(26), 11976–11997. doi:10.7150/thno.50663

Ziolkowska, S., Binienda, A., Jabłkowski, M., Szemraj, J., & Czarny, P. (2021). The interplay between insulin resistance, inflammation, oxidative stress, base excision repair and metabolic syndrome in nonalcoholic fatty liver disease. *Int J Mol Sci, 22*(20). doi:10.3390/ijms222011128

Chapter 2

Metal nanocomposites

Synthesis, characterization, and biomedical applications

Rachel Kaul, Surbhi Mishra, Varun Kumar, and Nitesh Kumar

LIST OF ABBREVIATIONS

ACE2	Angiotensin-Converting Enzyme 2
AFM	Atomic Force Microscopy
Al_2O_3/SiC	Aluminum Oxide, Silicon Carbide
ALD	Atomic Layer Deposition
ARCT-021	Arcturus Therapeutics
Au NPs	Gold Nanoparticles
B_4C/TiB_2	Boron Carbide/Titanium Diboride
BBB	Blood–Brain Barrier
BC	Bacterial Cellulose
BCA	Bio-Barcode Amplification Assay
BNT162b2	Pfizer-Biontech
CMNC	Ceramic Matrix Nanocomposites
CNPs	Cerium Oxide Nanoparticles
CNT	Carbon Nanotube
COVID	Coronavirus Disease
CVD	Chemical Vapor Deposition
DLS	Dynamic Light Scattering
DNA	Deoxyribose Nucleic Acid
EDX	Energy Dispersive X-Ray
FDA	Food Drug Administration
GPR120	G-Protein Receptor 120
HAV	Hepatitis A Virus
HBV	Hepatitis B Virus
HCV	Hepatitis C Virus
HDC	Hydrodynamic Chromatography
HEK	Human Embryonic Kidney
HeLa	Henrietta Lacks
HepG2	Hepatoblastoma Cell Line
HI	Hyperspectral Imaging
HIV	Human Immunodeficiency Virus
ICAM-1	Intercellular Adhesion Molecule-1
ICP-MS	Inductively Coupled Plasma Mass Spectroscopy
II	Ion Implantation

DOI: 10.1201/9781032621135-2

INPs	Iron Oxide Nanoparticles
LDL	Low-Density Lipoprotein
LIBD	Laser-Induced Breakdown Detection
LIF	Laser-Induced Fluorescence
MALDI	Matrix-Assisted Laser Desorption Ionization
MCF	Michigan Cancer Foundation
MF	Magnetic Fields
MMNC	Metal Matrix Nanocomposites
MN	Metal nanocomposites
MOPs	Metal Oxide Nanoparticles
MPER	Membrane Proximal External Region
MS	Mass Spectrometry
MT	Mycobacterium Tuberculosis
MTT	3-[4,5-Dimethylthiazol-2-Yl]-2,5 Diphenyl Tetrazolium Bromide
NCC	Nanocrystalline Cellulose
NFC	Nano-Fibrillated Cellulose
NIR	Near-Infrared
NMR	Nuclear Magnetic Resonance
NPs	Nanoparticles
NTA	Nanoparticle Tracking Analysis
PMNCs	Polymer-Metal Nanocomposites
PDT	Photodynamic Treatment
PLD	Pulsed Laser Deposition
PLGA	Poly Lactide-Glycolide, Copolymers
PS	Photosensitizer
PVD	Physical Vapor Deposition
PVD	Vapor Techniques
RNA	Ribose Nucleic Acid
ROS	Reactive Oxygen Species
RT-LAMP	Reverse Transcription Loop-Mediated Isothermal Amplification
SARS-CoV-2	Severe Acute Respiratory Syndrome Coronavirus 2
SAXS	Small Angle X-Ray Scattering
SEC	Size Exclusion Chromatography
SEM	Scanning Electron Microscopy
SERS	Surface-Enhanced Raman Scattering
SG	Sol-Gel
Si_3N_4/SiC	Silicon Nitride
SOD	Superoxide Dismutase
SP	Spray Pyrolysis
SpFN	Spike-Ferritin Nanoparticle
SQUID	Superconducting Quantum Interference Device
TB	Tuberculosis
TCSMPs	Triple Core-Shell Microparticles
TEM	Transmission Electron Microscopy
TGA	Thermogravimetric Analysis
$TiO2$	Titanium Oxide
VCAM-1	Vascular Cell Adhesion Molecule-1

VLPs   Virus-Like Particles
VSM   Vibrating Sample Magnetometry
WHO   World Health Organization
XPCS   X-Ray Photon Correlation Spectroscopy
XPS   X-Ray Photoelectron Spectroscopy
XRD   X-Ray Diffraction

2.1 INTRODUCTION

Nanomaterials are materials with at least one dimension on a nanoscale scale, particularly within the range of 100 nm (Jordan et al. 2005). Nanocomposites are multiphasic heterogeneous materials (NPs, nanotubes, nanoclays, or lamellar nanostructure) wherein the constituent materials, possessing different physical and chemical properties, integrate to develop new properties of the materials, ensuring that the constituent materials are within the size range of 1–100 nm (Ray and Bousmina 2007). Embedding materials engineer nanocomposites called the *reinforcing phase* into another constituent called the *matrix phase*; hence they comprise two parts, i.e., the reinforcing phase and the matrix phase, as shown in Figure 2.1. The continuous phase is matrix material, and it includes polymer, metallic/nonmetallic, and inorganic matrix materials, whereas the dispersed phase is reinforcing material, usually fibrous materials such as glass fiber, organic fiber, and so on (Camargo, Satyanarayana, and Wypych 2009; Tai, Kim, and Kim 2003).

During their formation process, each of the definite phases is unified in terms of structure and property to develop hybrid materials having multifunctionalities. Usually, the dispersed phase materials are strong with lower density, whereas the continuous phase is made up of malleable material. These composites possess the reinforcement strength in addition to the matrix toughness to accomplish an amalgamation of preferable characteristics not

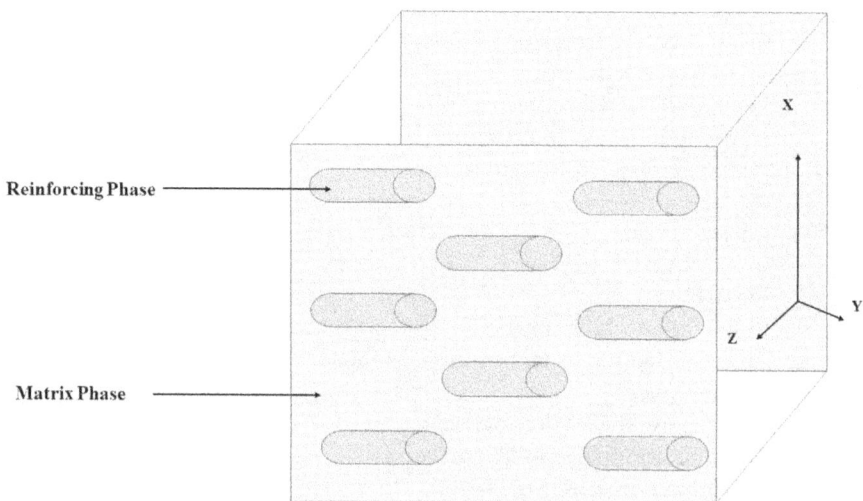

Figure 2.1 Constituents of nanocomposites (reinforcing phase and matrix phase).

obtainable in any singular constituent material upon accurate design and orchestration (Tai, Kim, and Kim 2003; Camargo, Satyanarayana, and Wypych 2009). Comprehensively, the principal rationale of making a nanocomposite is to generate an inherently diverse combined entity (nanocomposite) having synergistic or preferably properties superior to those of the constituents. However, the dispersion quality influences the interfaces among the phases, essentially governing the final properties of the nanocomposite; therefore, attaining NPs with homogeneous dispersion poses a major obstacle in nanocomposite processing (Sen 2020). Nanocomposites propound sparse properties from their comparable sizes and large surface area, making them suitable for various applications varying from the pharmaceutical, food packaging, and medical industries to the energy and electronics industries.

They hold unique electrical, mechanical, catalytic, thermal, and optical characteristics which are determined by several factors such as mobility, local chemistry, geometry, or crystallinity (Faupel et al. 2010; Camargo, Satyanarayana, and Wypych 2009). This section on nanocomposites describes their historical background and classification and lists their properties.

Nanocomposites have several advantages over conventional polymer composites characteristics that have shown significant improvements are as follows:

- Mechanical properties including bulk modules, strength, withstands limit, and so on
- Better durability
- Higher heat distortion temperature
- High smoothness
- Ease of availability
- Increased thermal stability
- Better flame retardancy
- Reduced permeability for water, solvents, and gases
- Better appearance of surface
- Increased electrical conductivity
- Improved chemical resistance
- Improved optical clarity compared to conventionally filled materials

2.2 HISTORICAL BACKGROUND

Nanomaterials have an immemorial existence as represented by the continuance of some nanostructured materials in the environment, like volcanic eruptions, seashells, skeletons, and early meteorites generating nanostructured materials. Nanocomposites have been deliberated for around 50 years though their first reference dates back to the 1950s. Coined by Theng in 1970, the term nanocomposite gained universal acceptance after 1992 (Theng 1970; Komarneni 1992).

The development of polymer nanocomposites commenced in the late 1980s in research and academics. The first decade of the 20th century witnessed the influence of reinforcing by carbon black filler on elastomer; nonetheless, it wasn't yet contemplated as a nanocomposite since the matrix filler size was in μm dimension (Noordermeer and Dierkes 2015). The idea of unifying silicate layers into polymer matrices is about half a century old. In 1966, Hess and Parker investigated the homogenous dispersion of cobalt NPs sized approximately 100 nm in the polymer (Hess and Parker Jr 1966).

Nevertheless, Toyota Company of Japan in 1988 marked the indisputable journey of polymer nanocomposites, as they used nanocomposites from silicate to manufacture their unconventional car models. Subsequently, many studies highlighting the increasing interest in clay minerals used in nanocomposites is montmorillonite, generically referred to as nano clay and sometimes as bentonite. According to Bouzouita (2016), a "natural clay" called bentonite is ordinarily prepared by the in-situ hydrothermal adaptation of volcanic rocks, and the modification of volcanic ash is extensively accessible and comparatively cost-effective, thus becoming universal clay in the applications of nanocomposite.

2.3 PROPERTIES OF NANOCOMPOSITES

Since the physiochemical and biological properties of nanomaterials vary from the properties of individual components, the resultant nanocomposites chiefly depend on the constituent materials' attributes, along with their morphology and interfacial characteristics. Although, the degree of incrementation substantially depended on geometry, aspect ratio, dispersion state, and the interfacial links of nanomaterial within the polymer matrix (Lin et al. 2007). The following subsection elaborates on a few significant properties of nanocomposites.

2.3.1 Physical properties of nanocomposites

One of the most pronounced accomplishments in nanocomposites (polymer) by the consolidation of nanomaterials is the preserved feature of lightweight polymers (Karak, Konwarh, and Voit 2010). This lightweight nature not only enhances sophistication but also facilitates ease of handling, also moderation in cost/unit volume. The influence of nanomaterials on the solubility and crystallinity of polymer nanocomposites is uneven, although insignificant.

2.3.2 Rheological properties of nanocomposites

From the standpoint of processing and exploiting them for industrial purposes, a thorough knowledge of the rheological property of polymer nanocomposites is extremely pivotal. Rheological behavior is associated with the fluidity and deformation of nanocomposites under external force, manifesting the flow properties and evaluation of viscoelastic behaviors such as the loss modulus and storage (Lin et al. 2007). Predominantly, viscosity upsurges proportionally to a certain concentration of the nanomaterial's loading. The alterations in loss of modulus and storage are analogous to the deformity's plastic and elastic response, respectively.

2.3.3 Mechanical properties of nanocomposites

Mechanical characteristics, particularly the pristine strength of the polymer, rise notably even at inadequate dose levels (<5% wt) upon the affiliation of appropriate nanomaterials. In contrast, the tensile strength typically elevates with the increasing nanomaterial loading up to a definite concentration, after which it could degenerate because of the accretion of nanomaterial in the polymer matrix. The exfoliated nanocomposites exhibit enhanced strength comparable to intercalated nanocomposites, and the strength in both nanocomposites is invariably higher than the macro-composites due to the possibility of the higher interface of nanomaterials with the matrix of polymers (Lagaly, Ogawa, and Dékány 2013).

2.3.4 Thermal properties of nanocomposites

Except for specifically engineered polymers with high thermostability, the thermostability of polymers at advanced temperatures represents one of their most striking imperfections. Integration of nanomaterials in the pristine polymers substantially ameliorates their thermal stability since these nanomaterials serve as the barrier of mass transport for volatiles produced during the decomposition and modify their decomposition process (Gilman 1999).

2.3.5 Barrier and chemical resistance

Developing appropriate polymer nanocomposites consequentially refines the chemical resistance and barrier characteristics of crude polymers. Nanomaterials bearing high aspect ratios impart large surface areas, which hinder the diffusion routes of various penetrating molecules. Furthermore, the diffusion rate of a penetrating molecule depends on the nanomaterials' degree of dispersion. Two-dimensional nanomaterials bestow a high aspect ratio which typically declines gas permeability adeptly. Therefore, the pronounced enhancement of barrier characteristics can be elucidated by employing a zigzag pathway. As the diffusion process becomes complicated, the chemical ions created from various chemical environments also face hitches in their interface with nanocomposites, and thus resistance (chemical) is also improved.

2.3.6 Biological properties of nanocomposites

The antimicrobial property, biodegradability, and biocompatibility concern the biological properties of the metal nanocomposites. Upon attuning their architecture and functionalization, polymer nanocomposites are shown to attain outstanding biodegradability and biocompatibility. Different nanocomposites of polyglycolic acid, polyurethane, polycaprolactone, polylactic acid, poly fumarate, chitosan-based polysaccharides, polyphosphazene, polyorthoester, poly(glycerol sebacate), polypyrrole, polyacrylates, poly (amido amine), etc., are used for targeting various cell types.

Biomedical research and material science have witnessed immense effectiveness using polymer nanocomposites. A predominant restraining effect is demonstrated by the most metal oxide nanoparticle-based nanocomposites against microorganisms (Wang et al. 2016). An amalgamation of NPs like copper and silver in the polymeric matrices leads to bactericidal activity against several pathogenic bacteria, like *Escherichia coli, Staphylococcus aureus, Trichophyton rubrum, Mycobacterium smegmatis, Bacillus subtilis, and Pseudomonas aeruginosa*, to name a few (Barua et al. 2014; De et al. 2015).

Polymer-based nanocomposites' biodegradability is yet another appealing and significant property due to their viability and tolerability with the environment. The constant environmental issues have resulted in the generation of improving demands for the establishment of composites that do not strain the environment impressively and deteriorate naturally. The magnitude of biodegradability is largely based on the chemical interactions within the polymer nanocomposites. The process of biodegradation essentially comprises four distinct stages: oxidation, water absorption, and hydrolysis, destruction of structural moieties, and integration of fragmented products by environmental microorganisms (De et al. 2015). Therefore, the formation of nanocomposites may govern at least any of these factors.

2.3.7 Catalytic properties of nanocomposites

Owing to their ease of separation, reusability, and long storability, polymer nanocomposites are capturing attention as selective catalysts for various transformations of organic material. These polymer-supported catalysts have improved efficiency over bare nanomaterials due to their uniform dispersity, stability, and enhanced surface activity of nanomaterials in the matrix.

2.4 CLASSIFICATION OF NANOCOMPOSITES

Nanocomposites are categorized into the following groups on the basis of the nature of matrices and the strength of interfacial interactions.

2.4.1 Based on the nature of matrices

Ceramic matrix nanocomposites (CMNCs)

CMNCs are composed of the matrix phase of a ceramic material enclosed by nanomaterials to improve the matrix's functionality. Their structure comprises a matrix containing the reinforcement components at the nanoscale (whiskers, fibers, particles, nanotubes). Generally fragile and easily fractured as a ramification of crack propagation, ceramics are rendered acceptable for applications in engineering via the integration of a matrix with a malleable metal phase, leading to enhanced durability, occurring due to the interactions between the various phases, reinforcements, and matrix at the interfaces. Examples of CMNC involve nanocomposites of carbide, alumina–zirconia nanocomposites, alumina–silicon, and carbon nanotube (CNT), wherein energy-dissipating components are integrated into the ceramic matrix to increase durability and decrease brittleness of the synthesized nanocomposites (Yu et al. 2017; Long et al. 2016) (Table 2.1).

Metal matrix nanocomposites (MMNCs)

MMNCs refer to the multiphasic combinational materials comprising ductile metals/alloys in the matrix phase that are amalgamated with the components of the reinforcement phase.

Table 2.1 Properties and processing methods of some commonly used CMNCs with their phases (Camargo, Satyanarayana, and Wypych 2009)

Matrix/reinforcement	Properties	Processing methods
Si_3N_4/SiC	Improves durability and strength	Powder process, sol-gel process
Al_2O_3/SiC	Improves durability and strength	Polymer precursor process
B_4C/TiB_2	Improves durability and strength	
$Al_2O_3/NdAlO_3$	Improved photoluminescence	Sol-gel process

Table 2.2 Some commonly used metal matrix nanocomposites with their method of fabrication

Method	System
Spray pyrolysis	MgO/Fe
	Cu/W
Liquid infiltration	Cu/Pb
	Cu/W, Nb/Cu
	Fe/Pb, Nb/Fe
	Al–C$_{60}$
Sol-gel	SiO$_2$/Fe
	Au/Ag
	Fe/Au/Au
Rapid solidification process	Al/X/Zr (X could be Si, Ni, and Cu,
	Pb/Al/Fe alloy
RSP with ultrasonics	Al/Si

Source: (Camargo, Satyanarayana, and Wypych 2009).

Mostly used metal matrices in their generation are Pb, Al, W, Mg, Sn, and Fe, while they have similar reinforcements to that of CMNC and polymer–metal nanocomposites (PMNC). Commonly used techniques for their preparation are liquid metal infiltration, vapor techniques (PVD), rapid solidification, spray pyrolysis, and chemical techniques, which involve sol-gel and colloidal processes (Table 2.2). High malleability, toughness, strength, and modulus represent some of the significant properties of MMNC (Kobayashi 2016; Dermenci et al. 2014).

Polymer–metal nanocomposites

Polymer–metal nanocomposites (PMNCs) comprise polymer-based nanomaterial as the matrix base and nano-scaled materials as the reinforcement phase. They can be described as a combination of two or more materials involving a polymer matrix and having discrete phases about a minimum of one dimension smaller than 100 nm. The last decades have exhibited substantial improvement in the thermal, barrier, mechanical, and explosive properties of polymers independent of their processability due to the incorporation of low concentrations of nanofillers into the polymer nanocomposites.

The reinforcing effect of filler is accredited to various factors, like nature, concentration, and forms of nanofillers and polymers; attributes of the polymer matrix; particle size; and the orientation and distribution of particles. PMNCs serve as typical candidates for the production of medical implants owing to their impressive properties, such as lightweight and high elasticity. Additionally, this class of nanocomposites possesses certain characteristics like high thermostability, improved resistance to abrasion, and higher barrier capacity. Lastly, clays, carbon nanotubes, graphene, halloysite, and nitrocellulose have been widely utilized in NPs to prepare nanocomposites with various polymers (Ogasawara et al. 2004). A simplified diagram illustrating the overall classification of nanocomposites is presented in Figure 2.2. Among these classes, PMNCs have gained noteworthy status within industrial and real-life applications owing to their ease of manufacturing; ductile nature; lightweight; high strength; better resistance to corrosion, fire, and acids; higher fatigue strength, and so on (Figure 2.3).

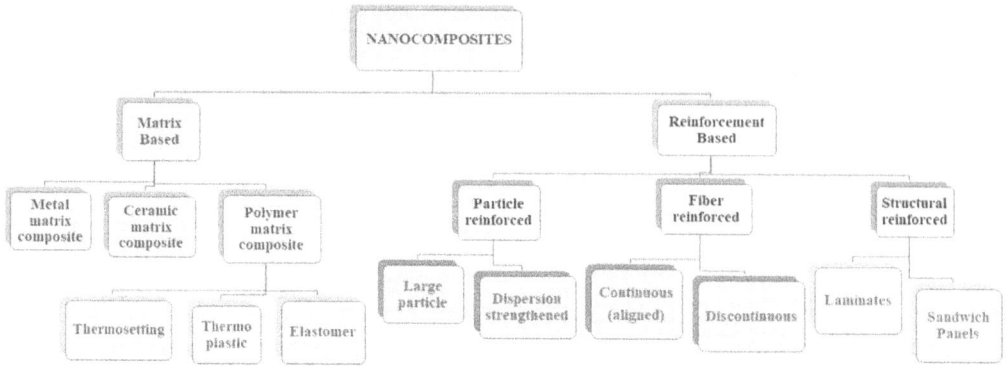

Figure 2.2 Classification of nanocomposites.

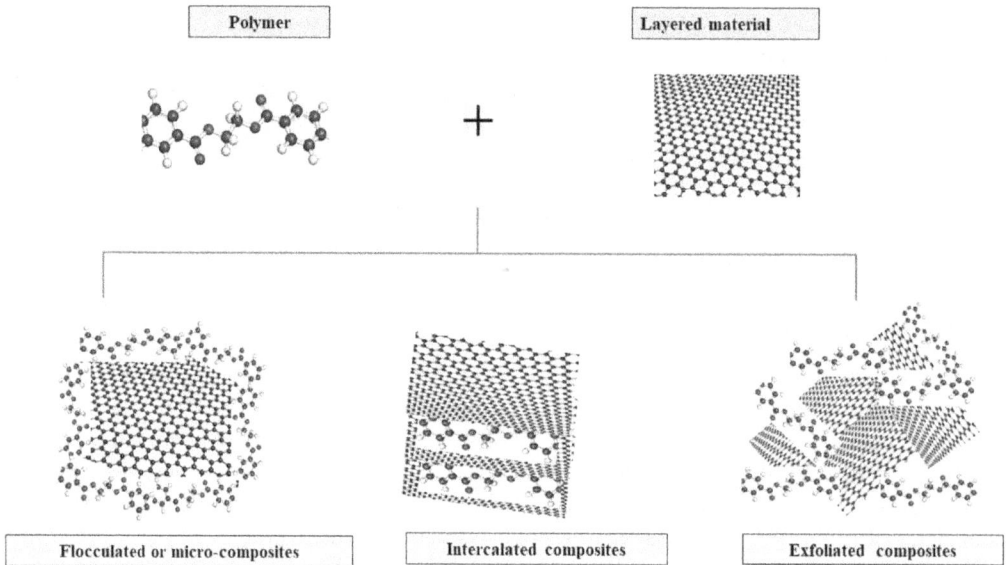

Figure 2.3 Illustration of various applications of polymeric nanocomposites.

2.4.2 Based on the strength of interfacial interactions

Furthermore, nanocomposites can be designated into the following three categories according to their strength of interfacial associations between the layered silicate and polymer matrix.

Intercalated

In this class of nanocomposites, the indigenous crystallographic structures of nanomaterials are sustained; however, the distance between the interlayers, sheets, or planes is more than the primary state due to the intercalation of the nanomaterials via a chain of polymers, thus resulting in a sandwich-like assembly. Polymer intercalation within the layers of inorganic molecules generates a nanocomposite comprising polymeric chains and interchanging layers of an inorganic molecule which generally creates an amplified interlayer spacing.

Exfoliated

These are apotheosized exfoliated structures comprising discrete layers of nanometer-scale suspended within a polymer matrix resulting from intensified polymer infiltration and delamination of the layer structure with disorder along the assembling axis of the nanocomposites. The principal difference between intercalated and exfoliated nanocomposites is that the former can weaken the van der Waals force by expanding the interlayer spacing. At the same time, the latter can degrade van der Waals forces between the adjacent layers by exercising some external strength.

Flocculated

These nanocomposites are analogous to intercalated nanocomposites. Occasionally, the inorganic layers in this class of nanocomposites are flocculated due to the silicate layers' hydroxylated edge–edge association (Figure 2.3).

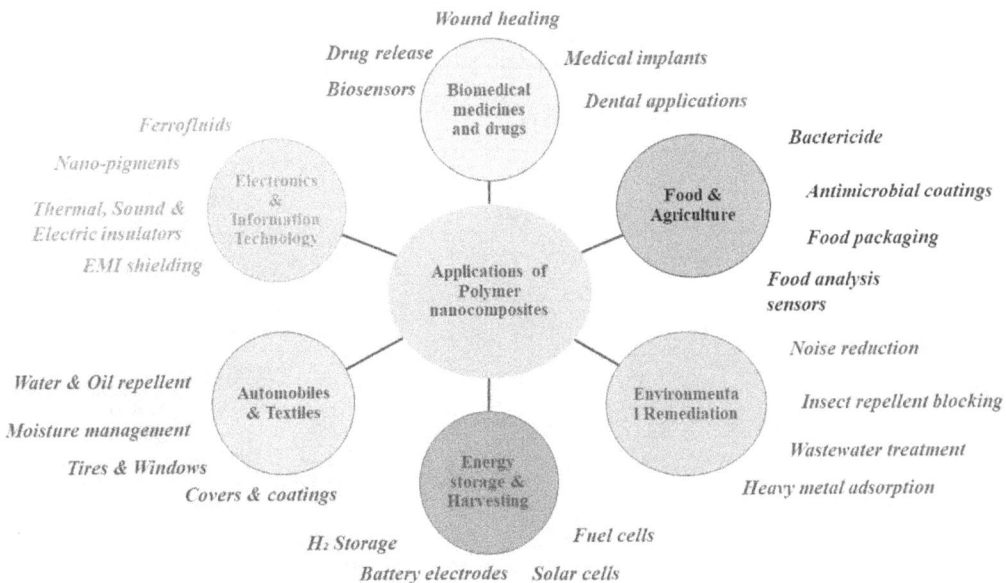

Figure 2.4 Demonstration/description of the various forms of composites generated from the interface between the polymers and layered compounds: (a) flocculated nanocomposites (b) intercalated nanocomposites; (c) exfoliated nanocomposites.

2.5 EMERGING TRENDS IN THE METAL OXIDE NANOCOMPOSITES

In recent years, nanotechnology has substantially metamorphosized healthcare strategies having a widespread impact on public health. This branch of science has captured extensive scientific attention due to the unique properties of nanocomposites and their concatenative application range in biological and medical sciences, particularly the evaluation of metal-based NPs and nanocomposites for biomedicine, biotechnology, and pharmaceutical applications (Bayda et al. 2019; Qasim et al. 2014). This section describes the recent and emerging trends in nanocomposites, specifically concerning diseases and drug delivery.

2.5.1 Nanocomposites as prospective therapeutic agents of major infectious diseases

Infectious diseases are responsible for causing millions of deaths worldwide, having a predominant global impact on healthcare and socioeconomic development. Furthermore, there is an intensifying need for the emergence of novel treatment strategies because of the increasing resistance to commercial drugs and undesired effects because of their prolonged application. Presently, acute respiratory infections like tuberculosis, COVID-19, hepatitis, and HIV infection are the chief public health issues (Mehendale, Joshi, and Patravale 2013) (Figure 2.5).

Coronavirus disease (COVID-19)

COVID-19 is a transmittable disease triggered by the novel coronavirus first identified in Wuhan, China, in December 2019. Belonging to the order *Nidovirales*, and *Coronaviridae*

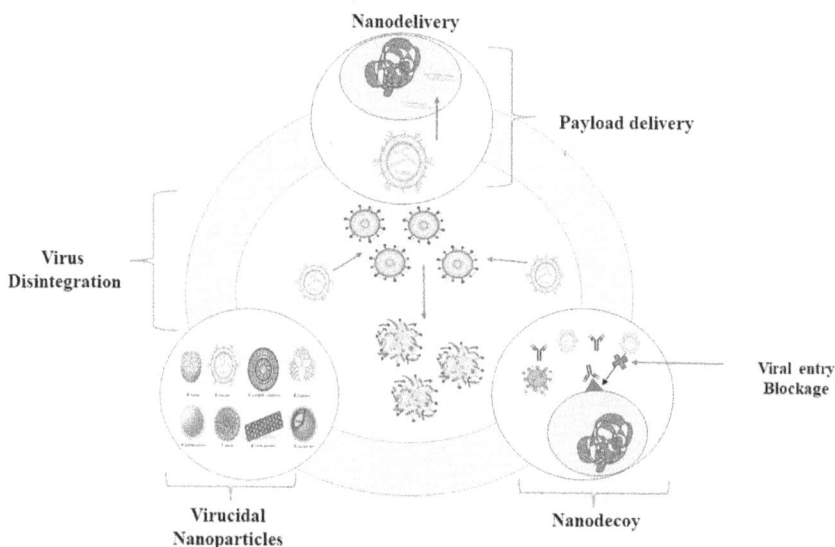

Figure 2.5 Nanotechnology-based methods for treating viral diseases: (1) NPs-based delivery of therapeutic cargoes to the infected cells, thus improving bioavailability; (2) virucidal properties of nanomaterials; and (3) nanodecoys to neutralize virus infectivity.

family, it is an enveloped virus consisting of a single-stranded positive (+ve) ribose nucleic acid (RNA) of about 25–33 kb as the genetic material. The virus uses ACE2 as the main receptor to bind with the epithelial cells in the nasal cavity and start its replication process. The main attributes of the virus involve sudden mutation, modification in tissue tropism, interspecies transmission, and viral adaptation to various epidemiological conditions. Essentially, a disease of the respiratory system affecting the lung parenchyma, COVID-19 presents fever, cough, and shortness of breath as its principal symptoms. However, the latest studies have shown the involvement of multiple organ systems, leading to extra-pulmonary symptoms (Zhu et al. 2020). Until now, SARS-CoV-2 has infected over 750 million people worldwide and caused over six million deaths.

Nanocomposites enfold great potential for preparing new therapeutics in identifying and treating COVID-19 infection. A recent study coupled NPs-based biosensors with transcription loop-mediated isothermal amplification (RT-LAMP) in diagnosing COVID-19 infection, proposing that this strategy is a potential means for COVID-19 infection. Itani et al. (2020) demonstrated NPs-based treatment along the intranasal route wherein therapeutic agents like siRNA, protein inhibitors (AAKI and JAK), or antibodies are conjugated to NPs. These NPs can then be administered smoothly through the nasal route to boost the efficacy against the COVID-19 virus.

HIV

Around 37 million people all over the globe are living with HIV, for which there is no empirical cure. Among the two major types of viruses, HIV-1, being more pathogenic and prevailing, primarily targets CD4+ T helper cells, causing a progressive loss in their number and many immunological abnormalities, eventually leading to AIDS. NPs are prospective agents that are largely employed to identify and treat HIV infection. Magnetic NPs were used to detect HIV-1-infected cells in an *in vitro* experiment. Correspondingly, bio-barcode amplification assay (BCA) utilizing Au NPs for detecting HIV-1 P24 antigens was used at lower concentrations. Apart from this, studies suggest antibody-NCs conjugation-mediated targeting as an efficient approach to combat HIV infection in astrocytes (das Neves et al. 2013; Tang and Hewlett 2010; Sharma and Garg 2010).

Hepatitis

Hepatitis refers to the inflammation of the liver. Hepatotropic viruses, i.e., hepatitis A virus (HAV), hepatitis B virus (HBV), and hepatitis C virus (HCV), are the most frequent causes of viral hepatitis leading to a wide array of liver diseases such as hemochromatosis, liver cirrhosis, and liver failure. The viruses invade hepatocytes, wherein they replicate affecting the functionality of the liver. Most of these viruses are acute and self-limiting even though types B, C, and E can become chronic (Ciupe et al. 2007).

Metal NPs are increasingly employed in diagnostic techniques owing to their higher sensitivity. These pioneering techniques include electrochemical assays of gold NPs for the detection of HBV DNA sequence effectively. Similarly, Tang et al. (2004) instantaneously detected the three hepatotropic viruses on a protein amplification chip by using gold NPs that provided strong signals for identifying viral antibodies following their stable and uniform shapes. A sandwich-type electrochemical immune sensor was developed using labeled graphene nanosheet over the polypyrrole-coupled gold NPs to detect HBV (Pei et al. 2019).

Tuberculosis

Tuberculosis (TB), one of the oldest diseases affecting humankind and a major cause of mortality across the world, is caused by the bacterium *Mycobacterium tuberculosis* (MT) (Luies and Du Preez 2020). Highly prevalent among the lower socioeconomic sections of the community, it is one of the deadliest diseases affecting one-third of the world's population, only, next to HIV/AIDS, as per WHO (WHO 2019). The different strains of MT can efficiently be detected using gold NPs conjugated with DNA (Cambier, Falkow, and Ramakrishnan 2014). Lately, electrochemical sensors containing DNA-coupled Au NPs and an aptamer H37Rv (as a probe) were developed to diagnose the H37Rv-type MT within a few hours (Zhang et al. 2019). Curdlan NPs, demonstrated by Basha et al., are used for the targeted delivery of drugs to host macrophages. Upon entry into the macrophage, these curdlan-cyclodextrin conjugate NPs block the dectin-1 receptor resulting in the degeneration of *M. smegmatis* (Basha, TS, and Doble 2019).

2.5.2 Nanocomposites application in drug delivery

Nanotechnology offers multiple benefits in delivering therapeutic drugs to precisely targeted locations in a controlled fashion, thereby having the caliber to revolutionize the therapy of several diseases such as diabetes, cancer, and neurodegenerative diseases (Petros and DeSimone 2010; Shrestha 2012). Nanostructures can be employed as delivery agents by enclosing or tethering therapeutic drugs and delivering them to intended tissues in a defined, regulated manner. This is accomplished due to their unique characteristics of decreased particle size that assists the delivery of drugs into anatomically privileged locations, improved surface area to volume ratios guaranteeing accommodation of high payloads, and alterable surface charge to help in cellular entry across the membrane, thus making them attractive tools for the treatment of viral diseases (Caron et al. 2010; Alexis et al. 2008; Mitchell et al. 2021). In addition to this, NPs have been reported to contain biomimetic properties leading to innate antiviral properties popularly demonstrated by silver NPs and dendrimers (Figure 2.6). Table 2.3 summarizes their characteristic advantages and limitations (Mitchell et al. 2021).

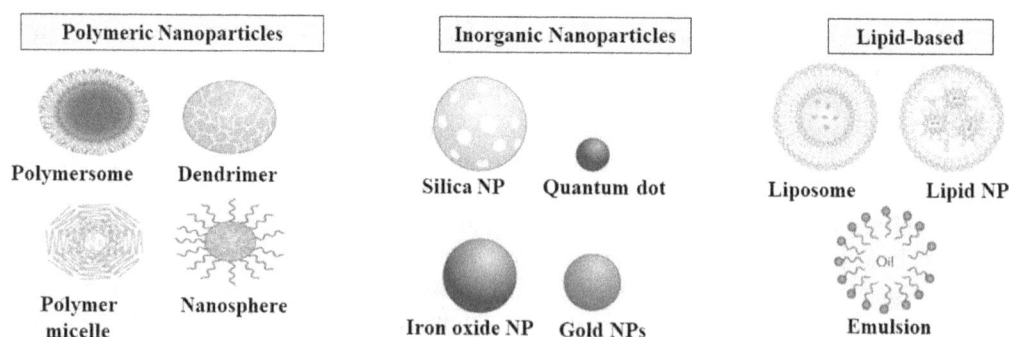

| Polymeric Nanoparticles | Inorganic Nanoparticles | Lipid-based |

Polymersome   Dendrimer   Silica NP   Quantum dot   Liposome   Lipid NP

Polymer micelle   Nanosphere   Iron oxide NP   Gold NPs   Emulsion

Figure 2.6 Different classes of NPs (NP) featuring multiple subclasses which serve essential roles in cargo delivery and patient response.

Table 2.3 Summary of NPs and their characteristics, advantages, and limitations

Nanoparticle types	Example	Properties	Advantage	Limitations	Reference
NPs of polymers: poly lactide-glycolide, copolymers (PLGA), poly (ε-caprolactone), alginate, polylactide (PLA), and chitosan	Decapeptyl®, Gonapeptyl Depot®, Enantone Depot®, Abraxane	Biodegradable aliphatic NPs, prepared from synthetic polymers	It helps in the modulation of drug release and protects functional moieties from damage, and prevents the toxic effects of drugs on healthy cells	Poor reproducibility	(Ram Prasad et al. 2017)
Nanogels: poly (ε-caprolactone), poly (lactic acid), polyacrylates poly (glycolic) copolymers, polymethacrylates	Aerogel AA, C16–catCA nanogel, PGMA Nano gels (PGED-NGs)	Biocompatible, improved surface area, and high moisture content	Polymer-based drug vehicle systems enable long-lasting stability in blood	Less stable	(Kesharwani et al. 2019)
Nanoemulsion: a) water in oil b) oil in water c) bi-continuous	Norvir, Restasis, Cyclosporin A, Etomidat, Diprivan, Flurbiprofenaxtil, Troypofol, Dexamethasone, Alprostadilpalmitate	Sustained and isotropic	The colloidal delivery carrier enables the prolonged effect of drugs and protection from oxidation	Less stable	(Halnor et al. 2018)
Solid-lipid NPs:	Ciprofloxacin (CIP)-loaded SLNs	Controlled-drug release	Cost-friendly raw materials, improved biocompatibility, protection to functional moieties from damage	The crystallinity of solid lipids NPs make them more prone to oxidative damage	(Jores et al. 2004)
Nonstructured lipid carriers: Solid lipids incorporated into liquid lipids	Fluconazole-loaded NLCs	Immobilization of therapeutic drugs and prevents coalescence	Low toxicity, controlled and targeted release, and drug protection.	More prone to gelation, minimal loading efficiency	(Wairkar, Patel, and Singh 2022)
Dendrimers: poly- polyamidoamine, propylenemine	PPI, AstromolR, DAB, PAMAM Starburstk, PAMAM dendrimers	Aqueous solubility, nano size, less polydispersity index	Investigated delivery via oral, transdermal, pulmonary and ocular routes.	Low manufacturing capacity	(Prasad et al. 2018)
Nanocapsules: Resveratrol-loaded lipid-core nanocapsules	SOLUDOTS-PTX	Biocompatible, chemically stable, and high reproducibility	Protect the incorporated drug and can be used in varieties of treatment	Delayed release of functionally active moieties	(Figueiró et al. 2013)
Nanosponges	Brexin, Glymasason, and Prostavastin		Potential to incorporate both lipophilic and hydrophilic moieties	Incorporate only low molecular weight drugs	(Sherje et al. 2017)

2.5.3 Nanocomposites in vaccine development

Vaccination refers to administering foreign pathogenic material into the body to protect against a specific disease by boosting the host immune system. Nevertheless, preparing potential drugs against several emerging diseases is a long quest due to the various opposition and restrictions like proper delivery of antigens, antigen selection, and adjuvant engineering (Ovsyannikova and Poland 2011; Rappuoli et al. 2011). The nanoscience has been explored to conquer these subjects by assisting in the preparation of next-generation drugs

Table 2.4 Summary of the nanoparticle-based vaccine candidates for infectious diseases in clinical trials

Disease	Vaccine agents	Nanoparticle types	Clinical identifier	Clinical phase	References
SARS-CoV-2	ARCT-021	Lipid	NCT04480957	Phase I/II trials completed	(Vu et al. 2021)
	BNT162b2	Lipid	NCT04760132	Recruiting for phase IV trials	(Vu et al. 2021)
	Covac 1	Lipid	ISRCTN17072692	Phase I trials completed	(Heitmann et al. 2022)
	ChulaCov19	Lipid	NCT04566276	NA	(Joyce et al. 2021)
	mRNA-1273	Lipid	NCT04760132	Recruiting in Phase IV trials	(Joyce et al. 2021)
	mRNA 1273.351	Lipid	NCT04785144	Phase I trial is active	(Joyce et al. 2021)
	SpFN	Ferritin	NCT04784767	Phase I trial is active	(Joyce et al. 2021)
Influenza	HA-Ferritin	Ferritin	NCT03186781	Phase I trials completed	(Houser et al. 2022)
HIV	MPER-565	Liposome	NCT03934541	Phase I trials completed	(Williams et al. 2021)

Table 2.5 The associated advantages and disadvantages of nanoparticle-based vaccine candidates

Nanotools	Advantages	Disadvantages	References
Liposomes	• Biocompatible, less toxic, and biodegradable • FDA-approved earliest nano-vaccine • Mimics structure of cell membrane, no safety issues • Enhanced drug entrapment for both hydrophobic and hydrophilic antigens • Enable controlled drug release and protection from GI-based destruction	• Not cost-efficient • Low aqueous solubility • Short residence period • Rather bigger size • Prone to GI-based destruction • Low encapsulation efficiency for hydrophobic agents because of the limited interior of the lipid domain	(Barenholz 2012; Tandrup Schmidt et al. 2016; Jazayeri et al. 2021)

Nanotools	Advantages	Disadvantages	References
PGLA NPs	• Biodegradable and biocompatible • American FDA approved • A broad range of encapsulation • Act as an adjuvant; excites maturation of dendritic cells • Ease of modification for broad functional and targeted deliveries	• Decreased encapsulation capacity for anionic antigens like pDNA; • Needs surface charge modification for quick uptake by the cell membrane • Nonspecific drug release profile • Comparatively short residence period, inability to cross the blood-brain barrier (BBB)	(Barenholz 2012; Nimesh 2013; Cappellano et al. 2019)
Chitosan NPs	• Biodegradable and biocompatible • Can be explored for mucosal delivery routes and longer antigen residential periods • Enable penetration of epithelium by tight junctions	• Bad aqueous solubility at basic or neutral pH • Degradation at low pH, such as in a gastric environment, causes cytotoxicity	(Aderibigbe and Naki 2019; Boroumand et al. 2021; Keong and Halim 2009)
Ferritin NPs	• Biocompatible • Able to display 24 separate surface molecules	• Heterogeneity of NPs • Inadequate interactions of antigen among the subunits results in reduced functionality	(Lee, Cho, and Kim 2022)
Calcium Phosphate NCs	• Biodegradable and biocompatible • Nontoxic • Storage stability • High affinity for the entrapment of nucleic acid	• Difficulties in commercialization • Agglomeration of the NPs • The synthesis method could use high-end laser equipment	(Heng, Yew, and Poh 2022; Sharma et al. 2015)

as exemplified during the COVID-19 pandemic, further underlining the obligation to design effective novel vaccine platforms that exploit nanoscale particulates mimicking the structural characteristics of the pathogens for control and prevention of infectious diseases (Liao et al. 2014; Liao et al. 2016). Until now, some of the nano vaccines listed in Table 2.4 have been accepted for human application, and many are being investigated in pre-clinical or clinical trials.

Nanocomposites of the next-generation vaccine development involve those prepared from organic polymers and inorganic NPs (NPs). Further, virus-like particles (VLPs), micelles, self-assembled proteins, inorganic NPs, liposomes, and polymers are also increasingly being investigated. Their advantages and limitations as vaccine candidates are discussed in Table 2.5.

2.6 FABRICATION OF METAL OXIDE NANOCOMPOSITES

Regarding nanocomposites, metal oxide NPs (MOPs) have received much attention in various industries, including environmental remediation, drug delivery, food and agriculture production, catalysis, sensing, and medicine. As a result, many synthesis routes for these metal oxide NPs have been developed. Chemical and physical methods have primarily synthesized MOPs. MOPs can be synthesized by both the bottom-up and top-down approaches (Bayda et al. 2019; Qasim et al. 2014; Hendricks et al. 2015; Rao, Tian, and Chen 2020). Top-down approaches are less popular than bottom-up approaches. Nowadays, several methods are used to synthesize MOPs; these are discussed in this section (Qasim et al. 2014; Mehendale, Joshi, and Patravale 2013).

Nanofabrication employs physical or chemical interactions at the nanoscale to arrange simple blocks into more complex assemblies (Wan et al. 2020; Rao, Tian, and Chen 2020). Bottom-up methods are an increasingly significant complement to top-down methods in nanofabrication as component sizes decrease. In biological systems, where nature has used chemical forces to make all the components required for life, bottom-up strategies find their inspiration. The following categories further split the bottom-up approach to synthesizing metal oxide nanocomposites.

2.6.1 Physiochemical methods

The physiochemical methods use several different physical and chemical processes to develop MOPs. Physical techniques like pulsed laser deposition (PLD), chemical vapor deposition (CVD), physical vapor deposition (PVD), ion implantation (II), atomic layer deposition (ALD), spray pyrolysis (SP), and chemical techniques like hydrothermal co-precipitation, and microemulsion routes, sol-gel, electrochemical, and photochemical processes are used to fabricate MOPs from aqueous solutions (Rao, Tian, and Chen 2020; Wan et al. 2020). These physiochemical production techniques are associated with certain demerits, like being exclusive, laborious, and unsafe for the environment.

2.6.2 Physical vapor deposition (PVD)

PVD refers to various techniques for vacuum deposition that can be utilized to develop MPOs, coatings, and thin films. PVD is a method in which the material in the condensed phase transitions to its vapor phase and then again condenses to a thin film (Zhu et al. 2020). The frequently used PVD methods are evaporation and sputtering. The steps involved in PVD are discussed here:

(i) Sputtering or evaporation of various precursors to create vapors
(ii) Supersaturation of the vapor phase in an inert condition to facilitate condensation of the MOPs
(iii) Thermal treatment in an inert atmosphere to consolidate the MOPs

2.6.3 Pulsed laser deposition (PLD)

The PLD technique has developed as a substitute and added benefit of maintenance of the target phase's stoichiometry. PLD holds great potential for creating MOPs. By maintaining the deposition parameters, the PLD method can develop NPs with the desired thickness, shape, and composition (Itani, Tobaiqy, and Al Faraj 2020). This method additionally enables the deposition of various target materials with distinct physicochemical and biological characteristics over one substrate for graded (functionally) coatings. There are three main steps in the PLD technique:

(i)   The target material is ablated;
(ii) A plume with high energy is created; and
(iii) The film is developed on the substrate.

2.6.4 Ion implantation (II)

An ambient temperature method is ion implantation by which ions from one element are accelerated onto the surface of a solid target, thereby altering the target surface's chemical, electrical, or physical properties. The ion implantation method is utilized in the fabrication of MOPs and in research related to materials science. Chemical and physical changes are induced when ions strike a high-energy target during ion implantation. The energetic collision cascades can potentially damage or even destroy the target's crystal structure, and ions with enough energy can lead to nuclear transmutation (Rao et al. 2020).

2.6.5 Atomic layer deposition (ALD)

ALD, a cutting-edge deposition technology, enables the highly controlled deposition of ultra-thin films of NPs with a thickness of just a few nanometers. In addition to great controlled thickness and homogeneity, ALD also enables the coating of 3D structures with conformal materials for a high aspect ratio. Due to its reliance on self-limiting surface reactions, ALD often offers extremely low particle and pin-hole levels, which might be good for various applications (Maartens, Celum, and Lewin 2014).

2.6.6 Spray pyrolysis (SP)

SP includes the process of thermal decomposition, which needs a short residence time and high temperatures. The droplets are produced using a nozzle and are further evaporated before being converted into a gaseous phase by flame. The gaseous molecules of precursor species are burnt, resulting in a self-sustainable combustion spray. The decomposition step comprises various reactions known as precursor pyrolysis, depending on the type of gases and precursor moiety used. Both metal oxide and metal showed extremely low saturation vapor pressures (133 Pa is needed for zinc at 500°C, for example). Therefore, the resultant gaseous vapors are significantly stochastically formed collisions and supersaturated to enable the creation and deformation of molecular structures, which are based on nucleation (heterogeneous and homogeneous) and process parameters to produce the required shape of the NPs (das Neves et al. 2013).

2.6.7 Sol-gel (SG) method

The process of sol-gel is more often known as a wet chemical method for developing different nanostructures, especially MOPs. In this procedure, the molecular precursor (often metal alkoxide) is mixed in water or alcohol, heated, and stirred until it gels. Since the gel formed during hydrolysis/alcoholysis is wet or moist, it needs to be dried appropriately depending on the gel's intended use and desired qualities. The developed gels are pulverized and then calcined after the drying stage. The cost-effectiveness of the sol-gel process allows for good control over the products' chemical composition because of the low reaction temperature (Tang and Hewlett 2010).

2.6.8 Hydrothermal synthesis

The frequently utilized technique for fabricating nanocomposites is hydrothermal synthesis. Essentially, it uses a solution-reaction-based methodology. The fabrication of MOPs during hydrothermal synthesis can take place at temperatures ranging from near ambient to extremely high temperatures. Based on the vapor pressure of the primary components in the fabrication, either high- or low-pressure conditions can regulate the geometry of the synthesized MOPs. This approach has been employed to develop a broad range of nanomaterials. The hydrothermal-based synthesis technique has several merits over other ones. It is possible for hydrothermal synthesis to fabricate nanomaterials that are sensitive to high temperatures (Tang and Hewlett 2010).

2.6.9 Co-precipitation

A quick and easy way to develop NPs of different sizes is through co-precipitation. Co-precipitation is the precipitate's removal of soluble materials under specific circumstances. The solution often experiences an abrupt appearance of nucleation when the substance concentration exceeds supersaturation. Diffusion will enable the nucleation to expand onto the surface and develop into NPs. To obtain uniform NPs, nucleation must be slowed down during the growth process. The pH and ion concentration in the solution could be changed to regulate the size of the nanocomposites (Sharma and Garg 2010).

2.6.10 Green synthesis of metal oxide nanocomposites

Although chemical and physical processes have historically dominated the synthesis of MOPs, an approach based on green chemistry to nanoparticle preparation is now receiving considerable attention, particularly in this environmental and human health era. The green synthesis of MOPs uses biomolecules like carbohydrates, amino acids, and proteins from microbes, plants, and living cells, along with mild reaction conditions like that of the sol-gel method. Green synthesis, which uses less or no harmful constituents to manufacture NPs, has generated interest because it is both economically and environmentally friendly. It is an emerging way of addressing nanoparticle toxicity that is typically linked to traditional physical and chemical fabrication methods (Ciupe et al. 2007; Pei et al. 2019; Hendricks et al. 2015). It employs non-hazardous chemicals, simple methods, and mild reaction conditions. Green chemistry-based synthesis methods involve synthesis using a variety of biomaterials like microorganisms and plants as either reducing or stabilizing agents or can be used as both in some circumstances. The fabrication of MOPs using plants and microorganisms is termed biosynthesis. Green synthesis and biosynthesis are terms that can be used interchangeably.

Various bioactive compounds are found in plant species that act as reducing agents for salts of metals, including alkaloids, phenols, ascorbic acid, flavonoids, citric acid, polyphenols, reductase, and terpenes. Furthermore, the ability of microorganisms to fabricate highly specialized inorganic MOPs is astounding. Several material scientists have turned their attention to biological systems to learn more about them and acquire the skills required for the precise fabrication of nanostructures due to their magnificent abilities (Ciupe et al. 2007; Pei et al. 2019; Hendricks et al. 2015).

Generally, biologically controlled and biologically induced synthesis have been used to describe MOP synthesis by living beings. It is well known that a few organisms engage in the biologically regulated synthesis of NPs naturally. The living organisms may regulate the composition, particle size, and surface area of the developed NPs during this sort of synthesis. Although the synthesis of MOPs via a biologically controlled manner demonstrates remarkable control over the composition and geometry of the synthesized NPs, it is limited to synthesizing a small number of NPs. It presents only in a small group of organisms.

Nonetheless, scientists have successfully used microbes to produce a variety of nanomaterials, including MOPs, from a simple metal ion precursor. Induced biological synthesis of nanomaterials has a huge variety of compositions than controlled biological synthesis (Ciupe et al. 2007; Pei et al. 2019). The microorganisms that can produce nanomaterials like metal oxides have been explored.

2.7 CHARACTERIZATION OF THE FABRICATED METAL OXIDE NANOCOMPOSITES

The synthesized nanocomposites present numerous characterization obstacles that can influence the detailing and suitability of characterization for the NPs. Therefore, it is crucial to comprehend the challenges of nanoparticle characterization and choose an appropriate characterization method (Hendricks et al. 2015; Rao, Tian, and Chen 2020; Tang et al. 2004; Wang et al. 2020). Several characterization tools may be utilized to assess parameters of nanocomposites, such as ultraviolet-visible spectroscopy (UV-Vis), scanning electron microscopy (SEM), transmission electron microscopy (TEM), atomic force microscopy (AFM), dynamic light scattering (DLS), thermogravimetric analysis (TGA), x-ray photoelectron spectroscopy (XPS), x-ray diffraction (XRD), nuclear magnetic resonance (NMR), dual-polarization interferometry, nanoparticle tracking analysis (NTA) for determination of Brownian movement, and measuring the size of the particles (Tang et al. 2004; Hendricks et al. 2015; AbdelAllah et al. 2020). Some of the crucial characterization techniques designed to measure specific parameters are described in detail here.

2.7.1 Spectroscopic analysis

Spectroscopic analysis is used to investigate hybrid composites, providing useful information on chemical composition, optical, elemental type, crystallinity, and electrical properties of the synthesized NPs.

UV-visible spectroscopy (UV-Vis)

UV-Vis is a crucial characterization tool that is frequently used for MOPs. It measures the amount of UV light a substance absorbs. This data can be applied to estimate nanocomposites' chemical composition and analyte concentrations. The wavelength of light between

300 nm and 800 nm could be used to determine the properties of various MOPs with sizes between 2 nm and 10 nm (Hendricks et al. 2015; Luies and Du Preez 2020).

FTIR spectroscopy

FTIR spectroscopy is performed to assess the chemical/functional groups present on the surface of nanocomposites. It is a non-destructive and rapid approach that can detect different groups (functional or chemical) and is very sensitive to changes in molecular geometry. FTIR analyses the amount of infrared radiation transmitted by the nanocomposite against wavelength. The infrared absorption bands then reveal the composition and structures of molecules. This is well-suited regarding surface chemistry and the detection of functional groups on the surface of NPs (e.g., ketones, amines). The absorption peaks in the FTIR spectrum, which are correlated to the intensities of vibrations among the bonds of an atom in the nanoparticle, serve as a fingerprint of the synthesized NPs (Cambier, Falkow, and Ramakrishnan 2014; Rao, Tian, and Chen 2020).

2.7.2 Microscopic analysis

The morphology of the synthesized NPs has a significant role in the properties of the synthesized NPs. The outcome of microscopic images provides the actual topology of each metal oxide and expresses distinct dimensions resulting from the growth and process of nucleation. Several microscopic analytical techniques, including TEM, SEM, and AFM, are frequently utilized to study the surface of nanocomposites (Rao, Tian, and Chen 2020; Hendricks et al. 2015; Zhang et al. 2019; Basha, TS, and Doble 2019).

Scanning electron microscopy (SEM)

SEM is one of the most widely utilized techniques for characterizing nanomaterials and nanocomposites. SEM reveals more information related to nanomaterials by employing a high beam of electrons on a very fine scale onto the sample surface. A specimen's surface is scanned by the primary beam of electrons, produced in a high vacuum condition. An image of the surface is produced by detecting changes in signals generated after the primary electrons strike the specimen. This imaging technique enables us to easily observe the size distribution, size, and morphology of the synthesized MOPs (Zhang et al. 2019; Basha, TS, and Doble 2019; Hendricks et al. 2015).

Transmission electron microscopy (TEM)

The TEM method is a potent instrument for an in-depth analysis of the structural properties of MOPs. The TEM uses electrons rather than light to perform the same fundamental operations as the light microscope. When a very thin sample is exposed to a high-energy electron beam, properties like the crystal structure and structural features of the nanocomposites can be seen because of interactions between the atoms and electrons. The best resolution feasible for TEM images is many orders of magnitude superior to that from a light microscope because the wavelength of electrons is significantly shorter than that of light. As a result, TEM may show even the finest details about the interior structure, even down to the level of individual atoms (Zhang et al. 2019; Basha, TS, and Doble 2019).

Atomic force microscopy (AFM)

The morphology of synthesized nanocomposites and biomolecules is investigated using AFM. AFM creates 3D images instead of SEM and TEM, allowing particle height and volume to be assessed. This technique, which relies on the physical scanning of analytes at the submicron level using the tip of a probe, can determine particle size with extremely high resolution. Software-based processing of the captured image can analyze quantitative data about individual NPs and groups of NPs using AFM, including size (height, length, and width), surface texture, and morphology. AFM can be carried out in a gas or a liquid medium. The main benefit of AFM is that it can be used to analyze nonconducting samples without any special preparation and delicate biological and polymeric nanocomposites (Basha, TS, and Doble 2019; Zhang et al. 2019).

2.7.3 Elemental analysis of the metal oxide nanocomposites

Several techniques can be employed for the elemental analysis of the synthesized NPs, and some of them are discussed here.

Energy dispersive X-ray (EDX)

EDX technique is employed to characterize nanocomposites depending on their elemental compositions. To perform this, an electron beam is made to bombard the NPs, which release X-rays. The X-rays emitted to balance out the energy variation among two electrons can be detected using an EDS detector connected to an SEM. Since the energy of X-rays (emitted) is a defining characteristic of the element, qualitative and quantitative analyses can be performed on it (Pi et al. 2020).

X-ray diffraction (XRD)

XRD is used to investigate the nanocomposite crystalline nature and is recognized by the unique diffraction patterns of a crystal structure. X-rays can pass through the nanocomposites to produce a specific diffraction pattern. Moreover, the following equation has been used to determine the mean crystal size and d-spacing of the synthesized NPs using the Debye-Scherrer formula and Bragg's law equations (Pi et al. 2020):

$D = K\lambda/\beta \cos\theta$-Debye-Scherrer formula
$D = \lambda \ 2\sin\theta$-Braggs law
Where K = Scherrer constant (0.98)
β = Full width at half maximum
λ = X-ray wavelength
D = Particle size
θ = Bragg's angle while d is the d-spacing

2.7.4 Mean particle size and surface area analysis

It is well-established that a high redox reaction is motivated by a wide surface area. Surface charge is yet another significant parameter in characterizing NPs. The strength and nature of the surface charge are crucial because they affect the interaction of nanocomposites with the environment (biological) and the bioactive substances produced by plants and microorganisms through electrostatic interactions (Pi et al. 2020; WHO 2019).

Dynamic light scattering (DLS)

DLS is a rapid and most often used technique for evaluating particle size distribution. DLS is frequently used to observe the size of Brownian particles in colloidal nanoformulations. When a beam of monochromatic light (laser) is incident onto a suspension of NPs in Brownian motion, a Doppler shift is observed, thus shifting the wavelength of the incident light by a value corresponding to the size of the particle. Therefore, DLS enables the size distribution estimation, and the NPs' motion in the nanoformulation can be evaluated by determining the diffusion coefficient (WHO 2019; Basha, TS, and Doble 2019).

Nanoparticle tracking analysis (NTA)

An upgraded approach called nanoparticle tracking analysis is used to distinguish various types of NPs according to their sizes, which can range from 30 nm to 1000 nm, along with a lower limit of detection based on the refractive index. This technique makes it possible to visualize the nanoparticle suspension directly, and it can be used to deliver drugs precisely to the targeted locations in a controlled manner using NPs (Basha, TS, and Doble 2019; Pi et al. 2020).

X-Ray photoelectron spectroscopy (XPS)

XPS, which stands for X-ray photoelectron spectroscopy, is a surface-sensitive quantitative technique that uses the photoelectric consequence to identify the presence of specific elements in a nanocomposite or on its surface, as well as their general electronic structure, chemical nature, and electron density within the material. Additionally, XPS can be used to determine the mechanism of the reaction on the exterior of magnetic MOPs, evaluate how well the various elements bond together, and confirm the structure and composition of the NPs (Amini et al. 2020).

Zeta potential analyzer

The surface potential and stability of colloidal nanocomposites are determined indirectly by employing a nanomachine called Zetasizer. Zeta potential analysis refers to the potential differences among the surface of the shear and outer Helmholtz plane. The stability of the colloidal dispersion during storage can be predicted by observing the zeta potential. High values of either the positive or negative zeta potential should be attained to ensure stability and prevent particle agglomeration. Also, the value of surface hydrophobicity can be measured. Based on zeta potential, the type of components entrapped within NPs or coated on their surface can also be examined (Basha, TS, and Doble 2019).

Thermogravimetric analysis (TGA)

The TGA technique is utilized to predict the chemical composition of surface coatings like polymers or surfactants to determine the binding capacity on the surface of nanocomposites (Kong et al. 2020).

2.7.5 Other characterization techniques

Apart from the techniques mentioned earlier for the characterization of MOPs, there are several other techniques that can be used to separate, characterize, and fractionate nano-composites for specific applications.

Assessment of surface hydrophobicity

The surface hydrophobicity of the synthesized nanocomposites can be determined utilizing various analytical techniques such as biphasic partitioning, probe adsorption, hydrophobic association, contact angle measurements, and chromatography. Recently, X-Ray photon correlation spectroscopy (XPCS) has been explored to determine specific functional/chemical groups on the exterior of nanocomposites. By using time correlations in the frequency of light scattered by an X-ray beam, XPCS can identify the structural dynamics of nanocomposites at the nanoscale scale. This skill has given researchers a special understanding of the complex fluid based on the microscopic origin of the rheological behavior (Basha, TS, and Doble 2019; Kong et al. 2020; Amini et al. 2020).

Magnetic properties

The magnetic properties of MOPs can be determined using superconducting quantum interference device (SQUID) magnetometry and vibrating sample magnetometry (VSM). Nevertheless, both techniques can measure only general magnetism and are not element-specific. SQUID magnetometry is usually utilized to evaluate the properties of magnetic nanocomposites. To achieve this, first, nanocomposites are cooled down, either with or without a magnetic field, and then warmed up through applying magnetic force. Magnetization is observed as a function of temperature (Prasad et al. 2018).

Chromatography and related methods

Various chromatography-based techniques have been explored to separate NPs in the colloidal solution. These techniques are also rapid, non-destructive, and sensitive (Dang and Guan 2020).

Size exclusion chromatography (SEC)

SEC is a well-known method for size-based separation of NPs, including polystyrene, single-walled carbon nanotubes, and quantum dots. Usually, the columns of SEC are packed with a small and rigid porous material of sizes between 3 μm and 20 μm with pore sizes ranging from 50 Å to 107 Å. The analytes in SEC separate based on their size in solution or hydrodynamic volume. The molecules with the larger size in a sample elute out first than the smaller ones because the larger molecules have access to fewer pores of the packed column matrix. SEC has been used over the two decades to fractionate a variety of NPs such as Ag NPs, Au NPs, and quantum dots (QDs). Since many NPs and QD features are size-dependent, NPs with a narrow size distribution are frequently obtained using a separation technique like SEC (Dang and Guan 2020; Bayda et al. 2019).

Hydrodynamic chromatography (HDC)

HDC divides particles based on their hydrodynamic size. Depending on the length of a column, the range of size separation for the HDC columns that are currently on the market is between 5 nm and 1200 nm. The relatively broad separation range of HDC enables nanocomposites of varied sizes to be evaluated in different suspensions and is specifically valuable in observing the development of aggregates (Bayda et al. 2019; Dang and Guan 2020).

Field-flow fractionation

Field-flow fractionation is another promising technique for separating elemental nanocomposites by size in complex materials. Similar to chromatographic methods, this technique separates NPs solely in an open platform without the involvement of a stationary phase. The particles are separated based on how they are impacted by an applied force (Dang and Guan 2020; Bayda et al. 2019).

Filtration and centrifuge techniques

Filtration and centrifugation are the key techniques utilized for the preparative size fractionation of MOPs. These techniques are very efficient and economical. On TEM and AFM substrates, preparative ultracentrifugation is utilized to settle down fine particulates and separate and extract aquatic colloids or NPs. Conventional membrane filtration separates MOPs with pore sizes ranging from 0.5 nm to 1 nm (Bayda et al. 2019; Dang and Guan 2020).

Hyperspectral imaging (HI)

HI is performed to determine the type of nanocomposites developed in an environmental system and enables studies of the fate and transition of these NPs in an aqueous environment. Additionally, the nanocomposites' functional groups and surface chemistry can be determined. The data is obtained expressing sample information which is observed as spatial distributions and spectral properties unique to each sort of nanocomposite at the sensitivity of a single nanoparticle (size < 10 nm) (Basha, TS, and Doble 2019).

Laser-induced breakdown detection (LIBD)

LIBD is a laser-based technique with low detection limits, which can analyze the concentration and colloid size based on the evaluated breakdown probability. It is a highly promising method for nanocomposite characterization. Several laser-based methods are used, such as mass spectrometry (MS), Raman spectroscopy, and laser-induced fluorescence (LIF). Fluorescent-labeled NPs are analyzed using MS-based techniques like LIF, MALDI (Matrix Assisted Laser Desorption Ionization), or ion trap mass spectrometry (Basha, TS, and Doble 2019; WHO 2019).

Small angle X-ray scattering (SAXS)

To explore the structural properties of the fluid and solid composites in the nanoscale, the SAXS-based approach is used. This non-destructive method can analyze solid, powdered, or liquid samples to identify elemental composition and determine their concentrations in a solution (Basha, TS, and Doble 2019; Amini et al. 2020).

Inductively coupled plasma mass spectroscopy (ICP-MS)

This technique is employed for the identification of ultra-stress metals in a variety of nanocomposites. To evaluate the cytotoxicity of synthesized nanocomposites for their possible application in biomedicines, a variety of NPs, including nickel, gold, and carbon nanotubes (single-walled) are quantified by using the ICP-MS (Basha, TS, and Doble 2019; Pi et al. 2020).

2.8 FACTORS INFLUENCING THE FABRICATION AND CHARACTERIZATION OF METAL OXIDE NANOCOMPOSITES

Several variables influence the nanocomposites' fabrication, characterization, and application. The key parameters which can influence nanoparticle synthesis are the pH of the reaction medium, concentration, temperature, raw materials concentration, size, and, most importantly, the protocols followed in the preparation of NPs (Keshvadi et al. 2019; Hasanzadeh et al. 2019). Some of the dominant parameters that influence the synthesis of NPs are mentioned in this section.

2.8.1 Particular method or technique

Nanocomposites can be fabricated using a variety of approaches, from physical ones involving mechanical processes to chemical or biological ones involving several organic or inorganic compounds and living things, respectively. Each method is associated with certain merits and downfalls. However, compared to conventional methods, biological approaches for the fabrication of nanocomposites involve nontoxic and environmentally safe components in combination with green chemistry-based technology and are thus eco-friendly and more acceptable than the conventional ones (Hasanzadeh et al. 2019; Keshvadi et al. 2019).

2.8.2 pH

pH has a significant influence on the synthesis of NPs. Researchers demonstrated that the pH of the reaction medium affects the particle size and texture of the fabricated nanocomposites. Consequently, the size of NPs can be tailored by changing the pH of the reaction media (Hasanzadeh et al. 2019; Keshvadi et al. 2019).

2.8.3 Temperature

Temperature is yet another parameter that influences the fabrication of nanocomposites via any method. The physical synthesis method uses a high temperature (>350°C), while the chemical synthesis methods use a temperature lower than 350°C. The synthesis process temperature determines the NPs' nature (Hasanzadeh et al. 2019; Keshvadi et al. 2019).

2.8.4 Pressure

In the fabrication of nanocomposites, the role of pressure is crucial. The size and shape of the fabricated nanocomposites are influenced by the pressure applied to the reaction media. At ambient pressure, it has been observed that the rate of metal ions reduction by biological systems is significantly faster (Hasanzadeh et al. 2019; Keshvadi et al. 2019).

2.8.5 Time

The reaction's time duration substantially influences the type and quality of the fabricated nanocomposite. Similarly, the properties of nanocomposites also change according to time and are significantly impacted by synthesis, light exposure, storage conditions, and many more (Hasanzadeh et al. 2019; Keshvadi et al. 2019).

2.8.6 Preparation cost

It is necessary to govern and control the cost involved in their synthesis to make it easier for NPs to be applied potentially in modern-day usage. Hence, the cost-effectiveness of the fabrication process is a key parameter that affects the synthesis of nanocomposites (Hasanzadeh et al. 2019; Keshvadi et al. 2019).

2.8.7 Pore size

The porosity of the fabricated nanocomposites has a significant impact on their quality and application. Immobilizing biomolecules on the surface of NPs has been made possible, which will expand their application in the biomedical and drug delivery fields (Hasanzadeh et al. 2019; Keshvadi et al. 2019).

2.8.8 Environment

The nature of the prepared NPs is significantly influenced by their surrounding environment. In many situations, a single nanoparticle quickly transforms into a core-shell nanoparticle by absorbing materials or interacting with others through oxidation or reduction processes. In a biological system, the fabricated NPs develop a coating that increases their size and thickness. In addition, the chemical and physical construction of the prepared NPs is significantly influenced by their environment (Hasanzadeh et al. 2019; Keshvadi et al. 2019).

2.8.9 Proximity

A majority of the time, the properties of the individual NPs change when they come close to some other NPs. Developing more tailored NPs can take advantage of the changing behavior of NPs. The proximity effect of NPs has numerous implications, including those related to the charge of a nanoparticle, substrate interactions, and the magnetic characteristics of the NPs (Hasanzadeh et al. 2019; Keshvadi et al. 2019).

2.9 BIOMEDICAL APPLICATIONS OF METAL OXIDE NANOCOMPOSITES IN OXIDATIVE STRESS-INDUCED DISORDERS

Oxidative stress, a condition of lost harmony between the oxidants and antioxidant systems of the cells, results in the upregulation of reactive oxygen species (ROS) and oxidative free radicals. ROS are usually considered to be unhealthy. Excessive ROS production may attack cellular proteins, nucleic acids, and lipids, causing cellular dysfunction, including reduced biological activity, inflammation, and immune activation. Oxidative stress acts as a pathogenic factor in developing many diseases, such as cancer, diabetes, and atherosclerosis (Pothipor et al. 2019; Ishida 2017; WHO 2022).

2.9.1 Metal oxide nanocomposites in *in vitro* diagnostics of obesity

Obesity jeopardizes people's health because it increases the risk of developing several chronic diseases such as atherosclerosis, hyperlipidemia, diabetes, hypertension, neurological diseases, and respiratory disorders, as shown in Figure 2.7. The identification and treatment of obesity and disorders related to obesity are facilitated by the identification of factors associated with obesity in the bloodstream and cells, including leptin and adiponectin. Adiponectin levels always fall, and serum leptin levels rise in obesity; however, the adiponectin level is severely raised in extremely obese patients because of its anti-inflammatory properties (Pothipor et al. 2019; WHO 2022).

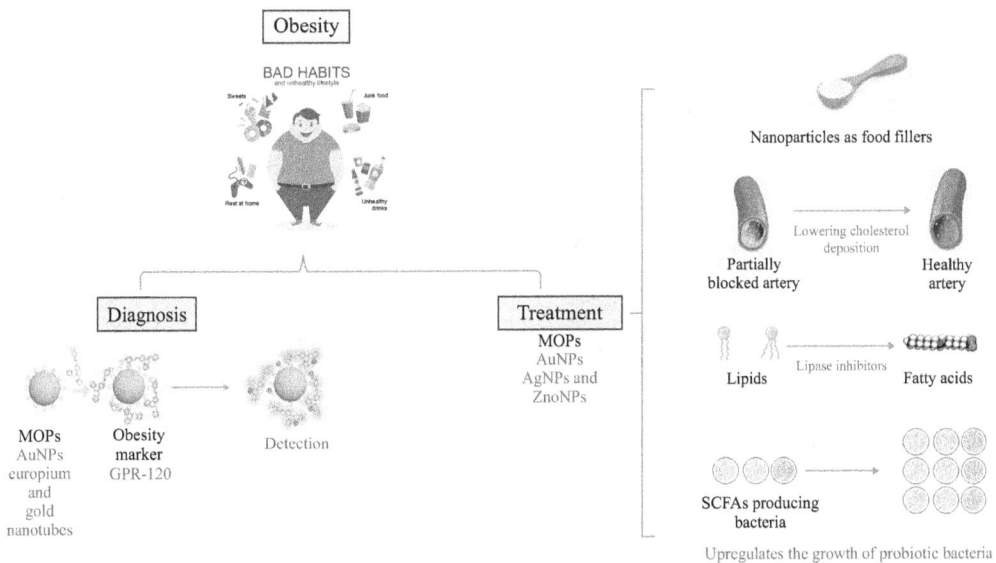

Figure 2.7 Biomedical application of MOPs in the identification and treatment of obesity and obesity-related issues.

Immunoassay termed as Surface-Enhanced Raman Scattering (SERS) is based on MPIO@ mesosilica@ silver (MPIO@SiO$_2$@Ag) triple core-shell microparticles (TCSMPs). It is effective in measuring serum adipokines levels in obese patients. Ag NPs on TCSMPs exerted SERS activity, and the MPIO provided TCSMPs magnetic characteristics that made them easy to separate. The G-protein receptor 120 (GPR120) can facilitate response to long-chain fatty acids, and GPR120 expression is enhanced as an increasingly long chain of fatty acids (Pothipor et al. 2019; Ishida 2017; WHO 2022). Evaluating GPR 120 could aid in diagnosing obesity because the concentration of fatty acids is relatively high in obese people, leading to a high level of GPR120 expression. Bimodal SERS-fluorescence microscopy was used to identify and image the GPR120 in individual live cells. The dual functional nanoprobe, composed of calcium molybdate doped with europium and gold nanorods (CaMoO$_4$:Eu^{3+}@ AuNR), was used for the coating to effectively detect GPR120 *in vitro* (Pothipor et al. 2019; Ishida 2017; WHO 2022).

Metal oxide nanocomposites as a food filler

Fat is the most prevalent nutrient in food, which also enhances the flavor, but abrupt consumption would raise the threat of getting obese, having hyperlipidemia, and hypertension. Food fillers, such as substitutes for fat, help people lose weight by consuming fewer calories and increasing satiety while preserving taste. Nanomaterials, particularly nanocellulose, provide a promising strategy for weight loss because of their distinctive physical and chemical characteristics (Pothipor et al. 2019; Ishida 2017).

The most prevalent macromolecule in nature is cellulose which has good biocompatibility and can stimulate intestinal peristalsis and lower caloric intake. Nanocellulose consists of bacterial cellulose (BC), nano-fibrillated cellulose (NFC), and nanocrystalline cellulose (NCC). Providing NCC and NFC would decrease the absorption of fat, which is also advantageous for lowering blood lipid levels and body weight. There are two basic processes of TG reduction as NFC might adsorb droplets of lipids and bile salt. The sequestration of bile salts might hinder the digestion of lipids, and the accumulation of lipid molecules on NFC fibers reduces the interacting surface area among lipids and lipase (Pothipor et al. 2019; Ishida 2017).

The fibers of NFC can develop a heterogeneous network and delay the diffusion of glucose and starch decomposition significantly without affecting the function of α-glucosidase and α-amylase. Moreover, NFC may make liquid foods viscous and prevent glucose from being absorbed into the body when starch-containing foods are consumed *in vivo*. In the rat model, blood glucose and lipid levels can be controlled by NCC to treat obesity-related disorders (Pothipor et al. 2019; Ishida 2017; WHO 2022).

Metal oxide nanocomposites in lowering cholesterol

Chitosan NPs may lower cholesterol and lessen gastrointestinal absorption of fat. Chitosan NPs are more effective than chitosan and water-soluble chitosan at lowering body weight, serum total cholesterol (TC), and LDL-C. Montmorillonite, a natural smectite extracted from bentonite, can be an absorbing substrate for dietary fats and TC. TC levels of mice could be significantly reduced, and obesity could be avoided by spray-drying montmorillonite with a 13.8 nm pore size. Using a different mechanism of action, spray-dried montmorillonite performed a similar function to orlistat (commercial drug) while having fewer side effects than currently available products (Pothipor et al. 2019; Ishida 2017; WHO 2022).

Metal oxide nanocomposites aid the growth of probiotic bacteria to regulate antioxidants

Probiotic bacteria like SCFAs-forming bacteria (short chain fatty acids), akkermansia, and bifidobacterium could aid in weight loss. Several nanomaterials, including those containing zinc oxide and Au NPs, boost the number of probiotic bacteria in the gut microflora. ZnO NPs boost the number of bacteria that produce SCFAs, which enhance anti-inflammatory activity and improve antioxidant status (Pothipor et al. 2019; Ishida 2017; WHO 2022).

Metal oxide nanocomposites as lipase inhibitors

Another strategy to treat obesity is inhibiting lipase activity, and various molecules have been employed in this regard. Orlistat is a small inhibitory drug to gastrointestinal lipase to treat obesity (Pothipor et al. 2019; Ishida 2017; WHO 2022). This drug has low bioavailability and poor delivery efficiency. Melt emulsification and high-pressure homogenization were employed to enhance the orlistat bioavailability by regulating the hydrodynamic size of the orlistat. Compared to the original form of orlistat and other similar commercial drugs, the efficiency of lipase inhibition of the emulsified nano-orlistat was higher by 2.4 times (Pothipor et al. 2019; Ishida 2017; WHO 2022).

2.9.2 Metal oxide nanocomposites in the diagnostics of carcinogenesis

MOPs have gained much interest due to their application in cancer therapeutics (Figure 2.8). Studies in the relevant area have demonstrated that different MOPs elicit cellular toxicity in cancer cells but not healthy ones. In some instances, it has been shown that the nanoparticle exhibits anticancer activity alone or in conjugation with other treatments involving photocatalytic therapy or some other drugs (anticancer) (Petros and DeSimone 2010).

Iron oxide MOPs in cancer therapy

Cancer cells could be directly targeted by radiation (nontoxic) like oscillating magnetic fields (MF) or near-infrared (NIR) that can be absorbed and transitioned into a detrimental signal of ROS production or hyperthermia by iron oxide NPs. The benefit of this method is that MOPs can be easily targeted to the desired location by covalent attachment of tissue-specific surface markers or with the aid of an external magnetic field in the case of magnetic MOPs (Petros and DeSimone 2010; Shrestha 2012).

Consequently, such MOPs can specifically target cancer cells, converting radiation energy into ROS or heat. When doxorubicin and iron oxide NPs are combined, they form magneto-sensitive nanoparticle complexes that have a stronger anticancer effect than doxorubicin alone. Its enhanced antitumor effect is most likely caused by the generation of hydroxyl ions, which destroy the mitochondria, lipids, DNA, proteins, and other structural components of

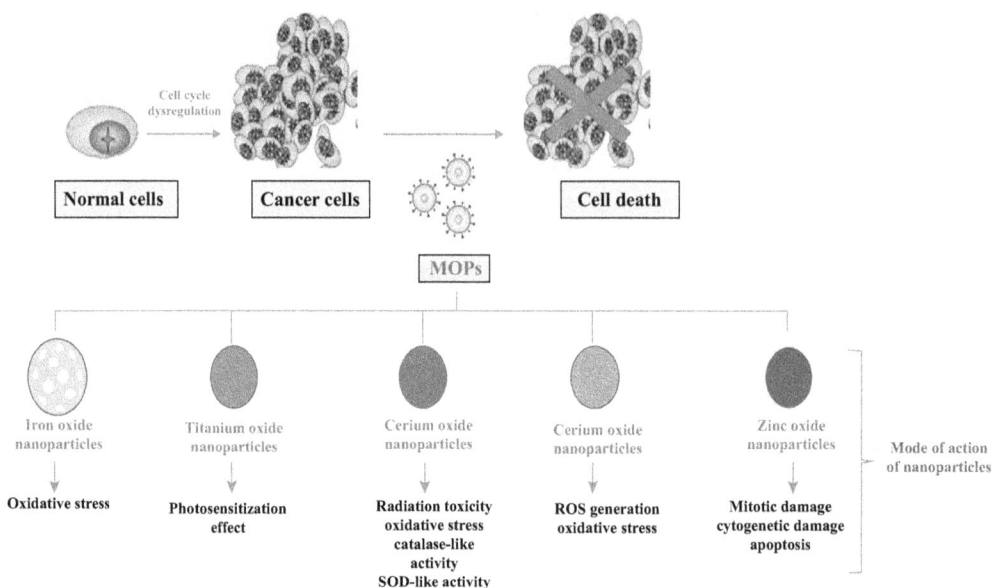

Figure 2.8 Modes of action of different MOPs employed in the treatment of cancer cells.

tumor cells and ultimately cause them to die or undergo necrosis. An innovative approach for potential cancer treatments in the future is made possible by the interaction of magnetic fields and an anticancer magneto-sensitive nanocomposite (Petros and DeSimone 2010; Shrestha 2012).

Titanium dioxide NPs in cancer therapy

The basic idea underlying photodynamic treatment (PDT) is that a hydrophobic organic photosensitizer (PS) is activated by electromagnetic rays in the visible or ultraviolet spectrum to give rise to cytotoxic ROS that cause apoptosis. An alternative strategy comprises the utilization of inorganic NPs like titanium oxide (TiO_2), a photo-sensitizing agent that can be utilized in place of PS molecules. It has been demonstrated that photocatalyzed TiO_2 NPs kill cancer cells. TiO_2 NPs were shown to be integrated into the cytoplasm and cell membrane of the cancer cell lines, including HeLa, U937, and T-24 cells. TiO_2 NPs are harmless and durable without exposure to light, and they can be maintained for a very long period inside the body (Petros and DeSimone 2010; Shrestha 2012).

Cerium oxide NPs (CNPs) in cancer therapy

CNPs are new and extremely important material for radiation-based therapies due to their "smart" capability to selectively irradiate cancer cells and induce death while preventing the surrounding healthy cells from damage induced by radiation and oxidative stress. Thus, CNPs have the exceptional property of functioning as radio-protecting and radio-sensitizing agents. These CNPs specifically induce oxidative stress and cell death in the cancer cells (irradiated) while preventing healthy cells. It has been postulated that the precise cytotoxicity of CNPs for cancer cells results from the inhibition of the catalase-like characteristic of CNPs that occur at low pH (pH 4.3); the SOD-like activity generated via ionizing radiations is preserved even at low pH and would result in the accumulation of H_2O_2, enhancing radiation-induced toxicity (Petros and DeSimone 2010; Shrestha 2012).

According to recent research, CNPs are cytotoxic to pancreatic cancer cells (L3.6pl) and are used in radiation therapy, but they exert little-to-no side effects on neighboring cells (hTERT-HPNE). The conclusion supports the application of CNPs as a stand-alone therapy in treating pancreatic cancer. It was demonstrated that the CNPs exhibit either antioxidant or pro-oxidant redox effects (Petros and DeSimone 2010; Shrestha 2012).

Zinc oxide NPs in cancer therapy

The NPs of ZnO are extremely toxic to T98G cancer cells, relatively less efficient to KB cells, and safe to normal cells (HEK). These findings have shown that treating T98G cells with ZnO NPs boosts both mitotic damage (cytogenetic) and interphase death (apoptosis). The ZnO NPs act as genotoxic drugs since they trigger micronuclei formation inside the cancer cells (Petros and DeSimone 2010; Shrestha 2012). These findings may be useful in developing anticancer drugs for therapeutic applications.

The mechanism of apoptosis is correlated with the synthesis of intracellular ROS and cancer cells with different doses of ZnO NPs. ZnO NPs were utilized at an extremely low dose and were found to show significant effect against liver cancer (HepG2) and breast cancer (MCF-7) cells in a dose-dependent way; cell viability was measured by the MTT cytotoxicity assay, demonstrating a dose-dependent death of cancer cells. Since ZnO has electrostatic characteristics, it can show surface charge differently in an acidic and basic environment. Moreover, ZnO NPs can also show photodynamic property that causes substantial production of ROS in cancer cells and may lead to cell death (Petros and DeSimone 2010; Shrestha 2012).

Copper oxide NPs in cancer therapy

Copper oxide NPs have shown cytotoxic activity on human hepatic (A549) and breast (MCF-7) cancer cells. The mechanism exhibited by copper oxide NPs in cytotoxicity was shown to be via the initiation of apoptosis with the enhanced production of ROS. The green chemistry-based fabrication of copper oxide NPs has been suggested as a reliable, nontoxic, eco-friendly, and easy process (Table 2.6). Importantly, the findings also concluded that

Table 2.6 Anticancer potential of MOPs in in vivo system

MOPs	Animal model (In Vivo)	Size of NPs (nm)	Reference
NPs of cerium oxide	Orthotopic administration of pancreatic tumor cells in mice (nude)	5–8	(Duncan and Gaspar 2011)
	Melanoma cells inoculation in xenograft mice	3–5	(Wang et al. 2012)
NPs of titanium dioxide	Human oral carcinoma in mice	50	(Pollock et al. 2008)
NPs of copper oxide	Subcutaneous melanoma in mice	40–110	(Jackman, Lee, and Cho 2016)

these NPs were eliminated from the cells and exhibited negligible or no systemic toxicity (Petros and DeSimone 2010; Shrestha 2012).

2.9.3 Metal oxide NPs in the treatment of cardiovascular diseases (CVD)

Several CVDs can be assessed, evaluated, and treated using new therapeutic, diagnostic, and prognostic techniques (Figure 2.9). Nanomedicine has attained increased interest over the last few decades. Indeed, it has been recognized as a pivotal, powerful, and harmless platform that can be used in the therapeutics of angiogenic, ischemic, metabolic, and inflammatory disorders like hyperlipidemia, atherosclerosis, and hypertension (Pothipor et al. 2019; Ishida 2017).

Iron oxide NPs (INPs) in the treatment of atherosclerosis and heart failure

INPs have been explored in biomedical applications because of their unique chemical and magnetic characteristics. These NPs have also been utilized to treat debilitating conditions, including multiple sclerosis and myocardial failure. To manage CVD, INPs have been studied only *in vitro* studies and animal models. Indeed, INPs are mainly fabricated for treating atherosclerotic plaque and stem cell usage. Moreover, these NPs were used to identify surface CD163, a biomarker of macrophages (M2), in atherosclerotic lesions (Pothipor et al. 2019; Ishida 2017).

Gold NPs (Au NPs) in the cardiac remodeling

Similar to INPs, Au NPs have undergone extensive research and are applied in various sectors, such as chemical, biological, and medical, because of their exceptional bio-optical

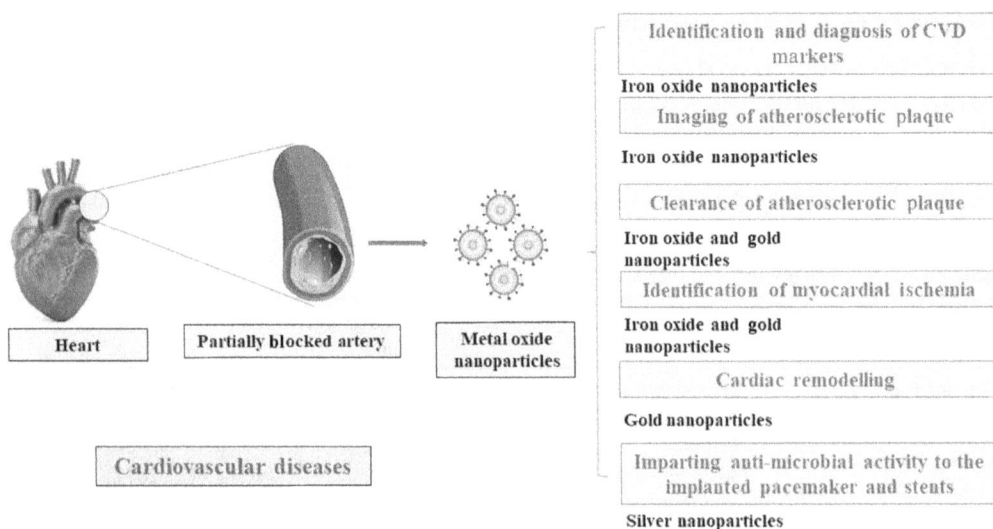

Figure 2.9 Identification of biomarkers and treatment of CVDs using MOPs.

characteristics. Current evidence, interpreted from the pre-clinical data and *in vitro* studies, emphasizes the potential application of Au NPs in both CVD treatment and imaging (Pothipor et al. 2019; Ishida 2017). Au NPs are useful for the radiological diagnosis and computation of atherosclerosis because they act as probes in targeting specific atheromarkers. For instance, the integration of these optically and biologically unique NPs in the imaging of atherosclerosis enables precise targeting of different inflammatory markers involving α5β3-integrin, vascular cell adhesion molecule-1 (VCAM-1), and intercellular adhesion molecule-1 (ICAM-1), Moreover, cardiac scarred tissues can be detected using Au NPs labeled with anti-collagen I peptides, which allows for an amazing identification of myocardial ischemia (Pothipor et al. 2019; Ishida 2017).

Silver NPs (Ag NPs) in the treatment of CVD

Ag NPs have drawn more attention over the last few decades and are frequently used in nonmedical and medical fields. Consequently, human exposure to these NPs is considered relatively common due to their widespread distribution. The coating of NPs over the pacemakers and cardiac stents has been used to minimize the risk of infections. The antimicrobial properties of Ag NPs are responsible for the beneficial reduction of infections (Pothipor et al. 2019; Ishida 2017).

2.10 CHALLENGES AND LIMITATIONS OF METAL OXIDE NANOCOMPOSITES

MOPs are now being used more often. Numerous challenges and new obstacles need to be resolved before using these MOPs in clinical or commercial applications for diagnosis, therapy, theranostics, cosmetics, sensors, etc. Differential toxicity profiles of the MOPs are associated with their altered optical, chemical, magnetic, and structural characteristics. These NPs can also experience changes in their physical state due to changes in pH and interactions with the cellular proteins, lipids, blood components, and genetic material. Similarly, MOPs tend to display lower stability. After being introduced to a biological environment, they are more prone to ion release and dissolution, following the production of ROS and oxidative stress inside the cells. Despite the significant potential of MOPs in the biomedical field, the main issue with ROS-based therapies is the incapability to control the ROS delivery to specified tissues or cells. However, some undefined mechanisms of MOP's extracellular and cellular functioning are unrelated to the production of ROS and have yet to be successfully demonstrated. So the assessment of the safety profile for MOPs should also involve *in vivo* and *in vitro* investigations of the mechanisms of MOPs' transmission, accumulation, long-term toxicity/safety; their interaction with receptors, neighboring cells, signaling pathways; and their overall phagocytic activity.

Both benefits and downfalls challenge biological scientists and material scientists who ought to create and investigate the relationship between a specific therapeutic MOP's structure and biological performance. The proper comprehension of all interactions occurring between regulatory and activation factors and toxicity mechanisms should serve as the foundation for the production of functional, safe, and biocompatible MOPs used in the nanomedicine-based therapy of human diseases. Improved control over the migration, cell cycle, proliferation, angiogenesis, and other processes could be used if the molecular regulatory mechanisms of the reaction and other signaling pathways prompted by MOPs are thoroughly known. Therefore, it is crucial to evaluate the short- and long-term toxicity of MOPs or their nano-formulations to assure the safety of the global biome and avoid a fiasco.

2.11 CONCLUSION AND FUTURE PERSPECTIVES

Nanocomposites are multiphasic heterogeneous materials (NPs, nanotubes, nano clays, or lamellar nanostructure) wherein the constituent materials, possessing different physical and chemical properties, integrate to develop new properties of the materials, ensuring that the constituent materials are within the size range of 1–100 nm. Since the properties of physical, chemical, and biological nanomaterials vary from individual atoms and molecules, the resultant properties of nanocomposites are highly dependent on the attributes of the constituent materials along with their morphology and interfacial characteristics. In recent decades, nanotechnology has substantially metamorphosized healthcare strategies having a widespread impact on public health. The term "nanotechnology" encompasses the development of particles with dimension(s) that fall into the nanometer range. This branch of science has captured extensive scientific attention due to its unique properties and concatenative application spectrum in the medical and biological sciences, particularly the evaluation of metal-based NPs and nanocomposites for the biotechnology, biomedicine, and pharmaceutical industry.

Regarding nanomaterials, MOPs have received much attention in various industries, including environmental remediation, drug delivery, food and agriculture production, catalysis, sensing, and medicine. As a result, many synthesis routes for these MOPs have been developed. Chemical and physical methods have primarily synthesized MOPs. These synthesized nanocomposites present a variety of characterization obstacles that influence the detailed and appropriate characterization of the nanocomposites. Therefore, it is crucial to comprehend the challenges involved in nanoparticle characterization and choose an appropriate characterization method. Herein, the techniques for the characterization of NPs are grouped into microscopic, spectroscopic, and elemental analyses and the analysis of crystalline phases utilizing the XRD technique. Specifically, nanocomposites are characterized to evaluate the porosity, aggregation, surface area, solubility, particle size distribution, zeta potential, shape, and intercalation and distribution within the nanocomposites. Several variables that can influence the fabrication, characterization, and application of the synthesized nanocomposites are the pH of the solution, concentration, temperature, raw materials concentration, size, and, most importantly, the protocols followed in preparing nanocomposites.

The MOPs have attained increased interest over the last few decades in dealing with ROS-related disorders. Indeed, it has been recognized as a pivotal, powerful, and harmless platform for therapeutics of angiogenic, ischemic, metabolic, and inflammatory disorders like hyperlipidemia, atherosclerosis, and hypertension. Reactive oxygen species (ROS) are usually considered to be unhealthy. Excessive ROS production may attack cellular proteins, nucleic acids, and lipids, causing cellular dysfunction, including reduced biological activity, inflammation, and immune activation. Oxidative stress is a pathogenic factor in the emergence of many diseases, including cancer, diabetes, and atherosclerosis.

Numerous challenges and new obstacles need to be resolved before using MOPs in clinical or commercial applications for diagnosis, therapy, theranostics, cosmetics, sensors, etc. Differential toxicity profiles of the MOPs are associated with their altered optical, chemical, magnetic, and structural characteristics. These NPs can also experience changes in their physical state due to changes in pH and interactions with the cellular proteins, lipids, blood components, and genetic material. Similarly, MOPs tend to display lower stability. After being introduced to a biological environment, they are more prone to ion release and dissolution, following the production of ROS and oxidative stress inside the cells. Therefore, the proper comprehension of all interactions occurring between regulatory and activation factors, as well as toxicity mechanisms, should serve as the foundation for the production of functional, safe, and biocompatible MOPs used in the nanomedicine-based therapy of human diseases to assure the safety of the global biome and avoid a fiasco.

REFERENCES

AbdelAllah, Nourhan H, Yasser Gaber, Mohamed E Rashed, Ahmed F Azmy, Heba A Abou-Taleb, and Sameh AbdelGhani. 2020. "Alginate-coated chitosan NPs act as an effective adjuvant for hepatitis: A vaccine in mice." *International Journal of Biological Macromolecules* 152:904–912.

Aderibigbe, Blessing Atim, and Tobeka Naki. 2019. "Chitosan-based nanocarriers for nose to brain delivery." *Applied Sciences* 9 (11):2219.

Alexis, Frank, Eric Pridgen, Linda K Molnar, and Omid C Farokhzad. 2008. "Factors affecting the clearance and biodistribution of polymeric NPs." *Molecular Pharmaceutics* 5 (4):505–515.

Amini, Parisa, Sina Nassiri, Alexandra Malbon, and Enni Markkanen. 2020. "Differential stromal reprogramming in benign and malignant naturally occurring canine mammary tumours identifies disease-modulating stromal components." *Scientific Reports* 10 (1):1–13.

Barenholz, Yechezkel Chezy. 2012. "Doxil®-The first FDA-approved nano-drug: Lessons learned." *Journal of Controlled Release* 160 (2):117–134.

Barua, Shaswat, Pronobesh Chattopadhyay, Mayur M Phukan, Bolin K Konwar, Johirul Islam, and Niranjan Karak. 2014. "Biocompatible hyperbranched epoxy/silver-reduced graphene oxide-curcumin nanocomposite as an advanced antimicrobial material." *RSC Advances* 4 (88):47797–47805.

Basha, Rubaiya Yunus, Sampath Kumar TS, and Mukesh Doble. 2019. "Dual delivery of tuberculosis drugs via cyclodextrin conjugated curdlan NPs to infected macrophages." *Carbohydrate Polymers* 218:53–62.

Bayda, Samer, Muhammad Adeel, Tiziano Tuccinardi, Marco Cordani, and Flavio Rizzolio. 2019. "The history of nanoscience and nanotechnology: From chemical-physical applications to nanomedicine." *Molecules* 25 (1):112.

Boroumand, Homa, Fereshteh Badie, Samaneh Mazaheri, Zeynab Sadat Seyedi, Javid Sadri Nahand, Majid Nejati, Hossein Bannazadeh Baghi, Mohammad Abbasi-Kolli, Bita Badehnoosh, and Maryam Ghandali. 2021. "Chitosan-based NPs against viral infections." *Frontiers in Cellular and Infection Microbiology* 11:643953.

Bouzouita, Amani. 2016. "Elaboration of polylactide-based materials for automotive application: Study of structure-process-properties interactions." Université de Valenciennes et du Hainaut-Cambresis; Université de Mons.

Camargo, Pedro Henrique Cury, Kestur Gundappa Satyanarayana, and Fernando Wypych. 2009. "Nanocomposites: Synthesis, structure, properties and new application opportunities." *Materials Research* 12:1–39.

Cambier, CJ, Stanley Falkow, and Lalita Ramakrishnan. 2014. "Host evasion and exploitation schemes of Mycobacterium tuberculosis." *Cell* 159 (7):1497–1509.

Cappellano, Giuseppe, Cristoforo Comi, Annalisa Chiocchetti, and Umberto Dianzani. 2019. "Exploiting PLGA-based biocompatible NPs for next-generation tolerogenic vaccines against autoimmune disease." *International Journal of Molecular Sciences* 20 (1):204.

Caron, Joachim, L Harivardhan Reddy, Sinda Lepêtre-Mouelhi, Séverine Wack, Pascal Clayette, Christine Rogez-Kreuz, Rahima Yousfi, Patrick Couvreur, and Didier Desmaële. 2010. "Squalenoyl nucleoside monophosphate nanoassemblies: New prodrug strategy for the delivery of nucleotide analogues." *Bioorganic & Medicinal Chemistry Letters* 20 (9):2761–2764.

Ciupe, Stanca M, Ruy M Ribeiro, Patrick W Nelson, and Alan S Perelson. 2007. "Modeling the mechanisms of acute hepatitis B virus infection." *Journal of Theoretical Biology* 247 (1):23–35.

Dang, Yu, and Jianjun Guan. 2020. "Nanoparticle-based drug delivery systems for cancer therapy." *Smart Materials in Medicine* 1:10–19.

das Neves, José, Francisca Araújo, Fernanda Andrade, Johan Michiels, Kevin K Ariën, Guido Vanham, Mansoor Amiji, Maria Fernanda Bahia, and Bruno Sarmento. 2013. "In vitro and ex vivo evaluation of polymeric NPs for vaginal and rectal delivery of the anti-HIV drug dapivirine." *Molecular Pharmaceutics* 10 (7):2793–2807.

De, Bibekananda, Kuldeep Gupta, Manabendra Mandal, and Niranjan Karak. 2015. "Biocide immobilized OMMT-carbon dot reduced Cu2O nanohybrid/hyperbranched epoxy nanocomposites: Mechanical, thermal, antimicrobial and optical properties." *Materials Science and Engineering: C* 56:74–83.

Dermenci, Kamil Burak, Bora Genc, Burçak Ebin, Tugba Olmez-Hanci, and Sebahattin Gürmen. 2014. "Photocatalytic studies of Ag/ZnO nanocomposite particles produced via ultrasonic spray pyrolysis method." *Journal of Alloys and Compounds* 586:267–273.

Duncan, Ruth, and Rogerio Gaspar. 2011. "Nanomedicine (s) under the microscope." *Molecular Pharmaceutics* 8 (6):2101–2141.

Faupel, Franz, Vladimir Zaporojtchenko, Thomas Strunskus, and Mady Elbahri. 2010. "Metal-polymer nanocomposites for functional applications." *Advanced Engineering Materials* 12 (12):1177–1190.

Figueiró, Fabrício, Andressa Bernardi, Rudimar L Frozza, Thatiana Terroso, Alfeu Zanotto-Filho, Elisa HF Jandrey, José Claudio F Moreira, Christianne G Salbego, Maria I Edelweiss, and Adriana R Pohlmann. 2013. "Resveratrol-loaded lipid-core nanocapsules treatment reduces in vitro and in vivo glioma growth." *Journal of Biomedical Nanotechnology* 9 (3):516–526.

Gilman, Jeffrey W. 1999. "Flammability and thermal stability studies of polymer layered-silicate (clay) nanocomposites." *Applied Clay Science* 15 (1–2):31–49.

Halnor, VV, VV Pande, DD Borawake, and HS Nagare. 2018. "Nanoemulsion: A novel platform for drug delivery system." *Journal of Materials Science and Nanotechnology* 6 (1):104.

Hasanzadeh, Mohammad, Mahsa Feyziazar, Elham Solhi, Ahad Mokhtarzadeh, Jafar Soleymani, Nasrin Shadjou, Abolghasem Jouyban, and Soltanali Mahboob. 2019. "Ultrasensitive immunoassay of breast cancer type 1 susceptibility protein (BRCA1) using poly (dopamine-beta cyclodextrine-cetyl trimethylammonium bromide) doped with silver NPs: A new platform in early stage diagnosis of breast cancer and efficient management." *Microchemical Journal* 145:778–783.

Heitmann, Jonas S, Tatjana Bilich, Claudia Tandler, Annika Nelde, Yacine Maringer, Maddalena Marconato, Julia Reusch, Simon Jäger, Monika Denk, and Marion Richter. 2022. "A COVID-19 peptide vaccine for the induction of SARS-CoV-2 T cell immunity." *Nature* 601 (7894):617–622.

Hendricks, Gabriel L, Lourdes Velazquez, Serena Pham, Natasha Qaisar, James C Delaney, Karthik Viswanathan, Leila Albers, James C Comolli, Zachary Shriver, and David M Knipe. 2015. "Heparin octasaccharide decoy liposomes inhibit replication of multiple viruses." *Antiviral Research* 116:34–44.

Heng, Wen Tzuen, Jia Sheng Yew, and Chit Laa Poh. 2022. "Nanovaccines against viral infectious diseases." *Pharmaceutics* 14 (12):2554.

Hess, Patrick H, and P Harold Parker Jr. 1966. "Polymers for stabilization of colloidal cobalt particles." *Journal of Applied Polymer Science* 10 (12):1915–1927.

Houser, Katherine V, Grace L Chen, Cristina Carter, Michelle C Crank, Thuy A Nguyen, Maria Claudia Burgos Florez, Nina M Berkowitz, Floreliz Mendoza, Cynthia Starr Hendel, and Ingelise

J Gordon. 2022. "Safety and immunogenicity of a ferritin nanoparticle H2 influenza vaccine in healthy adults: A phase 1 trial." *Nature Medicine* 28 (2):383–391.

Ishida, T. 2017. "Anticancer activities of silver ions in cancer and tumor cells and DNA damages by Ag+-DNA base-pairs reactions." *MOJ Tumor Res* 1 (1):8–16.

Itani, Rasha, Mansour Tobaiqy, and Achraf Al Faraj. 2020. "Optimizing use of theranostic NPs as a life-saving strategy for treating COVID-19 patients." *Theranostics* 10 (13):5932.

Jackman, Joshua A, Jaywon Lee, and Nam-Joon Cho. 2016. "Nanomedicine for infectious disease applications: Innovation towards broad-spectrum treatment of viral infections." *Small* 12 (9):1133–1139.

Jazayeri, Seyed Davoud, Hui Xuan Lim, Kamyar Shameli, Swee Keong Yeap, and Chit Laa Poh. 2021. "Nano and microparticles as potential oral vaccine carriers and adjuvants against infectious diseases." *Frontiers in Pharmacology* 12:682286.

Jordan, Jeffrey, Karl I Jacob, Rina Tannenbaum, Mohammed A Sharaf, and Iwona Jasiuk. 2005. "Experimental trends in polymer nanocomposites: A review." *Materials Science and Engineering: A* 393 (1–2):1–11.

Jores, Katja, Wolfgang Mehnert, Markus Drechsler, Heike Bunjes, Christoph Johann, and Karsten Mäder. 2004. "Investigations on the structure of solid lipid NPs (SLN) and oil-loaded solid lipid NPs by photon correlation spectroscopy, field-flow fractionation and transmission electron microscopy." *Journal of Controlled Release* 95 (2):217–227.

Joyce, M Gordon, Wei-Hung Chen, Rajeshwer S Sankhala, Agnes Hajduczki, Paul V Thomas, Misook Choe, Elizabeth J Martinez, William C Chang, Caroline E Peterson, and Elaine B Morrison. 2021. "SARS-CoV-2 ferritin nanoparticle vaccines elicit broad SARS coronavirus immunogenicity." *Cell Reports* 37 (12):110143.

Karak, Niranjan, Rocktotpal Konwarh, and Brigitte Voit. 2010. "Catalytically active vegetable-oil-based thermoplastic hyperbranched polyurethane/silver nanocomposites." *Macromolecular Materials and Engineering* 295 (2):159–169.

Keong, Lim Chin, and Ahmad Sukari Halim. 2009. "In vitro models in biocompatibility assessment for biomedical-grade chitosan derivatives in wound management." *International Journal of Molecular Sciences* 10 (3):1300–1313.

Kesharwani, Disha, Sandhya Mishra, Swarnali Das Paul, Rishi Paliwal, and Trilochan Satapathy. 2019. "The functional nanogel: An exalted carrier system." *Journal of Drug Delivery and Therapeutics* 9 (2-s):570–582.

Keshvadi, M, F Karimi, S Valizadeh, and A Valizadeh. 2019. "Comparative study of antibacterial inhibitory effect of silver NPs and garlic oil nanoemulsion with their combination." *Biointerface Research in Applied Chemistry* 9:4560–4566.

Kobayashi, Takaomi. 2016. *Applied environmental materials science for sustainability*: IgI Global.

Komarneni, Sridhar. 1992. "Nanocomposites." *Journal of Materials Chemistry* 2 (12):1219–1230.

Kong, Wenyan, Qi Wang, Guoying Deng, Hang Zhao, Linjing Zhao, Jie Lu, and Xijian Liu. 2020. "Se@ SiO 2@ Au-PEG/DOX NCs as a multifunctional theranostic agent efficiently protect normal cells from oxidative damage during photothermal therapy." *Dalton Transactions* 49 (7):2209–2217.

Lagaly, G, M Ogawa, and I Dékány. 2013. "Clay mineral-organic interactions." In *Developments in Clay Science*, 435–505: Elsevier.

Lee, Na Kyeong, Seongeon Cho, and In-San Kim. 2022. "Ferritin-a multifaceted protein scaffold for biotherapeutics." *Experimental & Molecular Medicine* 54 (10):1652–1657.

Liao, Wenzhen, Wen Li, Tiantian Zhang, Micheal Kirberger, Jun Liu, Pei Wang, Wei Chen, and Yong Wang. 2016. "Powering up the molecular therapy of RNA interference by novel NPs." *Biomaterials Science* 4 (7):1051–1061.

Liao, Wenzhen, Zhengxiang Ning, Luying Chen, Qingyi Wei, Erdong Yuan, Jiguo Yang, and Jiaoyan Ren. 2014. "Intracellular antioxidant detoxifying effects of diosmetin on 2, 2-azobis

(2-amidinopropane) dihydrochloride (AAPH)-induced oxidative stress through inhibition of reactive oxygen species generation." *Journal of Agricultural and Food Chemistry* 62 (34):8648–8654.

Lin, Bin, Genaro A Gelves, Joel A Haber, and Uttandaraman Sundararaj. 2007. "Electrical, rheological, and mechanical properties of polystyrene/copper nanowire nanocomposites." *Industrial & Engineering Chemistry Research* 46 (8):2481–2487.

Long, Xin, Changwei Shao, Hao Wang, and Jun Wang. 2016. "Single-source-precursor synthesis of SiBNC-Zr ceramic nanocomposites fibers." *Ceramics International* 42 (16):19206–19211.

Luies, Laneke, and Ilse Du Preez. 2020. "The echo of pulmonary tuberculosis: Mechanisms of clinical symptoms and other disease-induced systemic complications." *Clinical Microbiology Reviews* 33 (4):e00036–20.

Maartens, Gary, Connie Celum, and Sharon R Lewin. 2014. "HIV infection: Epidemiology, pathogenesis, treatment, and prevention." *The Lancet* 384 (9939):258–271.

Mehendale, Rujuta, Medha Joshi, and Vandana B Patravale. 2013. "Nanomedicines for treatment of viral diseases." *Critical Reviews™ in Therapeutic Drug Carrier Systems* 30 (1).

Mitchell, Michael J, Margaret M Billingsley, Rebecca M Haley, Marissa E Wechsler, Nicholas A Peppas, and Robert Langer. 2021. "Engineering precision NPs for drug delivery." *Nature Reviews Drug Discovery* 20 (2):101–124.

Nimesh, S. 2013. "Poly (D, L-lactide-co-glycolide)-based NPs." *Gene Therapy: Potential Applications of Nanotechnology: Woodhead Publishing Series in Biomedicine: Sawston, UK*:309–329.

Noordermeer, Jacobus WM, and Wilma K Dierkes. 2015. "Carbon black reinforced elastomers." *Encyclopedia of Polymeric Nanomaterials*: 287–299.

Ogasawara, Toshio, Yuichi Ishida, Takashi Ishikawa, and Rikio Yokota. 2004. "Characterization of multi-walled carbon nanotube/phenylethynyl terminated polyimide composites." *Composites Part A: Applied Science and Manufacturing* 35 (1):67–74.

Ovsyannikova, Inna G, and Gregory A Poland. 2011. "Systems biology approaches to new vaccine development Ann L Oberg, Richard B Kennedy 2, 3, Peter Li." *Current Opinion in Immunology* 23:436–443.

Pei, Fubin, Ping Wang, Enhui Ma, Qingshan Yang, Haoxuan Yu, Chunxiao Gao, Yueyun Li, Qing Liu, and Yunhui Dong. 2019. "A sandwich-type electrochemical immunosensor based on RhPt NDs/NH2-GS and Au NPs/PPy NS for quantitative detection hepatitis B surface antigen." *Bioelectrochemistry* 126:92–98.

Petros, Robby A, and Joseph M DeSimone. 2010. "Strategies in the design of NPs for therapeutic applications." *Nature reviews Drug Discovery* 9 (8):615–627.

Pi, Jiang, Ling Shen, Enzhuo Yang, Hongbo Shen, Dan Huang, Richard Wang, Chunmiao Hu, Hua Jin, Huaihong Cai, and Jiye Cai. 2020. "Macrophage-targeted isoniazid-selenium NPs promote antimicrobial immunity and synergize bactericidal destruction of tuberculosis bacilli." *Angewandte Chemie* 132 (8):3252–3260.

Pollock, Stephanie, Raymond A Dwek, Dennis R Burton, and Nicole Zitzmann. 2008. "N-Butyldeoxynojirimycin is a broadly effective anti-HIV therapy significantly enhanced by targeted liposome delivery." *Aids* 22 (15):1961–1969.

Pothipor, Chammari, Natta Wiriyakun, Thitirat Putnin, Aroonsri Ngamaroonchote, Jaroon Jakmunee, Kontad Ounnunkad, Rawiwan Laocharoensuk, and Noppadol Aroonyadet. 2019. "Highly sensitive biosensor based on graphene-poly (3-aminobenzoic acid) modified electrodes and porous-hollowed-silver-gold nanoparticle labelling for prostate cancer detection." *Sensors and Actuators B: Chemical* 296:126657.

Prasad, Minakshi, Upendra P Lambe, Basanti Brar, Ikbal Shah, J Manimegalai, Koushlesh Ranjan, Rekha Rao, Sunil Kumar, Sheefali Mahant, and Sandip Kumar Khurana. 2018. "Nanotherapeutics: An insight into healthcare and multi-dimensional applications in medical sector of the modern world." *Biomedicine & Pharmacotherapy* 97:1521–1537.

Qasim, Muhammad, Dong-Jin Lim, Hansoo Park, and Dokyun Na. 2014. "Nanotechnology for diagnosis and treatment of infectious diseases." *Journal of Nanoscience and Nanotechnology* 14 (10):7374–7387.

Ram Prasad, Ram Prasad, Rishikesh Pandey Rishikesh Pandey, Ajit Varma Ajit Varma, and Ishan Barman Ishan Barman. 2017. "Polymer-based NPs for drug delivery systems and cancer therapeutics." In *Natural Polymers for Drug Delivery*, 53–70: CABI Wallingford UK.

Rao, Lang, Rui Tian, and Xiaoyuan Chen. 2020. "Cell-membrane-mimicking nanodecoys against infectious diseases." *ACS nano* 14 (3):2569–2574.

Rao, Lang, Shuai Xia, Wei Xu, Rui Tian, Guocan Yu, Chenjian Gu, Pan Pan, Qian-Fang Meng, Xia Cai, and Di Qu. 2020. "Decoy NPs protect against COVID-19 by concurrently adsorbing viruses and inflammatory cytokines." *Proceedings of the National Academy of Sciences* 117 (44):27141–27147.

Rappuoli, Rino, Christian W Mandl, Steven Black, and Ennio De Gregorio. 2011. "Vaccines for the twenty-first century society." *Nature Reviews Immunology* 11 (12):865–872.

Ray, Suprakas Sinha, and Mosto Bousmina. 2007. *Polymer Nanocomposites and their Applications*: American Scientific Publishers.

Sen, Mousumi. 2020. "Nanocomposite materials." *Nanotechnology and the Environment*: 1–12.

Sharma, Puneet, and Sanjay Garg. 2010. "Pure drug and polymer based nanotechnologies for the improved solubility, stability, bioavailability and targeting of anti-HIV drugs." *Advanced drug delivery reviews* 62 (4–5):491–502.

Sharma, Shweta, Ashwni Verma, B Venkatesh Teja, Gitu Pandey, Naresh Mittapelly, Ritu Trivedi, and PR Mishra. 2015. "An insight into functionalized calcium based inorganic nanomaterials in biomedicine: Trends and transitions." *Colloids and Surfaces B: Biointerfaces* 133:120–139.

Sherje, Atul P, Bhushan R Dravyakar, Darshana Kadam, and Mrunal Jadhav. 2017. "Cyclodextrin-based nanosponges: A critical review." *Carbohydrate Polymers* 173:37–49.

Shrestha, Mohan. 2012. "Nanotechnology to revolutionize medicine." *Journal of Drug Delivery and Therapeutics* 2 (5).

Tai, Weon-Pil, Young-Sung Kim, and Jun-Gyu Kim. 2003. "Fabrication and magnetic properties of Al2O3/Co nanocomposites." *Materials Chemistry and Physics* 82 (2):396–400.

Tandrup Schmidt, Signe, Camilla Foged, Karen Smith Korsholm, Thomas Rades, and Dennis Christensen. 2016. "Liposome-based adjuvants for subunit vaccines: Formulation strategies for subunit antigens and immunostimulators." *Pharmaceutics* 8 (1):7.

Tang, DP, R Yuan, YQ Chai, X Zhong, Y Liu, JY Dai, and LY Zhang. 2004. "Novel potentiometric immunosensor for hepatitis B surface antigen using a gold nanoparticle-based biomolecular immobilization method." *Analytical Biochemistry* 333 (2):345–350.

Tang, Shixing, and Indira Hewlett. 2010. "Nanoparticle-based immunoassays for sensitive and early detection of HIV-1 capsid (p24) antigen." *Journal of Infectious Diseases* 201 (Supplement_1):S59-S64.

Theng, BKG. 1970. "Interactions of clay minerals with organic polymers: Some practical applications." *Clays and Clay Minerals* 18:357–362.

Vu, Mai N, Hannah G Kelly, Stephen J Kent, and Adam K Wheatley. 2021. "Current and future nanoparticle vaccines for COVID-19." *EBioMedicine* 74:103699.

Wairkar, Sarika, Dhrumi Patel, and Abhinav Singh. 2022. "Nanostructured lipid carrier based dermal gel of cyclosporine for atopic dermatitis-in vitro and in vivo evaluation." *Journal of Drug Delivery Science and Technology* 72:103365.

Wan, Yushun, Jian Shang, Rachel Graham, Ralph S Baric, and Fang Li. 2020. "Receptor recognition by the novel coronavirus from Wuhan: An analysis based on decade-long structural studies of SARS coronavirus." *Journal of Virology* 94 (7):10–1128.

Wang, Chun, Qianling Cui, Xiaoyu Wang, and Lidong Li. 2016. "Preparation of hybrid gold/polymer nanocomposites and their application in a controlled antibacterial assay." *ACS Applied Materials & Interfaces* 8 (42):29101–29109.

Wang, Lin, Alexandra Beumer Sassi, Dorothy Patton, Charles Isaacs, BJ Moncla, Phalguni Gupta, and Lisa Cencia Rohan. 2012. "Development of a liposome microbicide formulation for vaginal delivery of octylglycerol for HIV prevention." *Drug Development and Industrial Pharmacy* 38 (8):995–1007.

Wang, Wenjun, Xiaoxiao Zhou, Yingjie Bian, Shan Wang, Qian Chai, Zhenqian Guo, Zhenni Wang, Ping Zhu, Hua Peng, and Xiyun Yan. 2020. "Dual-targeting nanoparticle vaccine elicits a therapeutic antibody response against chronic hepatitis B." *Nature Nanotechnology* 15 (5):406–416.

WHO. 2019. "WHO guidelines on tuberculosis infection prevention and control: 2019 update." World Health Organization.

WHO. 2022. "Cancer." World Health Organization.

Williams, Wilton B, Kevin Wiehe, Kevin O Saunders, and Barton F Haynes. 2021. "Strategies for induction of HIV-1 envelope-reactive broadly neutralizing antibodies." *Journal of the International AIDS Society* 24:e25831.

Yu, Zhaoju, Yaxing Pei, Shuyi Lai, Shuang Li, Yao Feng, and Xinya Liu. 2017. "Single-source-precursor synthesis, microstructure and high temperature behavior of TiC-TiB2-SiC ceramic nanocomposites." *Ceramics International* 43 (8):5949–5956.

Zhang, Xiaoqing, Ye Feng, Shaoyun Duan, Lingling Su, Jialin Zhang, and Fengjiao He. 2019. "Mycobacterium tuberculosis strain H37Rv electrochemical sensor mediated by aptamer and AuNPs-DNA." *ACS Sensors* 4 (4):849–855.

Zhu, Xiong, Xiaoxia Wang, Limei Han, Ting Chen, Licheng Wang, Huan Li, Sha Li, Lvfen He, Xiaoying Fu, and Shaojin Chen. 2020. "Reverse transcription loop-mediated isothermal amplification combined with NPs-based biosensor for diagnosis of COVID-19." *MedRxiv*: March. 17.20037796.

Chapter 3

The general concept of metal nanocomposites in nanotherapeutics for oxidative stress-induced metabolic disorders

Jigar Vyas and Nensi Raytthatha

LIST OF ABBREVIATIONS

μM	micromolar
4MPBA	4-mercaptophenylboronic acid
ACE	angiotensin-converting enzyme
Ag	silver
ASK1	apoptotic signaling kinase 1
Au NFs	gold nanoflowers
Au NPs	gold nanoparticles
Au	gold
CeO_2	cerium oxide
CH	chitosan
ChOx	cholesterol oxidase
CNTs	carbon nanotubes
CO	carbon monoxide
COX	cyclooxygenase
CT	computed tomography
CTC	circulating tumor cells
CuO	cupric oxide
CVD	cardiovascular diseases
Fe_2O_3	ferric oxide
Fe_3O_4	black iron oxide
FFA	free fatty acids
fT3	free triiodothyronine
FTIR	Fourier transform infrared spectroscopy
Gd NPs	gadolinium nanoparticles
HBA_1C	glycated hemoglobin
hTSH	human thyroid stimulating hormone
ICAM-1	intercellular adhesion molecule-1
IDA	interdigitated microelectrode arrays
IONPs	iron oxide nanoparticles
IR	insulin resistance
ITO	indium-tin oxide
JNK	c-Jun N-terminal kinase
LFIA	lateral flow immunoassay

DOI: 10.1201/9781032621135-3

MD	metabolic disorders
MgO	magnesium oxide
MI	myocardial infarction
MNCs	metal nanocomposites
MTT	(3-[4,5-dimethylthiazol-2-yl]-2,5 diphenyl tetrazolium bromide)
NCs	nanocomposites
nM	nanomolar
OS	oxidative stress
Pd	palladium
PEG	polyethylene glycol
PET	positron emission tomography
PPy	polypyrrole
PSA	prostate-specific antigen
Pt	platinum
RGO	reduced graphene oxide
ROS	reactive oxygen species
RT	radiation therapy
SnO_2	tin (IV) oxide
SPION	superparamagnetic iron oxide
SPR	surface plasmon resonance
TEM	transmission electron microscopy
THs	thyroid hormones
TRX1	thioredoxin 1
VCAM-1	vascular cell adhesion molecule-1
WO_3	tungsten trioxide
XRD	X-ray diffraction
ZnO	zinc oxide

3.1 NANOTHERAPEUTICS AND NANOCOMPOSITES

Nanotherapeutics is a relatively new application of nanotechnology that significantly impacts the medical field (Freitas 2000). Research and development in the field of nanotherapeutics are very extensive. The drug is attached to nanoparticles (NPs), allowing it to act more correctly and efficiently with fewer side effects. Nanotherapeutics offers special possibilities to raise the safety and efficiency of traditional medicines. Nanocomposites and nanotherapeutics are directly interrelated as a framework for the formulation and administration of nanotherapeutics (Wagner et al. 2006). Nanocomposites are promising for use in various fields, including the biomedical, automotive, and other sectors.

Polymer nanocomposites, ceramic nanocomposites, and metal nanocomposites (MNCs) are three classes of nanocomposites. In an ideal situation, the polymer nanocomposites would have isolated nanoscale particles uniformly dispersed throughout the polymer. In application, it is employed to disperse aggregated nanoparticles. The synthesis of polymer nanocomposites using the sol-gel method appears to be the most promising method (Kuo et al. 2005). Before being added to the polymer gel, the nanoparticles are distributed at the molecular or nearly molecular scale. Aluminum is particularly promising, especially in ceramic matrix nanocomposites such as aluminum oxide–silicon carbide systems. All of the research has shown that adding silicon carbide with a moderate volume fraction (10%) and a certain volume to an aluminum oxide matrix improves the matrix's characteristics. Several methods have been used to prepare ceramic matrix nanocomposites, including the traditional powder method, the polymer precursor route, chemical methods, and spray pyrolysis, the most common of which are the colloidal and precipitation procedures, sol-gel process and template synthesis (Palmero 2015).

MNCs are composed of at least two physically and chemically distinct phases, which are distributed in such a way that they can coexist. They are made up of two phases separated by a metal matrix, a fibrous phase and a particulate phase. These MNCs are constructed from flexible alloys or metallic matrix and reinforcement material that is nanoscale in size. These materials have significant strength and modulus as well as ceramic and metal qualities like durability and toughness (Al-Mutairi, Mehdi, and Kadhim 2022). As a result, metal-based nanocomposites represent a distinct class of cutting-edge materials that aim to promote synergistic qualities, like the mechanical robustness of metal nanomaterials and the flexibility and processing of polymers made from the similar fusion of two distinct adaptable materials. Due to their special properties, which may be adjusted to fulfill specific criteria for drug administration, nanocomposites can be employed as a framework for forming nanotherapeutics. They enable design and characteristic choices that are impossible with conventional composites (Khan, Saeed, and Khan 2017). To achieve significant drug-loading capability, targeted release of drugs, and controlled drug release, for instance, nanocomposites can be developed. This can reduce side effects while increasing the effectiveness of pharmacological therapy (Safdari and Al-Haik 2018).

3.2 OXIDATIVE STRESS (OS)-INDUCED METABOLIC DISORDERS (MD)

The potential importance of oxidative stress in MD is gradually becoming apparent. The results support the hypothesis that symptoms of MD, such as atherosclerosis, type II diabetes, coronary heart disease, and hyperlipidemia, may be influenced by higher oxidative stress. Oxidative stress has also been associated with obesity and insulin resistance in men, suggesting that OS could represent an early marker in the pathogenesis of such chronic diseases. Figure 3.1 shows several metabolic disorders and the involvement of OS as one of the main causes.

The risk of cardiovascular diseases (CVD) and death from all causes is much greater in those with metabolic syndrome. For instance, according to Katzmarzyk et al. (Katzmarzyk, Church, and Blair 2004), middle-aged men having metabolic syndrome had a fourfold higher chance of passing away across an 11-year follow-up than those without the condition. A growing body of research suggests that OS, such as the oxidation of lipoproteins that include apolipoprotein-B, may be crucial to the atherogenicity of lipoproteins. OS indicators, such as 8-iso-erythrocyte GPX1, PGF2, oxLDL, MPO, and nitrotyrosine, independently predict a person's chance of developing CVD and myocardial infarction in their early years.

MPO, a hemoprotein highly expressed in neutrophils and has significant proatherogenic characteristics and serving as the enzymatic source of NO-derived oxidants, is of special interest. MPO can oxidize LDL cholesterol, boosting its absorption by macrophages and sustaining foam cell development. The oxidized purines enhance with the levels of damaged

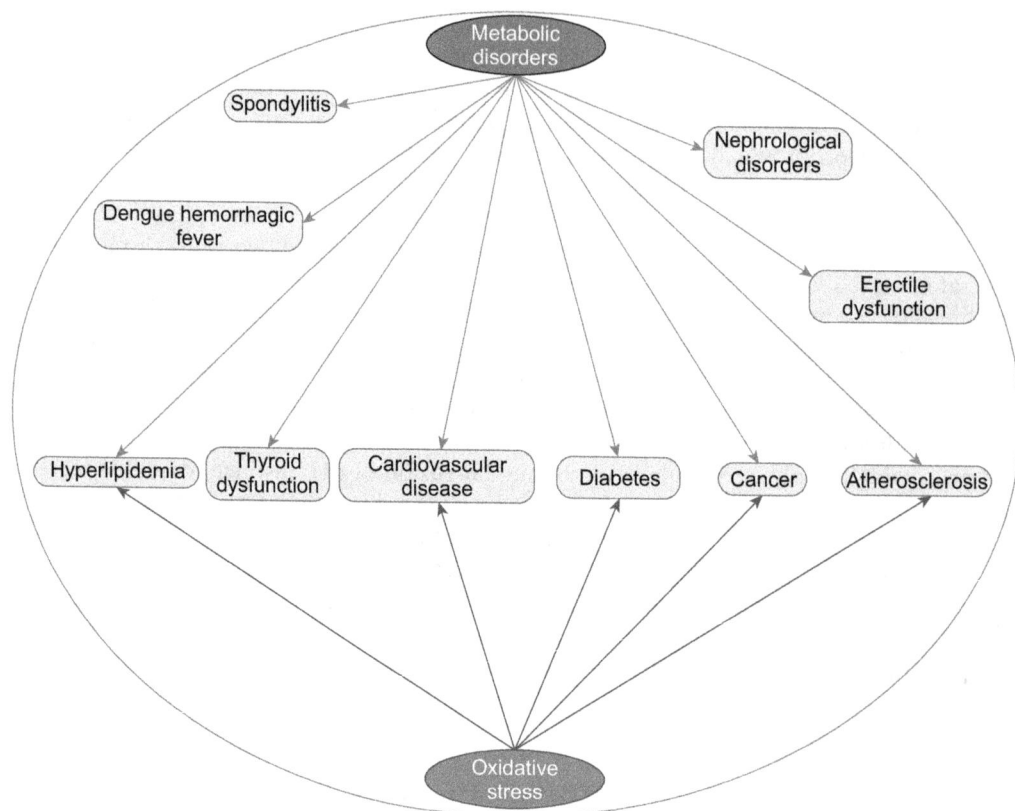

Figure 3.1 Involvement of oxidative stress in the development of various metabolic disorders.

arteries and are positively linked with the degree of CVD as determined by the number of coronary lesions. Moreover, diabetic and dyslipidemic individuals had comparable findings, indicating that the pathogenic processes underlying these linked disorders are identical (Botto et al. 2002).

Free fatty acid (FFA) concentrations in blood often increase in obese people and are associated with fat mass growth. FFA infusion induces higher levels of OS and insulin resistance (IR) in healthy persons, which can be restored by an antioxidant such as glutathione. Increased β-oxidation of cytoplasmic long-chain acyl-CoA ester, the physiologically active form of FFAs, occurs with obesity that increases OS. Moreover, FFAs can increase reactive oxygen species (ROS) generation in the aorta smooth muscle cells and endothelial cells via PKC-dependent stimulation of NADPH oxidase (Wojtczak and Schönfeld 1993).

People with a high BMI are more likely to acquire endometrial, colorectal, ovarian, and breast cancer. Obesity, according to Rehman et al., increases OS by raising the quantity of ROS, which is a primary factor in cancer formation (Rehman et al. 2020). One of the prominent reasons that OS exhibits its harmful consequences is by creating genomic instability. ROS cause DNA damage by inducing base/nucleotide changes and strand breaks.

Cancer development has been linked to a disruption in redox balance, which affects multiple signaling pathways that regulate the proliferation of cells, apoptosis, invasiveness, treatment resistance, and energy consumption. ROS accumulation may contribute to tumor formation by functioning as a signaling molecule or encouraging genomic DNA mutation. Histone alterations, in addition to DNA modification, are recognized to play an important role in genomic repair and carcinogenesis (Chen et al. 2020). Histone methylation and acetylation, catalyzed by histone methylation transferases and histone acetyltransferases, are major regulators of gene expression. ROS have been shown to have a role in histone modification and the consequent influence on cell viability (Cheng et al. 2020).

According to research, OS is a critical factor in atherogenesis (Witztum and Berliner 1998). It is widely defined as the redox state that occurs when there is an imbalance with antioxidant capabilities and activity species, including ROS and nitrogen (RNS) species and non-radical and free radical species. These circumstances harm the cells by directly oxidizing cell proteins, lipids, and DNA or by activating cell death pathways (Sinha et al. 2013). Ambient levels of specific ROS are utilized as signaling molecules in the cell to maintain essential cellulate processes. In contrast, reactive oxidants and free radicals are formed without a physiological stimulus, depleting small molecule antioxidants or overwhelming antioxidize systems. It causes an overall increase in OS and physiologically activated ROS. It is not only crucial in the pathophysiology of CVD but also has physiological roles that may modify cardiomyocytes (Santos et al. 2011).

Thyroid hormones (THs), namely tetraiodothyronine (thyroxin, T4) as well as a considerably lesser fraction of triiodothyronine, exert physiological functions via binding to a series of specialized receptors that connect to both genomic and nongenomic signaling pathways. While ROS generation mostly depends on the mitochondria, THs need not simply dictate the respiratory status of the mitochondria. They also influence the antioxidant status of the cell. The increased generation of free radicals reduced ability of the antioxidative defense, results in hypothyroidism-associated ROS. The excess of thyroid-stimulating hormone (TSH) may affect OS pathways (Nanda, Bobby, and Hamide 2008a). The respiratory chain of the mitochondria malfunction caused by hypothyroidism can increase free radical generation. Hyperlipidemia is associated with increased lipid peroxidation, a consistent biochemical finding. The prevalence of OS in thyroidism correlates with atherosclerosis lipid risk factors (Nanda, Bobby, and Hamide 2008b).

Hyperglycemia has been proven to be the main reason for diabetic tissue damage seen in clinical situations, such as neuropathy, retinopathy, nephropathy, or vascular damage. Crucially, endothelial cells maintain the development of non-insulin-dependent GLUTs, allowing internal glucose to grow concurrently with external glucose concentrations. Hyperglycemia increases adipocyte ROS production because high glucose levels trigger mitochondrial ROS production during 3T3-L1 adipocytes. Indeed, primary adipocytes subjected to hyperglycemic conditions produce more ROS *in vitro* and *in vivo* (Lin et al. 2004). Insulin sensitivity is lowered in hyperglycemic adipocytes. Hence, mounting data suggests that OS may play a role in IR.

3.3 ROLE OF MNCS IN THERANOSTICS AND MANAGEMENT OF OS-INDUCED METABOLIC DISORDERS

3.3.1 CVD

CVDs are conditions that affect the heart and blood arteries. Disorders include stroke, heart failure, coronary artery disease (CAD), and peripheral arterial disease, high BP, elevated lipid levels, hyperglycemia, smoking, diabetes, and obesity are common risk factors (Ruan et al. 2018).

MNCs in the diagnosis of CVDs

The diagnosis of CVDs typically involves biochemical testing, cardiograms, and radio-imaging testing. Biochemical testing includes lipid profiling, C-reactive protein, homocysteine, and troponin. Biochemical testing includes a wide range of methods, including both optical methods, like UV spectroscopy and Fourier transform infrared spectroscopy (FTIR) for the detection of biomarkers used and titration-based methods, like acid-base titrations and complexometric titrations, and can be used for the analysis of samples (Brewer, Svatikova, and Mulvagh 2015). Hence, there is no role for MNCs. Echocardiograms and electrocardiograms are medical tests used to diagnose heart-related conditions, and MNCs have no direct role in diagnosing CVD. MNCs have a role in radio-imaging testing and are discussed herewith.

MRI-based testing using gadolinium (Gd) nanoparticles (NPs) has indeed been employed in CVD diagnostic imaging such as MRI (magnetic resonance imaging). Gadolinium is a paramagnetic contrasting agent that helps to improve the MRI of the blood vessels and heart arteries by enhancing the MRI signal. Gadolinium nanoparticles (Gd NPs) may be employed in cardiac imaging to identify and diagnose vascular abnormalities, heart failure, and myocardial infarction (MI). It can also aid in determining the severity of certain conditions and monitoring disease progression. One advantage of employing Gd NPs in MRI is that it can produce high-resolution pictures of the heart and blood arteries while avoiding ionizing radiation, which is a problem with other imaging techniques such as CT scans. In summary, Gd NPs are an excellent contrast agent for medical testing of CVD, especially with MRI (Danila, Johnson, and Kee 2013).

Computed tomography (CT) is a strong imaging technique with several benefits, such as great spatial and temporal resolution and excellent quantitative capability. CT with NPs as external contrast agents has mostly been utilized to examine the coronary arteries in atherosclerosis. Several studies have examined the viability of utilizing NPs for CT to assess myocardial infarction (MI) pathways in small animals. Danila et al., for example, employed gold NPs functionalized with a collagen-recognition peptide to target the cardiac scar. Sawall et al. (2017) have established a preclinical micro-CT approach for longitudinally monitoring cardiac processes *in vivo* in normal and MI-inflicted mice utilizing a contrast dye depending on nanocapsules having high iodine content (ExiTronTM MyoC 8000). The investigators looked at NPs with a high alkaline-earth metal composition (ExiTronTM nano). ExiTron MyoC 8000 permitted the detection of MI even at extremely low doses (Chithrani and Chan 2007).

X-ray and CT X-ray-based testing using gold nanoparticles (Au NPs) with photothermal properties were used externally for plaque-specific administration of such NPs to detect photothermal rejuvenation of blocked arteries (Ghann et al. 2012). To develop Au NPs-based CT contrast agents, Ghann et al. incorporated lisinopril as an active agent. By ligand exchange reaction on citrate-coated Au NPs, Au NPs coupled to pure lisinopril, reduced thioctic acid-lisinopril, or thioctic acid-lisinopril was developed. X-ray-CT was utilized to investigate the target of angiotensin-converting enzyme (ACE) using higher stability features of thioctic lisinopril Au NPs. The scans showed strong contrast in the heart and lungs, indicating that ACE was being targeted, and ACE overexpression was associated with the onset of respiratory and heart fibrosis (Namdari et al. 2017). This new strategy may be useful for monitoring cardiovascular pathophysiology using CT. Au NPs used alone or in combination may offer large-scale treatment and diagnosis approaches in CVDs.

MNCs for treatment of CVD

Lifestyle modifications are well-known strategies, such as regular exercise, a healthy diet, not smoking, and managing stress. These lifestyle modifications are typically not related to the use of MNCs. Surgery is a well-established treatment option for certain types of CVD, such as coronary artery disease, heart valve disease, and congenital heart defects; hence, MNCs are not typically used in surgical interventions for CVD. Cardiac rehabilitation is a comprehensive program that involves exercise training, education, and counseling. Specifically, the role of MNCs in cardiac rehabilitation is not yet well-defined and requires further study and evaluation. Medication-based treatments include blood thinners, cholesterol-lowering drugs, and blood pressure medications; however, they have side effects, poor compliance, limited efficacy, and toxicity to reduce the drawbacks mentioned previously.

Au NPs are a common nanocarrier for the administration of cardioprotective drugs. The use of Au NPs in conjunction with existing medicinal medications has been proposed as a novel method with increased promise in the therapy of cardiac ailments. Gold nanocarriers have been widely employed because of their ease of manufacturing, stability, low toxicity, and immunogenicity. *Simdax* is a heart disease therapy that has received clinical approval. Simdax coupled on Au NPs has cardioprotective properties in rats with doxorubicin-induced heart failure. Gold nanoconjugates and simdax have greater cardioprotective qualities than their counterparts due to improved localization of gold nanoconjugates to wounded tissue (Xu et al. 2016).

Iron oxide NPs have been studied only in animals and *in vitro* studies for CVD treatment. Emerging research suggests that these iron nanoelements might be used in diagnosing, assessing, and treating various life-threatening CVDs. Iron oxide nanoparticles have been shown to have a role in the magnetic-mediated transport of mesenchymal stem cell precursors to the infarcted myocardium. It improves stem cell regeneration capability by increasing availability near the damaged location (Liu et al. 2017).

3.3.2 Hyperlipidemia

Hyperlipidemia is a medical condition characterized by abnormally high levels of lipids (fats) in the blood. This includes elevated levels of cholesterol, triglycerides, or both. Hyperlipidemia is a significant risk factor for CVD, as excess lipids can accumulate in the walls of blood vessels, leading to atherosclerosis (hardening and narrowing of the arteries), ultimately resulting in heart attacks or strokes.

MNCs in the diagnosis of hyperlipidemia

Currently, the diagnosis of hyperlipidemia primarily relies on biochemical testing of lipids, which measures the levels of different lipids (such as cholesterol and triglycerides) in the blood. Nanocomposites have been studied for their potential applications as biosensors in biochemical testing externally, which can be used to detect various biomolecules, including lipids. For example, nanocomposites could be incorporated into a biosensor externally that detects the levels of LDL cholesterol or triglycerides in the blood. Some of the tests that utilize metal-based biosensors are discussed here.

Biochemical testing using an Au-based biosensor was built to detect cholesterol electrochemically. The sensor was developed in two steps by integrating electrochemical and

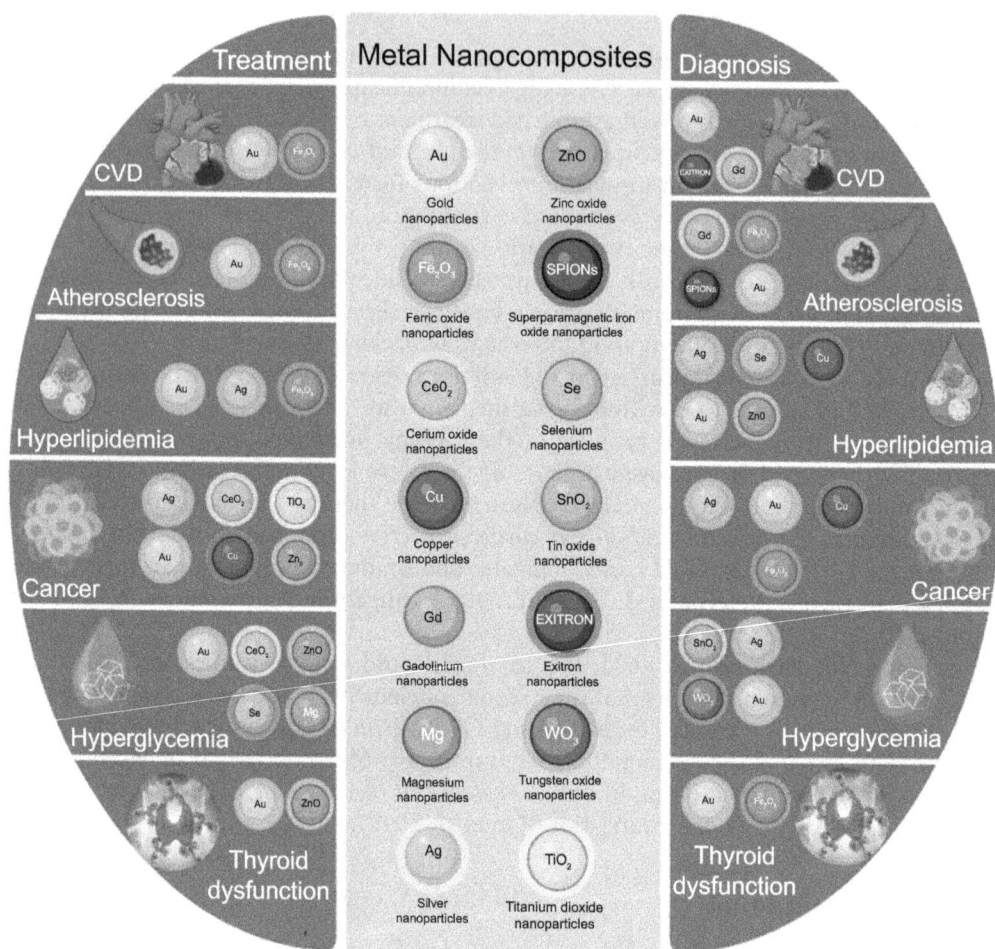

Figure 3.2 Role of metal nanocomposites in the diagnosis and treatment of various oxidative stress-induced metabolic disorders.

chemical methods, and it was tested using X-ray diffraction (XRD), transmission electron microscopy (TEM), and FTIR spectroscopy. The sensor demonstrated excellent selectivity, sensitivity, and repeatability. The sensor's high performance was due to the improved electrical conductivity caused by Au and the strong polypyrrole matrix used to retain the enzyme. As a result, the sensor may be used to detect lipids in biological specimens.

Recent research has revealed that metal oxide–chitosan nanocomposites may enhance the sensitivity and specificity of cholesterol biosensors. A cholesterol biosensor based on a ZnO–chitosan nanocomposite, with a sensitivity of 50.8 A/mM and a detection limit of 0.2 mM cerium oxide NPs (CeO$_2$), has indeed been effectively incorporated into chitosan (CH) matrices for detecting cholesterol, as well as lipid oxidase, which has been adsorbed on CH–CeO$_2$ nanocomposites. The cholesterol oxidase/chitosan–NanoCeO$_2$/

indium-tin-oxide (ChOx/CH-NanoCeO$_2$/ITO) bioelectrode demonstrated remarkable catalytic behavior toward cholesterol due to the enlarged active center of the nanocomposites produced by NanoCeO$_2$ as well as the suitable microenvironment it provides for high ChOx loading. This improves the transfer of electrons between the analyte and the surface of the ChOx/CH-NanoCeO$_2$/ITO electrode. The CeO$_2$NPs/PPy nanocomposites were used as a mediator to immobilize ChOx onto the electrode surface, resulting in a cholesterol biosensor. The results show that the ChOx/CeO$_2$ NPs/PPy/electrode does have a broad linear response range and good sensitivity. Furthermore, the cholesterol biosensor demonstrated high selectivity, repeatability, and storage stability. The findings suggest that the cholesterol biosensor developed can be useful for lipid monitoring in clinical diagnostics (Malhotra and Kaushik 2009).

A chemiluminescent cholesterol sensor with excellent selectivity and sensitivity has been developed using the peroxidase-like activity induced by cupric oxide CuO nanoparticles. CuO nanoparticles can catalyze the oxidation of luminol by H$_2$O$_2$, which is formed via a cholesterol-oxygen reaction catalyzed by ChOx. By combining these two processes, cholesterol oxidation might be converted into luminol chemiluminescence. The chemiluminescence intensity was proportionate to the cholesterol concentration throughout a range of 0.625–12.5 µM under optimal circumstances, with a detection limit of 0.17 µM. The suggested method's applicability has been tested by determining cholesterol in milk powder and human blood samples with favorable outcomes (Hong et al. 2013).

The wurtzite structure of CdSe/ZnS, core/shell nanoparticles, was effectively produced because of its great specificity, selectivity, fast response, and simple fabrication. This optical sensor employing enzyme-coupled quantum dots can be expanded for enzymatic detection of additional biomolecules. Because of cyclooxygenase's selective enzymatic oxidation process (COX), the bioconjugates demonstrated selectivity to cholesterol molecules. For precision cholesterol detection, simple quantum dot-based optic biosensors are suggested (Arshad et al. 2020).

A sensitive and specialized colorimetric technique for determining cholesterol is zinc oxide nanoparticles (ZnO NPs) on the surfaces of carbon nanotubes (CNTs). The suggested nanocomposite catalyzed the oxidation of 2,2'-azino-bis (3-ethylbenzthiazoline-6-sulfonic acid) in hydrogen peroxide, yielding a green product that can be measured at 405 nm. In the presence of ChOx, hydrogen peroxide is the oxidized by-product of cholesterol. By integrating these two processes, the oxidation of cholesterol may be quantitatively correlated to the colorimetric response. The colorimetric reaction was proportionate to the cholesterol concentration under ideal experimental circumstances (Li et al. 2019).

The use of Au@Ag core-shell nanoparticles in fabricating a chromogenic biosensor substrate for hyperglycemia and hyperlipidemia has demonstrated effectiveness. Because of the surface plasmon resonance (SPR) absorption of Ag NPs, Au@Ag NPs had a clear absorbance peak at 375 nm. H$_2$O$_2$ was produced in the presence of hyperglycemia and hyperlipidemia by the catalytic activity of enzymes such as cholesterol oxidase and glucose oxidase which etched the Ag NPs shells of Au@Ag NPs. As a result, their typical absorbance at 375 nm diminished, accompanied by a detectable color shift from orange to red, which may be utilized by the naked eye to identify hyperglycemia and hyperlipidemia (Rastogi, Dash, and Sashidhar 2021).

A chromogenic sensor based on gold-magnetic NPs with increased peroxidase-like characteristics for the measurement of hyperlipidemia with selectivity and accuracy was developed. Gold-magnetic particles, which have numerous advantages, including high catalytic properties, strong chemical stability, excellent monodispersity, biocompatibility, and optical characteristics, have been widely employed in biocatalysis, immunoassay, drug targeting, and other applications. Moreover, it facilitates quick enrichment, separation, and recovery (Guan et al. 2020).

MNCs in the treatment of hyperlipidemia

Treating hyperlipidemia, or high levels of lipids (fats) in the blood, typically involves lifestyle modifications and medications (Ezeh and Ezeudemba 2021). For example, lifestyle modifications can be difficult to maintain for some people. However, lifestyle modifications need to be followed. Medications such as statins can cause side effects, including muscle pain and liver damag; hence, metal-based nanocomposites are introduced to reduce the drawbacks.

MNCs have also been studied for their possible application in treating hyperlipidemia. Metal NPs as carriers for cholesterol-lowering drugs are one option. Metal nanoparticles with distinctive characteristics, such as Ag, Au, and Fe_2O_3, make them valuable as drug delivery vehicles. They have a high surface area-to-volume ratio, allowing for great loading of drugs, and they may be derivatized with targeting ligands, allowing them to deliver pharmaceuticals selectively to certain cells or tissues. In one study, the researcher formulated Au NPs coated with atorvastatin, a cholesterol-lowering drug. In rats with elevated cholesterol levels, the Au NPs were shown to lower cholesterol levels (Zamora-Justo et al. 2019).

Overall, MNCs can potentially be an innovative strategy for treating elevated cholesterol levels. However, further research is needed to assess their toxicity, efficacy, and clinical application prospects. While there is some encouraging investigation regarding the utilization of MNCs in treating hyperlipidemia, clinical use is still in its early phases.

3.3.3 Cancer

Cancer is a class of disorders defined by the uncontrollable development and dissemination of abnormal cells. Cells within the human body normally develop and divide to generate new cells in a regulated manner (Hamilton 2010).

MNCs in the diagnosis of cancer

Conventional cancer diagnosis typically involves blood tests, radio-imaging tests, biopsies, and genetic tests. Genetic cancer tests involve analyzing a patient's DNA to identify mutations or genetic variations that increase their risk of developing certain types of cancer or may indicate the presence of cancer, and MNCs have no role. MNCs have a role in biopsies, radio-imaging testing, and blood tests and have shown great potential in improving sensitivity and specificity compared to other testing methods (Patra et al. 2010).

Biopsies are modified using Au NPs and have been extensively explored for their possible use in cancer detection, including biopsy. Au NPs can be coupled with antibodies or other targeted ligands to adhere to cancer cells or tissues precisely. During a biopsy, this allows for precise detection and separation of cancerous cells from surrounding healthy tissues (Lim et al. 2011).

Radio-Imaging test

Computed tomographic (CT) scanning using Au NPs has been extensively studied for cancer diagnosis due to their high biocompatibility and ease of functionalization. They can be coated with various biomolecules, such as antibodies or peptides, to target specific cancer cells or biomarkers. Once bound to cancer cells, they can be imaged using a CT scan or optical imaging, providing high sensitivity and specificity. An important demonstration of multifunctional Au NPs usage for drug delivery was the use of 5-nm Au NPs as a delivery vehicle, covalently bound to cetuximab, as an active targeting agent and gemcitabine as a therapeutic payload in pancreatic cancer (Zanganeh et al. 2016). Patra et al. (2008) reported that this approach may reach high intra-tumoral Au concentrations (4500 µg/g) with minimal retention in the spleen or kidneys compared to 600 µg/g with untargeted Au NPs. Metal NPs, such as Au NPs, have indeed been demonstrated to enhance radiation absorption by cancer cells while decreasing absorption by healthy tissues. Metal nanoparticles can be employed in combination with contrast agents to increase the imaging of malignant tissue and the efficacy of detection. Au NPs have been employed as contrast agents in CT scanning to increase tumor imaging (Hamilton 2010).

MRI-based testing using iron oxide nanoparticles (IONPs) has also been investigated for cancer detection using MRI. IONPs may be functionalized with various targeting ligands and imaging agents to detect cancerous cells and their surroundings. An increasing amount of preclinical and clinical data shows that NPs, especially IONPs, and their interaction with host cells of the immune system promote immune recognition of tumors, potentially improving cancer detection (Wang, Wang, and Zha 2009).

Several metal NPs, such as Ag and Cu, have shown promise for cancer diagnosis by imaging systems and surface-enhanced Raman spectroscopy (Wang, Wang, and Zha 2009). Blood testing using MNCs, such as Au NPs and Fe_2O_3 NPs, has shown considerable promise in increasing the sensitivity and specificity of cancer blood tests. MNCs can be utilized in blood testing for cancer to improve the identification of circulating tumor cells (CTCs). CTCs are cells that have separated from a primary tumor and are migrating through the circulation, and their discovery gives valuable information regarding the aggressiveness of cancer and treatment success. CTCs, on the other hand, are highly infrequent and difficult to identify. CTCs may be labeled with particular biomarkers using MNCs, making identification simpler and more precise. MNCs are used alternatively in cancer blood tests to improve the identification of cancer biomarkers. MNCs are coupled with antibodies or other specific agents to precisely attach to tumor markers in the bloodstream, improving detection. Au NPs were coupled with antibodies against prostate-specific antigens (PSA) to detect prostate cancer. Overall, MNCs have a high sensitivity and specificity for cancer diagnosis and the capacity to specifically target tumor cells and markers (Zhang et al. 2019).

MNCs in the treatment of cancer

Cancer treatment includes lifestyle modifications, chemotherapy, surgery, and radiation-based treatments. Lifestyle modifications include diet and physical activities. MNCs play a crucial role in the modification of chemotherapy and radiation-based treatments. MNCs have been explored as a potential alternative or complementary treatment for cancer. Previous research and current development suggest MNCs' significant contribution in managing cancer (Wang, Wang, and Zha 2009).

Chemotherapy is a common treatment for cancer that involves the use of drugs to kill cancer cells. While chemotherapy is effective in treating many types of cancer, it has several drawbacks and side effects, such as nausea, vomiting, hair loss, fatigue, and increased risk of infections. These side effects significantly impact the quality of life and may require additional medications and supportive care for a patient. Additionally, chemotherapy drugs damage healthy cells in the body, leading to long-term health problems such as heart damage, nerve damage, and fertility issues (Zhang et al. 2019). Hence, MNCs reduce these side effects caused by chemotherapy and are discussed further.

Au NPs are one of the most extensively researched metal-based nanomedicines. Reduced cytotoxicity and immunogenicity, greater stability, improved permeability and retention, good biocompatibility, innate immune activation capabilities, and readily modifiable surface are among the interesting aspects of Au NPs (Shang et al. 2021). Moreover, Au NPs are of good use in cancer hyperthermia therapy. They can generate heat due to the contained mobile carriers that can be resonant at specific frequencies, depending on the structure of nanoplatforms. The heating effect increases under a plasmon resonance frequency when all mobile carriers present on the particle resonate (Szunerits and Boukherroub 2018). Green-synthesized Au NPs were shown to have intrinsic antitumor properties, with favorable outcomes once evaluated against many human cancer cell lines, including liver melanoma, lung melanoma (Chaturvedi et al. 2020), colon melanoma (Wang et al. 2019), pancreatic melanoma (López-Miranda et al. 2021), breast melanoma (Păduraru et al. 2022), cervix melanoma (Mikhailova 2020), and ovarian melanoma (Piktel et al. 2021).

Ag NPs have increased the interest of oncologists because of their inherent anticancer properties and effectiveness as antitumor drug delivery vehicles. Ag NPs have also been studied to modify cancer cell autophagy as cytotoxic agents on their own, in conjunction with transportable compounds, or in conjunction with other therapies. Regarding anticancer mechanisms, Ag NPs influence membrane fluidity, allowing for easy entrance and aggregation in cancerous cells, causing cancer cells to die or limiting their uncontrolled growth. Furthermore, Ag NPs can produce Ag^+ cations, which grab electrons, raise cellular OS, elevate ROS generation, decrease cancer cell ATP levels, and limit cell proliferation rates. According to reports, Ag^+ ions are produced primarily in mitochondria and secondarily in nuclei, interacting with DNA, causing disruption and inducing apoptosis (Perera et al. 2020).

ZnO is a ubiquitous metal NPs in the world. ZnO NPs can generate ROS if exposed to light. ZnO NPs can be chemically changed to improve their photocatalytic efficacy and capacity to create ROS in several ways, such as metal doping, polymeric alteration, and organic photosensitizing agents. The enhanced anticancer activity of altered ZnO NPs can be related to their higher efficacy in producing ROS (Gao et al. 2014). The MTT test was used to analyze the possible antitumor activity of the curcumin-loaded ZnO NPs on the rhabdomyosarcoma RD cultured cells. At the same time, the resazurin analysis was employed to assess their cytotoxic effects on the RD cell lines. The possible antitumor properties of ZnO NPs coated with curcumin were investigated using the MTT test on the rhabdomyosarcoma RD cell line, and their side effects were assessed using the resazurin test on human kidney embryonic cells. The high surface/volume ratio of ZnO NPs was thought to be a cause of the enhanced cytotoxicity of the NPs (Wason et al. 2018).

Cerium oxide nanoparticles (CeO_2 NPs) enclosed by an oxide lattice have shown the potential to cause cancer cell death. CeO_2 NPs render pancreatic cancer more susceptible to radiation therapy (RT) by oxidizing and activating the c-Jun N-terminal kinase (JNK) apoptotic pathway. CeO_2 NPs promote cancer cell ROS production. ROS has been shown to activate thioredoxin 1 (TRX1) oxidation, which culminates in the stimulation of apoptotic signaling kinase 1 (ASK1) (Prasad, Kanchi, and Naidoo 2016).

Copper is required for both animal and plant metabolism. Broccoli green extract has been suggested as a greener and ecologically safe precursor for one-pot biosynthesis of CuO NPs. The CuO NPs demonstrated to be successful in the therapy of prostatic cancers. A chitin-based Au and Cu nanocomposite was tested for cytotoxicity toward breast cancer (MCF-7) cells in humans. The MNCs' inhibitory concentration (IC_{50}) was determined to be 31 mg. Subsequent research demonstrated an increase in ROS generation, reduced antioxidant enzymatic activities, and membrane stability breakdown, validating the copper–silver NP-based nanocomposite's harmful action (Solairaj, Rameshthangam, and Arunachalam 2017).

Nontoxicity, inertness, superparamagnetism, biocompatibility, and readily adjustable surface characteristics of IONPs make them particularly suitable for biomedical applications (Aslam et al. 2021). For cancer care, IONPs have been FDA-approved for clinical testing in cancer detection, imaging, and magnetic hyperthermia therapy, also exhibiting possibilities in experimental contexts for photodynamic and photothermal treatments. Moreover, IONPs have raised the curiosity of researchers working on magnetic nanoparticle-based medication delivery systems. Specifically, by providing an outside magnetic field, injected IONP-based delivery mechanisms flow via blood capillaries to the target region, releasing the drugs into tumor cells and boosting therapeutic efficacy while causing no harm to the surrounding cells (Schuemann et al. 2020). Moreover, their magnetic features enable the translation of light radiation into heat or ROS after using a localized magnetic field from the outside, hence decreasing the negative consequences of cancer treatment.

Titanium dioxide nanoparticles (TiO_2 NPs) are promising nanomaterials for photodynamic treatment. Its mode of action depends on the stimulation of hydrophobic interaction with electromagnetic energy that reaches the visible or UV light spectrum, leading to the formation of ROS and the subsequent activation of apoptosis. Moreover, TiO_2 NPs are cytotoxic in various human cancer cell lines, including osteosarcoma, breast, and colon cancer (Tolkaeva et al. 2021).

Radiation-based treatment is a preventive therapy that uses high-intensity radiation capable of killing cancerous cells. While radiation therapy can effectively treat many types of cancer, it has several drawbacks. One major drawback of radiation therapy is that it can cause damage to healthy cells surrounding the cancerous area, leading to side effects such as skin irritation, fatigue, and nausea. Additionally, radiation therapy can increase some patients' risk of developing secondary cancers. MNCs have been explored as a potential alternative or complementary treatment for cancer (Rafieian-kopaei et al. 2012).

Metal-based nanoparticles are frequently used to enhance the precision of radiations to the target location, hence reducing radiation dosage and preventing toxicity and harm to normal tissues. This radiation causes the formation of ROS, and because of the unpaired electron, they cause severe DNA damage. Metal NPs increase radiation targeting in a variety of ways. Metals enhance OS in tumor cells, causing selective apoptosis and decreasing clonogenic survival. Since MNCs have a high atomic number, they absorb stronger radiations than surrounding tissues. This reduces the dose of radiation administered to the tumor tissue, improving therapy efficacy. MNCs can scatter radiations, which can enhance radiation distribution within malignant cells. Numerous studies have looked at the use of MNCs as radiation therapy enhancers, and while the outcomes are encouraging, further study is needed to completely understand the advantages and drawbacks (Bejarano et al.

2018). It should be noted that MNCs are still considered experimental and have not yet been approved for clinical use.

3.3.4 Atherosclerosis

It is a chronic disorder whereby the arteries constrict and harden due to plaque buildup. Plaques are composed of lipids, fats, calcium, and other constituents that can accumulate on the inner side of the artery walls, causing atherosclerosis. Over time, plaque buildup in the arteries can restrict blood flow and increase the incidence of CVD, stroke, and other ailments (Wang 2011).

MNCs in the diagnosis of atherosclerosis

Atherosclerosis can be diagnosed through radio-imaging tests, biochemical testing, and biopsy (Wang 2011). MNCs have no role in biopsy, and hence further research is required. The role of MNCs in biochemical and radio-imaging tests is discussed further.

Biochemical testing using Au NPs as biosensors to detect biomarkers of atherosclerosis, such as oxidized LDL cholesterol or inflammatory markers. Au NPs are functionalized with specific ligands that bind to the biomarker of interest, causing a change in the optical or electrical properties of the nanoparticles that can be detected and measured. This approach can potentially provide a sensitive and rapid diagnostic test for atherosclerosis that could be used in clinical settings (Tarin et al. 2015).

Radio-imaging testing with MRI combines great temporal and spatial resolution with various plaques and vessel morphological readouts. The NPs of superparamagnetic iron oxide (SPION) AMI-25, formed as a sustainable hydrophilic colloidal suspension of magnetite, were the initial magnetic nanoparticles employed for vessel imaging. The Fe_2O_3 was encapsulated for intravenous administration with a noncovalently bound low-molecular-weight dextran (Kim et al. 2007). Metal NPs have been used in recent studies for the imaging of atherosclerosis. Tarin et al. (2015) created Au-coated Fe_2O_3 NPs for MRI detection of CD163 in atherosclerosis. The enhanced expression of the membrane receptor CD163 in macrophages from intraplaque hemorrhagic areas or asymptomatic plaques is the basis for this targeted strategy. Au-coated Fe_2O_3 NPs coupled with anti-CD163 antibodies accumulate over time in apoE-deficient mice's atherosclerotic lesions (Tarin et al. 2015).

Computed tomography or computed tomographic scanning is the most clinically robust and reliable approach for assessing coronary arterial stenosis and identifying plaque calcification. Utilizing nanoparticles with unique features, specificity, and contrast can be increased in CT scanning. Kim et al. (2007) developed a revolutionary *in vivo* CT contrast dye using Au NPs coated with polyethylene glycol (PEG). CT scans of rats employing PEG-coated Au NPs revealed a clear distinction between heart ventricles and major arteries.

Photoacoustic imaging is another imaging approach that uses nanoparticles as contrast agents (*in vivo*), and it measures the spread of optical absorption of the contrast agents within organs. The contrast of a photoacoustic image can be increased by utilizing Au NPs due to the oscillations of free charge carriers on the Au NPs surface with the wavelength range that results in significant optical absorption and enhanced contrast. Kelly et al. (2005) demonstrated the possibility of using gold nanorods coupled to anti-intercellular adhesion molecule-1 (ICAM-1) to detect inflammation in endothelial cells, which might be connected to the progression of atherosclerotic plaques. Wang et al. infused macrophages with accumulated Au NPs to target atherosclerotic lesions (Wang et al. 2008).

Multimodal testing with nanosystems offers an interesting way to discuss the constraints of current diagnostic approaches. The goal is to increase detection by combining the characteristics of distinct nanoparticles in hybrid nanosystems using a mix of imaging methods. PET-MRI and PET-CT combine the sensitivity of positron emission tomography (PET) for various metabolic image analyses and monitoring labeled cells and receptors with exceptional functional and structural characterization of tissues offered by MRI and the anatomical precision provided by CT. IONPs functionalized with atherosclerosis-specific ligands, and Gd NPs were employed as contrast media in MRI to identify atherosclerotic plaques, two examples of multimodal techniques employing MNCs. As contrast agents in CT scans, Au NPs functionalized with peptides specifically bind to atherosclerotic lesions. Kelly et al. (2005) developed CVHSPNKKCGGSK(FITC) GK-modified magnetofluorescent NPs that demonstrated high affinity for endothelium exhibiting vascular cell adhesion molecule 1 (VCAM-1) and were visible by MRI and fluorescent imaging to solve this problem. Endothelial adhesion molecules are suitable targets for diagnostics due to their stringent spatial and temporal control and critical role in atherosclerosis (Wilson, Stem, and Bruehlman 2021).

MNCs in the treatment of atherosclerosis

Treatment of atherosclerosis is medication based on either lifestyle modifications or surgical procedures. Lifestyle modifications, such as changes to diet, exercise habits, and smoking cessation, are important in preventing and managing atherosclerosis. Surgical procedures such as angioplasty, stenting, or bypass surgery may be necessary to open blocked arteries and improve blood flow to the heart. Drug-based treatment includes statins to lower cholesterol levels, blood pressure medications to control hypertension, and antiplatelet medications to prevent blood clots. Drug-based treatment and the role of nanocomposites in drug-based treatment are discussed further.

Drug-based treatments include an effective approach to managing atherosclerosis but have some drawbacks. For example, some medications can have side effects, such as muscle pain, liver damage, and gastrointestinal issues, which can impact the quality of life of a patient. Additionally, long-term use of some medications can be costly and require ongoing monitoring and management. One option that has been investigated is using Au NPs in conjunction with cyclodextrin. When coupled with Au NPs, cyclodextrin forms a composite material that selectively removes lipids from plaque deposits in the arteries (Chaker et al. 2017).

Another strategy is to employ IONPs, which can be employed for imaging and medicinal applications. Such NPs can be coated with a material that specifically targets the cells implicated in atherosclerosis, allowing for tailored medication or other therapeutic agent administration (Baluta et al. 2023). While research on the application of MNCs for the therapy of atherosclerosis is still in its early phases, there is rising interest in this technique due to the promise of focused and successful therapy for this condition. However, additional research and clinical studies will be required to completely comprehend the safety and effectiveness of these compounds in the treatment of atherosclerosis.

3.3.5 Thyroid Dysfunction

Thyroid dysfunction is a term used to describe a range of disorders that affect the thyroid gland. Thyroid dysfunction is classified into two categories. Hypothyroidism happens whenever the thyroid gland does not generate enough thyroid hormones. In contrast, hyperthyroidism happens when the gland generates thyroid hormones in excess, which can cause an imbalance in the body's metabolic system (Liu et al. 2019).

MNCs in the diagnosis of thyroid dysfunction

Radio-imaging tests, biochemical testing, and biopsy are all part of primary diagnostic testing (Liu et al. 2019). The function of MNCs in biopsy testing has yet to be investigated and is thus not examined further. In contrast, the role of MNCs in biochemical testing and radio-imaging testing is discussed here.

Biochemical testing using gold and iron oxide NPs has been investigated as possible diagnostics for thyroid disorders. One method is incorporating NPs with thyroid hormone receptors, which may selectively attach to thyroid hormone in blood samples. This is a more accurate and precise approach to identifying thyroid hormones in circulation (Liu et al. 2019).

Radio-imaging is another option to utilize NPs as contrast media (*in vivo*) in imaging modalities, including CT scans, MRI, ultrasound, flow cytometry and colorimetric detection. The NPs in this approach are designed to specifically aggregate in the thyroid gland, which can increase gland visibility and perhaps assist with identifying hypothyroidism (Liu et al. 2019). Potentially useful NPs are described in this section. The GCE/Fe_3O_4 biosensor was designed around glass carbon electrodes modified with a semiconducting Fe_3O_4@graphene nanocomposite, an antibody (anti-PDIA3) having a significant affinity for free triiodothyronine (fT3), and laccase, which catalyzed the redox reaction of fT3. The electrode modification approach was studied using a cyclic voltammetric technique based on the peak current's reaction following alterations. All of the created biosensor's operating parameters were examined using differential pulse voltammetry. The obtained experimental findings demonstrated that the biosensor was sensitive to fT3 in a range of concentrations of 10–200 µM, a detection limit equal to 27 nM, and a limit of quantification equal to 45.9 nM.

Furthermore, the biosensor was selective for fT3 regardless of interfering chemicals such as tyrosine, levothyroxine, and ascorbic acid. The proposed biosensor also showed good stability and could be used for subsequent diagnostic purposes (Khayal et al. 2021). A gold core covered with europium (III)-chelate fluorophore-doped silica shells (Au NPs@SiO_2-Eu^{3+}) hybridized nanocomposite particles were developed and used as an antibody label in lateral flow immunoassay (LFIA) instrument for the detection of human thyroid stimulating hormone (TSH). The detection limits for observing through the naked eye of LFIA sensors are 5 IU/mL for colorimetric analysis and 0.1 IU/mL for fluorometric analysis, respectively (Preechakasedkit et al. 2018). A quantifiable linear association between the red fluorescence and the logarithm concentrations of TSH was obtained ($R^2 = 0.988$) with an LOD of 0.02 IU/mL using the fluorescence detection system in conjunction with a cellphone and computerized image augmentation. Lastly, LFIA devices were shown to be successful in identifying TSH in spiked dilute serum samples of humans, with recovery values ranging from 100% to 116%. As a result, Au NPs@SiO_2-Eu^{3+}-based LFIA sensors with dual signaling may offer an alternative way for straightforward testing of both hyperthyroidism and hypothyroidism (Cash and Clark 2010).

MNCs in the treatment of hypothyroidism

The therapy consists of both lifestyle modifications and medication. The drug-based therapies may include pharmaceuticals that inhibit thyroid hormone production, which includes antithyroid drugs (e.g., levothyroxine) or radioactive iodine therapy, which kills hyperactive thyroid tissue. In some circumstances, surgical excision of the thyroid gland may be required (Liu et al. 2019). The role of MNCs is examined in further detail.

Drug-based treatments include levothyroxine, the medication used to treat hypothyroidism, which causes side effects like headaches, irritability, and insomnia. Sometimes, the dosage may need to be adjusted over time, and it can take several weeks to months to achieve optimal thyroid hormone levels. Antithyroid drugs used to treat hyperthyroidism may cause side effects such as skin rash, joint pain, and liver damage. These medications also take several months to achieve optimal results, and there is a risk of relapse of hyperthyroidism once treatment is stopped. The use of metal NPs for treating hypothyroidism is still a relatively new area of research, and more studies are needed to understand their potential benefits and risks fully. Yet some researchers have shown that metal NPs enhance thyroid function. One study looked into the usage of Au NPs to treat thyroid dysfunction. The study discovered that Au NPs might boost thyroid hormone production and secretion in hypothyroid rats, implying that they could be a treatment agent for hypothyroidism (Li et al. 2020). Another research looked into the usage of ZnO NPs to treat thyroid dysfunction. The researchers discovered that ZnO NPs could raise thyroid hormone levels in hypothyroid rats, implying that they may be used to treat hypothyroidism. While these trials yield encouraging findings, additional study is required to evaluate the safety and efficacy of metal NPs in treating thyroid dysfunction in humans (Lin and Yi 2017).

3.3.6 Hyperglycemia

High blood sugar levels, often known as hyperglycemia, are typical signs of diabetes. High bloodstream glucose levels can occur if the body does not make enough insulin or does not utilize insulin adequately. Other hyperglycemia causes include sickness, stress, and a high-carbohydrate diet (Lin and Yi 2017). Biochemical testing is used to diagnose hyperglycemia. Measurement of glucose or biological fluids, measurement of glycosylated HbA_1c, including measurement of acetone in human breath are all part of biochemical testing. The role of MNCs in biochemical analysis is examined in further detail (Lin and Yi 2017).

MNCs for the diagnosis of hyperglycemia

Diagnosis of hyperglycemia includes biochemical testing. Biochemical testing involves monitoring glucose in biofluids, monitoring glycosylated HbA_1c in blood using electrochemical biosensors, and detecting acetone in human breath. The role of nanocomposites in the diagnosis of hyperglycemia is discussed further.

Monitoring glucose in biofluids: MNCs, such as metal NPs, have been studied for their possible application in hyperglycemic biochemical testing. These NPs may be functionalized with ligands or antibodies that target diabetes biomarkers, allowing them to be detected in plasma and other biological fluids. The use of Au NPs biosensors to monitor diabetes biomarkers like glucose or insulin is one strategy that has been investigated. Au NPs can be designed and synthesized with particular ligands bound to the biomarker of concern, leading to a shift in optical or electrical characteristics that may be recognized and evaluated. This method can develop an accurate and rapid diabetes diagnostic test that could be employed in primary care (Righettoni, Tricoli, and Pratsinis 2010).

Monitoring HbA1c in blood using electrochemical biosensors: Biosensors, such as wearables that monitor sweat glucose levels, can be applied to the surface for noninvasive monitoring. With these biosensors, a transducer is frequently employed to convert the glucose level to an electric signal that can be monitored and measured. Contrarily, biosensors can be inserted within the body and used for intrusive monitoring to measure the glucose levels in the bloodstream or interstitial fluid. These biosensors, frequently placed below the skin or in different tissues, may be programmed to measure glucose concentrations in real time continuously. Electrochemical biosensors offer a promising diagnosis strategy due to their high sensitivity, low cost, ease of use, and optimal specificity (Liu et al. 2021). Innovative electrochemical biosensors with fast electron/ion transmission rate, high adsorption abilities and loading, and optional immobilization of biomolecules were designed for diabetes detection in combination of nanostructural materials and particular biomolecules (e.g., antibodies or enzymes). For example, a gold nanoflowers (Au NFs) modified screen-printed carbon electrodes electrochemical sensor for label-free and accurate measurement of HbA1c has been created. A chemical ligand (4-mercaptophenylboronic acid, 4MPBA) was added to the electrochemical biosensor and interacted with the sugar component of HbA_1C. As a result, impedance-based electrochemical nanosensors for the identification and detection of HbA_1C were produced. Interdigitated Au microelectrode arrays (IDAs) were employed to boost signal sensitivity (Luo et al. 2006).

Acetone-based sensors are used to analyze acetone levels in exhaled breath, and chemoresistive detectors constructed with semiconductor tungsten trioxide (WO_3) nanofilms have been employed. This is achieved by interactions between the sensor and acetone with high dipole moments enhanced by the spontaneous electrical dipole moments of the WO_3 NPs phase. To guide deposition onto interdigitated electrodes, Righettoni et al. (2010) employed pure and Si-doped WO_3 nanoparticles with a gaseous phase. This resulted in a unique acetone detector. Ultrasmall WO_3-based NPs expand certain surface areas and improve sensors' ability to detect acetone molecules. They discovered that at 673.15 K and 90% relative humidity, a considerable sensor response gap might easily discriminate between healthy persons and diabetes patients (Shim et al. 2019). Tin (IV) oxide (SnO_2) has been examined as a semiconductor metal oxide. The selectivity and sensitivity of SnO_2-based acetone sensors are limited. Efforts have been made to tune their micro/nanostructures or include nanosized catalysts to improve their sensitivity to specific gas species. Among them, 1D SnO_2 nanotubes adorned with nanocatalysts show tremendous potential for improved sensing by optimizing the ratio of responsive groups at the inner and outer surfaces and nanocatalyst's electrical or chemical sensing characteristics of nanocatalysts. Ag-based NPs, which include Ag–Pd, Ag–Pt, and Ag–Au NPs, display exciting variations in their localized SPR, which aids in detecting H_2O_2 concentration changes. Based on these properties, researchers can mix Ag with many additional elements to produce a robust, enzyme-free, sensitive H_2O_2 biosensor. Liu et al. (2021) developed an innovative Ag–Au NPs alloy embedded in Cu_2O nanocrystals on reduced graphene oxide (RGO) nanosheets for H_2O_2 monitoring.

MNCs in the treatment of hyperglycemia

The primary treatments for hyperglycemia are lifestyle changes and medication. In general, the role of MNCs in lifestyle modifications has yet to be identified, while the role of nano-composites in drug-based therapy is described in detail in the following section.

Drug-based treatments do not address the underlying causes of hyperglycemia, which can result in long-term problems such as nerve damage, renal disease, and CVD. Metformin and sulfonylureas, used to manage hyperglycemia, might induce gastrointestinal problems, hypoglycemia, and weight gain. Yet a novel strategy for treating hyperglycemia is found in MNCs. Noble metals exhibit strong synergistic effects, and Au–Pt NPs have been widely utilized as catalysts in oxidation reactions. Pt, for example, can operate as the initiator of a dehydrogenation process, while Au and Pt can act together to enhance oxidative desorption and CO adsorption. The synergistic catalytic activity of Au–Pt NPs has been used in glucose detection for diabetes. Shim et al. (2019) synthesized core-shell organized Au–Pt NPs having Au core and Pt shell plated on a carbon electrode. These were produced synthetically and purified using centrifugation, with more Au inserted into the nanochannels. Pt nanoshells' sensitivity, selectivity, and stability were enhanced. At a potential of 0.35 V, the developed sensors had suitable variable range and detection limits in PBS saline solution.

Zinc is critical for insulin structure and is essential for insulin production, release, and storage. Many zinc transporters, including zinc transporter-8, are important in pancreatic beta-cell insulin secretion. Zinc also improves insulin sensitivity via various mechanisms, including enhanced insulin receptors' phosphorylation, greater phosphoinositide 3-kinase activities, or suppression of glycogen synthase kinase-3. ZnO NPs can also reverse alterations in pancreatic tissue caused by diabetes. Prior research has used ZnO NPs and conventional antidiabetic drugs to preserve cell structure and function in type II diabetic rat models and to investigate the efficacy of dipeptidyl peptidase-IV NPs with or without ZnO in type II diabetes models. Because of their extensive pharmacological and biological features, ZnO NPs and Ag NPs were shown to have more effective antidiabetic action than MC and CeO_2 NPs (Siddiqui et al. 2020).

Magnesium (Mg) is an essential ion that contributes to glucose control. Mg also plays a significant role in phosphorylation glucose metabolic rate by implicating many key enzymes in these reactions, and it may also play a role in insulin secretion. Mg deficiency caused insulin resistance, carbohydrate dyslipidemia, and diabetic side effects in mice. MgO NPs reduced blood sugar levels in diabetic mice by enhancing insulin sensitivity and removing lipid alterations such as high LDL and triglyceride levels and low HDL (Naghsh and Kazemi 2014). Thus, we can conclude that magnesium can effectively treat diabetes, especially type II diabetes, and this needs some other studies.

Cerium oxide nanoparticles (Ce_2O NPs) have the potential to relieve and reverse the parameters and negative effects of diabetes, such as elevated blood sugar levels, decreased insulin production, decreased pancreatic cell survival and activity, and raised OS biomarkers. Additionally, Ce_2O NPs have been shown to improve diabetes complications such as neurotoxicity, hepatotoxicity, nephritis, embryopathy, reproductive damage, tissue repair impairment, retinopathy, and cataract development. Moreover, Ce_2O NPs were able to modulate body fat and blood glucose levels in pregnant mice, effectively reducing the risk of gestational diabetes and the emergence of postnatal type II diabetes. Since Ce_2O NPs have antioxidant properties, they might be employed as a therapy option for diabetes control (Chai and Tang 2021).

Copper is a transitional element that is involved in many different metabolic reactions. CuO NPs have high antioxidant characteristics and radical scavenging activity in animals by blocking alpha-glucosidase and alpha-amylase, making them useful in treating type II diabetes. In conclusion, there may be a link between CuO NPs and diabetes.

Selenium (Se) is a trace element found in almost all plants. A lack of selenium in the body has been linked to various disorders, including diabetes. Se NPs acts as an antioxidant by scavenging different peroxides, protecting lipids and cell macromolecules from oxidative membrane damage, and boosting glutathione peroxidase and thioredoxin reductase levels. Se NPs demonstrated antidiabetic activity by preserving pancreatic beta cell integrity, lowering glucose levels, amplifying insulin, and maintaining the ratio of oxidative versus antioxidant production (Zhao et al. 2017). Se NPs is a new therapy strategy that cures most diabetic problems and IR synergistically.

3.4 CONCLUSION AND FUTURE PERSPECTIVES OF MNCS

Metal NPs have many advantages in medical care, including greater biocompatibility and stability, cheap operation, maintenance and capital costs, and lower environmental impacts. MNCs have demonstrated remarkable promise in diagnosing and treating several OS-induced metabolic disorders. The combination of artificial intelligence (AI) and MNCs has the potential to transform the diagnosis and treatment of various OS-induced metabolic diseases. AI systems can scan enormous volumes of data to uncover patterns and links humans may miss, enabling more precise diagnosis and personalized treatment choices. AI combined with MNCs can significantly improve patient outcomes in diagnosing and treating OS-induced metabolic disorders. However, further studies are required to fully comprehend the possibilities of this technique and ensure its reliability and effectiveness. Yet the possibilities for this approach's future are promising since it might lead to new and creative diagnostic and treatment alternatives for a broad spectrum of diseases.

REFERENCES

Al-mutairi, Nabeel Hasan, Atheer Hussain Mehdi, and Ban Jawad Kadhim. 2022. "Nanocomposites Materials Definitions, Types and Some of their Applications: A Review." *European Journal of Research Development and Sustainability* 3 (2): 102–108.

Arshad, Humaira, Madeeha Chaudhry, Shahid Mehmood, Ayesha Farooq, Minqiang Wang, and A.S. Bhatti. 2020. "The Electrochemical Reaction Controlled Optical Response of Cholestrol Oxidase (COx) Conjugated CdSe/ZnS Quantum Dots." *Scientific Reports* 10 (1).

Aslam, Hira, Shazia Shukrullah, Muhammad Yasin Naz, Hareem Fatima, Sami Ullah, and Abdullah G. Al-Sehemi. 2021. "Multifunctional Magnetic Nanomedicine Drug Delivery and Imaging-Based Diagnostic Systems." *Particle and Particle Systems Characterization* 38 (12): 2100179. https://doi.org/10.1002/ppsc.202100179.

Baluta, Sylwia, Marta Romaniec, Kinga Halicka-Stępień, Michalina Alicka, Aleksandra Piela, and Katarzyna Pala. 2023. "A Novel Strategy for Selective Thyroid Hormone Determination Based on an Electrochemical Biosensor with Graphene Nanocomposite." *Sensors* 23 (2): 602. https://doi.org/10.3390/s23020602.

Bejarano, Julian, Mario Navarro-Marquez, Francisco Morales-Zavala, Javier O. Morales, Ivonne Garcia-Carvajal, and Eyleen Araya-Fuentes. 2018. "Nanoparticles for Diagnosis and

Therapy of Atherosclerosis and Myocardial Infarction: Evolution toward Prospective Theranostic Approaches." *Theranostics* 8 (17): 4710–4732. https://doi.org/10.7150/thno.26284.

Botto, Nicoletta, Serena Masetti, Lucia Petrozzi, Cristina Vassalle, Samantha Manfredi, and Andrea Biagini. 2002. "Elevated Levels of Oxidative DNA Damage in Patients with Coronary Artery Disease." *Coronary Artery Disease* 13 (5): 269–274. https://doi.org/10.1097/00019501-200208000-00004.

Brewer, LaPrincess C., Anna Svatikova, and Sharon L. Mulvagh. 2015. "The Challenges of Prevention, Diagnosis and Treatment of Ischemic Heart Disease in Women." *Cardiovascular Drugs and Therapy* 29 (4): 355–368. https://doi.org/10.1007/s10557-015-6607-4.

Cash, Kevin J., and Heather A. Clark. 2010. "Nanosensors and Nanomaterials for Monitoring Glucose in Diabetes." *Trends in Molecular Medicine* 16 (12): 584–593. https://doi.org/10.1016/j.molmed.2010.08.002.

Chai, W. F., and K. San Tang. 2021. Protective potential of cerium oxide nanoparticles in diabetes mellitus. *Journal of Trace Elements in Medicine and Biology* 66: 126742.

Chaker, Layal, Antonio C. Bianco, Jacqueline Jonklaas, and Robin P. Peeters. 2017. "Hypothyroidism." *The Lancet* 390 (10101): 1550–1562. https://doi.org/10.1016/s0140-6736(17)30703-1.

Chaturvedi, V.K., N. Yadav, N.K. Rai, N.H.A. Ellah, R.A. Bohara, I.F. Rehan, N. Marraiki, G.E. Batiha, H.F. Hetta, M.P Singh. 2020. "Pleurotus sajor-caju-Mediated Synthesis of Silver and Gold Nanoparticles Active against Colon Cancer Cell Lines: A New Era of Herbonanoceutics." *Molecules*. 25: 3091.

Chen, Taiwei, Jinfu Qian, Wu Luo, Peiren Shan, Yan Cai, and Ke Lin. 2020. "Macrophage-Derived Myeloid Differentiation Protein 2 Plays an Essential Role in Ox-LDL-Induced Inflammation and Atherosclerosis." *SSRN Electronic Journal* 53: 102706. https://doi.org/10.2139/ssrn.3529432.

Cheng, Ming J., Ronodeep Mitra, Chinedu C. Okorafor, Alina A. Nersesyan, Ian C. Harding, Nandita N. Bal, and Rajiv Kumar. 2020. "Targeted Intravenous Nanoparticle Delivery: Role of Flow and Endothelial Glycocalyx Integrity." *Annals of Biomedical Engineering* 48 (7): 1941–1954. https://doi.org/10.1007/s10439-020-02474-4.

Chithrani, B. Devika, and Warren C.W. Chan. 2007. "Elucidating the Mechanism of Cellular Uptake and Removal of Protein-Coated Gold Nanoparticles of Different Sizes and Shapes." *Nano Letters* 7 (6): 1542–1550. https://doi.org/10.1021/nl070363y.

Danila, Delia, Evan Johnson, and Patrick Kee. 2013. "CT Imaging of Myocardial Scars with Collagen-Targeting Gold Nanoparticles." *Nanomedicine: Nanotechnology, Biology and Medicine* 9 (7): 1067–1076. https://doi.org/10.1016/j.nano.2013.03.009.

Ezeh, Kosisochukwu J., and Obiora Ezeudemba. 2021. "Hyperlipidemia: A Review of the Novel Methods for the Management of Lipids." *Cureus* 13 (7): e16412.

Freitas, R.A. 2000. "Nanomedicine, Volume 1: Basic Capabilities." *Kybernetes* 29 (9–10): 1333–1340. https://doi.org/10.1108/k.2000.29.9_10.1333.3.

Gao, Ying, Fei Gao, Kan Chen, and Jin-lu Ma. 2014. "Cerium Oxide Nanoparticles in Cancer." *OncoTargets and Therapy* 7 (May): 835. https://doi.org/10.2147/ott.s62057.

Ghann, William E., Omer Aras, Thorsten Fleiter, and Marie-Christine Daniel. 2012. "Syntheses and Characterization of Lisinopril-Coated Gold Nanoparticles as Highly Stable Targeted CT Contrast Agents in Cardiovascular Diseases." *Langmuir* 28 (28): 10398–10408.

Guan, Huanan, Yan Song, Bolin Han, Dezhuang Gong, and Na Zhang. 2020. "Colorimetric Detection of Cholesterol Based on Peroxidase Mimetic Activity of GoldMag Nanocomposites." *Spectrochimica Acta Part A: Molecular and Biomolecular Spectroscopy* 241: 118675.

Hamilton, William. 2010. "Cancer Diagnosis in Primary Care." *British Journal of General Practice* 60 (571): 121–128. https://doi.org/10.3399/bjgp10x483175.

Hong, L, A.L. Liu, G.W. Li, W. Chen, and X.H. Lin. 2013 May 15. "Chemiluminescent Cholesterol Sensor based on Peroxidase-Like Activity of Cupric Oxide Nanoparticles." *Biosens Bioelectron*. 43: 1–5. doi: 10.1016/j.bios.2012.11.031. Epub 2012 Dec 6. PMID: 23274189.

Katzmarzyk, Peter T., Timothy S. Church, and Steven N. Blair. 2004. "Cardiorespiratory Fitness Attenuates the Effects of the Metabolic Syndrome on All-Cause and Cardiovascular Disease Mortality in Men." *Archives of Internal Medicine* 164 (10): 1092. https://doi.org/10.1001/archinte.164.10.1092.

Kelly, Kimberly A., Jennifer R. Allport, Andrew T. sourkas, Vivek R. Shinde-Patil, Lee Josephson, and Ralph Weissleder. 2005. "Detection of Vascular Adhesion Molecule-1 Expression Using a Novel Multimodal Nanoparticle." *Circulation Research* 96 (3): 327–336. https://doi.org/10.1161/01.res.0000155722.17881.dd.

Khan, Ibrahim, Khalid Saeed, and Idrees Khan. 2017. "Nanoparticles: Properties, Applications and Toxicities." *Arabian Journal of Chemistry* 12 (7). https://doi.org/10.1016/j.arabjc.2017.05.011.

Khayal, Eman El-Sayed, Hanaa M. Ibrahim, Amany Mohamed Shalaby, Mohamed Ali Alabiad, and Arwa A. El-Sheikh. 2021. "Combined Lead and Zinc Oxide-Nanoparticles Induced Thyroid Toxicity through 8-OHdG Oxidative Stress-Mediated Inflammation, Apoptosis, and Nrf2 Activation in Rats." *Environmental Toxicology* 36 (12): 2589–2604. https://doi.org/10.1002/tox.23373.

Kim, Dongkyu, Sangjin Park, Jae Hyuk Lee, Yong Yeon Jeong, and Sangyong Jon. 2007. "Antibiofouling Polymer-Coated Gold Nanoparticles as a Contrast Agent for in Vivo X-Ray Computed Tomography Imaging." *Journal of the American Chemical Society* 129 (24): 7661–7665. https://doi.org/10.1021/ja071471p.

Kuo, M.C., C.M. Tsai, J.C. Huang, and M. Chen. 2005. "PEEK Composites Reinforced by Nano-Sized SiO2 and Al2O3 Particulates." *Materials Chemistry and Physics* 90 (1): 185–195. https://doi.org/10.1016/j.matchemphys.2004.10.009.

Li, Juanjuan, Ruitao Cha, Huize Luo, Wenshuai Hao, Yan Zhang, and Xingyu Jiang. 2019. "Nanomaterials for the Theranostics of Obesity." *Biomaterials* 223 (December): 119474. https://doi.org/10.1016/j.biomaterials.2019.119474.

Li, Yang, Zhao Yao, Wenjing Yue, Chunwei Zhang, Song Gao, and Cong Wang. 2020. "Reusable, Non-Invasive, and Ultrafast Radio Frequency Biosensor Based on Optimized Integrated Passive Device Fabrication Process for Quantitative Detection of Glucose Levels." *Sensors* 20 (6): 1565. https://doi.org/10.3390/s20061565.

Lim, Zhao-Zhin Joanna, Jia-En Jasmine Li, Cheng-Teng Ng, Lin-Yue Lanry Yung, and Boon-Huat Bay. 2011. "Gold Nanoparticles in Cancer Therapy." *Acta Pharmacologica Sinica* 32 (8): 983–990. https://doi.org/10.1038/aps.2011.82.

Lin, Hua, and Jun Yi. 2017. "Current Status of HbA1c Biosensors." *Sensors (Basel, Switzerland)* 17 (8). https://doi.org/10.3390/s17081798.

Lin, Ying, Anders H. Berg, Puneeth Iyengar, Tony K.T. Lam, Adria Giacca, and Terry P. Combs. 2004. "The Hyperglycemia-Induced Inflammatory Response in Adipocytes." *Journal of Biological Chemistry* 280 (6): 4617–4626. https://doi.org/10.1074/jbc.m411863200.

Liu, Fangzhou, Dawei Ma, Wei Chen, Xinyuan Chen, Yichun Qian, and Yanbin Zhao. 2019. "Gold Nanoparticles Suppressed Proliferation, Migration, and Invasion in Papillary Thyroid Carcinoma Cells via Downregulation of CCT3." *Journal of Nanomaterials* 2019 (February): 1–12. https://doi.org/10.1155/2019/1687340.

Liu, Yang, Qiangwei Kou, Dandan Wang, Lei Chen, Yantao Sun, and Ziyang Lu. 2017. "Rational Synthesis and Tailored Optical and Magnetic Characteristics of Fe_3O_4-Au Composite Nanoparticles." *Journal of Materials Science* 52 (17): 10163–10175.

Liu, Yuntao, Siqi Zeng, Wei Ji, Huan Yao, Lin Lin, and Haiying Cui. 2021. "Emerging Theranostic Nanomaterials in Diabetes and Its Complications." *Advanced Science* 9 (3): 2102466. https://doi.org/10.1002/advs.202102466.

López-Miranda, J. Luis, Gustavo A. Molina, Rodrigo Esparza, Marlen Alexis González-Reyna, Rodolfo Silva, and Miriam Estévez. 2021. "Green Synthesis of Homogeneous Gold Nanoparticles Using Sargassum Spp. Extracts and Their Enhanced Catalytic Activity for Organic Dyes." *Toxics* 9 (11): 280. https://doi.org/10.3390/toxics9110280.

Luo, Jin, Peter N. Njoki, Yan Lin, Derrick Mott, Lingyan Wang, and Chuan-Jian Zhong. 2006. "Characterization of Carbon-Supported AuPt Nanoparticles for Electrocatalytic Methanol Oxidation Reaction." *Langmuir: The ACS Journal of Surfaces and Colloids* 22 (6): 2892–2898. https://doi.org/10.1021/la0529557.

Malhotra, Bansi D., and Ajeet Kaushik. 2009. "Metal Oxide: Chitosan Based Nanocomposite for Cholesterol Biosensor." *Thin Solid Films* 518 (2): 614–620. https://doi.org/10.1016/j.tsf.2009.07.036.

Mikhailova, Ekaterina O. 2020. "Silver Nanoparticles: Mechanism of Action and Probable Bio-Application." *Journal of Functional Biomaterials* 11 (4): 84. https://doi.org/10.3390/jfb11040084.

Naghsh, N., and S. Kazemi. 2014. "Effect of Nano-Magnesium Oxide on Glucose Concentration and Lipid Profile in Diabetic Laboratory Mice." *Iranian Journal of Pharmaceutical Science* 10: 63–68.

Namdari, Mehrdad, Mostafa Cheraghi, Babak Negahdari, Ali Eatemadi, and Hadis Daraee. 2017. "Recent Advances in Magnetoliposome for Heart Drug Delivery." *Artificial Cells, Nanomedicine, and Biotechnology* 45 (6): 1051–1057. https://doi.org/10.1080/21691401.2017.1299159.

Nanda, Nivedita, Zachariah Bobby, and Abdoul Hamide. 2008a. "Association of Thyroid Stimulating Hormone and Coronary Lipid Risk Factors with Lipid Peroxidation in Hypothyroidism." *Clinical Chemistry and Laboratory Medicine* 46 (5). https://doi.org/10.1515/cclm.2008.139.

Nanda, Nivedita, Zachariah Bobby, and Abdoul Hamide. 2008b. "Oxidative Stress and Protein Glycation in Primary Hypothyroidism: Male/Female Difference." *Clinical and Experimental Medicine* 8 (2): 101–108. https://doi.org/10.1007/s10238-008-0164-0.

Păduraru, Dan Nicolae, Daniel Ion, Adelina-Gabriela Niculescu, Florentina Muşat, Octavian Andronic, and Alexandru Mihai Grumezescu. 2022. "Recent Developments in Metallic Nanomaterials for Cancer Therapy, Diagnosing and Imaging Applications." *Pharmaceutics* 14 (2): 435. https://doi.org/10.3390/pharmaceutics14020435.

Palmero, Paola. 2015. "Structural Ceramic Nanocomposites: A Review of Properties and Powders' Synthesis Methods." *Nanomaterials* 5 (2): 656–696. doi:10.3390/nano5020656.

Patra, Chitta Ranjan, Resham Bhattacharya, Debabrata Mukhopadhyay, and Priyabrata Mukherjee. 2010. "Fabrication of Gold Nanoparticles for Targeted Therapy in Pancreatic Cancer." *Advanced Drug Delivery Reviews* 62 (3): 346–361. https://doi.org/10.1016/j.addr.2009.11.007.

Patra, Chitta Ranjan, Resham Bhattacharya, E. Wang, A.K atarya, J.S. Lau, S. Dutta, et al. 2008. "Targeted Delivery of Gemcitabine to Pancreatic Adenocarcinoma using Cetuximab as a Targeting Agent." *Cancer Res* 68: 1970–1978.

Perera, W.P.T.D., Ranga K. Dissanayake, U.I. Ranatunga, N.M. Hettiarachchi, K.D.C. Perera, and Janitha M. Unagolla. 2020. "Curcumin Loaded Zinc Oxide Nanoparticles for Activity-Enhanced Antibacterial and Anticancer Applications." *RSC Advances* 10 (51): 30785–30795. https://doi.org/10.1039/d0ra05755j.

Piktel, Ewelina, Ilona Ościłowska, Łukasz Suprewicz, Joanna Depciuch, Natalia Marcińczyk, and Ewa Chabielska. 2021. "ROS-Mediated Apoptosis and Autophagy in Ovarian Cancer Cells Treated with Peanut-Shaped Gold Nanoparticles." *International Journal of Nanomedicine* 16 (March): 1993–2011. https://doi.org/10.2147/IJN.S277014.

Prasad, P. Reddy, S. Kanchi, and E.B.B. Naidoo. 2016. "In-Vitro Evaluation of Copper Nanoparticles Cytotoxicity on Prostate Cancer Cell Lines and Their Antioxidant, Sensing and Catalytic Activity: One-Pot Green Approach." *Journal of Photochemistry and Photobiology B: Biology* 161 (August): 375–382. https://doi.org/10.1016/j.jphotobiol.2016.06.008.

Preechakasedkit, Pattarachaya, Kota Osada, Yuta Katayama, Nipapan Ruecha, and Suzuki Koji. 2018. "Gold Nanoparticle Core-Europium(iii) Chelate Fluorophore-Doped Silica Shell Hybrid Nanocomposites for the Lateral Flow Immunoassay of Human Throid Stimulating Hormone with a Dual Signal Readout." *The Royal Society of Chemistry* 143 (2): 564–570.

Rafieian-kopaei, Mahmoud, Mahbubeh Setorki, Esfandiar Heidarian, Najmeh Shahinfard, and Roya Ansari. 2012. "Effect of Anethum Graveolens on Hypelipidemia Induced Hepatotoxicity." *Toxicology Letters* 211 (1): S167. https://doi.org/10.1016/j.toxlet.2012.03.604.

Rastogi, Lori, K. Dash, and R.B. Sashidhar. 2021. "Selective and Sensitive Detection of Cholesterol Using Intrinsic Peroxidase-like Activity of Biogenic Palladium Nanoparticles." *Current Research in Biotechnology* 3: 42–48. https://doi.org/10.1016/j.crbiot.2021.02.001.

Rehman, Tahniat, Muhammad Asim Shabbir, Muhammad Inam-Ur-Raheem, Muhammad Faisal Manzoor, Nazir Ahmad, Zhi-Wei Liu, and Muhammad Haseeb Ahmad. 2020. "Cysteine and Homocysteine as Biomarker of Various Diseases." *Food Science and Nutrition* 8 (9): 4696–4707. https://doi.org/10.1002/fsn3.1818.

Righettoni, Marco, Antonio Tricoli, and Sotiris E. Pratsinis. 2010. "Si: WO3Sensors for Highly Selective Detection of Acetone for Easy Diagnosis of Diabetes by Breath Analysis." *Analytical Chemistry* 82 (9): 3581–3587. https://doi.org/10.1021/ac902695n.

Ruan, Ye, Yanfei Guo, Yang Zheng, Zhezhou Huang, Shuangyuan Sun, and Paul Kowal. 2018. "Cardiovascular Disease (CVD) and Associated Risk Factors among Older Adults in Six Low-and Middle-Income Countries: Results from SAGE Wave 1." *BMC Public Health* 18 (1). https://doi.org/10.1186/s12889-018-5653-9.

Safdari, Masoud, and Marwan S. Al-Haik. 2018. "A Review on Polymeric Nanocomposites." *Carbon-Based Polymer Nanocomposites for Environmental and Energy Applications*, Elsevier Inc, 113–146.

Santos, Celio X.C., Narayana Anilkumar, Min Zhang, Alison C. Brewer, and Ajay M. Shah. 2011. "Redox Signaling in Cardiac Myocytes." *Free Radical Biology and Medicine* 50 (7): 777–793. https://doi.org/10.1016/j.freeradbiomed.2011.01.003.

Sawall S., D. Franke, A. Kirchherr, J. Beckendorf, J. Kuntz, and J. Maier. 2017. "In Vivo Quantification of Myocardial Infarction in Mice Using Micro-CT and a Novel Blood Pool Agent." *Contrast Media & Molecular Imaging*: Article ID 2617047.

Schuemann, Jan, Alexander F. Bagley, Ross Berbeco, Kyle Bromma, Karl T. Butterworth, and Hilary L. Byrne. 2020. "Roadmap for Metal Nanoparticles in Radiation Therapy: Current Status, Translational Challenges, and Future Directions." *Physics in Medicine and Biology* 65 (21): 21RM02. https://doi.org/10.1088/1361-6560/ab9159.

Shang, L., X. Zhou, J. Zhang, Y. Shi, L. Zhong. 2021. "Metal Nanoparticles for Photodynamic Therapy: A Potential Treatment for Breast Cancer." *Molecules.* 26: 6532. doi: 10.3390/molecules26216532.

Shim, Kyubin, Won Lee, Min-Sik Park, Mohammed Shahabuddin, Yusuke Yamauchi, and Md Shahriar Hossain. 2019. "Au Decorated Core-Shell Structured Au@Pt for the Glucose Oxidation Reaction." *Australian Institute for Innovative Materials-Papers* (January): 88–96. https://doi.org/10.1016/j.snb.2018.09.048.

Siddiqui, Shafayet Ahmed, Md. Mamun Or Rashid, Md. Giash Uddin, Fataha Nur Robel, Mohammad Salim Hossain, and Md. Azizul Haque. 2020. "Biological Efficacy of Zinc Oxide Nanoparticles against Diabetes: A Preliminary Study Conducted in Mice." *Bioscience Reports* 40 (4). https://doi.org/10.1042/bsr20193972.

Sinha, Krishnendu, Joydeep Das, Pabitra Bikash Pal, and Parames C. Sil. 2013. "Oxidative Stress: The Mitochondria-Dependent and Mitochondria-Independent Pathways of Apoptosis." *Archives of Toxicology* 87 (7): 1157–1180. https://doi.org/10.1007/s00204-013-1034-4.

Solairaj, Dhanasekaran, Palanivel Rameshthangam, and Gnanapragasam Arunachalam. 2017. "Anticancer Activity of Silver and Copper Embedded Chitin Nanocomposites against Human Breast Cancer (MCF-7) Cells." *International Journal of Biological Macromolecules* 105 (December): 608–619. https://doi.org/10.1016/j.ijbiomac.2017.07.078.

Szunerits, S., and R. Boukherroub. 2018. "Near-Infrared Photothermal Heating with Gold Nanostructures." In: Wandelt K., editor. *Encyclopedia of Interfacial Chemistry.* Elsevier; Oxford, UK:. pp. 500–510.

Tarin, Carlos, Monica Carril, Jose Luis Martin-Ventura, Irati Markuerkiaga, Daniel Padro, and Patricia Llamas-Granda. 2015. "Targeted Gold-Coated Iron Oxide Nanoparticles for CD163 Detection in Atherosclerosis by MRI." *Scientific Reports* 5 (1): 17135. https://doi.org/10.1038/srep17135.

Tolkaeva, Mariia, Kaushala Prasad Mishra, Ekaterina Evstratova, and Vladislav Petin. 2021. "Synergistic Interaction of Heavy Metal Salts with Hyperthermia or Ionizing Radiation." *Journal of Radiation and Cancer Research* 12 (1): 23. https://doi.org/10.4103/jrcr.jrcr_69_20.

Wagner, Volker, Anwyn Dullaart, Anne-Katrin Bock, and Axel Zweck. 2006. "The Emerging Nanomedicine Landscape." *Nature Biotechnology* 24 (10): 1211–1217. https://doi.org/10.1038/nbt1006-1211.

Wang, Bo, Evgeniya Yantsen, Timothy Larson, Andrei B. Karpiouk, Shriram Sethuraman, and Jimmy L. Su. 2008. "Plasmonic Intravascular Photoacoustic Imaging for Detection of Macrophages in Atherosclerotic Plaques." *Nano Letters* 9 (6): 2212–17.

Wang, Sujing, Zhiyong Wang, and Zhenggen Zha. 2009. "Metal Nanoparticles or Metal Oxide Nanoparticles, an Efficient and Promising Family of Novel Heterogeneous Catalysts in Organic Synthesis." *Dalton Transactions* 29 (1): 9363. https://doi.org/10.1039/b913539a.

Wang, Xiao-Ming, Yan-Yan Xu, Gang Yu, Zhen Rong, and Rui-Chao Geng. 2019. "Pure Transanal Total Mesorectal Excision for Rectal Cancer: Experience with 55 Cases." *Gastroenterology Report* 8 (1): 42–49. https://doi.org/10.1093/gastro/goz055.

Wang, Yi-Xiang J. 2011. "Superparamagnetic Iron Oxide Based MRI Contrast Agents: Current Status of Clinical Application." *Quantitative Imaging in Medicine and Surgery* 1 (1): 35–40. https://doi.org/10.3978/j.issn.2223-4292.2011.08.03.

Wason, Melissa, Heng Lu, Lin Yu, Satadru Lahiri, Debarati Mukherjee, and Chao Shen. 2018. "Cerium Oxide Nanoparticles Sensitize Pancreatic Cancer to Radiation Therapy through Oxidative Activation of the JNK Apoptotic Pathway." *Cancers* 10 (9): 303. https://doi.org/10.3390/cancers10090303.

Wilson, Stephen A., Leah A. Stem, and Richard D. Bruehlman. 2021. "Hypothyroidism: Diagnosis and Treatment." *American Family Physician* 103 (10): 605–613. https://pubmed.ncbi.nlm.nih.gov/33983002/.

Witztum, Joseph L., and Judith A. Berliner. 1998. "Oxidized Phospholipids and Isoprostanes in Atherosclerosis." *Current Opinion in Lipidology* 9 (5): 441–448. https://doi.org/10.1097/00041433-199810000-00008.

Wojtczak, Lech, and Peter Schönfeld. 1993. "Effect of Fatty Acids on Energy Coupling Processes in Mitochondria." Biochimica et Biophysica Acta (BBA)-Bioenergetics 1183 (1): 41–57. https://doi.org/10.1016/0005-2728(93)90004-y.

Xu, Shuangjiao, Yanqin Wang, Dayun Zhou, Meng Kuang, Dan Fang, and Weihua Yang. 2016. "A Novel Chemiluminescence Sensor for Sensitive Detection of Cholesterol Based on the Peroxidase-like Activity of Copper Nanoclusters." *Scientific Reports* 6 (1). https://doi.org/10.1038/srep39157.

Zamora-Justo, José Alberto, Paulina Abrica-González, Guillermo Rocael Vázquez-Martínez, Alejandro Muñoz-Diosdado, José Abraham Balderas-López, and Miguel Ibáñez-Hernández. 2019. "Polyethylene Glycol-Coated Gold Nanoparticles as DNA and Atorvastatin Delivery Systems and Cytotoxicity Evaluation." *Journal of Nanomaterials* 2019 (October): 1–11. https://doi.org/10.1155/2019/5982047.

Zanganeh, Saeid, Gregor Hutter, Ryan Spitler, Olga Lenkov, Morteza Mahmoudi, and Aubie Shaw. 2016. "Iron Oxide Nanoparticles Inhibit Tumour Growth by Inducing Pro-Inflammatory Macrophage Polarization in Tumour Tissues." *Nature Nanotechnology* 11 (11): 986–994.

Zhang, Xi, Zhenyue Tan, Kunjing Jia, Wenzhi Zhang, and Minyan Dang. 2019. "Rabdosia Rubescens Linn: Green Synthesis of Gold Nanoparticles and Their Anticancer Effects against Human Lung Cancer Cells A549." *Artificial Cells, Nanomedicine, and Biotechnology* 47 (1): 2171–2178. https://doi.org/10.1080/21691401.2019.1620249.

Zhao, Shao-Jun, De-Hua Wang, Yan-Wei Li, Lei Han, Xing Xiao, and Min Ma. 2017. "A Novel Selective VPAC2 Agonist Peptide-Conjugated Chitosan Modified Selenium Nanoparticles with Enhanced Anti-Type 2 Diabetes Synergy Effects." *International Journal of Nanomedicine* 12 (2): 2143–2160. https://doi.org/10.2147/IJN.S130566.

Chapter 4

Nanotherapeutics for leptin resistance and obesity using metal nanocomposites

Ipsa Padhy, Biswajit Banerjee, and Tripti Sharma

LIST OF ABBREVIATIONS

5-HT	5-hydroxy tryptamine
AMPK	Adenosine monophosphate activated kinase
AP-1	Activated protein-1
BAT	Brown adipose tissue
BBB	Blood–brain barrier
BMI	Body mass index
DNA	Deoxyribonucleic acid
ER	Endoplasmic reticulum
FFA	Free fatty acid
GERD	Gastroesophageal reflux disease
GI	Gastrointestinal
GLP-1	Glucagon-like peptide-1
GLPR1	Glucagon-like peptide-1 receptor
GPx	Glutathione peroxidase
GR	Glutathione reductase
GSH	Glutathione
ICAM-1	Intracellular adhesion molecule-1
IL-1	Interleukin 1
IL-6	Interleukin 6
JNK	c-Jun-n terminal kinase
KCl	Potassium chloride
LDL	Low-density lipoprotein
NADPH	Nicotinamide adenine dinucleotide phosphate
NAFLD	Non-alcoholic fatty liver disease
NF-κB	Nuclear factor kappa B
NIR	Near infrared
NPs	Nanoparticles
PAI-1	Plasminogen activator inhibitor-1
PEG	Poly ethylene glycol
PGMC	Propylene glycol monocaprylate
PLGA	Poly (lactic-o-glycolic acid)
PYY	Peptide YY
ROS	Reactive oxygen species

DOI: 10.1201/9781032621135-4

SOD   Superoxide dismutase
SPIONs   Superparamagnetic iron oxide nanoparticles
TGA   Triglyceride
TNF-α   Tumor necrosis factor alpha
UCP-1   Uncoupling protein 1
VCAM-1   Vascular cell adhesion molecule 1
WAT   White adipose tissue

4.1 INTRODUCTION

Obesity indeed has enticed far-reaching attention globally and has been the root cause of several medical problems (Ogden et al, 2012). Copious philosophies have been proposed to illustrate the multifaceted etiology of obesity. Two significant findings in the elucidation of the obesity network are leptin coded by an obese gene and physiologically expressed via binding to leptin receptor b (Wauman, 2011). Leptin regulates neurons of the hypothalamus, midbrain, and brainstem, thereby centrally governing body weight by inducing anorexic signals and enhanced energy disbursement (Myers et al, 2010). Most individuals with diet-induced obesity show high levels of circulating leptin (Myers et al, 2012). Interestingly, even injecting extra leptin in patients with obesity has reportedly failed to thwart weight gain (Maffei et al, 1995). Subsequently, the theory of leptin resistance arose to clarify raised levels of leptin that transpire in obese individuals. Leptin resistance is both a symbol and a risk factor for obesity. Leptin also regulates reproduction, bone homeostasis, and immune functions (Gonzalez-Bulnes et al, 2012; Wauman, 2011; Wong et al, 2013). As a result, leptin resistance also partakes in various pathological progressions of other cardiovascular disorders (Patel et al, 2008), osteoporosis (Odabasi et al., 2000), systemic inflammations (DeLany, 2008), and depression (Yamada-Goto et al., 2012).

The pervasiveness, impact, and economic costs of obesity management demand grander therapeutics and better thoughtful analysis of the physiological progressions. In this context, nanocomposites have the repertoire to potentiate currently available medications and approaches for averting, monitoring, and alleviating diseases through their multitasking traits (Hauser et al., 2015). Nanocomposites are multiphasic nanosystems entrenched in ceramic or metal or polymer matrices (Sanchez et al., 2001; Sanchez et al., 2005). The nanocomposites are synthesized by numerous techniques like sol-gel route, hydrothermal, self-assembly process, and chemical co-precipitation. Nanocomposite materials are classified as inorganic nanostructures, organic nanostructures, and organic–inorganic hybrid nanostructures. Inorganic nanocomposite materials usually include carbon nanotubes, metal oxides of gold, silver, platinum, zinc, iron, cerium, tin, titanium, zirconium, and palladium. The organic nanocomposite units comprise polyaniline polypyrrole, poly (3,4-ethylene dioxythiophene), chitosan, and a self-assembled monolayer of 8-amino polythiophene. Fabrication of nanocomposites involves different intermolecular forces like van der Waal's interactions, weak electrostatic interactions, hydrogen bonding, or covalent bonds (Ajayan et al., 2004; Camargo et al., 2009). Carbon nanotubes, nanofibers, nanoparticles, and nanorods are basic inorganic components of nanocomposites. The nanocomposites exhibit an excellent surface-to-volume ratio enabling the stuffing of biomolecules, higher tensile strength, greater electric conductivity, redox potential, and catalytic action (Luo et al., 2005; Xian et al., 2006). Nanocomposites have immense applications in imaging, targeted drug delivery, and artificial implants. Owing to their flexibility as biologically functionalized systems and unique characteristics, nanocomposites are used for the diagnosis of cancers and microbial infections as well (An et al., 2004; Croce et al., 1998; Kubacka et al., 2014; Ma et al., 2012; Wang et al., 2011; Zhang et al., 2012). Nanocomposites fabricated from metal oxides, metal nanohybrid materials, metal–carbon nanomaterials, and polymers are utilized in the designing of enzyme-based biosensors (Rajesh & Kumar., 2009). Explicitly, nanocomposites aid as transducers for immobilizing protein molecules on their surfaces by physical interactions, entrapment, or covalent interfaces in between them (Rege et al., 2003).

Numerous studies have started to reconnoiter approaches for obesity therapy by addressing leptin resistance. The chapter reviews the major developments explaining the present-day understanding of phenomena underlying leptin resistance as well as its meticulous correlation with obesity. Furthermore, theranostic applications of the metal nanocomposites to leptin resistance and obesity as a whole are also discussed in the following sections.

4.2 PATHOPHYSIOLOGY OF LEPTIN RESISTANCE AND OBESITY

Obesity is a consequence of extra energy consumed in calories than spent. Nonetheless, continuing explorations indicate that energy homeostasis is a more complex course than just reflexive accretion of extra calories. The principal pathophysiological manifestation of obesity is the body's imperfect regulation of energy balance maneuvered by an intricate interplay of multiple metabolic pathways, both central and peripheral (Ritten & LaManna, 2017; Schwartz et al., 2017).

4.2.1 Central pathway for energy regulations

Hypothalamus

Energy homeostasis of the body is regulated within the arcuate nucleus located within the hypothalamus via orchestral antagonistic (Figure 4.1) signal transductions (Srivastava & Apovian, 2017). The orexigenic agouti-related peptide/neuropeptide Y expressing neurons increase appetite and food intake and reduce energy investment in response to low plasma levels of leptin and insulin. The pro-opiomelanocortin/cocaine and amphetamine-regulated transcript-expressing neurons upon stimulation by leptin release neuropeptide α-melanocyte-stimulating hormone that reduces appetite and food intake thereby triggering fat loss in a fasting state (Srivastava & Apovian, 2017). Neuronal signals from adipocytes, gastrointestinal tract, liver, and pancreas also control the functioning of both neuron systems thereby forming a complex energy homeostatic environment. Orexigenic hormone ghrelin and anorexigenic hormones GLP-1, PYY, and leptin majorly constitute a negative response circuit between the brain and the periphery primarily acting upon the cerebellum and nucleus of the tractus solitaries (Apovian et al., 2015). Moreover, the hedonic pathways operating in the corticolimbic system of the brain reportedly also regulate eating behavior and food palatability in individuals in tandem with the hypothalamus (Berthoud et al., 2017). Neurotransmitters like serotonin or 5-HT, dopamine, and norepinephrine regulating via reward pathways (Figure 4.1) also regulate food intake.

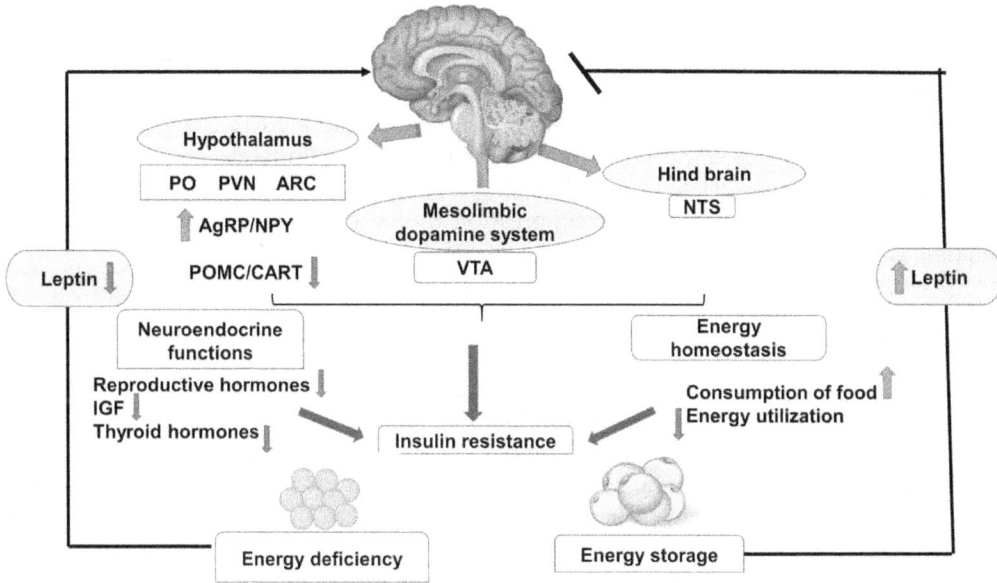

Figure 4.1 Role of leptin in energy homeostasis.

4.2.2 Peripheral pathways for energy regulations

Leptins

The white adipose tissue primarily secretes the peptide hormone, leptin. Undoubtedly, the body fat mass dictates the elevated leptin plasma levels, which are usually high in obese individuals. As mentioned earlier, leptin regulates hypothalamic neurons (Figure 4.1) to maintain energy homeostasis by transducing orexigenic and anorexigenic signals. The effect of the ghrelin hormone is also countered by leptin, although maximum obese persons are leptin resistant. According to the most prevalent hypotheses for obesity-related leptin resistance, leptin resistance is caused by a complex set of mechanisms that involve several different parameters such as a problem with the transportation of leptin across the blood–brain barrier (BBB) (Banks et al., 1996; Banks et al., 1999; Caro et al., 1996; Schwartz et al., 1996), deterioration of leptin signaling (Bjørbæk et al., 1998; Loh et al., 2011; White et al., 2009; Wilsey & Scarpace, 2004), anxiety of endoplasmic reticulum (ER) (Ozcan et al., 2009; Ramírez & Claret, 2015; Williams et al., 2014), inflammation (Kleinridders et al., 2009; Milanski et al., 2009; Valdearcos et al., 2014; Zhang et al., 2008), inadequate autophagy (Kaushik et al., 2011; Quan et al., 2012), and other associated problems. The functions of leptin might be interrupted due to epigenetic intonation of the leptin signaling loop (Crujeiras et al., 2015). Leptin may bind to extracellular circulating substances like C-reactive protein, changing its biological effects (Hribal et al., 2014). According to recent research, mice with circadian disorder also developed leptin resistance (Kettner et al., 2015). Another study also observed that hyperleptinemia itself may potentially contribute to leptin resistance (Knight et al., 2010).

Gut Hormones

Gut hormones regulate GI motility, digestion, and food intake and thereby maintain energy homeostasis (Oussaada et al., 2019). Cholecystokinin, PYY, and GLP-1 are anorexigenic hormones secreted by the intestine that suppress appetite and increase satiety. The hunger hormone ghrelin is orexigenic and promotes food intake. Ghrelin levels are elevated during fasting and are inversely related to BMI. Reduced postprandial suppression of ghrelin is a hallmark of obesity. Decreased ghrelin suppression and/or increased anorexigenic hormonal signaling can affect the energy balance.

Adipocytes

Adipocytes are the peripheral regulators of energy balance. The adipose tissue comprises white and brown adipose tissues with varying morphology, distribution, gene expression, and physiological functions. The white adipose tissue (WAT) is the energy reservoir along with a secretory hub for many hormones, adipokines, and cytokines. Besides adipocytes, the WAT comprises macrophages, leucocytes, and fibroblasts which account for its broad metabolic activities, insulin resistance, inflammatory role, and other vascular complications in obesity and comorbidities (Coon et al., 1992; Mathieu et al., 2010). The types of adipokine secretion for visceral and subcutaneous adipose tissues are way different which also defines their role in obesity pathogenesis (Fain et al., 2004). Peripheral obesity is common in women subjected to the accumulation of subcutaneous fat and is not associated with an increased risk of other pathological conditions (Snijder et al., 2003). Epidemiological studies have explored that central and abdominal obesity, which is more common in men, is subjected to accumulation of visceral fat.

Both central and abdominal obesity are associated with insulin resistance, diabetes mellitus, and hypertension, thereby heightening the risk of cardiovascular diseases (Fox et al., 2007). Lipid accumulation in WAT triggers inflammatory responses via hyperactivated JNK and NF-κB signaling pathways releasing proinflammatory cytokines, leptin, resistin, chemokines (monocyte chemoattractant protein-1), proatherogenic mediators (PAI-1), endothelial adhesion molecules (ICAM-1 and VCAM-1), and chemoattractant molecules. The binding of chemoattractant molecules to integrins and chemokine receptors recruits monocytes and macrophages into adipose tissue thereby amplifying the inflammatory response (Chawla et al., 2011). Conversely, brown adipose tissue (BAT) is inversely associated with BMI in humans (Cypess et al., 2009). Above and beyond thermogenesis, the latest reports revealed BAT possibly has a prominent role in lipid and carbohydrate metabolism. Elevated levels of cholesterol and triglycerides are reduced by activated BAT, thereby lessening obesity (Bartelt et al., 2011; Bartelt et al., 2012; Hoeke et al., 2016; Ouellet et al., 2012; Williams, 2008).

4.2.3 Other pathways of energy regulation

Microbiomes

The term microbiome refers to the bacterial flora of the gut forming a complex and dynamic ecosystem. The gut microbiota acts as an organ system of one's own body with specific metabolic interplay, immunologic and endocrine actions greatly influencing health. Any disturbance in the gut microflora disburses negative health impact triggering gut upset, cardiovascular malfunctioning, and obesity (Gérard, 2015). Roughly, 80–90% of the gut microflora comprise firmicutes and Bacteroides strains. In obese individuals, the firmicutes population is comparably high as compared to the Bacteroides population, eventually creating a microbiome imbalance. The use of excessive antibiotics in early life might amend the microbiome ecosystem and therefore contributes to obesity (Chelimo et al., 2020).

Muscles

Striated and heart muscles mediate a crucial part in metabolic distress, regulation of body weight, and energy ingestion. Sarcopenia declines energy expenditure and triggers the accumulation of adipose tissue often termed sarcopenic obesity. Raised serum TGA, FFA, and glucose, together with a lack of physical activity, affect the metabolism of voluntary and heart muscles. As voluntary skeletal muscles can acclimatize to augmented substrate availability, energy balance homeostasis is changed, therefore triggering obesity and diabetes mellitus (Baskin et al., 2015).

4.3 RELATION BETWEEN OXIDATIVE STRESS AND OBESITY

Oxidative stress is a consequence of the imbalance between the generation of reactive free radicals and the capability of the antioxidant defense system to quench the free radicals (Takaki et al., 2013). Excessive fat accumulation in visceral organs and tissues as in the case of obesity increases the levels of free fatty acids in the portal circulation and consequently disturbing glucose metabolism. Increased lipid and glucose levels in the serum recruit energy substrates to different cellular metabolic pathways and induce ROS generation. Obesity-induced oxidative stress is associated with the activation of the innate immune system within the adipose tissue. The adipose tissue releases the pro-inflammatory cytokines, TNF-α, IL-1, and IL-6, which further increase ROS production and lipid peroxidation rates. Moreover, the ROS induces the further release of inflammatory cytokines and expression of adhesion molecules via activation of redox-sensitive transcription factors like NF-κB and AP-1 as well as NADPH oxidase pathway oxidation (Bryan et al., 2013). The adipokines also generate ROS. Leptin secreted from the WAT is pro-oxidative and pro-inflammatory, increase the macrophage phagocytic activity, and induces the release of several endothelial cell dysfunction and activation markers (Hukshorn et al., 2004). Similarly, visfatin, another adipokine dictates many pro-oxidative and pro-inflammatory processes via the NF-κB signaling pathway (Kim et al., 2008). Mitochondrial and peroxisomal lipid peroxidation rates are enhanced due to oxidative stress, which leads to mitochondrial DNA injury and ATP depletion (Rzheshevsky, 2013). Mitochondrial dysfunction disturbs the insulin signaling pathways and ultimately induces insulin resistance (Wang et al., 2013).

The piling of excess fat in the liver reduces its antioxidant functional machinery. Activities of CAT, GPx, and SOD (antioxidant enzymes) are inversely related to BMI in adults and children (Mittal & Kant, 2009). Antioxidant enzyme activities are impaired in obese individuals, especially with mineral deficiencies (Via, 2012). Several *in vivo* and *in vitro* studies have also demonstrated the interrelationship between obesity and activity levels of antioxidant enzymes. In the 3T3-L1 differentiated adipocytes co-cultured with macrophages demonstrated higher levels of extracellular SODs (Adachi et al., 2009). The upregulated SOD expressions could be related to a defense action in response to infiltering macrophages.

In a genetically obese Zucker rat model with characteristic insulin resistance and hyperlipidemia, the plasma samples and heart tissue showed decreased levels of SOD while GPx expressions remained unaltered (Martinelli et al., 2020). In another study, it was reported that a fructose-enriched diet enhanced the plasma levels of free fatty acids and triglycerides in the male Wistar rat model.

The elevated levels of free fatty acids induced ROS generation and insulin resistance. Subsequently, the activities of CAT, GPx, and SODs were significantly reduced (Malaisse, 2011). Ample documented pieces of evidence suggest that impairment in the antioxidant enzyme machinery leads to further complications in obese subjects with insulin resistance. A significant reduction in GSH levels was observed in obese male and female human subjects with non-alcoholic fatty liver disease. It was further observed that reduced serum GSH levels had impaired the activity of GPx and GRs (Irie et al., 2016). In obese children of both sexes, reduced gene expressions of Mn-SOD and CAT were prevalent in comparison to normal-weight children (Mohseni et al., 2018). Other clinical investigations have reported reduced levels of serum paraoxonase-1 in obese individuals with cardiovascular complications and diabetes mellitus (D'Archivio et al., 2011). Serum paraoxonase-1 are antioxidant enzyme that prevents the oxidation of lipoproteins as well as protects circulating cells from oxidative damage (Savini et al., 2013).

4.4 PHARMACOLOGICAL TREATMENT OF OBESITY AND CHALLENGES

Pharmacotherapy for obesity is aimed at reducing appetite, increasing energy expenditure, or a combination of both approaches (Klobučar Majanović et al., 2016). The investigated drug candidates against obesity range from cannabinoid receptor antagonists, (Pi-Sunyer et al., 2006), gastrointestinal lipase inhibitors, (Kwon et al., 2021), serotonergic agonists (Lucchetta et al., 2017; Mathus-Vliegen et al., 1992; Müller et al., 2018), sympathomimetic agents (Kilian et al., 2015) to peptides optimized for therapeutic use (Ambery et al., 2018; Bhat et al., 2013; Hasib et al., 2018; Mack et al., 2009; Tillner et al., 2018). The century-old history of obesity pharmacotherapy has witnessed several drugs (Table 4.1) being approved by regulatory authorities for therapeutic use. Predominantly, all the anti-obesity medications exert their pharmacological role by regulating either peripheral or central pathways governing energy balance but not both (Cercato and Fonseca, 2019). For instance, the centrally acting drugs enhance satiety by regulating signaling cascades in the noradrenergic, serotonergic, or dopaminergic pathways either by inhibiting catecholamine reuptake or by influencing satiety receptors in the limbic system and hypothalamus. On the contrary, peripherally acting drugs like orlistat, a pancreatic lipase inhibitor, inhibits dietary fat uptake from the gut. The etiology of obesity is heterogeneous and rarely monogenetic and more commonly polygenic dictated by behavioral, metabolic, and endocrine causes and epigenetic factors (Hebebrand et al., 2001; Huypens et al., 2016). This heterogeneity in genetic and epigenetic processes in addition to environmental risk factors stands responsible for varied responses of individuals to developed anti-obesity medications (Melvin et al., 2018; O'Rahilly, 2009).

Many drug candidates have been withdrawn with the sobering realization of serious side effects. Fenfluramine and its stereoisomer, dexfenfluramine (Cannistra et al., 1997 Connolly et al., 1997), lorcaserin (a 5HT2c agonist) (Sharretts et al., 2020; Smith et al., 2010), first-generation endocannabinoid receptor 1 antagonist taranabant

Table 4.1 Currently used anti-obesity medications.

Drug	Drug type	Mean weight reduction Placebo/drug	Adverse effects
Diethylpropion	Sympathomimetic	7%/10% in studies ranging from 6–52 weeks	Dizziness, dry mouth, nausea, headache, hallucinations, seizures, insomnia
Phentermine		1.6%/6.6–7.4% in studies ranging from 2–24 weeks	Tachycardia, pulmonary hypertension, insomnia, dizziness, dry mouth, constipation, nervousness
Cathine		2.4%/6.6–9.9% in studies ranging from 2–24 weeks	Tachycardia, hypertension, insomnia, depression
Phendimetrazine		0.80%/6.96% in studies ranging from 2–14 weeks	Dry mouth, nausea, diarrhea, rhabdomyolysis, pulmonary infarction, persistent psychosis, addiction
Orlistat	Pancreatic lipase inhibitor	6.1%/10.2% in studies carried out for 12–24 months	Oily stool, diarrhea, abdominal pain, nausea, vomiting, reduced absorption of fat-soluble vitamins, and reports of macrocytic anemia and thrombocytopenia
Phentermine/ topiramate	Sympathomimetic/ anticonvulsant	1.2%/9.8% in studies carried out for 12–24 months (dose-dependent)	Paresthesia, dizziness, tachycardia, dry mouth, sleep and mood disorders, constipation, dysgeusia, metabolic acidosis
Naltrexone/ bupropion	Opioid antagonist/ nor-epinephrine and dopamine reuptake inhibitor	1.3%/6.2% in studies carried out for 12–24 months (dose-dependent)	Seizures, manic episodes, blurred vision, mood alternations, skin problems, muscle or joint pain, dry mouth, headache, constipation, nausea, vomiting, insomnia
Liraglutide	GLPR1 agonist	2.6%/8.2% in studies carried out for 12–24 months	Nausea, vomiting, dyspepsia, gastroenteritis, fatigue, hypoglycemia, constipation, GERD diarrhea, and abdominal pain
Semaglutide	GLPR1 agonist	2.4%/14.9% in studies carried out for 12–24 weeks	Nausea, vomiting, abdominal pain, constipation, diarrhea

(Addy et al., 2008; Aronne et al., 2010; Proietto et al; 2010), rimonabant (Després et al., 2005; Pi-Sunyer et al., 2006; Van Gaal et al., 2005), SLV-319 (Bristol Meyers Squibb), AZD-2207 (AstraZeneca), V-23434 (Vernalis), E-6776 (Esteve) (Colca, 2009; Onakpoya Igho et al., 2016), and sibutramine (Bosello et al., 2002; Harrison-Woolrych et al., 2010; Torp-Pedersen et al., 2007) were withdrawn eventually from anti-obesity drug research owing to severe abnormalities reported in various clinical studies. Multifactorial disease-producing pathways, side effects of available pharmacotherapies, and lesser efficacy have evoked many challenges in the drug development process. These bottlenecks have also led to the underuse of drugs for weight management (Chao et al., 2020; Kanj & Levine, 2020).

4.5 THERANOSTIC APPLICATIONS OF METAL NANOCOMPOSITES IN LEPTIN RESISTANCE AND OBESITY

Nano theranostics, which integrates diagnostic and therapeutic functions into a single system using the benefits of nanotechnology, is extremely attractive for personalized medicine. Several types of nanocarriers have been developed so far for nano theranostics, which include dendrimers, liposomes, micelles, polymer conjugations, metal and inorganic nanoparticles (NPs), solid–lipid NPs, and carbon nanotubes. Metal matrix-based nanocomposites own novel mechanical functionalities in terms of their high elastic modulus, specific strength, thermal and electrocatalytic properties in comparison with respective monolithic alloys (Mortensen & Llorca, 2010; Zhang & Chen, 2008). Larger surface area, robust adsorption, elevated bioavailability, worthy tissue targeting, and modifiable releasing rates enable the nanocomposites to improvise the diagnosis and management of obesity/associated disorders.

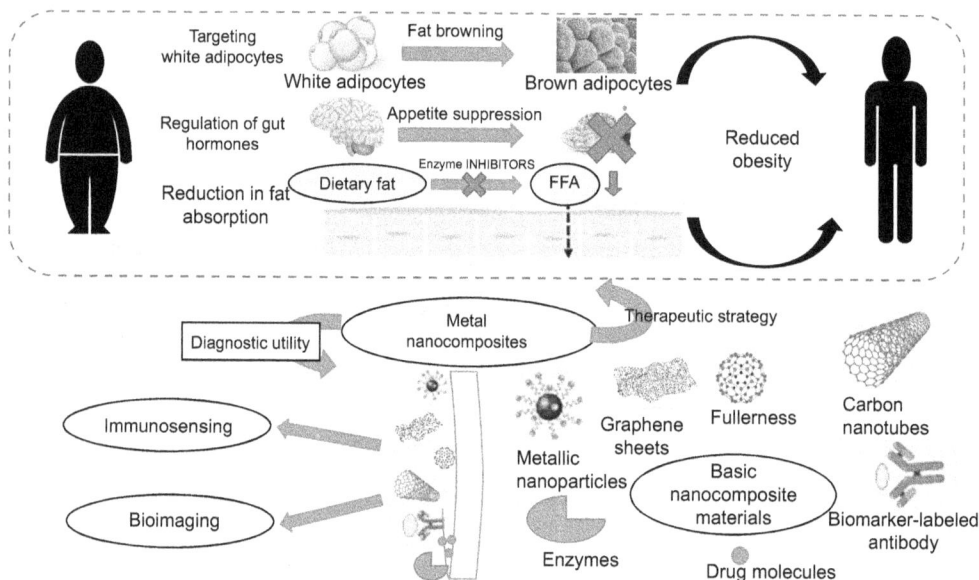

Figure 4.2 Therapeutic and diagnostic applications of metal nanocomposites.

The following sections of the chapter have mainly discussed the use of metal nanocomposites in the diagnosis of obesity by the detection of specific biochemical markers, especially leptin (Figure 4.2). The therapeutic ability and applicability of these nanoengineered metal composites in countering obesity (Figure 4.2) have also been reviewed.

4.5.1 Diagnosis of obesity using metal nanocomposites

In recent years, obesity has indeed evolved into a serious health problem worldwide. Detection of adiponectin and leptin levels in serum and cells proves crucial in the diagnosis of obesity and related disorders. Leptin levels are usually elevated while adiponectin levels are reduced in obese individuals. Leptin hormone secreted from adipocytes is primarily responsible for proper maintenance of body weight (Tjong, 2013; Zhang et al., 2013). Leptin deficiency majorly disturbs physiological processes such as reproduction, pubertal development, bone metabolism, and hypothalamic amenorrhea (Zhang et al., 1994). The average serum levels of leptin in obese persons are 31 ng mL^{-1}, whereas the normal leptin level is around 7.5 ng mL^{-1} in healthy normal-weight persons (Procaccini et al., 2012). It is quite clear that precise monitoring of serum concentrations would be beneficial to define the physiological role of leptin. Many methods have been reported for the detection of leptin in serum, including capillary electrophoresis, immunocapture techniques, and enzyme-linked immunosorbent assays (Considine et al., 1996; Imagawa et al., 1998; Richards et al., 1999). Despite being reliable, these methods are a bit expensive and time-consuming. Yet several chemiluminescence-based immunosensors have been designed of late for the detection of leptin and other obesity-related biomarkers in serum. The reported nanocomposite-based immunosensors not only show great sensitivity, selectivity, and wider linear ranges for leptin but are also instrumental in the detection of susceptible plaques, recruited macrophages of adipose tissue, fatty liver, and other physiological obesity markers as well (Wang & Heilig, 2012).

A regenerable, impedimetric electrochemical sensor for serum detection of leptin was fabricated by co-electropolymerizing pyrrole and pyrrole propylic acid with gold nanoparticles conjugated with protein G to capture antibody as a probe for the immunoassay. The hybrid nanocomposite immunosensor was porous, hydrophilic, stable, and possessed good conductivity. The sensor can be regenerated by rinsing with 0.1 M glycine buffer (pH 2.7). The immunosensor effectively detected leptin in a variable range of 10–100,000 ng mL^{-1} with a limit of detection of 10 ng mL^{-1} in PBS + 1% serum solution. Besides biochemical analysis, it is feasible to fabricate such regenerable immunosensor for the routine environment and food analysis as well (Chen et al., 2010).

Dong and his research group devised an interesting novel sandwich electrochemical immunosensor for the detection of human leptin in serum samples. They used glassy carbon electrodes modified with single-walled chitosan film-based carbon nanotubes. The fabrication involved a specific anti-human leptin antibody (capture antibody) covalently immobilized on the modified glassy carbon electrode surface followed by incubation with target human leptin and reaction with biotinylated anti-human leptin. The immunosensor surface contained also streptavidin-alkaline phosphatase to catalyze the hydrolysis of substrate α-naphthyl phosphate. The immunosensor generated a good amperometric response in leptin detection (0.05–500 ng mL^{-1}) with a limit of detection of 30 ng mL^{-1}. The immunosensor was successfully applied in the detection of leptin in spiked serum (Dong et al., 2014).

A sandwich-type chemoluminescence immunosensor was developed with hermin/G-quadruplex DNAzymes and functional supramagnetic nanocomposites. The sensing platform was built up on polydopamine as a surface adhesive layered with ferric oxide/gold nanoparticles. The immunosensor showed high sensitivity and good specificity and had a wide linear range of leptin detection (1.08×10^3 pg mL^{-1}) with a limit of detection confined to 0.3 pg mL^{-1}. The developed immunosensor was found to be the most sensitive in detecting leptin owing to efficient catalysis of DNAzymes and analyte enriched by magnetic capture (He et al., 2015).

GPR120, a member of the rhodopsin family of G-protein-coupled receptors are sensor for long-chain fatty acids. In obese individuals, the serum free fatty acids remain elevated and so also the expression of the GPR120. Thereby detection of high levels of GPR120 could be instrumental in the diagnosis of obesity. Xiao and coworkers reported for the first time a novel SERS-fluorescence bimodal microscopy method for the detection and imaging of GPR120s in individual cells. They synthesized europium-doped calcium molybdate/gold nanocomposites encapsulated in 4-mercaptobenzoic acid as imaging probes. The hybrid nanocomposites were conjugated with antibodies to evaluate their biocompatibility on HEK293 cell lines by transfection assay. GPR120s were successfully observed in single cells by SERS mapping. A linear relationship was observed between linoleic acid concentration and GPR120 activity in the range of 0–60 µM. The fabricated novel hybrid nanocomposite-based imaging probe could prove beneficial in multiple signaling-based detections of many other fat-responsive receptors directed toward obesity diagnosis (Xiao et al., 2018).

Similarly, Cai and coworkers fabricated an eco-friendly and label-free immunosensor based on porous graphene-functionalized black phosphorus hybrid nanocomposite for leptin detection. The nanoprobes were further coated with gold nanoparticles conjugated with glutaraldehyde and cysteamine for the effective fixation of anti-leptin antibodies. Such fabrication uplifted the protein loading capacity and detective sensitivity of the immunoprotein. The immunosensor displayed sensitive detection of leptin in a linear range of 0.150–2500 pg mL^{-1} with a limit of detection of 0.036 pg mL^{-1}. The immunoprobe was selective and exhibited good anti-interference ability. The fabricated probe could be used for screening for obesity and NAFLD (Cai et al., 2019).

Square wave voltammetry is an efficient analytical technique to examine the antigen–antibody binding on a solid–liquid interface. Very recently, Shukla and co-research workers used an effective voltammetry method for detecting leptin in serum samples. They devised a polymer–metallic hybrid nanocomposite electrode comprising antibody-conjugated micellar gold nanoparticles layered on a 3-aminopropyl trimethoxysilane base. Bradford assay was used to determine the ratio of bound antibody molecules. The antigen coupling rate was found to be consistent with binding affinity as well as variation in antibody density. Leptin could be detected in a very wide range (6.2 ng mL^{-1}–0.12 fg mL^{-1}) while the limit of detection was kept nearer to 0.25 fM mL^{-1} (Shukla et al., 2022).

A novel impedimetric hybrid nanocomposite aptasensor was designed to detect human leptin in serum and plasma samples. The electrode was coated with gold and titanium dioxide nanoparticles by electrode position. The aptasensor was fabricated by immobilizing the modified electrode surface with thiol-tethered DNA. Leptin detection was possible in two linear ranges (1.0–100.0 pg mL^{-1} and 100.0–1,000.0 pg mL^{-1}), with a limit of detection of 0.312 pg mL^{-1} (Erkmen et al., 2022).

Using a high-temperature pyrolysis technique a porous carbon-based ferric/nickel bimetallic nanocomposite system was fabricated as an electrochemical immunosensor for detecting leptin. The sensor detected leptin in serum samples in a wide linear range (500 fg mL^{-1}–80 mL^{-1}) with a limit of detection varying between 155 fg mL^{-1} and 185 fg mL^{-1}. The sensor showed good stability and feasibility in detecting leptin *in vivo* (Islam et al., 2022).

Another innovative ultrasensitive electrochemical immunosensor was fabricated using the immobilization method by Uludag and Sezginturk. Indium tin oxide-coated polyethylene terephthalate nanosheets were used for the fabrication of working electrodes. The anti-leptin antibodies were immobilized on the electrode surface by enhancing covalent interactions using cyanogen bromide. The designed immunosensor had a wider detection range (0.05–100 pg mL^{-1}) as well as a low limit of detection (0.0086 pg mL^{-1}). The designed immunosensor though disposable, retained activity after repeated uses (Uludağ & Sezgintürk, 2022).

Ghrelin is a hunger hormone secreted in the small intestine and stomach. It stimulates gastric acid secretion and regulates gastric contractility and motor activity. Ghrelin plays a major role in carbohydrate metabolism and regulates gluconeogenesis. In human serum, ghrelin occurs in deacylated form and a varying range (in pg mL^{-1}). Generally, in healthy individuals, the ghrelin levels are higher in comparison to obese persons. Therefore, the detection of the hunger hormone ghrelin as a biomarker for obesity becomes important. On such grounds, Li and his research group developed a bimetal oxide–polymer hybrid nanocomposite-based electrochemical immunusensor for the detection of ghrelin in serum samples. The nanocomposite is composed of hafnium oxide–praseodymium oxide nanoflakes layered on chitosan and polyethyleneimine polymer base. The fabricated immunosensor displayed good sensitivity, electrical conductivity, and a larger surface area for antibody loading. The constructed immunosensor detected ghrelin in a wider linear range of 0.01 pg mL^{-1}–50 ng mL^{-1} with a limit of detection around 0.006 pg mL^{-1} (Li et al., 2021).

4.5.2 Therapy of obesity with metal nanocomposites

Nanotechnology-based therapeutic strategies for obesity are focused on refining bowel strength, plummeting food intake, healing cellular anomalies, and amending disturbed metabolic imbalances (Li et al., 2019). Many metal matrix nanocomposites have been designed and evaluated for their anti-obesity potential. Obesity and associated comorbidities are characterized by aberrations of functional adipocytes, macrophages, and vascular endothelial cells. Nanocomposites possess admirable drug-loading and targeting modalities, seeming as better alternatives to curb adverse effects and increase the bioavailability of potent anti-obesity drugs (Li et al., 2019).

El-seidy et al., (2022) designed and synthesized KCl matrix-based ZnO nanocomposites for targeting and reducing diet-induced obesity in the high-fat diet-induced rat model. The designed ZnO nanocomposites showcased appreciable anti-obesity effects via depressing body weight increase, free radical-induced oxidative stress, body mass index, lipids, and insulin resistance.

Adipocytes more recently have been considered as potential targets for anti-obesity drug discovery. Tumbling the dimensions and multiplication of adipose tissue cells might offer resolutions for obesity management. Nanocomposites can promote fat browning, lessen lipid content in adipocytes, and check their multiplication.

Alsenousy et al. (2022) studied the fat-reducing effects of superparamagnetic iron oxide nanoparticles (SPIONs) in high-fat diet-induced rat models. The SPIONs reduced body weight, high blood sugar, and levels of adiponectin, leptin, and dyslipidemia in obese rats. The SPIONs depressed the expressions of proinflammatory cytokines and markers involved in mitochondrial biogenesis. The PEG-conjugated SPIONs induced fat browning in WAT by inducing the expression of UCP-1.

In another study, the anti-obesity effects of hollow gold nanoshells coated with polypyrrole were explored. The synthesized polypyrrole-coated hollow gold nanoshells presented admirable photothermal permanence in near-infrared (NIR) laser radiation in repeated five sets of experimental cycles. Histopathological examinations *ex vivo* revealed adipocyte necrosis subjected to near infra-red laser beam radiation (Han et al., 2019). Besides nanocomposites, other nanosystems have been designed to augment the therapeutic effectiveness of synthetic drugs via targeted delivery (Table 4.2).

Drug-based treatment of obesity should be prolonged therapy, subsequently as most individuals recapture weight after discontinuing medicine. Consequently, we must contemplate whether or not nanostructures unaided promote weight loss and improve healthier lipid profiles. Nanostructures, such as superparamagnetic iron oxide grafted with carboxyethyl silane triol (Sharifi et al., 2013), cerium oxide (Rocca et al., 2015), chitosan and water-soluble chitosan (Zhang et al., 2012), silica (Kupferschmidt et al., 2014), nanorods based on gold (Sheng et al., 2014) and nanospheres (Lee et al., 2017) based on gold and hyaluronate, have therapeutic roles in the management of obesity.

Obesity treatment with herbal nanotherapeutics poses suitable alternatives to currently available anti-obesity drugs. Many nanostructures (Table 4.3) of potential plant extracts have been grafted and evaluated for anti-obesity and anti-dyslipidemic potential.

Furthermore, numerous metal and polymer-based nanostructures have been fabricated for phytoconstituents like resveratrol (Zu et al., 2021), curcumin (Rajabzadeh-Khosroshahi et al., 2022), hydroxy citric acid (Ezhilarasi et al., 2016), chlorogenic acid (Nallamuthu et al., 2015) and ϒ-oryzanol (Kozuka et al., 2017). The pharmacological outcomes underlined their effects on the

Table 4.2 Types of nanosystems for obesity management

Designed nanostructure	Size range	Role in countering obesity	Reference
Gold nanoparticles-loaded adipose-homing peptide	Less than 50 nm	Targeted delivery to WAT vasculature *in vivo*	(Thovhogi et al., 2015)
Rosiglitazone-loaded polymeric nanoparticles	200 nm	Reduction in inflammatory responses of WAT	(Di Mascolo et al., 2013)
Dextran and dextran-PEG grafted nanocarriers for dexamethasone	4–30 nm	Decreased obesity-induced inflammatory actions	(Ma et al., 2016)
Capryol PGMC and Cremophor RH40 grafted orlistat nano emulsion	139–150 nm	Improved bioavailability and enhanced pancreatic lipase inhibition	(Sangwai et al., 2012)
Endothelial targeted peptides conjugated to PLGA-PEG nanoparticles	100 nm	Reduced diet-induced weight gain in a mouse model	(Xue et al., 2016)

Table 4.3 Herbal nanotherapeutics for obesity and dyslipidemia

Plant extract	Nano-formulation	Pharmacological effects	Reference
Salacia chinensis	Gold nanoparticles	Activated AMPK kinase, enhanced levels of adiponectin, decreased leptin levels, proinflammatory cytokines, and reduced body weight.	(Gao, et al., 2020)
Smilax glabra	Hollow gold nanospheres	Reduced body weight, decreased insulin resistance, and glucose intolerance. The normalized activity of liver marker enzymes.	(Ansari et al., 2019)
Poria cocos	Polydispersed gold nanoparticles	Enhance satiety hormone expressions, regulate glucose and lipid metabolism, reduce WAT-induced inflammation, and reduce body weight.	(Li et al., 2020)
Argyreia nervosa	Silver nanoparticles	Potent alpha-amylase and alpha-glucosidase inhibitor.	(Saratale et al., 2017)
Ziziphus jujube	Gold nanoparticles	Counter dyslipidemia and scavenge free radicals, lower cholesterol.	(Javanshir et al., 2020)
Ribes nigrum (black currant)	Selenium nanoparticles	Anti-hyperlipidemia and free radical scavenger.	(Al-Kurdy & Khudair, 2020)
Nigella sativa	Silver nanoparticles	Decreased cholesterol, triglyceride, and LDL levels.	(Ali & Khudair, 2019)
Dendropanax morbifera	Gold nanoparticles	Decreased triglyceride levels, suppressed the expressions of adipogenetic master regulator proteins in 3T3-L1 preadipocytes. Fatty acid synthase and acetyl co-enzyme carboxylase levels in HepG2 cells were also curbed.	(YI et al., 2020)

metabolic signaling cascades, lipid profiles, free radicals, physiological markers, and so on. The explorations pose a greater arena where these nanostructured systems can be used as alternative clinical remedies to curb obesity and related complications.

4.6 CONCLUSION AND FUTURE PERSPECTIVES

Within a decade or two, obesity has evolved drastically as a global health issue. Besides modification in lifestyle and surgical procedures, clinical management of obesity with drugs is also important. Currently available drugs for the clinical management of obesity have poor bioavailability, narrow therapeutic windows, and are not target specific. The implementation of nanotechnology in the field of medicine ensures a wide array of innovative methodologies and approaches thus facilitating value-added theranostic features. In comparison to conventional therapy, metal-based nanocomposite systems have fewer adverse effects and superior efficiency in terms of drug delivery, drug loading, biocompatibility, and bioavailability. As discussed earlier, several metal-based nanocomposite systems have presented encouraging preclinical activities about weight reduction by clampdown of digestion or augmentation of energy expenditure. Even though there are numerous advantages, the fruitful clinical trials and bringing drugs to the therapeutic market remain perplexing by quite a few facets. First, diverse drug delivery transporters and administration routes might have an impact on therapeutic effectiveness. Second, reducing the adverse effects of these nano medications remains challenging. Molecules targeting adipocytes might be integrated into nano delivery systems to attain accurate therapeutic effects, but the blemishes of some designs must be considered, such as targeting proficiency and prospective immune response. Meanwhile, as weight reduction demands prolonged treatment, a comprehensive assessment of biocompatibility should be prudently executed.

REFERENCES

Adachi, Tetsuo, Taisuke Toishi, Haoshu Wu, Tetsuro Kamiya, and Hirokazu Hara. 2009. "Expression of Extracellular Superoxide Dismutase during Adipose Differentiation in 3T3-L1 Cells." *Redox Report* 14 (1): 34–40. doi:10.1179/135100009x392467.

Addy, Carol, Hamish Wright, Koen Van Laere, Ira Gantz, Ngozi Erondu, Bret J. Musser, Kaifeng Lu, et al. 2008. "The Acyclic CB1R Inverse Agonist Taranabant Mediates Weight Loss by Increasing Energy Expenditure and Decreasing Caloric Intake." *Cell Metabolism* 7 (1): 68–78. doi:10.1016/j.cmet.2007.11.012.

Ajayan, P. M., L. S. Schadler, and P. V. Braun. 2004. *Nanocomposite Science and Technology*. Weinheim, Germany: Wiley-VCH.

Ali, Zainab Sattar, and Khalisa Khadim Khudair. 2019. "Synthesis, Characterization of Silver Nanoparticles Using Nigella Sativa Seeds and Study Their Effects on the Serum Lipid Profile and DNA Damage on the Rats' Blood Treated with Hydrogen Peroxide." *The Iraqi Journal of Veterinary Medicine* 43 (2): 23–37. doi:10.30539/iraqijvm.v43i2.526.

Al-Kurdy, Masar Jabbar, and Khalisa Khadim Khudair. 2020. "The Effect of Black Currant Selenium Nanoparticles on Dyslipidemia and Oxidant- Antioxidant Status in D-Galactose Treated Rats." *Kufa Journal For Veterinary Medical Sciences* 11 (1): 23–38. doi:10.36326/kjvs/2020/v11i13300.

Alsenousy, Aisha H., Rasha A. El-Tahan, Nesma A. Ghazal, Rafael Piñol, Angel Millán, Lamiaa M. Ali, and Maher A. Kamel. 2022. "The Anti-Obesity Potential of Superparamagnetic Iron Oxide Nanoparticles against High-Fat Diet-Induced Obesity in Rats: Possible Involvement of Mitochondrial Biogenesis in the Adipose Tissues." *Pharmaceutics* 14 (10): 2134. doi:10.3390/pharmaceutics14102134.

Ambery, Philip, Victoria E. Parker, Michael Stumvoll, Maximilian G. Posch, Tim Heise, Leona Plum-Moerschel, Lan-Feng Tsai, et al. 2018. "MediO382, a GLP-1 and Glucagon Receptor Dual Agonist, in Obese or Overweight Patients with Type 2 Diabetes: A Randomised, Controlled, Double-Blind, Ascending Dose and Phase 2a Study." *The Lancet* 391 (10140): 2607–2618. doi:10.1016/s0140–6736(18)30726–8.

An, K. H., S. Y. Jeong, H. R. Hwang, and Y. H. Lee. 2004. "Enhanced Sensitivity of a Gas Sensor Incorporating Single-Walled Carbon Nanotube: Polypyrrole Nanocomposites." *Advanced Materials* 16 (12): 1005–1009. doi:10.1002/adma.200306176.

Ansari, Siddique Akber, Ahmed Bari, Riaz Ullah, Maghimaa Mathanmohun, Vishnu Priya Veeraraghavan, and Zhongwei Sun. 2019. "Gold Nanoparticles Synthesized with Smilax Glabra Rhizome Modulates the Anti-Obesity Parameters in High-Fat Diet and Streptozotocin Induced Obese Diabetes Rat Model." *Journal of Photochemistry and Photobiology B: Biology* 201: 111643. doi:10.1016/j.jphotobiol.2019.111643.

Apovian, Caroline M., Louis J. Aronne, Daniel H. Bessesen, Marie E. McDonnell, M. Hassan Murad, Uberto Pagotto, Donna H. Ryan, and Christopher D. Still. 2015. "Pharmacological Management of Obesity: An Endocrine Society Clinical Practice Guideline." *The Journal of Clinical Endocrinology & Metabolism* 100 (2): 342–362. doi:10.1210/jc.2014-3415.

Aronne, L. J., S. Tonstad, M. Moreno, I. Gantz, N. Erondu, S. Suryawanshi, C. Molony, et al. 2010. "A Clinical Trial Assessing the Safety and Efficacy of Taranabant, a CB1R Inverse Agonist, in Obese and Overweight Patients: A High-Dose Study." *International Journal of Obesity* 34 (5): 919–935. doi:10.1038/ijo.2010.21.

Banks, William A., Christopher R. DiPalma, and Catherine L. Farrell. 1999. "Impaired Transport of Leptin across the Blood-Brain Barrier in Obesity." *Peptides* 20 (11): 1341–1345. doi:10.1016/s0196-9781(99)00139-4.

Banks, William A., Abba J. Kastin, Weitao Huang, Jonathan B. Jaspan, and Lawrence M. Maness. 1996. "Leptin Enters the Brain by a Saturable System Independent of Insulin." *Peptides* 17 (2): 305–311. doi:10.1016/0196-9781(96)00025-3.

Bartelt, Alexander, Oliver T. Bruns, Rudolph Reimer, Heinz Hohenberg, Harald Ittrich, Kersten Peldschus, Michael G. Kaul, et al. 2011. "Brown Adipose Tissue Activity Controls Triglyceride Clearance." *Nature Medicine* 17 (2): 200–205. doi:10.1038/nm.2297.

Bartelt, Alexander, Martin Merkel, and Joerg Heeren. 2012. "A New, Powerful Player in Lipoprotein Metabolism: Brown Adipose Tissue." *Journal of Molecular Medicine* 90 (8): 887–893. doi:10.1007/s00109-012-0858-3.

Baskin, Kedryn K., Benjamin R. Winders, and Eric N. Olson. 2015. "Muscle as a 'Mediator' of Systemic Metabolism." *Cell Metabolism* 21 (2): 237–248. doi:10.1016/j.cmet.2014.12.021.

Berthoud, Hans-Rudolf, Heike Münzberg, and Christopher D. Morrison. 2017. "Blaming the Brain for Obesity: Integration of Hedonic and Homeostatic Mechanisms." *Gastroenterology* 152 (7): 1728–1738. doi:10.1053/j.gastro.2016.12.050.

Bhat, Vikas K., Barry D. Kerr, Peter R. Flatt, and Victor A. Gault. 2013. "A Novel Gip-Oxyntomodulin Hybrid Peptide Acting through GIP, Glucagon and GLP-1 Receptors Exhibits Weight Reducing and Anti-Diabetic Properties." *Biochemical Pharmacology* 85 (11): 1655–1662. doi:10.1016/j.bcp.2013.03.009.

Bjørbæk, Christian, Joel K. Elmquist, J. Daniel Frantz, Steven E. Shoelson, and Jeffrey S. Flier. 1998. "Identification of Socs-3 as a Potential Mediator of Central Leptin Resistance." *Molecular Cell* 1 (4): 619–625. doi:10.1016/s1097-2765(00)80062-3.

Bosello, O., M. O. Carruba, E. Ferrannini, and C. M. Rotella. 2002. "Sibutramine Lost and Found." *Eating and Weight Disorders: Studies on Anorexia, Bulimia and Obesity* 7 (3): 161–167. doi:10.1007/bf03327453.

Bryan, Sean, Boran Baregzay, Drew Spicer, Pawan K. Singal, and Neelam Khaper. 2013. "Redox-Inflammatory Synergy in the Metabolic Syndrome." *Canadian Journal of Physiology and Pharmacology* 91 (1): 22–30. doi:10.1139/cjpp-2012-0295.

Cai, Jinying, Xiaodan Gou, Bolu Sun, Wuyan Li, Dai Li, Jinglong Liu, Fangdi Hu, and Yingdong Li. 2019. "Porous Graphene-Black Phosphorus Nanocomposite Modified Electrode for Detection of Leptin." *Biosensors and Bioelectronics* 137: 88–95. doi:10.1016/j.bios.2019.04.045.

Camargo, Pedro Henrique, Kestur Gundappa Satyanarayana, and Fernando Wypych. 2009. "Nanocomposites: Synthesis, Structure, Properties and New Application Opportunities." *Materials Research* 12 (1): 1–39. doi:10.1590/s1516-14392009000100002.

Cannistra, Lauralyn B., Steven M. Davis, and Anne G. Bauman. 1997. "Valvular Heart Disease Associated with Dexfenfluramine." *New England Journal of Medicine* 337 (9): 636. doi:10.1056/nejm199708283370912.

Caro, José F., Jerzy W. Kolaczynski, Mark R. Nyce, Joanna P. Ohannesian, Irina Opentanova, Warren H. Goldman, Richard B. Lynn, Pei-Li Zhang, Madhur K. Sinha, and Robert V. Considine. 1996. "Decreased Cerebrospinal-Fluid/Serum Leptin Ratio in Obesity: A Possible Mechanism for Leptin Resistance." *The Lancet* 348 (9021): 159–161. doi:10.1016/s0140-6736(96)03173-x.

Cercato, C., and F. A. Fonseca. 2019. "Cardiovascular Risk and Obesity." *Diabetology & Metabolic Syndrome* 11 (1). doi:10.1186/s13098-019-0468-0.

Chao, Ariana M., Thomas A. Wadden, Robert I. Berkowitz, Kerry Quigley, and Frank Silvestry. 2020. "The Risk of Cardiovascular Complications with Current Obesity Drugs." *Expert Opinion on Drug Safety* 19 (9): 1095–1104. doi:10.1080/14740338.2020.1806234.

Chawla, Ajay, Khoa D. Nguyen, and Y. P. Goh. 2011. "Macrophage-Mediated Inflammation in Metabolic Disease." *Nature Reviews Immunology* 11 (11): 738–749. doi:10.1038/nri3071.

Chelimo, Carol, Carlos A. Camargo, Susan M. Morton, and Cameron C. Grant. 2020. "Association of Repeated Antibiotic Exposure up to Age 4 Years with Body Mass at Age 4.5 Years." *JAMA Network Open* 3 (1). doi:10.1001/jamanetworkopen.2019.17577.

Chen, Wei, Yu Lei, and Chang Ming Li. 2010. "Regenerable Leptin Immunosensor Based on Protein G Immobilized Au-Pyrrole Propylic Acid-Polypyrrole Nanocomposite." *Electroanalysis* 22 (10): 1078–1083. doi:10.1002/elan.200900536.

Colca, Jerry R. 2009. "Discontinued Drugs in 2008: Endocrine and Metabolic." *Expert Opinion on Investigational Drugs* 18 (9): 1243–1255. doi:10.1517/13543780903132673.

Connolly, Heidi M., Jack L. Crary, Michael D. McGoon, Donald D. Hensrud, Brooks S. Edwards, William D. Edwards, and Hartzell V. Schaff. 1997. "Valvular Heart Disease Associated with Fenfluramine: Phentermine." *New England Journal of Medicine* 337 9 (1997): 581–588. https://doi.org/10.1056/nejm199708283370901.

Considine, Robert V., Madhur K. Sinha, Mark L. Heiman, Aidas Kriauciunas, Thomas W. Stephens, Mark R. Nyce, Joanna P. Ohannesian, et al. 1996. "Serum Immunoreactive-Leptin Concentrations in Normal-Weight and Obese Humans." *New England Journal of Medicine* 334 (5): 292–295. doi:10.1056/nejm199602013340503.

Coon, P. J., E. M. Rogus, D. Drinkwater, D. C. Muller, and A. P. Goldberg. 1992. "Role of Body Fat Distribution in the Decline in Insulin Sensitivity and Glucose Tolerance with Age." *The Journal of Clinical Endocrinology & Metabolism* 75 (4): 1125–1132. doi:10.1210/jcem.75.4.1400882.

Croce, F., G. B. Appetecchi, L. Persi, and B. Scrosati. 1998. "Nanocomposite Polymer Electrolytes for Lithium Batteries." *Nature* 394 (6692): 456–458. doi:10.1038/28818.

Crujeiras, Ana B., Marcos C. Carreira, Begoña Cabia, Sara Andrade, Maria Amil, and Felipe F. Casanueva. 2015. "Leptin Resistance in Obesity: An Epigenetic Landscape." *Life Sciences* 140: 57–63. doi:10.1016/j.lfs.2015.05.003.

Cypess, Aaron M., Sanaz Lehman, Gethin Williams, Ilan Tal, Dean Rodman, Allison B. Goldfine, Frank C. Kuo, et al. 2009. "Identification and Importance of Brown Adipose Tissue in Adult Humans." *New England Journal of Medicine* 360 (15): 1509–1517. doi:10.1056/nejmoa0810780.

D'Archivio, Massimo, Giovanni Annuzzi, Rosaria Varì, Carmelina Filesi, Rosalba Giacco, Beatrice Scazzocchio, Carmela Santangelo, Claudio Giovannini, Angela A. Rivellese, and Roberta Masella. 2011. "Predominant Role of Obesity/Insulin Resistance in Oxidative Stress Development." *European Journal of Clinical Investigation* 42 (1): 70–78. doi:10.1111/j.1365-2362.2011.02558.x.

DeLany, Judith. 2008. "Leptin Hormone and Other Biochemical Influences on Systemic Inflammation." *Journal of Bodywork and Movement Therapies* 12 (2): 121–132. doi:10.1016/j.jbmt.2007.11.006.

Després, Jean-Pierre, Alain Golay, and Lars Sjöström. 2005. "Effects of Rimonabant on Metabolic Risk Factors in Overweight Patients with Dyslipidemia." *New England Journal of Medicine* 353 (20): 2121–2134. doi:10.1056/nejmoa044537.

Di Mascolo, Daniele, Christopher J. Lyon, Santosh Aryal, Maricela R. Ramirez, Jun Wang, Patrizio Candeloro, Michele Guindani, Willa A. Hsueh, and Paolo Decuzzi. 2013. "Rosiglitazone-Loaded Nanospheres for Modulating Macrophage-Specific Inflammation in Obesity." *Journal of Controlled Release* 170 (3): 460–468. doi:10.1016/j.jconrel.2013.06.012.

Dong, Fang, Rong Luo, Heng Chen, Wei Zhang, and Shijia Ding. 2014. "Amperometric Immunosensor Based on Carbon Nanotubes/Chitosan Film Modified Electrodes for Detection of Human Leptin." *International Journal of Electrochemical Science* 9 (2014): 6924–6935.

Elseidy, Ahmed, Samir Bashandy, Fatma Ibrahim, Sahar Abd El-Rahman, Omar Farid, Sherif Moussa, and Marwan El-Baset. 2022. "Zinc Oxide Nanoparticles Characterization and Therapeutic Evaluation on High Fat/Sucrose Diet Induced-Obesity." *Egyptian Journal of Chemistry.* doi:10.21608/ejchem.2022.112166.5113.

Erkmen, Cem, Gözde Aydoğdu tiğ, and Bengi Uslu. 2022. "First Label-Free Impedimetric Aptasensor Based on Au NPS/tio2 NPS for the Determination of Leptin." *Sensors and Actuators B: Chemical* 358: 131420. doi:10.1016/j.snb.2022.131420.

Ezhilarasi, P. N., S. P. Muthukumar, and C. Anandharamakrishnan. 2016. "Solid Lipid Nanoparticle Enhances Bioavailability of Hydroxycitric Acid Compared to a Microparticle Delivery System." *RSC Advances* 6 (59): 53784–53793. doi:10.1039/c6ra04312g.

Fain, John N., Atul K. Madan, M. Lloyd Hiler, Paramjeet Cheema, and Suleiman W. Bahouth. 2004. "Comparison of the Release of Adipokines by Adipose Tissue, Adipose Tissue Matrix, and Adipocytes from Visceral and Subcutaneous Abdominal Adipose Tissues of Obese Humans." *Endocrinology* 145 (5): 2273–2282. doi:10.1210/en.2003–1336.

Fox, Caroline S., Joseph M. Massaro, Udo Hoffmann, Karla M. Pou, Pal Maurovich-Horvat, Chun-Yu Liu, Ramachandran S. Vasan, et al. 2007. "Abdominal Visceral and Subcutaneous Adipose Tissue Compartments." *Circulation* 116 (1): 39–48. doi:10.1161/circulationaha.106.675355.

Gao, Lei, Yangxi Hu, Desheng Hu, Ying Li, Songpeng Yang, Xing Dong, Sulaiman Ali Alharbi, and Hansong Liu. 2020. "Anti-Obesity Activity of Gold Nanoparticles Synthesized from Salacia Chinensis Modulates the Biochemical Alterations in High-Fat Diet-Induced Obese Rat Model via AMPK Signaling Pathway." *Arabian Journal of Chemistry* 13 (8): 6589–6597. doi:10.1016/j.arabjc.2020.06.015.

Gérard, Philippe. 2015. "Gut Microbiota and Obesity." *Cellular and Molecular Life Sciences* 73 (1): 147–162. doi:10.1007/s00018-015-2061-5.

Gonzalez-Bulnes, A., L. Torres-Rovira, C. Ovilo, S. Astiz, E. Gomez-Izquierdo, P. Gonzalez-Añover, P. Pallares, M. L. Perez-Solana, and R. Sanchez-Sanchez 2012. "Reproductive, Endocrine and Metabolic Feto-Maternal Features and Placental Gene Expression in a Swine Breed with Obesity/Leptin Resistance." *General and Comparative Endocrinology* 176 (1): 94–101doi:10.1016/j.ygcen.2011.12.038.

Han, Soomin, and Younghun Kim. 2019. "Polypyrrole-Coated Hollow Gold Nanoshell Exerts Anti-Obesity Effects via Photothermal Lipolysis." *Colloids and Surfaces A: Physicochemical and Engineering Aspects* 570: 414–419. doi:10.1016/j.colsurfa.2019.03.063.

Harrison-Woolrych, Mira, Janelle Ashton, and Peter Herbison. 2010. "Fatal and Non-Fatal Cardiovascular Events in a General Population Prescribed Sibutramine in New Zealand." *Drug Safety* 33 (7): 605–613. doi:10.2165/11532440–000000000–00000.

Hasib, Annie, Ming T. Ng, Neil Tanday, Sarah L. Craig, Victor A. Gault, Peter R. Flatt, and Nigel Irwin. 2018. "Exendin-4(Lys27Pal)/Gastrin/Xenin-8-Gln: A Novel Acylated GLP-1/Gastrin/Xenin Hybrid Peptide That Improves Metabolic Status in Obese-Diabetic (OB/Ob) Mice." *Diabetes/Metabolism Research and Reviews* 35 (3). doi:10.1002/dmrr.3106.

Hauser, Anastasia K., Robert J. Wydra, Nathanael A. Stocke, Kimberly W. Anderson, and J. Zach Hilt. 2015. "Magnetic Nanoparticles and Nanocomposites for Remote Controlled Therapies." *Journal of Controlled Release* 219: 76–94. doi:10.1016/j.jconrel.2015.09.039.

He, Yuezhen, Jian Sun, Xiaoxun Wang, and Lun Wang. 2015. "Detection of Human Leptin in Serum Using Chemiluminescence Immunosensor: Signal Amplification by Hemin/G-Quadruplex DNAzymes and Protein Carriers by FE3O4/Polydopamine/AU Nanocomposites." *Sensors and Actuators B: Chemical* 221: 792–798. doi:10.1016/j.snb.2015.07.022.

Hebebrand, J., C. Sommerlad, F. Geller, T. Görg, and A. Hinney. 2001 "The Genetics of Obesity: Practical Implications." *International Journal of Obesity* 25: S1 doi:10.1038/sj.ijo.0801689.

Hoeke, Geerte, Sander Kooijman, Mariëtte R. Boon, Patrick C. N. Rensen, and Jimmy F. P. Berbée. 2016. "Role of Brown Fat in Lipoprotein Metabolism and Atherosclerosis." *Circulation Research* 118 (1): 173–182. doi:10.1161/circresaha.115.306647.

Hribal, Marta, Teresa Fiorentino, and Giorgio Sesti. 2014. "Role of C Reactive Protein (CRP) in Leptin Resistance." *Current Pharmaceutical Design* 20 (4): 609–615. doi:10.2174/138161281 13199990016.

Hukshorn, Chris J., Jan H. Lindeman, Karin H. Toet, Wim H. Saris, Paul H. Eilers, Margriet S. Westerterp-Plantenga, and Teake Kooistra. 2004. "Leptin and the Proinflammatory State Associated with Human Obesity." *The Journal of Clinical Endocrinology & Metabolism* 89 (4): 1773–1778. doi:10.1210/jc.2003–030803.

Huypens, Peter, Steffen Sass, Moya Wu, Daniela Dyckhoff, Matthias Tschöp, Fabian Theis, Susan Marschall, Martin Hrabě de Angelis, and Johannes Beckers. 2016 "Epigenetic Germline Inheritance of Diet-Induced Obesity and Insulin Resistance." *Nature Genetics* 48 (5): 497–499. doi:10.1038/ng.3527.

Imagawa, Keiichi, Yayoi Matsumoto, Yoshito Numata, Atsushi Morita, Shino Kikuoka, Mikio Tamaki, Chie Higashikubo, et al. 1998. "Development of a Sensitive Elisa for Human Leptin, Using Monoclonal Antibodies." *Clinical Chemistry* 44 (10): 2165–2171. doi:10.1093 /clinchem/44.10.2165.

Irie, Makoto, Tetsuro Sohda, Akira Anan, Atsushi Fukunaga, Kazuhide Takata, and Takashi Tanaka. 2016. "Reduced Glutathione Suppresses Oxidative Stress Innonalcoholic Fatty Liver Disease." *Euroasian Journal of Hepato-Gastroenterology* 6 (1): 13–18. doi:10.5005/ jp-journals-10018–1159.

Islam, Tamanna, Md Ariful Ahsan, Masud Hassan, Humayra Afrin, Jaqueline Pena-Zacarias, Ali Aldalbahi, Bonifacio Alvarado-Tenorio, Juan C. Noveron, and Md Nurunnabi. 2022. "Detection of Leptin Using Electrocatalyst Mediated Impedimetric Sensing." *ACS Biomaterials Science & Engineering.* doi:10.1021/acsbiomaterials.2c00642.

Javanshir, Reyhane, Moones Honarmand, Mehran Hosseini, and Mina Hemmati. 2020. "Anti-Dyslipidemic Properties of Green Gold Nanoparticle: Improvement in Oxidative Antioxidative Balance and Associated Atherogenicity and Insulin Resistance." *Clinical Phytoscience* 6 (1): 460–468. doi:10.1186/s40816-020-00224-6.

Kanj, Amjad, and Diane Levine. 2020. "Overcoming Obesity: Weight-Loss Drugs Are Underused." *Cleveland Clinic Journal of Medicine* 87 (10): 602–604. doi:10.3949/ccjm.87a.19102.

Kaushik, Susmita, Jose Antonio Rodriguez-Navarro, Esperanza Arias, Roberta Kiffin, Srabani Sahu, Gary J. Schwartz, Ana Maria Cuervo, and Rajat Singh. 2011. "Autophagy in Hypothalamic AgRP Neurons Regulates Food Intake and Energy Balance." *Cell Metabolism* 14 (2): 173–183. doi:10.1016/j.cmet.2011.06.008.

Kettner, Nicole M., Sara A. Mayo, Jack Hua, Choogon Lee, David D. Moore, and Loning Fu. 2015. "Circadian Dysfunction Induces Leptin Resistance in Mice." *Cell Metabolism* 22 (3): 448–459. doi:10.1016/j.cmet.2015.06.005.

Kilian, Tom-Marten, Nora Klöting, Ralf Bergmann, Sylvia Els-Heindl, Stefanie Babilon, Mathieu Clément-Ziza, Yixin Zhang, Annette G. Beck-Sickinger, and Constance Chollet. 2015. "Rational Design of Dual Peptides Targeting Ghrelin and Y2 Receptors to Regulate Food Intake and Body Weight." *Journal of Medicinal Chemistry* 58 (10): 4180–4193. doi:10.1021 /jm501702q.

Kim, Su-Ryun, Yun-Hee Bae, Soo-Kyung Bae, Kyu-Sil Choi, Kwon-Ha Yoon, Tae Hyeon Koo, Hye-Ock Jang, et al. 2008. "Visfatin Enhances ICAM-1 and VCAM-1 Expression through Ros-Dependent NF-KB Activation in Endothelial Cells." *Biochimica Et Biophysica Acta (BBA): Molecular Cell Research* 1783 (5): 886–895. doi:10.1016/j.bbamcr.2008.01.004.

Kleinridders, André, Dominik Schenten, A. Christine Könner, Bengt F. Belgardt, Jan Mauer, Tomoo Okamura, F. Thomas Wunderlich, Ruslan Medzhitov, and Jens C. Brüning. 2009. "Myd88 Signaling in the CNS Is Required for Development of Fatty Acid-Induced Leptin Resistance and Diet-Induced Obesity." *Cell Metabolism* 10 (4): 249–259. doi:10.1016/j.cmet.2009.08.013.

Klobučar Majanović, Sanja, Željka Crnčević Orlić, and Davor Štimac. 2016. "Current Trends in the Pharmacotherapy for Obesity." *Endocrine Oncology and Metabolism* 2 (1): 50–59. doi:10.21040/eom/2016.2.6.

Knight, Zachary A., K. Schot Hannan, Matthew L. Greenberg, and Jeffrey M. Friedman. 2010. "Hyperleptinemia Is Required for the Development of Leptin Resistance." *PLoS ONE* 5 (6). doi:10.1371/journal.pone.0011376.

Kozuka, Chisayo, Chigusa Shimizu-Okabe, Chitoshi Takayama, Kaku Nakano, Hidetaka Morinaga, Ayano Kinjo, Kotaro Fukuda, et al. 2017. "Marked Augmentation of PLGA Nanoparticle-Induced Metabolically Beneficial Impact of γ-Oryzanol on Fuel Dyshomeostasis in Genetically Obese-Diabetic OB/Ob Mice." *Drug Delivery* 24 (1): 558–568. doi:10.1080/10717544.2017.1279237.

Kubacka, Anna, María Suárez Diez, David Rojo, Rafael Bargiela, Sergio Ciordia, Inés Zapico, Juan P. Albar, et al. 2014. "Understanding the Antimicrobial Mechanism of tio2-Based Nanocomposite Films in a Pathogenic Bacterium." *Scientific Reports* 4 (1): 1–9. doi:10.1038/srep04134.

Kupferschmidt, Natalia, Robert I. Csikasz, Lluís Ballell, Tore Bengtsson, and Alfonso E Garcia-Bennett. 2014. "Large Pore Mesoporous Silica Induced Weight Loss in Obese Mice." *Nanomedicine* 9 (9): 1353–1362. doi:10.2217/nnm.13.138.

Kwon, Yu-Jin, Hyangkyu Lee, Chung Mo Nam, Hyuk-Jae Chang, Young-Ran Yoon, Hye Sun Lee, and Ji-Won Lee. 2021. "Effects of Orlistat/Phentermine versus Phentermine on Vascular Endothelial Cell Function in Obese and Overweight Adults: A Randomized, Double-Blinded, Placebo-Controlled Trial." *Diabetes, Metabolic Syndrome and Obesity: Targets and Therapy* 14: 941–950. doi:10.2147/dmso.s300342.

Lee, Jung Ho, Hyeon Seon Jeong, Dong Hyun Lee, Songeun Beack, Taeyeon Kim, Geon-Hui Lee, Won Chan Park, Chulhong Kim, Ki Su Kim, and Sei Kwang Hahn. 2017. "Targeted Hyaluronate: Hollow Gold Nanosphere Conjugate for Anti-Obesity Photothermal Lipolysis." *ACS Biomaterials Science & Engineering* 3 (12): 3646–3653. doi:10.1021/acsbiomaterials.7b00549.

Li, Juanjuan, Ruitao Cha, Huize Luo, Wenshuai Hao, Yan Zhang, and Xingyu Jiang. 2019. "Nanomaterials for the Theranostics of Obesity." *Biomaterials* 223: 119474. doi:10.1016/j.biomaterials.2019.119474.

Li, Wansen, Hong Wan, Shuxun Yan, Zhao Yan, Yalin Chen, Panpan Guo, Thiyagarajan Ramesh, Ying Cui, and Lei Ning. 2020. "Gold Nanoparticles Synthesized with Poria Cocos Modulates the Anti-Obesity Parameters in High-Fat Diet and Streptozotocin Induced Obese Diabetes Rat Model." *Arabian Journal of Chemistry* 13 (7): 5966–5977. doi:10.1016/j.arabjc.2020.04.031.

Li, Xiaohua, Lu-Yin Lin, Kai-Yi Wang, Jian Li, Li Feng, Lijun Song, Xinke Liu, Jr-Hau He, Rajalakshmi Sakthivel, and Ren-Jei Chung. 2021. "Streptavidin-Functionalized-Polyethyleneimine/Chitosan/HfO$_2$-Pr$_6$O$_{11}$ Nanocomposite Using Label-Free Electrochemical Immunosensor for Detecting The Hunger Hormone Ghrelin." *Composites Part B: Engineering* 224: 109231. doi:10.1016/j.compositesb.2021.109231.

Loh, Kim, Atsushi Fukushima, Xinmei Zhang, Sandra Galic, Dana Briggs, Pablo J. Enriori, Stephanie Simonds, et al. 2011. "Elevated Hypothalamic TCPTP in Obesity Contributes to Cellular Leptin Resistance." *Cell Metabolism* 14 (5): 684–699. doi:10.1016/j.cmet.2011.09.011.

Lucchetta, Rosa Camila, Bruno Salgado Riveros, Roberto Pontarolo, Rosana Bento Radominski, Michel Fleith Otuki, Fernando Fernandez-Llimos, and Cassyano Januário Correr. 2017. "Systematic Review and Meta-Analysis of the Efficacy and Safety of Amfepramone and Mazindol as a Monotherapy for the Treatment of Obese or Overweight Patients." *Clinics* 72 (5): 317–324. doi:10.6061/clinics/2017(05)10.

Luo, Xi-Liang, Jing-Juan Xu, Jin-Li Wang, and Hong-Yuan Chen. 2005. "Electrochemically Deposited Nanocomposite of Chitosan and Carbon Nanotubes for Biosensor Application." *Chemical Communications* 16: 2169. doi:10.1039/b419197h.

Ma, Liang, Tzu-Wen Liu, Matthew A. Wallig, Iwona T. Dobrucki, Lawrence W. Dobrucki, Erik R. Nelson, Kelly S. Swanson, and Andrew M. Smith. 2016. "Efficient Targeting of Adipose Tissue Macrophages in Obesity with Polysaccharide Nanocarriers." *ACS Nano* 10 (7): 6952–6962. doi:10.1021/acsnano.6b02878.

Ma, Xinxing, Huiquan Tao, Kai Yang, Liangzhu Feng, Liang Cheng, Xiaoze Shi, Yonggang Li, Liang Guo, and Zhuang Liu. 2012. "A Functionalized Graphene Oxide-Iron Oxide Nanocomposite for Magnetically Targeted Drug Delivery, Photothermal Therapy, and Magnetic Resonance Imaging." *Nano Research* 5 (3): 199–212. doi:10.1007/s12274–012–0200-y.

Mack, C. M., C. J. Soares, J. K. Wilson, J. R. Athanacio, V. F. Turek, J. L. Trevaskis, J. D. Roth, et al. 2009. "Davalintide (AC2307), a Novel Amylin-Mimetic Peptide: Enhanced Pharmacological Properties over Native Amylin to Reduce Food Intake and Body Weight." *International Journal of Obesity* 34 (2): 385–395. doi:10.1038/ijo.2009.238.

Maffei, M., J. Halaas, E. Ravussin, R. E. Pratley, G. H. Lee, Y. Zhang, H. Fei, et al. 1995. "Leptin Levels in Human and Rodent: Measurement of Plasma Leptin and OB RNA in Obese and Weight-Reduced Subjects." *Nature Medicine* 1(11): 1155–1161. doi:10.1038/nm1195–1155.

Malaisse, Willy. 2011. "Dietary Sardine Protein Lowers Insulin Resistance, Leptin and TNF-α and Beneficially Affects Adipose Tissue Oxidative Stress in Rats with Fructose-Induced Metabolic Syndrome." *International Journal of Molecular Medicine.* doi:10.3892/ijmm.2011.836.

Martinelli, Ilenia, Daniele Tomassoni, Michele Moruzzi, Proshanta Roy, Carlo Cifani, Francesco Amenta, and Seyed Khosrow Tayebati. 2020. "Cardiovascular Changes Related to Metabolic Syndrome: Evidence in Obese Zucker Rats." *International Journal of Molecular Sciences* 21 (6): 2035. doi:10.3390/ijms21062035.

Mathieu, P., I. Lemieux, and J. P. Després. 2010. "Obesity, Inflammation, and Cardiovascular Risk." *Clinical Pharmacology & Therapeutics* 87 (4): 407–416. doi:10.1038/clpt.2009.311.

Mathus-Vliegen, E. M., K. Voorde, A. M. Kok, and A. M. Res. 1992. "Dexfenfluramine in the Treatment of Severe Obesity: A Placebo-Controlled Investigation of the Effects on Weight Loss, Cardiovascular Risk Factors, Food Intake and Eating Behaviour." *Journal of Internal Medicine* 232 (2): 119–127. doi:10.1111/j.1365–2796.1992.tb00560.x.

Melvin, A., S. O'Rahilly, and D. B. Savage. 2018 "Genetic Syndromes of Severe Insulin Resistance." *Current Opinion in Genetics & Development* 50: 60–67. doi:10.1016/j.gde.2018.02.002.

Milanski, Marciane, Giovanna Degasperi, Andressa Coope, Joseane Morari, Raphael Denis, Dennys E. Cintra, Daniela M. Tsukumo, et al. 2009. "Saturated Fatty Acids Produce an Inflammatory Response Predominantly through the Activation of TLR4 Signaling in Hypothalamus: Implications for the Pathogenesis of Obesity." *The Journal of Neuroscience* 29 (2): 359–370. doi:10.1523/jneurosci.2760–08.2009.

Mittal, Poonam C., and Ruchi Kant. 2009. "Correlation of Increased Oxidative Stress to Body Weight in Disease-Free Post Menopausal Women." *Clinical Biochemistry* 42 (10–11): 1007–1011. doi:10.1016/j.clinbiochem.2009.03.019.

Mohseni, Roohollah, Zahra Arab Sadeghabadi, Mohammad Taghi Goodarzi, Maryam Teimouri, Mitra Nourbakhsh, and Maryam Razzaghy Azar. 2018. "Evaluation of MN-Superoxide Dismutase and Catalase Gene Expression in Childhood Obesity: Its Association with Insulin Resistance." *Journal of Pediatric Endocrinology and Metabolism* 31 (7): 727–732. doi:10.1515/jpem-2017–0322.

Mortensen, Andreas, and Javier Llorca. 2010. "Metal Matrix Composites." *Annual Review of Materials Research* 40 (1): 243–270. doi:10.1146/annurev-matsci-070909–104511.

Müller, T. D., C. Clemmensen, B. Finan, R. D. DiMarchi, and M. H. Tschöp. 2018. "Anti-Obesity Therapy: From Rainbow Pills to Polyagonists." *Pharmacological Reviews* 70 (4): 712–746. doi:10.1124/pr.117.014803.

Myers, Martin G., Steven B. Heymsfield, Carol Haft, Barbara B. Kahn, Maren Laughlin, Rudolph L. Lei-bel, Matthias H. Tschöp, and Jack A. Yanovski. 2012. "Challenges and Opportunities of Defining Clinical Leptin Resistance." *Cell Metabolism* 15 (2): 150–156. doi:10.1016/j.cmet.2012.01.002.

Myers, Martin G., Rudolph L. Leibel, Randy J. Seeley, and Michael W. Schwartz. "Obesity and Leptin Resistance: Distinguishing Cause from Effect." *Trends in Endocrinology & Metabolism* 21 11 (2010): 643–651. https://doi.org/10.1016/j.tem.2010.08.002.

Nallamuthu, Ilaiyaraja, Aishwarya Devi, and Farhath Khanum. 2015. "Chlorogenic Acid Loaded Chitosan Nanoparticles with Sustained Release Property, Retained Antioxidant Activity and Enhanced Bioavailability." *Asian Journal of Pharmaceutical Sciences* 10 (3): 203–211. doi:10.1016/j.ajps.2014.09.005.

Odabasi, E., M. Ozata, M. Turan, N. Bingol, A. Yonem, B. Cakir, M. Kutlu, and I. C. Ozdemir. 2000. "Plasma Leptin Concentrations in Postmenopausal Women with Osteoporosis." *European Journal of Endocrinology* 142 (2): 170–173. doi:10.1530/eje.0.1420170.

Ogden, Cynthia L., Margaret D. Carroll, Brian K. Kit, and Katherine M. Flegal 2012. "Prevalence of Obesity and Trends in Body Mass Index among Us Children and Adolescents, 1999–2010." *JAMA* 307(5): 483–490. doi:10.1001/jama.2012.40.

Onakpoya, Igho J., Carl J. Heneghan, and Jeffrey K. Aronson. 2016. "Post-Marketing Withdrawal of Anti-Obesity Medicinal Products Because of Adverse Drug Reactions: A Systematic Review." *BMC Medicine* 14 (1). doi:10.1186/s12916-016-0735-y.

O'Rahilly, Stephen. 2009. "Human Genetics Illuminates the Paths to Metabolic Disease." *Nature* 462 (7271): 307–314. doi:10.1038/nature08532.

Ouellet, Véronique, Sébastien M. Labbé, Denis P. Blondin, Serge Phoenix, Brigitte Guérin, François Haman, Eric E. Turcotte, Denis Richard, and André C. Carpentier. 2012. "Brown Adipose Tissue Oxidative Metabolism Contributes to Energy Expenditure during Acute Cold Exposure in Humans." *Journal of Clinical Investigation* 122 (2): 545–552. doi:10.1172/jci60433.

Oussaada, Sabrina M., Katy A. van Galen, Mellody I. Cooiman, Lotte Kleinendorst, Eric J. Haze-broek, Mieke M. van Haelst, Kasper W. ter Horst, and Mireille J. Serlie. 2019. "The Pathogenesis of Obesity." *Metabolism* 92: 26–36. doi:10.1016/j.metabol.2018.12.012.

Ozcan, Lale, Ayse Seda Ergin, Allen Lu, Jason Chung, Sumit Sarkar, Duyu Nie, Martin G. Myers, and Umut Ozcan. 2009. "Endoplasmic Reticulum Stress Plays a Central Role in Development of Leptin Resistance." *Cell Metabolism* 9 (1): 35–51. doi:10.1016/j.cmet.2008.12.004.

Patel, Sanjeev B., Garry P. Reams, Robert M. Spear, Ronald H. Freeman, and Daniel Villarreal. 2008. "Leptin: Linking Obesity, the Metabolic Syndrome, and Cardiovascular Disease." *Current Hypertension Reports* 10 (2): 131–137. doi:10.1007/s11906-008-0025-y.

Pi-Sunyer, F. Xavier, Louis J. Aronne, Hassan M. Heshmati, Jeanne Devin, Julio Rosenstock, and for the RIO-North America Study Group. 2006. "Effect of Rimonabant, a Cannabinoid-1 Receptor Blocker, on Weight and Cardiometabolic Risk Factors in Overweight or Obese Patients." *JAMA* 295 (7): 761. doi:10.1001/jama.295.7.761.

Procaccini, Claudio, Emilio Jirillo, and Giuseppe Matarese. 2012. "Leptin as an Immunomodulator." *Molecular Aspects of Medicine* 33 (1): 35–45. doi:10.1016/j.mam.2011.10.012.

Proietto, J., A. Rissanen, J. B. Harp, N. Erondu, Q. Yu, S. Suryawanshi, M. E. Jones, et al. 2010. "A Clinical Trial Assessing the Safety and Efficacy of the CB1R Inverse Agonist Taranabant in Obese and Overweight Patients: Low-Dose Study." *International Journal of Obesity* 34 (8): 1243–1254. doi:10.1038/ijo.2010.38.

Quan, Wenying, Hyun-Kyong Kim, Eun-Yi Moon, Su Sung Kim, Cheol Soo Choi, Masaaki Komatsu, Yeon Taek Jeong, et al. 2012. "Role of Hypothalamic Proopiomelanocortin Neuron Autophagy in the Control of Appetite and Leptin Response." *Endocrinology* 153 (4): 1817–1826. doi:10.1210/en.2011–1882.

Rajabzadeh-Khosroshahi, Maryam, Mehrab Pourmadadi, Fatemeh Yazdian, Hamid Rashedi, Mona Navaei-Nigjeh, and Bita Rasekh. 2020. "Chitosan/Agarose/Graphitic Carbon Nitride Nanocomposite as an Efficient Ph-Sensitive Drug Delivery System for Anticancer Curcumin Releasing." *Journal of Drug Delivery Science and Technology* 74 (2022): 103443. doi:10.1016/j.jddst.2022.103443.

Ramírez, Sara, and Marc Claret. 2015. "Hypothalamic Er Stress: A Bridge between Leptin Resistance and Obesity." *FEBS Letters* 589 (14): 1678–1687. doi:10.1016/j.febslet.2015.04.025.

Rege, Kaushal, Nachiket R. Raravikar, Dae-Yun Kim, Linda S. Schadler, Pulickel M. Ajayan, and Jonathan S. Dordick. 2003. "Enzyme-Polymer-Single Walled Carbon Nanotube Composites as Biocatalytic Films." *Nano Letters* 3 (6): 829–832. doi:10.1021/nl034131k.

Richards, Mark P., Christopher M. Ashwell, and John P. McMurtry. 1999. "Analysis of Leptin Gene Expression in Chickens Using Reverse Transcription Polymerase Chain Reaction and Capillary Electrophoresis with Laser-Induced Fluorescence Detection." *Journal of Chromatography A* 853 (1–2): 321–335. doi:10.1016/s0021-9673(99)00576-2.

Ritten, Angela, and Jacqueline LaManna. 2017. "Unmet Needs in Obesity Management." *Journal of the American Association of Nurse Practitioners* 29 (S1). doi:10.1002/2327-6924.12507.

Rocca, Antonella, Stefania Moscato, Francesca Ronca, Simone Nitti, Virgilio Mattoli, Mario Giorgi, and Gianni Ciofani. 2015. "Pilot in Vivo Investigation of Cerium Oxide Nanoparticles as a Novel Anti-Obesity Pharmaceutical Formulation." *Nanomedicine: Nanotechnology, Biology and Medicine* 11 (7): 1725–1734. doi:10.1016/j.nano.2015.05.001.

Rzheshevsky, A. V. 2013. "Fatal 'Triad': Lipotoxicity, Oxidative Stress, and Phenoptosis." *Biochemistry (Moscow)* 78 (9): 991–1000. doi:10.1134/s0006297913090046.

Sanchez, Clément, Beatriz Julián, Philippe Belleville, and Michael Popall. 2005. "Applications of Hybrid Organic-Inorganic Nanocomposites." *Journal of Materials Chemistry* 15 (35–36): 3559. doi:10.1039/b509097k.

Sanchez, Clément, G. J. Soler-Illia, F. Ribot, T. Lalot, C. R. Mayer, and V. Cabuil. 2001. "Designed Hybrid Organic-Inorganic Nanocomposites from Functional Nanobuilding Blocks." *Chemistry of Materials* 13 (10): 3061–3083. doi:10.1021/cm011061e.

Sangwai, Mayur, Surendra Sardar, and Pradeep Vavia. 2012. "Nanoemulsified Orlistat-Embedded Multi-Unit Pellet System (MUPS) with Improved Dissolution and Pancreatic Lipase Inhibition." *Pharmaceutical Development and Technology* 19 (1): 31–41. doi:10.3109/10837450.2012.75 1404.

Saratale, Ganesh Dattatraya, Rijuta Ganesh Saratale, Giovanni Benelli, Gopalakrishnan Kumar, Arivalagan Pugazhendhi, Dong-Su Kim, and Han-Seung Shin. 2017. "Anti-Diabetic Potential of Silver Nanoparticles Synthesized with Argyreia Nervosa Leaf Extract High Synergistic Antibacterial Activity with Standard Antibiotics against Foodborne Bacteria." *Journal of Cluster Science* 28 (3): 1709–1727. doi:10.1007/s10876-017-1179-z.

Savini, Isabella, Maria Catani, Daniela Evangelista, Valeria Gasperi, and Luciana Avigliano. 2013. "Obesity-Associated Oxidative Stress: Strategies Finalized to Improve Redox State." *International Journal of Molecular Sciences* 14 (5): 10497–10538. doi:10.3390/ijms140510497.

Schwartz, Michael W., Elaine Peskind, Murray Raskind, Edward J. Boyko, and Daniel Porte. 1996. "Cerebrospinal Fluid Leptin Levels: Relationship to Plasma Levels and to Adiposity in Humans." *Nature Medicine* 2 (5): 589–593. doi:10.1038/nm0596-589.

Schwartz, Michael W., Randy J. Seeley, Lori M. Zeltser, Adam Drewnowski, Eric Ravussin, Leanne M. Redman, and Rudolph L. Leibel. 2017. "Obesity Pathogenesis: An Endocrine Society Scientific Statement." *Endocrine Reviews* 38 (4): 267–296. doi:10.1210/er.2017-00111.

Sharifi, S., S. Daghighi, M. M. Motazacker, B. Badlou, B. Sanjabi, A. Akbarkhanzadeh, A. T. Rowshani, S. Laurent, M. P. Peppelenbosch, and F. Rezaee. 2013. "Superparamagnetic Iron Oxide Nanoparticles Alter Expression of Obesity and T2D-Associated Risk Genes in Human Adipocytes." *Scientific Reports* 3 (1). doi:10.1038/srep02173.

Sharretts, John, Ovidiu Galescu, Shanti Gomatam, Eugenio Andraca-Carrera, Christian Hampp, and Lisa Yanoff. 2020. "Cancer Risk Associated with Lorcaserin: The FDA's Review of the Camellia-Timi 61 Trial." *New England Journal of Medicine* 383 (11): 1000–1002. doi:10.1056/nejmp2003873.

Sheng, Wangzhong, Ali H. Alhasan, Gabriella DiBernardo, Khalid M. Almutairi, J. Peter Rubin, Barry E. DiBernardo, and Adah Almutairi. 2014. "Gold Nanoparticle-Assisted Selective Photothermolysis of Adipose Tissue (NanoLipo)." *Plastic and Reconstructive Surgery Global Open* 2 (12). doi:10.1097/gox.0000000000000251.

Shukla, Shubhangi, Pratik Joshi, Parand Riley, and Roger J. Narayan. 2022. "Square Wave Voltammetric Approach to Leptin Immunosensing and Optimization of Driving Parameters with Chemometrics." *Biosensors and Bioelectronics* 216: 114592. doi:10.1016/j.bios.2022.114592.

Smith, Steven R., Neil J. Weissman, Christen M. Anderson, Matilde Sanchez, Emil Chuang, Scott Stubbe, Harold Bays, and William R. Shanahan. 2010. "Multicenter, Placebo-Controlled Trial of Lorcaserin for Weight Management." *New England Journal of Medicine* 363 (3): 245–256. doi:10.1056/nejmoa0909809.

Snijder, Marieke B., Jacqueline M. Dekker, Marjolein Visser, Lex M. Bouter, Coen D. A. Stehouwer, Piet J. Kostense, John S. Yudkin, Robert J. Heine, Giel Nijpels, and Jacob C. Seidell. 2003. "Associations of Hip and Thigh Circumferences Independent of Waist Circumference with the Incidence of Type 2 Diabetes: The Hoorn Study,." *The American Journal of Clinical Nutrition* 77 (5): 1192–1197. doi:10.1093/ajcn/77.5.1192.

Srivastava, Gitanjali, and Caroline M. Apovian. 2017. "Current Pharmacotherapy for Obesity." *Nature Reviews Endocrinology* 14 (1): 12–24. doi:10.1038/nrendo.2017.122.

Takaki, Akinobu, Daisuke Kawai, and Kazuhide Yamamoto. 2013. "Multiple Hits, Including Oxidative Stress, as Pathogenesis and Treatment Target in Non-Alcoholic Steatohepatitis (NASH)." *International Journal of Molecular Sciences* 14 (10): 20704–20728. doi:10.3390/ijms141020704.

Thovhogi, Ntevheleni, Nicole Sibuyi, Mervin Meyer, Martin Onani, and Abram Madiehe. 2015. "Targeted Delivery Using Peptide-Functionalised Gold Nanoparticles to White Adipose Tissues of Obese Rats." *Journal of Nanoparticle Research* 17 (2): 114–120. doi:10.1007/s11051-015-2904-x.

Tillner, Joachim, Maximilian G. Posch, Frank Wagner, Lenore Teichert, Youssef Hijazi, Christine Einig, Stefanie Keil, et al. 2018. "A Novel Dual Glucagon-like Peptide and Glucagon Receptor Agonist SAR425899: Results of Randomized, Placebo-Controlled First-in-Human and First-in-Patient Trials." *Diabetes, Obesity and Metabolism* 21 (1): 120–128. doi:10.1111/dom.13494.

Tjong, Sie Chin. 2013. "Recent Progress in the Development and Properties of Novel Metal Matrix Nanocomposites Reinforced with Carbon Nanotubes and Graphene Nanosheets." *Materials Science and Engineering: R: Reports* 74 (10): 281–350. doi:10.1016/j.mser.2013.08.001.

Torp-Pedersen, Christian, Ian Caterson, Walmir Coutinho, Nick Finer, Luc Van Gaal, Aldo Maggioni, Arya Sharma, et al. 2007. "Cardiovascular Responses to Weight Management and Sibutramine in High-Risk Subjects: An Analysis from the Scout Trial." *European Heart Journal* 28 (23): 2915–2923. doi:10.1093/eurheartj/ehm217.

Uludağ, İnci, and Mustafa Kemal Sezgintürk. 2022. "A Direct and Simple Immobilization Route for Immunosensors by CNBR Activation for Covalent Attachment of Anti-Leptin: Obesity Diagnosis Point of View." *3 Biotech* 12 (1): 33. doi:10.1007/s13205-021-03096-w.

Valdearcos, Martin, Megan M. Robblee, Daniel I. Benjamin, Daniel K. Nomura, Allison W. Xu, and Suneil K. Koliwad. 2014. "Microglia Dictate the Impact of Saturated Fat Consumption on Hypothalamic Inflammation and Neuronal Function." *Cell Reports* 9 (6): 2124–2138. doi:10.1016/j.celrep.2014.11.018.

Van Gaal, Luc F., A. Rissanen, A. J. Scheen, O. Ziegler, and S. Rössner. 2005. "Effect of Rimonabant on Weight Reduction and Cardiovascular Risk." *The Lancet* 366 (9483): 369–370. doi:10.1016/s0140-6736(05)67020-1.

Via, Michael. 2012. "The Malnutrition of Obesity: Micronutrient Deficiencies That Promote Diabetes." *ISRN Endocrinology* 2012: 1–8. doi:10.5402/2012/103472.

Wang, Chensu, Jingyuan Li, Christian Amatore, Yu Chen, Hui Jiang, and Xue-Mei Wang. 2011. "Gold Nanoclusters and Graphene Nanocomposites for Drug Delivery and Imaging of Cancer Cells." *Angewandte Chemie* 123 (49): 11848–11852. doi:10.1002/ange.201105573.

Wang, Chih-Hao, Ching-Chu Wang, Hsin-Chang Huang, and Yau-Huei Wei. 2013. "Mitochondrial Dysfunction Leads to Impairment of Insulin Sensitivity and Adiponectin Secretion in Adipocytes." *The FEBS Journal* 280 (4): 1039–1050. doi:10.1111/febs.12096.

Wang, Yan, and Joseph S. Heilig. 2012. "Differentiation and Quantification of Endogenous and Recombinant-Methionyl Human Leptin in Clinical Plasma Samples by Immunocapture/Mass Spectrometry." *Journal of Pharmaceutical and Biomedical Analysis* 70: 440–446. doi:10.1016/j.jpba.2012.06.018.

Wauman, Joris. 2011. "Leptin Receptor Signaling: Pathways to Leptin Resistance." *Frontiers in Bioscience* 16(1): 2771–2793. doi:10.2741/3885.

White, Christy L., Amy Whittington, Maria J. Barnes, Zhong Wang, George A. Bray, and Christopher D. Morrison. 2009. "HF Diets Increase Hypothalamic PTP1B and Induce Leptin Resistance through Both Leptin-Dependent and -Independent Mechanisms." *American Journal of Physiology-Endocrinology and Metabolism* 296 (2): E291–E299. doi:10.1152/ajpendo.90513.2008.

Williams, Kevin Jon. 2008. "Molecular Processes That Handle-and Mishandle: Dietary Lipids." *Journal of Clinical Investigation* 118 (10): 3247–3259. doi:10.1172/jci35206.

Williams, Kevin W., Tiemin Liu, Xingxing Kong, Makoto Fukuda, Yingfeng Deng, Eric D. Berglund, Zhuo Deng, et al. 2014. "Xbp1s In POMC Neurons Connects ER Stress with Energy Balance and Glucose Homeostasis." *Cell Metabolism* 20 (3): 471–482. doi:10.1016/j.cmet.2014.06.002.

Wilsey, J., and P. J. Scarpace. 2004. "Caloric Restriction Reverses the Deficits in Leptin Receptor Protein and Leptin Signaling Capacity Associated with Diet-Induced Obesity: Role of Leptin in the Regulation of Hypothalamic Long-Form Leptin Receptor Expression." *Journal of Endocrinology* 181 (2): 297–306. doi:10.1677/joe.0.1810297.

Wong, Iris P. L., Amy D. Nguyen, Ee Cheng Khor, Ronaldo F. Enriquez, John A. Eisman, Amanda Sainsbury, Herbert Herzog, and Paul A. Baldock. 2013. "Neuropeptide Y is a Critical Modulator of Leptin's Regulation of Cortical Bone." *Journal of Bone and Mineral Research* 28 (4): 886–898. https://doi.org/10.1002/jbmr.1786.

Xian, Yuezhong, Yi Hu, Fang Liu, Yang Xian, Haiting Wang, and Litong Jin. 2006. "Glucose Biosensor Based on Au Nanoparticles: Conductive Polyaniline Nanocomposite." *Biosensors and Bioelectronics* 21 (10): 1996–2000. doi:10.1016/j.bios.2005.09.014.

Xiao, Lifu, Abdul K. Parchur, Timothy A. Gilbertson, and Anhong Zhou. 2018. "SERS-Fluorescence Bimodal Nanoprobes for in vitro imaging of the Fatty Acid Responsive Receptor GPR120." *Analytical Methods* 10 (1): 22–29. doi:10.1039/c7ay02039b.

Xue, Yuan, Xiaoyang Xu, Xue-Qing Zhang, Omid C. Farokhzad, and Robert Langer. 2016. "Preventing Diet-Induced Obesity in Mice by Adipose Tissue Transformation and Angiogenesis Using Targeted Nanoparticles." *Proceedings of the National Academy of Sciences* 113 (20): 5552–5557. doi:10.1073/pnas.1603840113.

Yamada-Goto, Nobuko, Goro Katsuura, Yukari Ochi, and Kazuwa Nakao. 2012. "An approach toward CNS dysfunction associated with metabolic syndrome; implication of leptin, which is a key molecule of obesity, in depression associated with obesity." *Japanese Journal of Psychopharmacology* 32 (5): 245–250.

Yi, Myoung Hi, Shakina Yesmin Simu, Sungeun Ahn, Verónica Castro Aceituno, Chao Wang, Ramya Mathiyalagan, Joon Hurh, et al. 2020. "Anti-Obesity Effect of Gold Nanoparticles from *Dendropanax Morbifera Léveille* by Suppression of Triglyceride Synthesis and Downregulation of PPARΓ and Cebpα Signaling Pathways in 3t3-L1 Mature Adipocytes and hepg2 Cells." *Current Nanoscience* 16 (2): 196–203. doi:10.2174/1573413716666200116124822.

Zhang, Hong-liang, Tao Zhong, and Wu Su. 2012. "Effects of Chitosan and Water-Soluble Chitosan Micro- and Nanoparticles in Obese Rats Fed a High-Fat Diet." *International Journal of Nanomedicine* 7 (2012): 4069–4076. doi:10.2147/ijn.s33830.

Zhang, Hong-liang, Xiao-bin Zhong, Yi Tao, Si-hui Wu, and Zheng-quan Su. 2012. "Effects of Chitosan and Water-Soluble Chitosan Micro- and Nanoparticles in Obese Rats Fed a High-Fat Diet." *International Journal of Nanomedicine* 7 (2012): 4069–4076. doi:10.2147/ijn.s33830.

Zhang, Jinying, Zihui Deng, Jie Liao, Cuihong Song, Chen Liang, Hui Xue, Luhuan Wang, Kai Zhang, and Guangtao Yan. 2013. "Leptin Attenuates Cerebral Ischemia Injury through the Promotion of Energy Metabolism via the Pi3K/Akt Pathway." *Journal of Cerebral Blood Flow & Metabolism* 33 (4): 567–574. doi:10.1038/jcbfm.2012.202.

Zhang, Xiaoqing, Guo Zhang, Hai Zhang, Michael Karin, Hua Bai, and Dongsheng Cai. 2008. "Hypothalamic Ikkβ/NF-KB and ER Stress Link Overnutrition to Energy Imbalance and Obesity." *Cell* 135 (1): 61–73. doi:10.1016/j.cell.2008.07.043.

Zhang, Yiying, Ricardo Proenca, Margherita Maffei, Marisa Barone, Lori Leopold, and Jeffrey M. Friedman. 1994. "Positional Cloning of the Mouse Obese Gene and Its Human Homologue." *Nature* 372 (6505): 425–432. doi:10.1038/372425a0.

Zhang, Z., and D.L. Chen. 2008. "Contribution of Orowan Strengthening Effect in Particulate-Reinforced Metal Matrix Nanocomposites." *Materials Science and Engineering: A* 483–484: 148–152. doi:10.1016/j.msea.2006.10.184.

Zu, Yujiao, Ling Zhao, Lei Hao, Yehia Mechref, Masoud Zabet-Moghaddam, Peter A. Keyel, Mehrnaz Abbasi, et al. 2021. "Browning White Adipose Tissue Using Adipose Stromal Cell-Targeted Resveratrol-Loaded Nanoparticles for Combating Obesity." *Journal of Controlled Release* 333: 339–351. doi:10.1016/j.jconrel.2021.03.022.

Chapter 5

Nanotherapeutics for insulin resistance and diabetes mellitus using metal nanocomposites

Jigar Vyas and Isha Shah

LIST OF ABBREVIATIONS

4-MPBA	4-mercapto phenylboronic acid
Ag	Silver
Ag NPs	Silver nanoparticles
Au	Gold
Au NFs	Gold nanoflowers
Au NPs	gold nanoparticles
ALT	Alanine aminotransferase
AST	Aspartate aminotransferase
CEB	Chlorophyll a/b binding
CeO_2	Cerium oxide
CeO_2 NPs	Cerium dioxide nanoparticles
CER	Ceramides
CNFs	Carbon nanofibers
Co_3O_4	Cobalt oxide
CTS-Se-NPs	Chitosan-stabilized nanoparticles
DAG	Diacylglycerols
DIDM	Diseases-induced diabetes mellitus
DIO	Diosmin
DM	Diabetes mellitus
DOM	Dimensionally ordered macroporous
ETC	Electron transport chain
FBS	Fasting blood sugar test
GDM	Gestational diabetes mellitus
GLUTs	Glucose transporters
GOx	Glucose oxidase
GS	graphite sheet electrode
GSH	Glutathione
GSK-3	Glycogen synthase kinase 3
GSSG	oxidized glutathione
GSH	Glutathione
GPx	Glutathione peroxidase

DOI: 10.1201/9781032621135-5

HbA$_1$c	Glycated hemoglobin
HIF-1α	Hypoxia-inducible factors alpha
HNC	Hollow nanocage
HRP	Horseradish peroxidase
HFD	High-fat diet
In$_2$O$_3$	Indium (III) oxide
IRS-1	Insulin receptor substrate 1
IRS-2	Insulin receptor substrate 1
IST	Insulin signal transduction
JNK	Jun amino-terminal kinases
LU	Luteolin
LDL-c	Low-density lipoprotein-cholesterol
LPO	Lipid peroxidation
MAPK	Mitogen-activated protein kinase
MOF	Metal-organic framework
MRC	Mitochondrial respiratory chains
MET	Metformin
NADPH	Nicotinamide adenine dinucleotide phosphate
NF-B/Nrf2	Nuclear factor erythroid-2-related factor 2
NF-κB	Nuclear factor kappa B
NOX	Nicotinamide adenine dinucleotide phosphate oxidase
NPs	Nanoparticles
NO	Nitric oxide
NEFAs	Non-esterified fatty acids
OGTT	Oral glucose tolerance test
PdO	Palladium oxide
PI3K	Phosphoinositide 3-kinase
PKC	Protein kinase C
PPAR-γ	Peroxisome proliferator-activated receptor gamma
Pt	Platinum
PVP	Polyvinyl pyrrolidone
QDs	Quantum dots
RAAS	Renin–angiotensin–aldosterone system
RBS	Random blood sugar test
rGO	Reduced graphene oxide
ROS	Reactive oxygen species
SAPK	Stress-activated protein kinases
Se NPs	Selenium nanoparticles
SGLT2	Sodium-glucose cotransporters inhibitors 2
SnO$_2$	Tin oxide
STZ	Streptozotocin
SFP	Sapodilla fruit peel
T1DM	Type 1 diabetes mellitus
T2DM	Type 2 diabetes mellitus
TiO$_2$	Titanium dioxide

TG   Triacylglycerols
VLDL   Very low-density lipoprotein cholesterol
WO₃   Tungsten oxide
ZnO   Zinc oxide
ZnO NPs   Zinc oxide nanoparticles

5.1 DIABETES MELLITUS (DM) AND INSULIN RESISTANCE

Diabetes mellitus (DM), usually recognized as diabetes, is a complicated physiologic ailment characterized by hyperglycemia, a functionally impaired state caused by consistently high blood sugar levels. Insulin signal transduction (IST) can be impaired by oxidative stress, increasing the risks of insulin resistance and diabetes (Freeman and Pennings 2019).

The pathophysiology of diabetic complications is influenced by this severe condition, which has a detrimental impact on a majority of metabolic processes. Oxidative stress is among the main causes underlying the emergence of diabetic symptoms. As the amount of free radical formation surpasses the antioxidant defense mechanisms, oxidative stress occurs, leading to harmful consequences of free radicals. In metabolic balance, free radical entities play a crucial role in the pathophysiology (Habtamu Wondifraw 2015; Yaribeygi et al. 2018a). However, oxidative stress develops if free radicals generation exceeds the antioxidant capacity of the body. The carriers of oxidative stress include reactive species, particularly reactive oxygen species (ROS) such as hydrogen peroxide, superoxide, and hydroxyl radical ions (Schieber and Chandel 2014), which are released at minimal physiological levels primarily in the peroxisomes and mitochondria. Both the onset of insulin resistance and challenges from diabetes are significantly influenced by oxidative stress (Yaribeygi, Mohammadi, and Sahebkar 2018b; Hurrle and Hsu 2017), which causes pathophysiologic molecular processes and sets off a chain reaction of harmful mechanisms that result in insulin resistance.

Diabetes is categorized into four types: T1DM, T2DM, GDM, and DIDM (diabetes-induced or related to certain diseases, pathologies, or disorders). T1DM affects approximately 5–10% of all diabetic patients and is caused by β-cell failure decreased insulin production and downregulation of circulatory insulin. T2DM is the most common form of diabetes, comprising about 90–95% of diabetics and is characterized by inadequate insulin release (reduced insulin sensitivity) and peripheral insulin resistance (Sreenivasamurthy 2021). The term gestational diabetes mellitus (GDM) refers to any extent of glucose intolerance or diabetes that is identified early on or during pregnancy, often in the second or third trimester. Apart from T1DM, T2DM, and GDM, diabetes is correlated to a variety of particular diseases, comprising various pathologies or multiple problems, however mostly in smaller numbers than the global diabetes occurrence paradigm (O'Neal, Johnson, and Panak 2016).

5.2 MECHANISMS OF INSULIN RESISTANCE AND DIABETES CAUSED BY OXIDATIVE STRESS

Insulin signal transduction (IST) can be impaired by oxidative stress, increasing the risks of insulin resistance and diabetes. It is emphasized that oxidative stress and diabetes have complicated interrelations that aggravate each other. The following sections discuss potential pathways through which reactive oxygen species (ROS) affect normal glucose metabolism and contribute to DM development.

5.2.1 Insulin signaling pathways

Insulin resistance and DM may occur as a result of any deficiencies in the insulin signaling pathways (Mackenzie and Elliott 2014). The PI3K enzyme, IRs (IRS-1 and IRS-2), Akt signaling pathways, and oxidative stress contribute to the impairment of functional IST. Oxidative stress causes IRS-1 and IRS-2 serine phosphorylation, which disrupts the IST (Paz et al. 1997). Through stimulation of the JNK/SAPK signaling pathways, free radicals can promote serine phosphorylation of IRS-1 and decrease normal IST (Aguirre et al. 2000). Oxidative stress can potentially affect IST by downregulating proteins responsible for normal IST. IRS, IRS-1, Akt, and GSK-3 are essential IST mechanisms that govern ROS and are dysregulated under oxidative stress, leading to insulin resistance and diabetic complications (Balbaa, Abdulmalek, and Khalil 2017).

5.2.2 β-cell dysfunction/insulin production and secretion

Developing DM is significantly influenced by a steady decrease in β-cell number and functioning. Glucose-induced insulin production from β-cells became uncontrolled and diminished within those environments, causing postprandial glucose to rise beyond usual (White, Shaw, and Taylor 2016). Many pathogenic mechanisms, in addition to oxidative stress, cause β-cell failure (Drews, Krippeit-Drews, and Düfer 2010). The primary source of free radicals in pancreatic β-cells is the mitochondrial respiratory chains (MRC) and NADPH oxidase/NOX enzyme activity. In β-cells, the MRC and NOX enzymes primarily produce superoxide anion (O_2^-) as a species of free radicals. Due to the limited antioxidant defense system capability of β-cells, oxidative stress in β-cells is common in diabetes and significantly contributes to the impairment of activity in both T1DM and T2DM (Gerber and Rutter 2017).

5.2.3 GLUT-4 expression or localization

GLUTs, or glucose transporters, are transmembrane proteins that carry glucose through cellular membranes. In addition, via facilitating glucose entrance into the skeletal and adipose tissues, insulin stimulates glucose absorption. It is assumed to be undertaken by insulin-induced GLUT4 translocation across intracellular compartments toward the cell membrane, which enhances the total amount of glucose flux to a cell (Wang et al. 2020). Any factor that lowers the expression of GLUT-4 significantly affects insulin sensitivity because it results in less glucose reaching target cells and lowers insulin sensitivity in such tissues (Reno et al. 2016). The transcriptional regulators of GLUT-4 expression, including CEB/Ps (CCAAT enhancer-binding proteins), PPAR-γ (peroxisome proliferator-activated receptor gamma), p85, nuclear factor-1, MEF2 (myocyte enhancer factor 2), NF-κB, and HIF-1α (hypoxia-inducible factors alpha) can be suppressed by chronic oxidative stress (Pessler, Rudich, and Bashan 2001). Furthermore, oxidative stress activates several oxidative stress-induced factors and metabolites, including p38 MAPK, JNK/SAPK, PKC (protein kinase C), sorbitol, and hexosamine all of which can decrease GLUT-4 expression. As a result, lowering GLUT-4 expression/localization is one of the key underlying pathways whereby oxidative stress promotes insulin resistance and results in the onset of DM (Hurrle and Hsu 2017).

5.2.4 Systemic mitochondrial dysfunction

Mitochondrial dysfunction can be caused by a reduction in mitochondrial biogenesis, a reduction in mitochondrial content, and a reduction in the protein levels and efficiency of oxidative proteins 'per unit of mitochondria' (for example, a change in electron transport chain complexes) (ETC). The majority of instances of mitochondrial dysfunction are caused by oxidative stress (Rose et al. 2014), which impairs mitochondrial functioning by changing the normal action of the mitochondrial respiratory chain (MRC), decreasing mitochondrial respiratory capacity, elevating proton leakage in the MRC, changing the potential difference throughout the inner mitochondrial membrane, thereby weakening mitochondrial membrane integrity (Wada and Nakatsuka 2016). All of these alterations could probably result in a reduction in substrate oxidation. Fuel oxidation is inhibited, resulting in lipid deposition, along with the accumulation of metabolically active lipid mediators like diacylglycerols (DAG) and ceramides (CER). Both DAG and CER were shown to decrease insulin signaling: DAG inhibits insulin signaling by translocating to the plasma membrane and inhibiting the insulin receptor via protein kinase C stimulation (Samuel, Petersen, and Shulman 2010), and CER inhibits insulin signaling via protein kinase AKT downregulation (Bruce et al. 2012). Deposition of DAG and CER is thus a potential correlation between mitochondrial malfunction and insulin resistance.

5.2.5 Inflammatory processes

The inflammatory response is among the major basic molecular mechanisms implicated in insulin resistance pathophysiology, diabetes, and its consequences (Goldberg 2009). These negative processes may potentially trigger other DM pathophysiologic pathways, such as IST impairment and β-cell impairment (Eizirik, Colli, and Ortis 2009). Cytokines can activate Janus kinase pathways (JNKs), leading to IRS-1 serine phosphorylation and IST dysfunction (Richardson et al. 2009). Several pro-inflammatory indicators, including IL-1, IL-6, TNF-α, CRP, and numerous chemokines, are indirectly or directly connected to IR (Akash, Rehman, and Chen, 2013), and they are most frequently associated with excessively increased amounts of pro-inflammatory cytokines, hypertension, obesity, and glucolipotoxicity (Akash, Rehman, and Chen 2013).

5.3 DIAGNOSIS OF DIABETES MELLITUS

The general diagnosis of diabetes is based on glucose level in biofluids, level of glycosylated hemoglobin in the blood and acetone gas level in breath.

5.3.1 Test based on glucose level in biofluids

Many biofluids, including blood, urine, saliva, sweat, and tears, can be used to measure the body's glucose level. The blood glucose level is examined using three different tests: fasting blood sugar test (FBS), random blood sugar test (RBS), and oral glucose tolerance test (OGTT). A person is considered diabetic if the FBS range is 126 mg/dL or above, and the RBS and OGTT ranges are 200 mg/dL or above. Normal urine glucose levels range from 0 mmol/L to 0.8 mmol/L (millimoles per liter), saliva glucose levels (8 M–0.21 mM), sweat glucose level (5–20 mg/dL), and fasting glucose levels in tears (3.6 mg/100 mL–16.6 mg/100 mL).

5.3.2 Test based on glycosylated hemoglobin in the blood

Glycated hemoglobin (HbA1c) in blood analysis provides information on an individual's average blood glucose levels all through the last 2–3 months, which is the estimated RBCs (Sherwani et al. 2016). For diagnosing and managing diabetes, especially T2DM, the HbA1c is currently indicated as a standard of care (SOC) (Florkowski 2013). An HbA1c of <5.7% is considered normal, approximately 5.7% and 6.4% suggest prediabetes, and 6.5% or more signifies diabetes.

5.3.3 Test based on acetone gas level in breath

Acetone, a volatile organic substance exhaled, has been utilized as a biomarker for diabetes mellitus, particularly T1DM (Salehi et al. 2014). Human breath includes hundreds of volatile organic chemicals (VOCs) at proportions ranging from ppt to ppm. The level of acetone in the breath ranges from 300 ppb to 900 ppb in healthy people to more than 1800 ppb in diabetics (Righettoni et al. 2012). Acetone can thus serve as a blood-based biomarker for metabolic (diabetic) disorders. The acetone content in the exhaled breath is then measured by the breath analyzer, which may be used to determine blood glucose levels.

5.4 METAL-BASED NANOSENSORS IN THE DIAGNOSIS OF DIABETES MELLITUS

Metal oxides are used in diabetes diagnosis as they interact with glucose and give a measurable indication. The metal oxides that can be used include tungsten oxide (WO_3), cerium oxide (CeO_2), zinc oxide (ZnO), platinum (Pt), gold (Au), silver (Ag), and others. There are various advantages of using metal oxide-based sensors for the diagnosis of diabetes (Figure 5.1). These sensors are simple to use and can enable fast, adequate output. Nanotechnology is involved in the emergence of biosensors through the inclusion of nanoparticles into biosensors. To identify underlying conditions, biosensor-based nanomaterials are used (Gholami, Farjami, and Younes 2021; Yoon et al. 2019). These materials typically comprise metals, nanocomposites, carbon nanomaterials, metal oxides, and newly synthesized nanomaterials. Biosensors based on nanotechnology advancements were developed to combat the limitations of conventional diagnostic tests such as false positives, false negatives, inconvenience, and inaccuracy. Metal

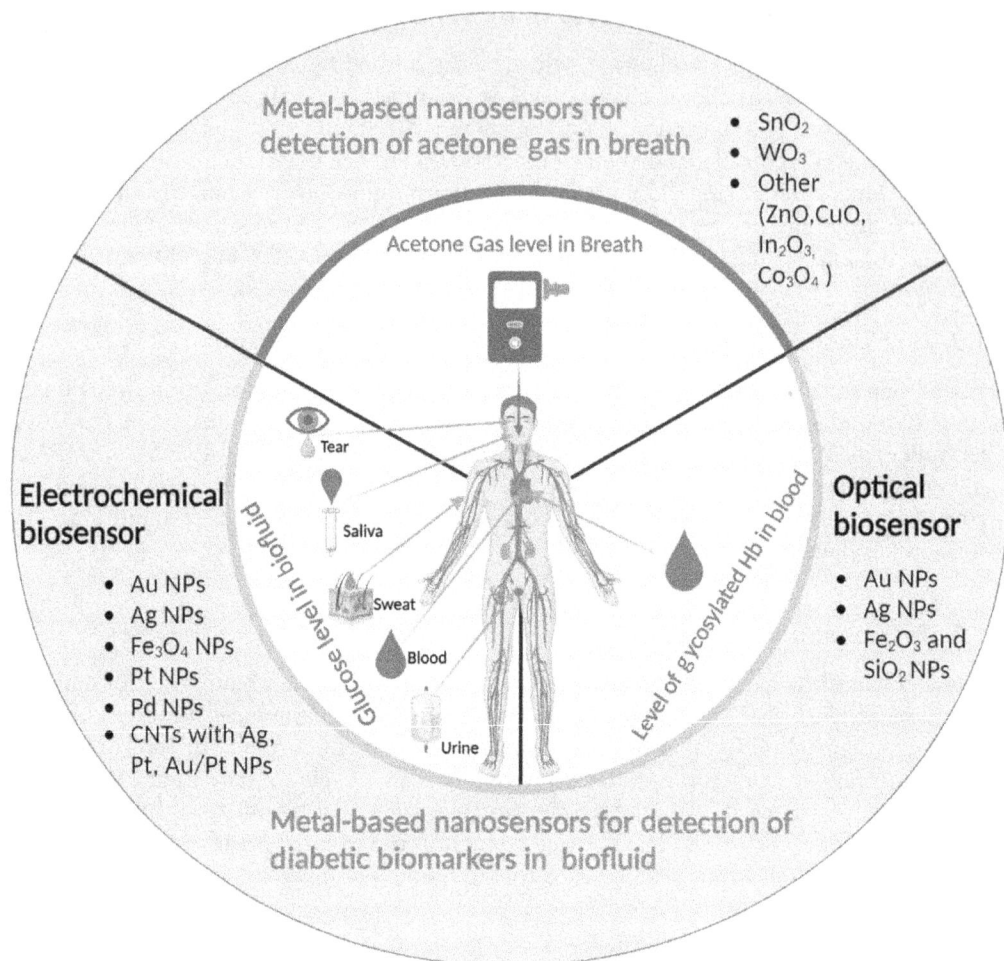

Figure 5.1 Diagnosis of diabetes mellitus based on glucose level in biofluids, glycosylated hemoglobin (Hb) in blood, and acetone gas level in breath with metal-based nanosensors.

nanosensors, in general, have the potential to increase the precision and rapidity of diabetes diagnosis; thus, research is done to diagnose diabetes biomarkers in human biofluids and acetone gas in the human breath which is discussed in the following section.

5.4.1 Metal-based nanosensors for detection of diabetic biomarkers in human biofluids

Diabetes is diagnosed using biofluids (such as tears, blood, and urine) that include markers for the disease such as insulin, glucose, and glycated hemoglobin (HbA1c). According to the specific signal source, emerging nanomaterial-based biosensors for diabetes biomarkers often rely on the underlying diagnostic procedures. According to particular applications for diagnosis, glucose biosensors coupled with metal nanoparticles can be placed both inside and outside the body. Biosensors can be placed externally to the skin for noninvasive moni-toring, such as wearable devices that assess sweat glucose levels. A transducer is often used in these biosensors to transform the glucose concentration to an electrical signal, which may be monitored and assessed. Biosensors, on the contrary, can be implanted inside the body to assess levels of glucose in the blood or interstitial fluid for invasive monitoring. These biosensors are often implanted beneath the skin or in other tissues and may be customized to continually assess glucose levels in real time.

Electrochemical detection of glucose

Because of the minimal price, ease of use, high sensitivity, and optimum specificity of elec-trochemical biosensors, it enables a viable point-of-care technique for disease detection. For instance, for accurate HbA1c monitoring, an electrochemical sensor based on altered screen-printed carbon electrodes and gold nanoflowers (Au NFs) has been developed (Wang et al. 2019). Using a chemical ligand (4-mercapto phenylboronic acid, 4-MPBA), the electro-chemical biosensor was improved. The phenylboronic acid of 4-MPBA combines with the sugar component of HbA1c. Certain electrochemical reactions were triggered when HbA1c was catalytically reduced to H_2O_2 and collected by 4-MPBA on the surface of the electrode (Wang et al. 2019). This approach might potentially be used in a human serum system, which has a lot of potential for monitoring actual blood samples from diabetic patients. To develop novel nanosensors with higher electrochemical sensitivity, carbon nanotubes/conductive graphene nanosheets had been coupled with nanoparticles of noble metals (Ag/Au). Shajaripour et al. (2019) made an HbA1c sensor by electrochemically deposited nano-materials of reduced graphene oxide (rGO) and Au upon a flexible and relatively affordable graphite sheet electrode (GS), which was later modified with aptamers that are specific to measuring HbA1c. Electrode surface area was increased, electron transport was acceler-ated, and a signal was activated by the rGO–Au nanocomposite. The rGO–Au on the elec-trode surface provided an adequate surface for a covalently attached DNA aptamer acting like a receptor of HbA1c. For electrochemical detection and diagnosis of diabetes, insulin blood levels, glucose, and HbA1c had all been taken into consideration. Gold–graphene nanocomposites on the surface of the dielectric Ag/Au nanoparticle-based sensors and Ag nanoflower–rGO composite-based microdisk electrodes have been developed for glucose and insulin monitoring (Ge et al. 2019; Yagati, Ngoc Le, and Cho 2020).

The sensors, which are made of metal oxides and nanostructured metals, electro-oxidize glucose directly using electrodes, resulting in significant electrocatalytic activity toward glu-cose (Bai et al. 2010). Because of their exceptional catalytic characteristics, noble metal nanoparticles (NPs) having a reduced activation barrier can catalyze a wide range of pro-cesses. A plethora of studies have been conducted on metals like Pt, Au, and Pd, in addition to their nanoparticles, along with metal oxides and sulfides of Zn, Ni, Mn, Co, Fe, and Cu for their improved catalytic characteristics and for oxidation of glucose specifically (Mei et al. 2016). Generally, the development of glucose electrochemical sensors has been divided into two categories: Enzymatic sensors and non-enzymatic sensors.

Enzymatic sensors

Glucose oxidase enzyme (GOx) is utilized to diagnose glucose in an enzymatic biosensor. The specificity of the enzyme allows the electrochemical glucose enzymatic sensor to operate with high sensitivity and high selectivity. The majority of enzymatic sensors require the utilization of various intermediates or additional catalysts to perform the efficient electrochemical interaction of the enzyme active center with electrodes (Patolsky, Weizmann, and Willner 2004). Clark developed the first enzymatic amperometric glucose sensor by immobilizing a platinum (Pt) electrode with glucose oxidase (GOx). GOx has been actively researched and employed for glucose biosensors since these early studies because of its excellent bioactivity, selectivity, inexpensive cost, and stability (Bankar et al. 2009).

CNTs in combination with silver (Lin et al. 2017), platinum (Wen, Ci, and Li 2009), or gold/platinum (Cash and Clark 2010) nanoparticles were developed as enzymatic biosensors. Nanostructured electrodes, like titanium dioxide (TiO_2) nanotube arrays or alumina-coated-silica-modified electrodes, can be used to deposit CNT membranes. It is possible to encapsulate GOx on an electrode surface using nanocomposites comprising nanotubes and polymers like cellulose (Wu et al. 2009). For glucose sensors, magnetic nanoparticles, typically formed of iron oxide, have also been employed. Both these particles and other systems, such CNTs, can be utilized separately or in combination (Yang et al. 2009).

Yang et al. (2009) designed a glucose biosensor that combines the intrinsic peroxidase-like activity of Fe_3O_4 NPs with the anti-interference capacity of a Nafion film. The new glucose biosensor has great sensitivity, a wide linear range, a quick response, good repeatability, a minimum limit of detection, and long-term stability. Because Fe_3O_4 NPs may catalyze the H_2O_2 reaction, the broad monitoring range and significant sensitivity can be attributed to the amplitude amplification of the current response.

Non-enzymatic sensor

The glucose detection technique in non-enzymatic biosensors is dependent on electrocatalytic activities. Non-enzymatic glucose biosensors are not affected by enzyme immobilization. A substrate is employed in non-enzymatic biosensors for inserting the catalyst. Carbon nanofibers (CNFs) are the useful substrate for non-enzymatic biosensors due to their simple method of fabrication, large surface area, and high electrical conductivity. Furthermore, free-standing CNF-based electrodes can be made without the utilization of a binder/glassy carbon electrode (GCE) for usage as a strip with the original glucometer. Noble metals, copper-based NPs, and nickel-based NPs have all been bonded to the CNF surface as electro-catalysts (Xie et al. 2020; Liu et al. 2015). Utilizing nanoporous Pt as an electrode material, Lee et al. (2018) developed an early version of a disposable non-enzymatic blood glucose sensor strip. Non-enzymatic biosensors are more stable than enzymatic sensors as enzymes are not present. Functionalized nanomaterials serving as catalysts or immobilization systems can improve detection specificity and sensitivity (Gamessa, Suman, and Tadesse 2018).

5.4.2 Metal-based nanosensors for the detection of glucose by optical method

Optical sensors relying on fluorescence, colorimetry, and photonic materials exhibiting analyte-triggered color-changing features were broadly explored to diagnose diabetes biomarkers. Following the particular application for diagnosis, optical biosensors coupled with metal nanoparticles can be placed both inside and outside the body. These biosensors can be mostly designed to be non-invasive, which may increase patient comfort and compliance.

Fluorescence-based nanosensors for the detection of glucose

The glucose fluorescent biosensor functions by processing a glucose signal into a corresponding fluorescence signal. In addition to electrochemical-based systems, fluorescence-based sensors have various benefits. They include high sensitivity with very low detection limits (Wang and Lee 2015); relatively no disruption to the host organism because of the NIR light's high penetration depth into tissues, which permits the monitoring of glucose content without any invasive process (Pickup et al. 2005). Nanomaterials can serve as effective carriers and detection devices for glucose in target cells/biological fluids. Biomolecules with enhanced biological activity are considerably stabilized on nanoparticles via conjugation or physical adsorption. The fluorescence-emitting nanomaterials that have received the greatest scientific attention include fluorescent semiconductor quantum dots (QDs), upconversion nanoparticles (UCNPs), lanthanide-doped nanomaterials, dye-doped silica NP (DDSNs), and fluorescent Au/Ag metal nanoclusters (Zhang and Wang 2014). Fluorescence tests have been performed on other nanomaterials, like fluorescence-interacting quenchers. Plasmonic Au/Ag NPs are employed as fluorescence quenchers.

Spectroscopic methods

Semiconductor quantum dots (QDs) exhibit many significant optoelectronic properties, comprising a broad excitation range, great photostability, narrow and symmetric emission, high quantum yields, and prolonged fluorescence lifetimes (Chen et al. 2020). NPs complexes with GOx-like characteristics tend to detect glucose, even if H_2O_2 may suppress the photoluminescence signal of QDs. As a GOx mimic, Au NPs may catalytically oxidize blood sugar and form H_2O_2 as a biomimetic catalyst. As a result, Au NPs had been altered, and a new enzyme-mimicking nanosensor for the detection of glucose was developed using QD–Au NP@silica, a mesoporous silica microsphere (Li et al. 2014).

Colorimetric method

The colorimetric technique, which is focused on enzyme cascade catalysis of GOx and horse-radish peroxidase (HRP), seems to be an optical analytic method for diabetes monitoring. The glucose-sensing capabilities of GOx and HRP were examined as the initial enzymatic set. Lu et al. (2015) developed colorimetric glucose detection utilizing GOx-attached Janus NPs composed of Fe_2O_3 and SiO_2. The inclusion of dual enzyme systems to Janus particles resulted in the formation of a single, reusable blood glucose sensor that could track glucose in complex samples like a serum. According to this, a hybrid multienzyme apparatus was developed by self-assembly, and CeO_2/GOx nano-complex was reported to demonstrate outstanding catalytic properties for the colorimetric detection of glucose (Liu et al. 2019a).

5.4.3 Metal-based nanosensors for the detection of acetone gas in the breath

Chemiresistive sensors that identify acetone are being utilized to diagnose diabetes. Semiconductor metal oxides such as tin oxide (SnO_2), tungsten oxide (WO_3), cerium oxide (CeO_2), and zinc oxide (ZnO) have been extensively used in chemiresistive detection of gases (Wang et al. 2019). These metal oxides are generally included in a breath analyzer device. The acetone content in the exhaled breath is then measured by the breath analyzer, which may be used to determine blood glucose levels. Nevertheless, the nanotubes or nanoparticles are normally employed in external breath analyzer equipment for diagnosis of diabetes rather than being injected within the body. The diagnostic methodology has relied on changes in electrical conductivity which emerge when gases and oxygen are available as they initiate catalytic oxidation or reduction processes on the surface of metal oxide.

Tin oxide (SnO$_2$)-based nanosensors

SnO$_2$ has been studied as a semiconductor metal oxide for detecting acetone gas. Acetone sensors dependent on SnO$_2$ have low selectivity and sensitivity. Efforts have been made to modify its micro/nanostructures or include nanosized catalysts in addition to improving sensitivity to certain gas species. As the electrical or chemical sensing characteristics of nanocatalysts, 1D SnO$_2$ nanotubes (high surface area) show significant potential for improved detection by optimizing the proportion of reactive sites on the outer and inner surfaces. Righettoni et al. (2015) used an SnO$_2$-based nanosensor based on Pt NP-immobilized thin-walled SnO$_2$ nanotubes and the electrical and chemical sensitizing characteristics of NPs. The hollow protein nanocage (e.g., Pt-encapsulated apoferritin) template approach was used to make the acetone nanosensor, which was then effectively complexed on the inner and exterior walls of electrospun SnO$_2$ nanotubes. At 350°C, Pt-loaded SnO$_2$ nanotubes demonstrated good sensitivity and selectivity to acetone with negligible cross-sensitivity to traces of interfering gases such as H$_2$S, NO, toluene, CO, pentane, and NH$_3$. As a result, the Pt-decorated 1D SnO$_2$ nanotubes may recognize diabetic gas indicators for diagnosis of diabetes (Jang et al. 2015).

Tungsten oxide (WO$_3$)-based nanosensors

Chemo-resistive detectors relying on semiconductor WO$_3$ nanofilms were employed for the study of exhaled breath acetone (Wang et al. 2008). The WO$_3$ nanoparticles' spontaneous electric dipole moments promote high dipole moment interactions between the sensor and acetone. Righettoni et al. (2010) utilized a gas phase to control the deposition of pure and Si-doped WO$_3$ NPs on interdigitated electrodes, leading to an innovative acetone detector. Ultrasmall nanoparticles, which increase certain surface areas, improve sensor accessibility to acetone molecules. At 400°C and 90% relative humidity, they discovered that diabetes patients (1800 ppb) and healthy individuals (900 ppb) could be easily separated from one another by a notable difference in sensor responses.

Other oxide semiconductor-based nanosensors

In addition to WO$_3$ and SnO$_2$, several kinds of semiconductor nanomaterials oxide with porous geometries and catalytic loadings were extensively studied as potential acetone gas detection alternatives. A three-dimensionally ordered microporous (DOM) ZnO nanomaterial complexed with graphene quantum dots (GQDs) was developed by Liu et al. (2019b) to detect acetone. The higher electrical conductivity of GQDs and the p-n heterojunction among p-type GQDs and n-type ZnO NPs significantly increased resistivity fluctuation caused by variations in oxygen adsorption, as reported by researchers. The hierarchical pore size distribution (macroscale and mesoscale), high surface area, and the electrical property notably increased the gas-detecting potential of the ZnO-based nanosensor.

A similar team of researchers additionally developed many 3DOM ZnO–CuO sensors that can detect acetone even in extreme humidity. They developed an entirely novel 3DOM indium (III) oxide (In_2O_3) film having Au NPs loaded on it to improve electrical responses after choosing Au NPs as surface modifying catalysts and appropriate oxygen adsorbent (Xing et al. 2015a). Hollow nanostructures have also been used in various semiconductor oxide-based acetone nanosensors. Koo et al. (2017) used metal–organic framework (MOF) templates to develop a nanoscale PdO catalyst-loaded Co_3O_4 hollow nanocage (HNC) for monitoring acetone. The novel structure offered a large surface area as well as robust catalytic activities, which enhanced acetone sensor outputs. Metal oxide semiconductor nanosensors show great promise in detecting diabetes-related trace acetone gas.

5.5 TREATMENT OF DIABETES MELLITUS AND INSULIN RESISTANCE

The subsequent step after a precise diagnosis of diabetes is to determine effective diabetic treatments. This chapter focuses on the three types of diabetes therapies, namely lifestyle modifications, conventional therapies, and nanotherapeutics.

5.5.1 Lifestyle modifications

Weight loss is indicated to avoid and prevent T2DM in obese people, as well as for treatment. All therapies are based on lifestyle changes such as a calorie-limited diet, reduced

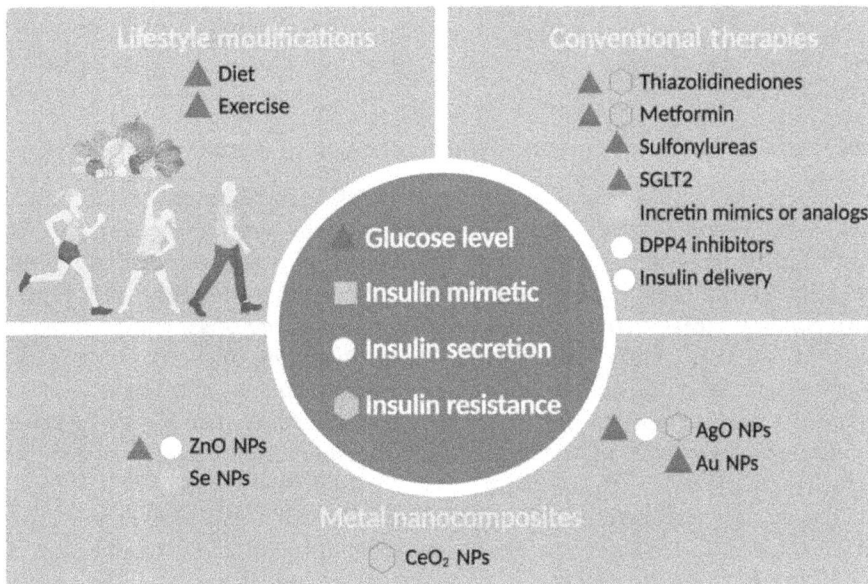

Figure 5.2 Symbolic depiction of the mechanism of action of various therapies in diabetes mellitus.

sedentary behavior, and increased exercise. Weight loss combined with lifestyle changes such as calorie control and exercising results in the normalization of the stimulated renin–angiotensin aldosterone system (RAAS), sympathetic activation, insulin resistance, and hyperleptinemia that are all common in T2DM and obesity (Masuo 2013). In addition, several dietary supplements are required. Micronutrients such as minerals and vitamins are needed by our body in minute amounts for various functions. They are often used as cofactors and coenzymes in metabolic pathways, assisting essential physiological functions in the process. Micronutrients have been studied as possible preventive and therapeutic agents for type I and type II diabetes, in addition to common diabetic symptoms (Franz 2001).

5.5.2 Conventional therapies for diabetes

Diabetes treatment and management have posed a significant problem for researchers and clinical individuals as a result of the surge in diabetic patients. Because T1DM is primarily caused by insulin insufficiency, the primary therapy option is insulin delivery with daily injections or an insulin pump. T2DM therapy, on the other hand, is significantly simpler because physical activity and diet can be effective therapies, particularly in the beginning. Certain drugs may additionally be given to patients as supplemental therapy. Metformin is commonly utilized as the first-line treatment for T2DM followed by sulfonylureas, which are typically utilized as the second-line treatment but can occasionally serve as the first-line in certain circumstances. For instance, if the patient does not tolerate metformin or when the person has a normal body weight, then thiazolidinediones, a third-line therapy is administered as the second-line treatment. Moreover, GDM patients can take metformin and glyburide. Incretin mimics or analogs DPP4 inhibitors which target the incretin axon (second-line therapy), and sodium–glucose cotransporters inhibitors 2 (SGLT2) that affects the kidneys' reabsorption of glucose may serve as monotherapy (if metformin is not tolerated) as well as second- and third-line drugs are among the newer treatments (Tasyurek et al. 2014; Wang et al. 2018).

5.5.3 Nanotherapeutics for the management of diabetes

In T2DM, there are multiple therapy options. Patients must begin treatment with antidiabetic drugs when exercise and diet are unable to regulate hyperglycemia. Current oral treatments for T2DM are primarily restricted by their limited bioavailability and rapid release of drugs, which necessitates increasing the frequency of doses. The development of therapeutic nanocarriers can increase patient compliance by reducing the dosage, prolonging release, and increasing bioavailability. Several nanocarriers, including liposomes, niosomes, nanospheres, polymeric micelles, ceramic nanoparticles, dendrimers, and metal nanoparticles, are employed to deliver antidiabetic drugs (Simos et al. 2020). Among all of these nanocarriers, metal nanoparticles of zinc, silver, gold, selenium, and cerium are an innovative approach that greatly enhances the quality of life for diabetics. Many studies have emphasized the efficacy of these metal NPs in physiological function in glucometabolic disorders. These metals have all been linked to blood sugar management and are utilized in diabetes treatment. They function as cofactors in numerous biochemical enzymatic processes. In the following section, these metal nanoparticles are discussed in detail.

Zinc oxide nanoparticles (ZnO NPs)

Zinc activates several enzymes in the body which are essential in many metabolic processes, particularly glucose metabolism (Haase, Overbeck, and Rink 2008). In addition to maintaining the structure of insulin, zinc is essential for the storage, synthesis, and secretion of insulin. Many studies have shown that zinc transporters in pancreatic β-cells, such as zinc transporter-8, play an important role in insulin secretion (Smidt et al. 2009). Zinc can improve insulin signaling through a variety of methods, involving increased phosphorylation of the insulin receptor, increased phosphoinositide 3-kinase activity, and inhibition of glycogen synthase kinase-3 (Jansen, Karges, and Rink 2009). As a result, there is a complicated interplay among zinc, diabetes, diabetic complications, and associated diabetic symptoms. Although zinc supplements have demonstrated a better impact in preclinical investigations (Tang 2019), the production of zinc-based therapeutics would be intriguing in the treatment of diabetes and its related problems. Siddiqui et al. (2020) employed mice models to investigate the hypoglycemic and glucose tolerance effects of ZnO NPs to confirm their impact on diabetes. The treatment of ZnO NPs in diabetic mice reduced fasting blood glucose levels significantly. Increased production (hepatic glycogenolysis and gluconeogenesis) and reduced use of glucose by tissues are the primary mechanisms of hyperglycemia. In mice with diabetes induced by STZ, the results showed that ZnO NPs dramatically lowered glucose levels (Siddiqui et al. 2020).

In type II diabetic rats, Gadoa et al. (2022) investigated the potential antidiabetic effects of ZnO NPs and pyrazolopyrimidine. The outcomes showed that the levels of the enzymes alanine aminotransferase (ALT) and aspartate aminotransferase (AST), low-density lipoprotein-cholesterol (LDL-c), malondialdehyde, and peroxisome proliferator-activated receptor gamma coactivator 1-alpha PGC-1 were higher in the diabetic group compared to the control group, whereas serum insulin and high-density lipoprotein cholesterol in the treatment groups (ZnO NPs, pyrazolopyrimidine, and ZnO NPs + pyrazolopyrimidine), these results were maintained. The results of their research established a beneficial impact of ZnO NPs and pyrazolopyrimidine on the profile of lipids, blood glucose levels, antioxidant status, mRNA expression of hepatic genes, and liver function enzymes.

According to Vinotha et al. (2019), the *Costus igneus* leaf extract was used in a unique way to form ZnO nanoparticles. An efficient and low-cost method was used to develop Ci-ZnO NPs, ZnO NPs coated with *C. igneus*. In comparison to the *C. igneus* leaf extract, the synthesized Ci-ZnO NPs demonstrated more potent antioxidant and antidiabetic effects. Additionally, against selected harmful bacteria, Ci-ZnO NPs have shown potential antibacterial and antibiofilm capabilities. Additionally, Ci-ZnO NPs showed their biocompatibility in goat RBCs, with an acceptable level of hemolysis. As a consequence, the findings of their research showed that Ci-ZnO NPs may have prospective uses in the biomedical and pharmaceutical industries.

Abd El-Aziz et al. (2021) investigated the antidiabetic effects of curcumin nanoparticles Curc-NPs, ZnO NPs, and Curc/ZnO-NC on streptozotocin (STZ)-induced diabetic rats. The outcomes correlate with both normal nondiabetic rats and animals given standard antidiabetic Diamicron. Although all treatment groups had substantial reductions in blood sugar, increased insulin levels, and altered GLUT-2 and GK genes, Curc/ZnO-NC animals demonstrated the most antidiabetic effect when compared to control rats.

Al-Radadi, Najlaa et al. (2022) employed zinc sulphate salt as a source for ZnO NPs. ZnO NPs were stabilized and decreased by the bioactive components of *Zingiber officinale*.

It has been performed to characterize ZnO NPs more. At high concentrations of NPs, ZnO NPs demonstrated notable antioxidant activity and RBC hemocompatibility, and no noticeable hemolysis was seen. Because of its considerable percentage inhibition of α-amylase, ZnO NPs have shown potential for treating diabetes.

Selenium nanoparticles (Se NPs)

Humans require selenium, which can improve the functions of antioxidant enzymes and immune system-related selenoenzymes like GPx (Srivastava, Braganca, and Kowshik 2014) and other selenium-dependent enzymes, hence enhancing both acquired and innate immunological responses (Hashem, Hassnin, and AbdEl-Kawi 2013). Gutiérrez et al. (2022) investigated the effect of luteolin (LU) and diosmin (DIO) in Se NPs in streptozotocin (STZ)-induced diabetic rats. When selenium was given to rats, it stimulated glucose transport and showed insulin-like actions in adipocytes and insulin-sensitive cyclic adenosine monophosphate phosphodiesterase (cAMP-PDE). The synergistic combination of selenium and flavonoids in the nanoparticle resulted in a potent antidiabetic efficacy (Gutiérrez et al. 2022).

Amany Abdel-Rahman Mohamed et al. (2020) investigated a novel approach alongside the hypothesis that the combination of the antidiabetic drug metformin (MET) and the chitosan-stabilized nanoparticles (CTS-Se-NPs) may have an effect on insulin level, liver impairment and cell death, and cardiac injury markers T2DM in rat model. Their findings have shown that HFD/STZ caused harmful effects on serum, hepatic, and cardiac tissues, comprising a notable increase in oxidative and inflammatory mediators, as well as increased levels of the expression of genes associated with apoptosis (Bax, Caspase-3, Fas, and Fas-L). Combining MET and CTS-Se-NP treatment had a more notable antidiabetic impact, as seen by significantly lower fasting blood glucose and insulin levels as well as enhanced anti-apoptotic gene (BCL-2) and anti-apoptotic gene expression levels. In comparison to a monotherapeutic approach, the combination method used in the research may have more impact on limiting the effects of diabetes and restoring insulin resistance, and it may represent an intriguing treatment option in T2DM rat models.

El-Borady et al. (2020) assessed the effectiveness of manufactured Se NPs capped with glucose and polyvinyl pyrrolidone (PVP) on the hyperglycemia and prooxidants/antioxidants imbalance observed in model streptozotocin (STZ)-induced diabetic rats. The treatment of Se NPs significantly decreased hyperglycemia, increased plasma and pancreatic insulin levels, and repaired the injured pancreatic tissue. By lowering the levels of pancreatic lipid peroxidation (LPO) and nitric oxide (NO), Se NPs also demonstrated improved clearance of the oxidative stress damage carried by diabetes. Additionally, both the glutathione (GSH) and glutathione peroxidase (GPx) activities and levels were elevated in the diabetic rats.

Abdel Maksoud et al. (2020) examined the reduction in blood sugar of selenium cleome droserifolia nanoparticles (Se-CNPs) and/or Galvus met® therapy on streptozotocin-induced diabetes mellitus in male rats. The findings revealed a substantial rise in serum glucose concentration, Aminotransferase (ALT) and Aspartate Aminotransferase (AST) activities, Triacylglycerols (TG), Very Low-Density Lipoprotein cholesterol (VLDL-c), Low Density Lipoprotein cholesterol (LDL-c), and Non-Esterified Fatty Acids (NEFAs), urea and creatinine levels in untreated diabetic rats compared to control, whereas a significant decrease in serum insulin and HDL-c concentration. Se-CNPs and/or Galvus met® were given every day to diabetic rats, and this significantly improved these parameters.

Omayma A.R. AboZaid et al. (2022) evaluated the antioxidant and antidiabetic properties of chitosan-encapsulated selenium nanoparticles in a streptozotocin-induced diabetes animal. As a standard antidiabetic drug, glibenclamide was utilized. By substantially modifying the examined parameters and physical characteristics of pancreatic tissue, findings have shown that STZ caused both diabetes and oxidative stress in normal rats. In comparison to STZ-induced diabetic rats, oral treatment of CTSSe NPs or Glib significantly improves the amount of serum fasting blood glucose, insulin, IGF-1, AST, ATL, and CK-MB.

Fan et al. (2020) carried out green synthesis of Se NPs through the combination of *Hibiscus sabdariffa* (roselle plant) leaf extract with a solution of selenious acid (H_2SeO_3) during continuous stirring conditions, yielding roselle plant secondary metabolites-conjugated Se NPs. Se NPs' antioxidative and protective activities were also investigated in diabetes-induced rats by streptozotocin (STZ). Se NPs or/and insulin therapy were given to these STZ-induced diabetic rats every day, and the impact of Se NPs on the variables linked to oxidative damage in the rat testes was assessed. According to biochemical tests, Se NPs can improve the drop in blood testosterone carried by STZ-induced diabetes. Se NPs can also considerably lower the testicular tissue's oxidative stress markers, including nitric oxide and lipid peroxidation. However, Se NPs treatment boosted glutathione content and antioxidant enzyme activity in testicular tissues in STZ-induced diabetic mice.

Cerium dioxide nanoparticles (CeO$_2$ NPs)

CeO$_2$ NPs, also known as nanoceria, typically display substantial antidiabetic benefits due to their antioxidant activity alongside low toxicity (Choudhury et al. 2018). To effectively combat diabetes, CeO$_2$ NPs modulate the nuclear factor kappa-light-chain-enhancer of activated B cells/nuclear factor erythroid-2-related factor 2 (NF-κB/Nrf2) pathways and significantly reduce apoptotic cell death (Khurana, Tekula, and Godugu 2018). The efficiency to resemble superoxide dismutase, function as effective reactive oxygen species (ROS) scavengers (Ce^{+3} to Ce^{+4}), and change the oxidation state to resemble catalase function that decreases H$_2$O$_2$ and releases protons and O$_2$ (Ce^{+4} to the initial Ce^{+3}) are all linked with the therapeutic efficacy. As a result of their self-regenerative characteristic, the nanoparticles are particularly helpful for diabetes treatment (Solgi et al. 2021).

Ayodhya et al. (2022) developed an easy, reliable, and environmentally friendly process for synthesizing CeO$_2$ NPs and Sapodilla fruit (*Manilkara zapota*) peel (SFP) extract CeO$_2$ NPs. Utilizing a paper disc approach instead of CeO$_2$ NPs, the synthesized SFP extract-CeO$_2$ NPs exhibit strong antibacterial efficacy against a number of diseases (bacteria and fungus). Additionally, both amylase and glucosidase enzymes strongly decrease the antidiabetic action of SFP extract-CeO2 NPs with IC50 values in comparison to the reference (acarbose).

Jan et al. (2020) have shown the method to synthesize highly biocompatible CeO$_2$ NPs with a variety of biological uses using the medicinally significant plant *Aquilegia pubiflora*. The NPs were examined for a variety of biological applications, including antimicrobial (antifungal, antibacterial, and antileishmanial), protein kinase inhibition, anticancer, antioxidant, antidiabetic, and biocompatibility activities. According to their research, CeO$_2$ NPs made synthetically from *Aquilegia pubiflora* are extremely biocompatible and have great prospects as possible treatments for leishmaniasis and cancer. Furthermore, these CeO$_2$ NPs may be cutting-edge nanotools for a variety of biological applications due to their strong antibacterial and antioxidant properties.

Saravanakumar et al. (2021) formed ceria oxide nanoparticles (CeO_2 NPs) from the plant extract of *Stachys japonica* Miq and tested their antioxidant and antidiabetic properties. CeO_2 NPs were shown to reduce staurosporine (STS)-induced oxidative stress in NIH3T3 cells by reducing MMP loss and nucleus disruption in a cell antioxidant study. In IR-HepG2 cells, administration with CeO_2 NPs reduced insulin resistance and enhanced glucose absorption by protecting the mitochondria from oxidative stress.

Silver nanoparticles (Ag NPs)

Silver is essential for a wide range of metabolic functions. Silver nanoparticles may be a source of insulin sensitivity because they increase the concentrations of cytosolic calcium ions and stimulate AMPK by phosphorylating it through the CAMKK pathway in SH-SY5Y cells in rats (Li et al. 2019). By enhancing its activity, AMPK stimulation might normalize the effects of insulin by improving insulin sensitivity (Zhang et al. 2018). The phosphorylation cascade from IRS1 is activated when insulin links to its receptor, causing glucose to be transported into the cells. Raising the protein levels of IRS1 may eventually lessen the difficulties associated with hyperglycemia because research has demonstrated that animal models without IRS1 exhibited hyperglycemia or T2DM. By elevating the IRS1 and GLUT2 expression levels, Ag NPs contribute to a decrease in blood glucose levels. Moreover, Ag NPs increase insulin expression and secretion (Wahab, Bhatti, and John 2022).

It is generally recognized that phytochemical elements are therapeutically active compounds and that they have an impact on a variety of illnesses and diseases through a number of different biological processes. Ul Haq, Shah, and Menaa (2022) employed a *T. couneifolia* extract encapsulated with silver nanoparticles as a model for diabetes mellitus study. Ullah et al. (2021) developed Ag NPs from the leaves of *Emblica phyllanthus*. Ag NPs made from phytosynthesized Ag NPs had been administered to alloxan-induced diabetic rats. When compared to diabetic control rats, the dyslipidemia status of the treated diabetic rats significantly improved. It also reduced down blood sugar levels over the days. The level of blood glucose dropped, body weight increased, and the lipid, liver, and renal profiles all significantly improved. According to the research, plant-mediated silver nanoparticles could potentially be utilized to treat diabetes since they greatly reduced the alloxan-induced diabetic modifications in the rats that underwent different treatments.

Essghaier et al. (2022) demonstrated the green production of Ag NPs from the extremophile plant *Aeonium haworthii*. There have been reports of the antioxidant, antidiabetic, and antibacterial effects. Ag NPs were shown to have better antioxidant action than ascorbic acid. The Ag NPs also have a bactericidal action, which inhibits bacterial growth. In comparison to normal Acarbose, the silver nanoparticles have strong antidiabetic action as measured by the inhibitory effect on α-amylase.

Elekofehinti (2022) used silver nitrate to make *Momordica charantia* nanoparticles to test the antidiabetic characteristics of the extract and the nanoparticles *in vivo*. In rats with streptozotocin-induced diabetes, the antidiabetic efficacy of *M. charantia* nanoparticle was evaluated *in vivo*. Blood sugar levels were significantly reduced in rats injected with *M. charantia* nanoparticles. When compared to diabetic rats that were left untreated, a decrease in pancreatic alanine transaminase, aspartate aminotransferase, and alkaline phosphatase was seen. When in comparison with diabetic rats not receiving treatment, *M. charantia* nanoparticles significantly increase the antioxidant enzymes in the rats. Comparing diabetic control rats to diabetic untreated rats, it was found that cholesterol, triglyceride, and low-density lipoprotein levels decreased. It was also found that the expression of Takeda-G-protein-receptor-5, superoxide dismutase, insulin, catalase, glucagon-like peptide-1, and NFE2-related factor 2 genes significantly increased.

Gold nanoparticles (Au NPs)

Au NPs had been shown to function as antioxidants by inhibiting the release of ROS, scavenging free radicals, and increasing antioxidant levels (Bednarski et al. 2015). Many *in vivo* investigations have revealed that Au NPs have antihyperglycemic properties. The Au NPs of 21 nm size synthesized by Opris et al. (2017) showed enhanced antioxidant activity in the liver, muscle, and blood, lowered blood sugar levels, and lessened oxidative stress induced by diabetes. Selim, Abd-Elhakim, and Al-Ayadhi. (2015) investigated the therapeutic benefits of Au NPs on autistic diabetic rats and observed that 50 nm Au NPs, 2.5 mg/kg markedly altered mostly all liver redox properties, such as glutathione (GSH) and oxidized glutathione (GSSG) levels, SOD and GPx activities, and oxygen radical absorbance capacity. The levels of glucose and lipids were also elevated, and pancreatic β cells seem to have the capability for healing after injury.

Sanaa Ayyoub et al. (2022) investigate the *in vivo* antidiabetic effects of Au NPs derived from the leaf extract of *Dittrichia viscosa* in rats with high-fat diet (HFD)/streptozotocin (STZ)-induced diabetes. Au NPs were formed using *D. viscosa* leaf extract, and the resulting materials were subsequently described. Au NP therapy dramatically decreased liver blood glucose levels, PEPCK activities, and gene expression in comparison to the diabetic group not receiving treatment. According to their studies, Au NPs made from *D. viscosa* leaf extract can lower hyperglycemia in diabetic rats given the HFD/STZ diet. This could be achieved by reducing hepatic gluconeogenesis by inhibiting the PEPCK gene's expression and function.

Mahmoudi et al. (2021) synthesized and characterized Au NPs from a plant extract of *Eryngium thyrsoideum Boiss*. Synthesized Au NPs were tested for their antidiabetic effectiveness on T2DM mice by observing how they affected inflammatory indicators and blood-related variables. The results demonstrated that Au NPs decreased white blood cell count, liver enzyme concentration, and blood sugar in diabetic rats. The Au NPs also drastically decreased the expression of the TNF-α and IL-6 genes in the visceral adipose tissue of T2DM patients. Furthermore, the Au NPs have no negative effects on the hematology and liver functions of rats. The results show that Au NPs are useful nanomaterials for the treatment of diabetes.

Guo et al. (2020) investigated the antidiabetic activity of Au NPs derived from *Fritillaria cirrhosa* in diabetic preclinical rats induced by streptozotocin (STZ). The standard drug glibenclamide had been compared with the formed Au NPs. Abdel-Halim et al. (2020) investigated the impact of *Bauhinia variegata (B. variegata)* extract on DM carried on by streptozotocin (STZ) in rats prior to and following integrating Au NPs. According to the study, Au NPs treatment restores serum, hepatic, and renal markers in diabetic rats that have been subjected to STZ. The Au NPs therapy decreases lipid peroxidation and changes the quantity of antioxidants due to its antioxidant properties. The delivery of Au NPs to experimental rats induces the pancreatic islet cells to regenerate, based on the pathology results. The study's findings showed that *B. variegata* and *Fritillaria cirrhosa* Au NPs have antidiabetic characteristics.

5.6 CONCLUSION AND FUTURE PERSPECTIVES

Because of the rising numbers of diabetes patients, there is an urgent requirement to develop novel drugs in addition to better drug delivery techniques with even better accurate efficacies and fewer adverse effects. Nanotechnology has enabled breakthroughs in both glucose sensors and self-regulated insulin delivery devices. Furthermore, innovative developments in fluorescence glucose detection can provide continuous *in vivo* glucose monitoring. Further improved approaches that fully exploit nanotechnology breakthroughs in the fight against the increased occurrence of T2DM are expected to be implemented in the upcoming years to achieve this goal.

Current nanosensors for glucose or HbA1c detection, on the other hand, are likely to be better. By altering the nanosensors with phenylboronic acid or glucose oxidase, the issue of acetone sensors detecting non-specificity may be solved. Nevertheless, more research is required to improve diagnosis because existing diabetes sensors require specialized devices to analyze the output. After all, advancements have been done in this domain, and paper-based glucose and acetone gas test strips and color-changing contact lenses are expected to be among the more promising options for diabetes detection. Future research in this area needs to target developing semiconductor oxide acetone sensors capable of being gathered, in addition to modifying clinical testing to help accommodate its actual aspects.

Despite functional nanosystems having made significant advances in the diagnosis, treatment, and management of diabetes and its consequences, this research subject is yet in its development and needs more interaction and collaboration among the fields of biology, medicine, chemistry, material science, and computer science. Further research must be undertaken to identify novel diabetes biomarkers, develop enhanced monitoring methods and point-of-care diagnostic technologies, and develop highly effective drug nanocarriers and efficient nanocomposites having distinct antidiabetic characteristics, as well as reduce the biotoxicity of existing nanocomposites. The specific characteristics of metal nanocomposites, like their large surface area, biocompatibility, and capacity for regulated drug release, have the possibility of transforming the treatment of DM. Furthermore, the application of artificial intelligence (AI) can enhance the manufacturing of these materials by enabling even more specific regulations regarding their characteristics and function.

REFERENCES

Abd El-Aziz, S. M., M. Raslan, M. Afify, M. D. E. Abdelmaksoud, and K. A. El-Nesr. 2021. "Anti-diabetic Effects of Curcumin/Zinc Oxide Nanocomposite in Streptozotocin-Induced Diabetic Rats." *IOP Conference Series: Materials Science and Engineering* 1046 (1): 012023. https://doi.org/10.1088/1757-899x/1046/1/012023.

Abdel-Halim, and Abeer Hamed et al. 2020. "Assessment of the Anti-Diabetic Effect of Bauhinia Variegata Gold Nano-Extract against Streptozotocin Induced Diabetes Mellitus in Rats." *Journal of Applied Pharmaceutical Science* 10 (5): 77–91. https://doi.org/10.7324/japs.2020.10511.

Abdel Maksoud, H. A., Omayma A. R. Abou Zaid, Mohamed G. Elharrif, M. A. Omnia, and E. A. Alaa. 2020. "Selenium Cleome Droserifolia Nanoparticles (Se-CNPs) and It's Ameliorative Effects in Experimentally Induced Diabetes Mellitus." *Clinical Nutrition ESPEN* 40 (December): 383–391. https://doi.org/10.1016/j.clnesp.2020.07.016.

Aguirre, Vincent, Tohru Uchida, Lynne Yenush, Roger Davis, and Morris F. White. 2000. "The C-Jun NH2-Terminal Kinase Promotes Insulin Resistance during Association with Insulin Receptor

Substrate-1 and Phosphorylation of Ser307." *Journal of Biological Chemistry* 275 (12): 9047–9054. https://doi.org/10.1074/jbc.275.12.9047.

Akash, Muhammad Sajid Hamid, Kanwal Rehman, and Shuqing Chen. 2013. "Role of Inflammatory Mechanisms in Pathogenesis of Type 2 Diabetes Mellitus." *Journal of Cellular Biochemistry* 114 (3): 525–531. https://doi.org/10.1002/jcb.24402.

Al-Radadi, Najlaa S., and Shah Faisal et al. 2022. "Zingiber Officinale Driven Bioproduction of ZnO Nanoparticles and Its Anti-Inflammatory, Anti-Diabetic, Anti-Alzheimer, Anti-Oxidant, and Anti-Microbial Applications." *Inorganic Chemistry Communications* 140 (April): 109274. https://doi.org/10.1016/j.inoche.2022.109274.

Ayodhya, D., A. Ambala, G. Balraj, M.P. Kumar, and P. Shyam. 2022. "Green Synthesis of CeO2 NPs Using Manilkara Zapota Fruit Peel Extract for Photocatalytic Treatment of Pollutants, Antimicrobial, and Antidiabetic Activities." *Results in Chemistry* 4: 100441.

Amany Abdel-Rahman Mohamed, Safaa I. Khater, Islam M. Saadeldin, and Mohamed M. M. Metwally et al. 2021. "Chitosan-Stabilized Selenium Nanoparticles Alleviate Cardio-Hepatic Damage in Type 2 Diabetes Mellitus Model via Regulation of Caspase, Bax/Bcl-2, and Fas/FasL-Pathway." *Gene* 768 (20): 145288 https://doi.org/10.1016/j.gene.2020.145288.

Bai, Hongyan, Min Han, Yuezhi Du, Jianchun Bao, and Zhihui Dai. 2010. "Facile Synthesis of Porous Tubular Palladium Nanostructures and Their Application in a Nonenzymatic Glucose Sensor." *Chemical Communications* 46 (10): 1739. https://doi.org/10.1039/b921004k.

Balbaa, Mahmoud, Shaymaa A. Abdulmalek, and Sofia Khalil. 2017. "Oxidative Stress and Expression of Insulin Signaling Proteins in the Brain of Diabetic Rats: Role of Nigella Sativa Oil and Antidiabetic Drugs." Edited by Christian Holscher. *PLOS ONE* 12 (5): e0172429. https://doi.org/10.1371/journal.pone.0172429.

Bankar, Sandip B., Mahesh V. Bule, Rekha S. Singhal, and Laxmi Ananthanarayan. 2009. "Glucose Oxidase: An Overview." *Biotechnology Advances* 27 (4): 489–501. https://doi.org/10.1016/j.biotechadv.2009.04.003.

Bednarski, Marek, Magdalena Dudek, Joanna Knutelska, Leszek Nowiński, Jacek Sapa, Małgorzata Zygmunt, Gabriel Nowak, et al. 2015. "The Influence of the Route of Administration of Gold Nanoparticles on Their Tissue Distribution and Basic Biochemical Parameters: In Vivo Studies." *Pharmacological Reports: PR* 67 (3): 405–409. https://doi.org/10.1016/j.pharep.2014.10.019.

Bruce, Clinton R., Steve Risis, Joanne R. Babb, Christine Yang, Greg M. Kowalski, Ahrathy Selathurai, Robert S. Lee-Young, et al. 2012. "Overexpression of Sphingosine Kinase 1 Prevents Ceramide Accumulation and Ameliorates Muscle Insulin Resistance in High-Fat Diet-Fed Mice." *Diabetes* 61 (12): 3148–3155. https://doi.org/10.2337/db12-0029.

Cash, Kevin J., and Heather A. Clark. 2010. "Nanosensors and Nanomaterials for Monitoring Glucose in Diabetes." *Trends in Molecular Medicine* 16 (12): 584–593. https://doi.org/10.1016/j.molmed.2010.08.002.

Chen, Sha, Danlian Huang, Piao Xu, Wenjing Xue, Lei Lei, Min Cheng, Rongzhong Wang, Xigui Liu, and Rui Deng. 2020. "Semiconductor-Based Photocatalysts for Photocatalytic and Photoelectrochemical Water Splitting: Will We Stop with Photocorrosion?" *Journal of Materials Chemistry A* 8 (5): 2286–2322. https://doi.org/10.1039/c9ta12799b.

Choudhury, Hira, Manisha Pandey, Chua Kui Hua, Cheah Shi Mun, Jessmie Koh Jing, Lillian Kong, Liang Yee Ern, et al. 2018. "An Update on Natural Compounds in the Remedy of Diabetes Mellitus: A Systematic Review." *Journal of Traditional and Complementary Medicine* 8 (3): 361–376. https://doi.org/10.1016/j.jtcme.2017.08.012.

Drews, Gisela, Peter Krippeit-Drews, and Martina Düfer. 2010. "Oxidative Stress and Beta-Cell Dysfunction." *Pflügers Archiv: European Journal of Physiology* 460 (4): 703–718. https://doi.org/10.1007/s00424-010-0862-9.

Eizirik, Décio L., Maikel L. Colli, and Fernanda Ortis. 2009. "The Role of Inflammation in Insulitis and β-Cell Loss in Type 1 Diabetes." *Nature Reviews Endocrinology* 5 (4): 219–226. https://doi.org/10.1038/nrendo.2009.21.

El-Borady, Ola M., Mohamed S. Othman, Heba H. Atallah, and Ahmed E. Abdel Moneim. 2020. "Hypoglycemic Potential of Selenium Nanoparticles Capped with Polyvinyl-Pyrrolidone in Streptozotocin-Induced Experimental Diabetes in Rats." *Heliyon* 6 (5): e04045. https://doi. org/10.1016/j.heliyon.2020.e04045.

Elekofehinti, Olusola Olalekan. 2022. "Momordica Charantia Nanoparticles Potentiate Insulin Release and Modulate Antioxidant Gene Expression in Pancreas of Diabetic Rats." *Egyptian Journal of Medical Human Genetics* 23 (1). https://doi.org/10.1186/s43042-022-00282-0.

Essghaier, Badiaa, Rihab Dridi, and Filomena Mottola et al. 2022. "Biosynthesis and Characterization of Silver Nanoparticles from the Extremophile Plant Aeonium Haworthii and Their Antioxidant, Antimicrobial and Anti-Diabetic Capacities." *Nanomaterials (Basel, Switzerland)* 13 (1): 100. https://doi.org/10.3390/nano13010100.

Fan, Dabei, Li Li, Zhizhen Li, and Ying Zhang. 2020. "Biosynthesis of Selenium Nanoparticles and Their Protective, Antioxidative Effects in Streptozotocin Induced Diabetic Rats." *Science and Technology of Advanced Materials* 21 (1): 505–514. https://doi.org/10.1080/14686996.2020. 1788907.

Florkowski, Chris. 2013. "HbA1c as a Diagnostic Test for Diabetes Mellitus: Reviewing the Evidence." *The Clinical Biochemist. Reviews* 34 (2): 75–83. www.ncbi.nlm.nih.gov/pmc/articles/PMC3799221/.

Franz, Marion J. 2001. "Medical Nutrition Therapy for Diabetes." *Nutritional Health*, 167–193. https://doi.org/10.1007/978-1-59259-226-5_12.

Freeman, Andrew M., and Nicholas Pennings. 2019. "Insulin Resistance." *Nih.gov. StatPearls Publishing*. 2019. www.ncbi.nlm.nih.gov/books/NBK507839/.

Gadoa, Zahraa Alaaeldein, Ahmed Hussein Moustafa, Samir Mohamed El Rayes, Ahmed A. Arisha, and Mohamed Fouad Mansour. 2022. "Zinc Oxide Nanoparticles and Synthesized Pyrazolopyrimidine Alleviate Diabetic Effects in Rats Induced by Type II Diabetes." *ACS Omega* 7 (41): 36865–36872. https://doi.org/10.1021/acsomega.2c05638.

Gamessa, Tadesse Waktola, Dabbu Suman, and Zerihun Ketema Tadesse. 2018. "Blood Glucose Monitoring Techniques: Recent Advances, Challenges and Future Perspectives." *International Journal of Advanced Technology and Engineering Exploration* 5 (46): 335–344. https://doi. org/10.19101/ijatee.2018.546008.

Ge, Yueping, Thangavel Lakshmipriya, Subash C.B. Gopinath, Periasamy Anbu, Yeng Chen, Firdaus Hariri, and Lu Li. 2019. "Glucose Oxidase Complexed Gold-Graphene Nanocomposite on a Dielectric Surface for Glucose Detection: A Strategy for Gestational Diabetes Mellitus." *International Journal of Nanomedicine* 14 (October): 7851–7860. https://doi.org/10.2147/ijn.s222238.

Gerber, Philipp A., and Guy A. Rutter. 2017. "The Role of Oxidative Stress and Hypoxia in Pancreatic Beta-Cell Dysfunction in Diabetes Mellitus." *Antioxidants & Redox Signaling* 26 (10): 501–518. https://doi.org/10.1089/ars.2016.6755.

Gholami, Ahmad, Fatmeh Farjami, and Ghasemi Younes. 2021. "The Development of an Amperometric Enzyme Biosensor Based on a Polyaniline-Multiwalled Carbon Nanocomposite for the Detection of a Chemotherapeutic Agent in Serum Samples from Patients." *Journal of Sensors*. (July). www.researchgate.net/publication/353250348_The_Development_of_an_Amperometric_Enzyme_Biosensor_Based_on_a_Polyaniline-Multiwalled_Carbon_Nanocomposite_for_the_Detection_of_a_Chemotherapeutic_Agent_in_Serum_Samples_from_Patients.

Goldberg, Ronald B. 2009. "Cytokine and Cytokine-like Inflammation Markers, Endothelial Dysfunction, and Imbalanced Coagulation in Development of Diabetes and Its Complications." *The Journal of Clinical Endocrinology and Metabolism* 94 (9): 3171–3182. https://doi.org/10.1210/jc.2008-2534.

Guo, Ying, Nan Jiang, Li Zhang, and Min Yin. 2020. "Green Synthesis of Gold Nanoparticles from Fritillaria Cirrhosa and Its Anti-Diabetic Activity on Streptozotocin Induced Rats." *Arabian Journal of Chemistry* 13 (4): 5096–5106. https://doi.org/10.1016/j.arabjc.2020.02.009.

Gutiérrez, Rosa Martha Pérez, Julio Téllez Gómez, Raúl Borja Urby, José G. Contreras Soto, and Héctor Romo Parra. 2022. "Evaluation of Diabetes Effects of Selenium Nanoparticles Synthesized from a Mixture of Luteolin and Diosmin on Streptozotocin-Induced Type 2 Diabetes in Mice." *Molecules* 27 (17): 5642. https://doi.org/10.3390/molecules27175642.

Haase, Hajo, Silke Overbeck, and Lothar Rink. 2008. "Zinc Supplementation for the Treatment or Prevention of Disease: Current Status and Future Perspectives." *Experimental Gerontology* 43 (5): 394–408. https://doi.org/10.1016/j.exger.2007.12.002.

Habtamu Wondifraw. 2015. "Classification, Pathophysiology, Diagnosis and Management of Diabetes Mellitus." *ResearchGate*. OMICS International. www.researchgate.net/publication/279274191_Classification_Pathophysiology_Diagnosis_and_Management_of_Diabetes_Mellitus.

Hashem, Khalid, Kamel M. A. Hassnin, and Samraa H. AbdEl-Kawi. 2013. "The Prospective Protective Effect of Selenium Nanoparticles against Chromium-Induced Oxidative and Cellular Damage in Rat Thyroid." *International Journal of Nanomedicine* (May): 1713. https://doi.org/10.2147/ijn.s42736.

Hurrle, Samantha, and Walter H. Hsu. 2017. "The Etiology of Oxidative Stress in Insulin Resistance." *Biomedical Journal* 40 (5): 257–262. https://doi.org/10.1016/j.bj.2017.06.007.

Jan, Hasnain, Muhammad Aslam Khan, and Hazrat Usman et al. 2020. "The Aquilegia Pubiflora (Himalayan Columbine) Mediated Synthesis of Nanoceria for Diverse Biomedical Applications." *RSC Advances* 10 (33): 19219–19231. https://doi.org/10.1039/d0ra01971b.

Jang, Ji-Soo, Sang-Joon Kim, Seon-Jin Choi, Nam-Hoon Kim, Meggie Hakim, Avner Rothschild, and Il-Doo Kim. 2015. "Thin-Walled SnO2 Nanotubes Functionalized with Pt and Au Catalysts via the Protein Templating Route and Their Selective Detection of Acetone and Hydrogen Sulfide Molecules." *Nanoscale* 7 (39): 16417–16426. https://doi.org/10.1039/C5NR04487A.

Jansen, Judith, Wolfram Karges, and Lothar Rink. 2009. "Zinc and Diabetes: Clinical Links and Molecular Mechanisms." *The Journal of Nutritional Biochemistry* 20 (6): 399–417. https://doi.org/10.1016/j.jnutbio.2009.01.009.

Khurana, Amit, Sravani Tekula, and Chandraiah Godugu. 2018. "Nanoceria Suppresses Multiple Low Doses of Streptozotocin-Induced Type 1 Diabetes by Inhibition of Nrf2/NF-KB Pathway and Reduction of Apoptosis." *Nanomedicine* 13 (15): 1905–1922. https://doi.org/10.2217/nnm-2018-0085.

Koo, Won-Tae, Sunmoon Yu, Seon-Jin Choi, Ji-Soo Jang, Jun Young Cheong, and Il-Doo Kim. 2017. "Nanoscale PdO Catalyst Functionalized Co3O4 Hollow Nanocages Using MOF Templates for Selective Detection of Acetone Molecules in Exhaled Breath." *ACS Applied Materials & Interfaces* 9 (9): 8201–8210. https://doi.org/10.1021/acsami.7b01284.

Lee, S., J. Lee, S. Park, H. Boo, H. C. Kim, and T. D. Chung. 2018. "Disposable Non-Enzymatic Blood Glucose Sensing Strip Based on Nanoporous Platinum Particles." *Applied Materials Today* 10: 24–29. https://doi.org/10.1016/j.apmt.2017.11.009.

Li, Lin, Lu Li, Xuejiao Zhou, Yang Yu, Zengqiang Li, Daiying Zuo, and Yingliang Wu. 2019. "Silver Nanoparticles Induce Protective Autophagy via Ca2+/CaMKKβ/AMPK/MTOR Pathway in SH-SY5Y Cells and Rat Brains." *Nanotoxicology* 13 (3): 369–391. https://doi.org/10.1080/17435390.2018.1550226.

Li, Yang, Qiang Ma, Ziping Liu, Xinyan Wang, and Xingguang Su. 2014. "A Novel Enzyme-Mimic Nanosensor Based on Quantum Dot-Au Nanoparticle@Silica Mesoporous Microsphere for the Detection of Glucose." *Analytica Chimica Acta* 840 (August): 68–74. https://doi.org/10.1016/j.aca.2014.05.027.

Lin, Jiehua, Chunyan He, Yue Zhao, and Shusheng Zhang. 2017. "One-Step Synthesis of Silver Nanoparticles/Carbon Nanotubes/Chitosan Film and Its Application in Glucose Biosensor." *Sensors and Actuators B-Chemical*. www.semanticscholar.org/paper/One-step-synthesis-of-silver-nanoparticles-carbon-Lin-He/6c75e814cac1fc03f881b023bbf2a-60ca3639e5f.

Liu, Dong, Qiaohui Guo, Xueping Zhang, Haoqing Hou, and Tianyan You. 2015. "PdCo Alloy Nanoparticle-Embedded Carbon Nanofiber for Ultrasensitive Nonenzymatic Detection of Hydrogen Peroxide and Nitrite." *Journal of Colloid and Interface Science* 450 (July): 168–173. https://doi.org/10.1016/j.jcis.2015.03.014.

Liu, Meng, Zhihao Li, Yingxue Li, Jiajia Chen, and Quan Yuan. 2019a. "Self-Assembled Nanozyme Complexes with Enhanced Cascade Activity and High Stability for Colorimetric Detection of Glucose." *Chinese Chemical Letters* 30 (5): 1009–1012. https://doi.org/10.1016/j.cclet.2018.12.021.

Liu, Wei, Xiangyu Zhou, Lin Xu, Shidong Zhu, Shuo Yang, Xinfu Chen, Biao Dong, Xue Bai, Geyu Lu, and Hongwei Song. 2019b. "Graphene Quantum Dot-Functionalized Three-Dimensional Ordered Mesoporous ZnO for Acetone Detection toward Diagnosis of Diabetes." *Nanoscale* 11 (24): 11496–11504. https://doi.org/10.1039/c9nr00942f.

Lu, Chang, Xiangjiang Liu, Yunfeng Li, Fang Yu, Longhua Tang, Yanjie Hu, and Yibin Ying. 2015. "Multifunctional Janus Hematite: Silica Nanoparticles: Mimicking Peroxidase-like Activity and Sensitive Colorimetric Detection of Glucose." *ACS Applied Materials & Interfaces* 7 (28): 15395–15402. https://doi.org/10.1021/acsami.5b03423.

Mackenzie, Richard, and Bradley Elliott. 2014. "Akt/PKB Activation and Insulin Signaling: A Novel Insulin Signaling Pathway in the Treatment of Type 2 Diabetes." *Diabetes, Metabolic Syndrome and Obesity: Targets and Therapy* 7 (February): 55. https://doi.org/10.2147/dmso.s48260.

Mahmoudi, Fariba, Farzaneh Mahmoudi, Khadijeh Haghighat Gollo, and Mostafa M. Amini. 2021. "Novel Gold Nanoparticles: Green Synthesis with Eryngium Thyrsoideum Boiss Extract, Characterization, and in Vivo Investigations on Inflammatory Gene Expression and Biochemical Parameters in Type 2 Diabetic Rats." *Biological Trace Element Research* 200 (5): 2223–2232. https://doi.org/10.1007/s12011-021-02819-7.

Masuo, Kazuko. 2013. "Lifestyle Modification Is the First Line Treatment for Type 2 Diabetes." *Www. intechopen.com.* IntechOpen. www.intechopen.com/chapters/45238.

Mei, He, Wenqin Wu, Beibei Yu, Huimin Wu, Shengfu Wang, and Qinghua Xia. 2016. "Nonenzymatic Electrochemical Sensor Based on Fe@Pt Core: Shell Nanoparticles for Hydrogen Peroxide, Glucose and Formaldehyde." *Sensors and Actuators B: Chemical* 223 (February): 68–75. https://doi.org/10.1016/j.snb.2015.09.044.

Omayma, A. R. AboZaid, Sawsan M. El-Sonbaty, Neama M. A. Hamam, Mostafa A. Farrag, and Ahmad S. Kodous. 2022. "Chitosan-Encapsulated Nano-Selenium Targeting TCF7L2, PPARγ, and CAPN10 Genes in Diabetic Rats." *Biological Trace Element Research* 201 (1): 306–323. https://doi.org/10.1007/s12011-022-03140-7.

O'Neal, Katherine S., Jeremy L. Johnson, and Rebekah L. Panak. 2016. "Recognizing and Appropriately Treating Latent Autoimmune Diabetes in Adults." *Diabetes Spectrum* 29 (4): 249–452. https://doi.org/10.2337/ds15-0047.

Opris, Razvan, Corina Tatomir, Diana Olteanu, Remus Moldovan, Bianca Moldovan, Luminita David, Andras Nagy, Nicoleta Decea, Mihai Ludovic Kiss, and Gabriela Adriana Filip. 2017. "The Effect of Sambucus Nigra L. Extract and Phytosinthesized Gold Nanoparticles on Diabetic Rats." *Colloids and Surfaces B: Biointerfaces* 150 (February): 192–200. https://doi.org/10.1016/j.colsurfb.2016.11.033.

Patolsky, Fernando, Yossi Weizmann, and Itamar Willner. 2004. "Long-Range Electrical Contacting of Redox Enzymes by SWCNT Connectors." *Angewandte Chemie (International Ed. In English)* 43 (16): 2113–2117. https://doi.org/10.1002/anie.200353275.

Paz, Keren, Rina Hemi, Derek LeRoith, Avraham Karasik, Eytan Elhanany, Hannah Kanety, and Yehiel Zick. 1997. "A Molecular Basis for Insulin Resistance." *Journal of Biological Chemistry* 272 (47): 29911–29918. https://doi.org/10.1074/jbc.272.47.29911.

Pessler, D., A. Rudich, and N. Bashan. 2001. "Oxidative Stress Impairs Nuclear Proteins Binding to the Insulin Responsive Element in the GLUT4 Promoter." *Diabetologia* 44 (12): 2156–2164. https://doi.org/10.1007/s001250100024.

Pickup, John C., Faeiza Hussain, Nicholas D. Evans, Olaf J. Rolinski, and David J. S. Birch. 2005. "Fluorescence-Based Glucose Sensors." *Biosensors and Bioelectronics* 20 (12): 2555–2565. https://doi.org/10.1016/j.bios.2004.10.002.

Reno, Candace M., Erwin C. Puente, Zhenyu Sheng, Dorit Daphna-Iken, Adam J. Bree, Vanessa H. Routh, Barbara B. Kahn, and Simon J. Fisher. 2016. "Brain GLUT4 Knockout Mice Have Impaired Glucose Tolerance, Decreased Insulin Sensitivity, and Impaired Hypoglycemic Counterregulation." *Diabetes* 66 (3): 587–597. https://doi.org/10.2337/db16-0917.

Richardson, S. J., A. Willcox, A. J. Bone, A. K. Foulis, and N. G. Morgan. 2009. "Islet-Associated Macrophages in Type 2 Diabetes." *Diabetologia* 52 (8): 1686–1688. https://doi.org/10.1007/s00125-009-1410-z.

Righettoni, Marco, Anton Amann, and Sotiris E. Pratsinis. 2015. "Breath Analysis by Nanostructured Metal Oxides as Chemo-Resistive Gas Sensors." *Materials Today* 18 (3): 163–171. https://doi.org/10.1016/j.mattod.2014.08.017.

Righettoni, Marco, Antonio Tricoli, Samuel Gass, Alex Schmid, Anton Amann, and Sotiris E. Pratsinis. 2012. "Breath Acetone Monitoring by Portable Si: WO3 Gas Sensors." *Analytica Chimica Acta* 738 (August): 69–75. https://doi.org/10.1016/j.aca.2012.06.002.

Righettoni, Marco, Antonio Tricoli, and Sotiris E. Pratsinis. 2010. "Si:WO3Sensors for Highly Selective Detection of Acetone for Easy Diagnosis of Diabetes by Breath Analysis." *Analytical Chemistry* 82 (9): 3581–3587. https://doi.org/10.1021/ac902695n.

Rose, S., R. E. Frye, J. Slattery, R. Wynne, M. Tippett, S. Melnyk, and S. J. James. 2014. "Oxidative Stress Induces Mitochondrial Dysfunction in a Subset of Autistic Lymphoblastoid Cell Lines." *Translational Psychiatry* 4 (4): 1–8. https://doi.org/10.1038/tp.2014.15.

Salehi, Sara, Ehsan Nikan, Abbas Ali Khodadadi, and Yadollah Mortazavi. 2014. "Highly Sensitive Carbon Nanotubes: SnO2 Nanocomposite Sensor for Acetone Detection in Diabetes Mellitus Breath." *Sensors and Actuators B: Chemical* 205 (December): 261–267. https://doi.org/10.1016/j.snb.2014.08.082.

Samuel, Varman T., Kitt Falk Petersen, and Gerald I. Shulman. 2010. "Lipid-Induced Insulin Resistance: Unravelling the Mechanism." *The Lancet* 375 (9733): 2267–2277. https://doi.org/10.1016/s0140-6736(10)60408-4.

Sanaa Ayyoub, Bahaa Al-Trad, Alaa A. A. Aljabali, and Walhan Alshaer. 2022. "Biosynthesis of Gold Nanoparticles Using Leaf Extract of Dittrichia Viscosa and in Vivo Assessment of Its Anti-Diabetic Efficacy." *Drug Delivery and Translational Research* 12 (12): 2993–2999. https://doi.org/10.1007/s13346-022-01163-0.

Saravanakumar, Kandasamy, Anbazhagan Sathiyaseelan, Arokia Vijaya Anand Mariadoss, and Myeong-Hyeon Wang. 2021. "Antioxidant and Antidiabetic Properties of Biocompatible Ceria Oxide (CeO2) Nanoparticles in Mouse Fibroblast NIH3T3 and Insulin Resistant HepG2 Cells." *Ceramics International* 47 (6): 8618–8626. https://doi.org/10.1016/j.ceramint.2020.11.230.

Schieber, Michael, and Navdeep S. Chandel. 2014. "ROS Function in Redox Signaling and Oxidative Stress." *Current Biology* 24 (10): R453–R462. https://doi.org/10.1016/j.cub.2014.03.034.

Selim, Manar E., Yasmina M. Abd-Elhakim, and Laila Y. Al-Ayadhi. 2015. "Pancreatic Response to Gold Nanoparticles Includes Decrease of Oxidative Stress and Inflammation in Autistic Diabetic Model." *Cellular Physiology and Biochemistry* 35 (2): 586–600. https://doi.org/10.1159/000369721.

Shajaripour Jaberi, Seyedeh Yasaman, Ali Ghaffarinejad, and Eskandar Omidinia. 2019. "An Electrochemical Paper Based Nano-Genosensor Modified with Reduced Graphene Oxide-Gold Nanostructure for Determination of Glycated Hemoglobin in Blood." *Analytica Chimica Acta* 1078 (October): 42–52. https://doi.org/10.1016/j.aca.2019.06.018.

Sherwani, Shariq I., Haseeb A. Khan, Aishah Ekhzaimy, Afshan Masood, and Meena K. Sakharkar. 2016. "Significance of Hba1c Test in Diagnosis and Prognosis of Diabetic Patients." *Biomarker Insights* 11 (11): BMI.S38440. https://doi.org/10.4137/bmi.s38440.

Siddiqui, Shafayet Ahmed, Md. Mamun Or Rashid, Md. Giash Uddin, Fataha Nur Robel, Mohammad Salim Hossain, Md. Azizul Haque, and Md. Jakaria. 2020. "Biological Efficacy of Zinc

Oxide Nanoparticles against Diabetes: A Preliminary Study Conducted in Mice." *Bioscience Reports* 40 (4). https://doi.org/10.1042/bsr20193972.

Simos, Yannis V., Konstantinos Spyrou, Michaela Patila, Niki Karouta, Haralambos Stamatis, Dimitrios Gournis, Evangelia Dounousi, and Dimitrios Peschos. 2020. "Trends of Nanotechnology in Type 2 Diabetes Mellitus Treatment." *Asian Journal of Pharmaceutical Sciences* 16 (1). https://doi.org/10.1016/j.ajps.2020.05.001.

Smidt, Kamille, Niels Jessen, Andreas Brønden Petersen, Agnete Larsen, Nils Magnusson, Johanne Bruun Jeppesen, Meredin Stoltenberg, et al. 2009. "SLC30A3 Responds to Glucose-and Zinc Variations in ß-Cells and Is Critical for Insulin Production and in Vivo Glucose-Metabolism during ß-Cell Stress." Edited by Kathrin Maedler. *PLOS ONE* 4 (5): e5684. https://doi.org/10.1371/journal.pone.0005684.

Solgi, Torab, Iraj Amiri, Sara Soleimani Asl, Massoud Saidijam, Banafsheh Mirzaei Seresht, and Tayebe Artimani. 2021. "Antiapoptotic and Antioxidative Effects of Cerium Oxide Nanoparticles on the Testicular Tissues of Streptozotocin-Induced Diabetic Rats: An Experimental Study." *International Journal of Reproductive BioMedicine (IJRM)* 19 (7). https://doi.org/10.18502/ijrm.v19i7.9465.

Sreenivasamurthy, L. 2021. "Evolution in Diagnosis and Classification of Diabetes." *Journal of Diabetes Mellitus* 11 (05): 200–207. https://doi.org/10.4236/jdm.2021.115017.

Srivastava, Pallavee, Judith M. Braganca, and Meenal Kowshik. 2014. "In Vivo synthesis of Selenium Nanoparticles By Halococcus Salifodinae BK18 and Their Anti-Proliferative Properties against HeLa Cell Line." *Biotechnology Progress* 30 (6): 1480–1487. https://doi.org/10.1002/btpr.1992.

Tang, Kim San. 2019. "The Current and Future Perspectives of Zinc Oxide Nanoparticles in the Treatment of Diabetes Mellitus." *Life Sciences* 239 (December): 117011. https://doi.org/10.1016/j.lfs.2019.117011.

Tasyurek, Hale M., Hasan Ali Altunbas, Mustafa Kemal Balci, and Salih Sanlioglu. 2014. "Incretins: Their Physiology and Application in the Treatment of Diabetes Mellitus." *Diabetes/Metabolism Research and Reviews* 30 (5): 354–371. https://doi.org/10.1002/dmrr.2501.

Ul Haq, Muhammad Nisar, Ghulam Mujtaba Shah, and Farid Menaa. 2022. "Green Silver Nanoparticles Synthesized from Taverniera Couneifolia Elicits Effective Anti-Diabetic Effect in Alloxan-Induced Diabetic Wistar Rats." *Nanomaterials* 12 (7): 1035. https://doi.org/10.3390/nano12071035.

Ullah, Salim, Ali Shah, and Muhammad Imran Qureshi et al. 2021. "Antidiabetic and Hypolipidemic Potential of Green AgNPs against Diabetic Mice." *ACS Applied Bio Materials* 4 (4): 3433–3442. https://doi.org/10.1021/acsabm.1c00005.

Vinotha, Viswanathan, Arokiadhas Iswarya, Rajagopalan Thaya, and Marimuthu Govindarajan. 2019. "Synthesis of ZnO Nanoparticles Using Insulin-Rich Leaf Extract: Anti-Diabetic, Anti-biofilm and Anti-Oxidant Properties." *Journal of Photochemistry and Photobiology B: Biology* 197 (August): 111541. https://doi.org/10.1016/j.jphotobiol.2019.111541.

Wada, Jun, and Atsuko Nakatsuka. 2016. "Mitochondrial Dynamics and Mitochondrial Dysfunction in Diabetes." *Acta Medica Okayama* 70 (3): 151–158. https://okayama.pure.elsevier.com/en/publications/mitochondrial-dynamics-and-mitochondrial-dysfunction-in-diabetes.

Wahab, Maryam, Attya Bhatti, and Peter John. 2022. "Evaluation of Antidiabetic Activity of Biogenic Silver Nanoparticles Using Thymus Serpyllum on Streptozotocin-Induced Diabetic BALB/c Mice." *Polymers* 14 (15): 3138. https://doi.org/10.3390/polym14153138.

Wang, Hui-Chen, and An-Rong Lee. 2015. "Recent Developments in Blood Glucose Sensors." *Journal of Food and Drug Analysis* 23 (2): 191–200. https://doi.org/10.1016/j.jfda.2014.12.001.

Wang, Keke, Yansong Zhang, Chunyang Zhao, and Mingyan Jiang. 2018. "SGLT-2 Inhibitors and DPP-4 Inhibitors as Second-Line Drugs in Patients with Type 2 Diabetes: A Meta-Analysis of Randomized Clinical Trials." *Hormone and Metabolic Research* 50 (10): 768–777. https://doi.org/10.1055/a-0733-7919.

Wang, L., A. Teleki, S. E. Pratsinis, and P. I. Gouma. 2008. "Ferroelectric WO_3 Nanoparticles for Acetone Selective Detection." *Chemistry of Materials* 20 (15): 4794–4796. https://doi.org/10.1021/cm800761e.

Wang, Tiannan, Jing Wang, Xinge Hu, Xian-Ju Huang, and Guo-Xun Chen. 2020. "Current Understanding of Glucose Transporter 4 Expression and Functional Mechanisms." *World Journal of Biological Chemistry* 11 (3): 76–98. https://doi.org/10.4331/wjbc.v11.i3.76.

Wang, Xiao, Jing Su, Dongdong Zeng, Gang Liu, Lizhuang Liu, Yi Xu, Chenguang Wang, Xinxin Liu, Lu Wang, and Xianqiang Mi. 2019. "Gold Nano-Flowers (Au NFs) Modified Screen-Printed Carbon Electrode Electrochemical Biosensor for Label-Free and Quantitative Detection of Glycated Hemoglobin." *Talanta* 201 (August): 119–125. https://doi.org/10.1016/j.talanta.2019.03.100.

Wen, Zhenhai, Suqin Ci, and Jinghong Li. 2009. "Pt Nanoparticles Inserting in Carbon Nanotube Arrays: Nanocomposites for Glucose Biosensors." *The Journal of Physical Chemistry C* 113 (31): 13482–13487. https://doi.org/10.1021/jp902830z.

White, Michael G., James A. M. Shaw, and Roy Taylor. 2016. "Type 2 Diabetes: The Pathologic Basis of Reversible β-Cell Dysfunction." *Diabetes Care* 39 (11): 2080–2088. https://doi.org/10.2337/dc16-0619.

Wu, Xuee, Feng Zhao, John R Varcoe, Alfred E. Thumser, Claudio Avignone-Rossa, and Robert C. T. Slade. 2009. "Direct Electron Transfer of Glucose Oxidase Immobilized in an Ionic Liquid Reconstituted Cellulose-Carbon Nanotube Matrix." *Bioelectrochemistry (Amsterdam, Netherlands)* 77 (1): 64–68. https://doi.org/10.1016/j.bioelechem.2009.05.008.

Xie, Hui, Guiling Luo, Yanyan Niu, Wenju Weng, Yixing Zhao, Zhiqiang Ling, Chengxiang Ruan, Guangjiu Li, and Wei Sun. 2020. "Synthesis and Utilization of Co3O4 Doped Carbon Nanofiber for Fabrication of Hemoglobin-Based Electrochemical Sensor." *Materials Science & Engineering: C, Materials for Biological Applications* 107 (February): 110209. https://doi.org/10.1016/j.msec.2019.110209.

Xing, Ruiqing, Qingling Li, Lei Xia, Jian Song, Lin Xu, Jiahuan Zhang, Yi Xie, and Hongwei Song. 2015a. "Au-Modified Three-Dimensional in_2O_3 Inverse Opals: Synthesis and Improved Performance for Acetone Sensing toward Diagnosis of Diabetes." *Nanoscale* 7 (30): 13051–13060. https://doi.org/10.1039/c5nr02709h.

Yagati, Ajay Kumar, Hien T. Ngoc Le, and Sungbo Cho. 2020. "Bioelectrocatalysis of Hemoglobin on Electrodeposited Ag Nanoflowers toward H2O2 Detection." *Nanomaterials* 10 (9): 1628. https://doi.org/10.3390/nano10091628.

Yang, Liuqing, Xiangling Ren, Fangqiong Tang, and Lin Zhang. 2009. "A Practical Glucose Biosensor Based on Fe_3O_4 Nanoparticles and Chitosan/Nafion Composite Film." *Biosensors & Bioelectronics* 25 (4): 889–895. https://doi.org/10.1016/j.bios.2009.09.002.

Yaribeygi, Habib, Farin R. Farrokhi, Ramin Rezaee, and Amirhossein Sahebkar. 2018a. "Oxidative Stress Induces Renal Failure: A Review of Possible Molecular Pathways." *Journal of Cellular Biochemistry* 119 (4): 2990–2998. https://doi.org/10.1002/jcb.26450.

Yaribeygi, Habib, Mohammad Mohammadi, and Amirhossein Sahebkar. 2018b. "PPAR-α Agonist Improves Hyperglycemia-Induced Oxidative Stress in Pancreatic Cells by Potentiating Antioxidant Defense System." *Drug Research* 68 (06): 355–360. https://doi.org/10.1055/s-0043-121143.

Yoon, Jinho, Sang Nam Lee, Min Kyu Shin, Hyun-Woong Kim, Hye Kyu Choi, Taek Lee, and Jeong-Woo Choi. 2019. "Flexible Electrochemical Glucose Biosensor Based on GOx/Gold/MoS2/Gold Nanofilm on the Polymer Electrode." *Biosensors and Bioelectronics* 140 (September): 111343. https://doi.org/10.1016/j.bios.2019.111343.

Zhang, Libing, and Erkang Wang. 2014. "Metal Nanoclusters: New Fluorescent Probes for Sensors and Bioimaging." *Nano Today* 9 (1): 132–157. https://doi.org/10.1016/j.nantod.2014.02.010.

Zhang, Ruixin, Xuze Qin, Ting Zhang, Qian Li, Jianxin Zhang, and Junxing Zhao. 2018. "Astragalus Polysaccharide Improves Insulin Sensitivity via AMPK Activation in 3T3-L1 Adipocytes." *Molecules* 23 (10): 2711. https://doi.org/10.3390/molecules23102711.

Chapter 6

Nanotherapeutics for diabetic retinopathy using metal nanocomposites

Mitali Patel and Rutvi Vaidya

LIST OF ABBREVIATIONS

8-OHdG	8-hydroxy-2′-deoxyguanosine
ACE	angiotensin-converting enzyme
ACEIs	angiotensin-converting enzyme inhibitors
AGEs	advanced glycation end products
Ag NPs	silver nanoparticles
ALR2	aldose reductase
anti-VEGF	anti-vascular endothelial growth factor
ARBs	angiotensin receptor blockers
ARL2	ADP-ribosylation factor-like protein 2
Au NPs	gold nanoparticles
BBB	blood–brain barrier
BRB	blood–retinal barrier
Cu	copper
DM	diabetes mellitus
DME	diabetic macular edema
DMI	diabetic macular ischemia
DR	diabetic retinopathy
HDL	high-density lipoprotein
HDL-C	high-density lipoprotein cholesterol
HRECs	human retinal endothelial cells
ICAM-1	intercellular adhesion molecule-1
IL-6	interleukin-6
LDH	lactate dehydrogenase
LDL-C	low-density lipoprotein cholesterol
MCP-1	monocyte chemotactic proteins-1
MNPs	metal NPs
NPDR	nonproliferative diabetic retinopathy
NPs	nanoparticles
NSAIDs	nonsteroidal anti-inflammatory drugs
OCT	analog octreotide
PDR	proliferative diabetic retinopathy
PEG	polyethylene glycol

DOI: 10.1201/9781032621135-6

PKC	protein kinase C
PPAR-α	proliferator-activated receptor alpha
RAS	renin–angiotensin system
ROS	reactive oxygen species
RPE	retinal pigment epithelium
SOD	superoxide dismutase
SPR	surface plasmon resonance
TLR4	toll-like receptor 4
TNFα	tumor necrosis factor
VEGF	vascular endothelial growth factor
Zn	zinc

6.1 INTRODUCTION

Diabetic retinopathy (DR) is a potentially blinding microvascular complication of diabetes that affects the eyes, particularly the blood vessels in the retina. It is characterized by changes in retinal blood vessels, such as microaneurysms, hemorrhages, exudates, and neovascularization that can lead to macular edema and vision loss if left untreated. The severity of diabetic retinopathy can range from mild nonproliferative to severe proliferative forms, which can cause retinal detachment and blindness (American Diabetes Association, 2022). The retina performs a constant, consistent, and heavy amount of visual processing while maintaining homeostasis being an intricately designed part of the nervous system. It has neuronal, glial, and vascular cells working simultaneously to maintain the anatomy of the retina and perform the physiology of vision. These functional cells get affected by diabetes, and understanding which physiological processes are most sensitive can help pave the way to prevent some of the irreversible changes brought on by DR and find therapeutic interventions including surgery to reduce the damage or ultimate fate.

6.2 ANATOMY AND PHYSIOLOGY OF RETINA

6.2.1 Anatomy

The retina is a complex layered structure that lines the inner surface of the eye and is responsible for detecting and processing visual information. The retina is composed of several layers of cells, including photoreceptor cells, bipolar cells, and ganglion cells, among others (Reese, 2011). These cells work together to convert light into electrical signals that are transmitted to the brain, allowing us to see. It is composed of several layers of cells forming a delicate membrane (Manchanda et al., 2018). The photoreceptor cells are further divided into two types: rods which are highly organized and cones which are concentrated in the center (Burns and Baylor, 2001). Rods are responsible for detecting low levels of light and are most sensitive to light at the blue-green end of the spectrum, while cones are responsible for color vision and are most sensitive to light at the red-green and blue-violet ends of the spectrum (Masland, 2012). The bipolar cells are responsible for transmitting signals from the photoreceptor cells to the ganglion cells, which then send the signals to the brain via the optic nerve (Snell et al., 2013) and transmit visual information to the brain (Sanes and Masland, 2015; Kolb et al., 2014). The interneurons of the retina are responsible for processing the visual information before it is transmitted to the brain (Kolb et al., 2014; Masland, 2012). These cells include amacrine cells, horizontal cells, and Müller cells, which modulate the signals between the photoreceptor cells and bipolar cells (Levin et al., 2011; Dowling, 2011). The retina is a complex and dynamic structure that undergoes continuous changes throughout life, including synaptic remodeling, cell death and replacement, and alterations in gene expression (Kolb et al., 2014; Masland, 2012). These changes are critical for maintaining normal vision and responding to environmental challenges.

The neurovascular unit is composed of several key components, including endothelial cells, pericytes, astrocytes, and neurons (Abbott et al., 2010; Daneman et al., 2015), and plays a critical role in the function of the retina. The retina is one of the most metabolically active tissues in the body and requires a constant supply of nutrients and oxygen to function properly. The neurovascular unit regulates blood flow to the retina and maintains the integrity of the blood–retinal barrier (BRB), which is similar in function to the BBB (Cunha-Vaz, 2004). Astrocytes and Müller's cells are types of glial cells that are critical for regulating blood flow and maintaining the integrity of the BRB (Reese, 2011). Dysfunction of the neurovascular unit in the retina has been implicated in many retinal disorders, including diabetic retinopathy, age-related macular degeneration, and retinopathy of prematurity. The BRB restricts the entry of substances into the retina, including many drugs, and limits the effectiveness of systemic drug administration (Cunha-Vaz, 2004). This has led to the development of several approaches for delivering drugs to the retina, including intravitreal injection, subconjunctival injection, and topical administration.

6.2.2 Physiology

The retinal neurovascular unit plays a crucial role in maintaining retinal homeostasis and function (Kurihara et al., 2012). This unit is composed of various cell types, including endothelial cells, pericytes, astrocytes, and Müller cells, which interact with each other to regulate blood flow, nutrient delivery, and waste removal (Reese, 2011). The endothelial cells of the retinal vessels form tight junctions and regulate the transport of nutrients and metabolites between the blood and the neural tissue (Romeo et al., 2002). Pericytes, on the other hand, control blood flow by contracting or relaxing their processes in response to local signals (Ozdemir et al., 2012). Astrocytes provide structural support to the blood vessels and modulate blood flow by releasing vasoactive molecules (Puro, 2007). Müller cells, which span the entire thickness of the retina, maintain ionic and osmotic homeostasis and play a role in the clearance of metabolic waste products (Bringmann et al., 2004). Dysfunction of the retinal neurovascular unit has been implicated in various retinal diseases, including diabetic retinopathy, age-related macular degeneration, and retinal vein occlusion (Campochiaro, 2015; Kaur et al., 2008). In these conditions, the BRB is compromised, leading to increased vascular permeability and the accumulation of fluid and lipids in the retina. Neurovascular coupling is the coordination between neural activity and blood flow, which ensures an adequate supply of oxygen and nutrients to the retinal neurons. Disruption of neurovascular coupling can lead to neuronal dysfunction and vision loss in the case of DR (Riva et al., 2005).

6.3 PATHOPHYSIOLOGY OF DIABETIC RETINOPATHY

6.3.1 Types of DR

DR is a serious complication of diabetes mellitus and is one of the leading causes of blindness worldwide. DR can be broadly classified into two types: nonproliferative diabetic retinopathy (NPDR) and proliferative diabetic retinopathy (PDR). In NPDR, the earliest stage of DR, small retinal blood vessels become damaged, leading to the development of microaneurysms, intraretinal hemorrhages, and hard exudates. As NPDR progresses, capillary closure and ischemia can cause areas of retinal non-perfusion, leading to the development of macular edema and vision loss. In PDR, new abnormal blood vessels grow on the surface of the retina and into the vitreous, which can cause vitreous hemorrhage and tractional retinal detachment (Yau et al., 2012). NPDR is characterized by the presence of microaneurysms, intraretinal hemorrhages, hard exudates, and cotton wool spots. In contrast, PDR is characterized by the growth of new and fragile blood vessels on the surface of the retina, which can lead to vitreous hemorrhage, tractional retinal detachment, and neovascular glaucoma.

6.3.2 Epidemiology of DR

DR is a significant public health concern worldwide, affecting millions of people globally. According to recent epidemiological studies, it is estimated that approximately one-third of individuals with DM have DR (Solomon et al., 2017) with higher rates in individuals with type I DM compared to those with type II DM. A meta-analysis of global DR prevalence found that the overall prevalence of any DR was 35.4% (95% CI 30.4–40.8) among individuals with diabetes, with higher rates reported in Asia (45.8%; 95% CI 37.3–54.5) compared to Europe (33.5%; 95% CI 24.9–43.0) and North America (34.3%; 95% CI 26.5–42.9) (Yau et al., 2012). The prevalence of vision-threatening DR, defined as severe NPDR, PDR, or diabetic macular edema (DME), was estimated to be 10.2% (95% CI 7.8–13.3) globally (Yau et al., 2012). It is a leading cause of visual impairment and blindness among working-age adults globally (Bourne et al., 2013) and high in low- and middle-income countries, where access to eye care services may be limited (Kyari et al., 2014). An estimated 191 million people are living with the condition in 2021 (International Diabetes Federation, 2021). In addition, certain populations, such as Indigenous peoples and racial/ethnic minorities, are disproportionately affected by DR and its associated vision loss (Sivaprasad et al., 2012). A review analyzed a total of 48 meta-analyses and found that DR is significantly associated with an increased risk of mortality, cardiovascular disease, kidney disease, and neuropathy in individuals with diabetes. Moreover, the review found that DR is also associated with poorer mental health outcomes, including depression, anxiety, and lower quality of life (Trott et al., 2022).

6.3.3 Pathogenesis of DR

The pathogenesis of DR involves a complex interplay of various biochemical, molecular, and cellular processes, including chronic hyperglycemia, oxidative stress, inflammation, and angiogenesis (Antonetti et al., 2012). Chronic hyperglycemia is considered the primary trigger for DR development, as it leads to the formation of advanced glycation end products (AGEs) and the activation of various signaling pathways, including the polyol, hexosamine, and protein kinase C (PKC) pathways (Antonetti et al., 2012). These pathways contribute to retinal capillary basement membrane thickening, pericyte loss, and endothelial dysfunction, which are key features of DR.

Oxidative stress, inflammation, and dyslipidemia

Oxidative stress promotes the generation of reactive oxygen species (ROS), which cause retinal vascular injury and endothelial cell dysfunction (Du et al., 2013). Inflammation stimulates the production of pro-inflammatory cytokines and chemokines, leading to leukocyte infiltration and capillary closure (Antonetti et al., 2012). Angiogenesis is a hallmark of advanced DR, as it promotes the growth of new blood vessels, which are often leaky and prone to hemorrhage (Ferrara et al., 1997). Dyslipidemia is another factor for DR, characterized by elevated levels of triglycerides (TGA) and low-density lipoprotein cholesterol (LDL-C) and decreased levels of high-density lipoprotein cholesterol (HDL-C), leading to the accumulation of lipids in retinal capillaries and the development of retinal microaneurysms and hemorrhages. Due to elevated ROS vicious cycle of oxidative stress begins which involves the following factors, biomarkers, and genetic components too.

Vascular endothelial growth factor (VEGF)

VEGF is a potent angiogenic factor that stimulates the growth of new blood vessels, particularly in hypoxic conditions, such as those found in the retina of diabetic patients (Antonetti et al., 2012). Overexpression of VEGF in the retina of diabetic patients leads to the breakdown of the BRB and the development of DME (Aiello et al., 1994). VEGF promotes the proliferation and migration of endothelial cells, leading to the formation of new fragile and leaky blood vessels in the retina causing hemorrhages and scarring (Antonetti et al., 2012). VEGF may also be involved in the early stages of DR, including the development of diabetic macular ischemia (DMI). Enhanced VEGF expression can also cause DME, neovascular glaucoma, and PDR (Gupta et al., 2013).

Genetic factors

Several susceptibility genes involved in inflammation, oxidative stress, angiogenesis, and the regulation of glucose metabolism have been identified that increase the risk of developing DR in individuals with DM (Cho et al., 2014). Genetic variations in the genes encoding for VEGF, angiotensin-converting enzyme (ACE), and aldose reductase (ALR2) have been associated with an increased risk of DR with their polymorphism (Abhary et al., 2009; Jankovic et al., 2021) and for toll-like receptor 4 (TLR4) may contribute to the development and progression of DR. Several studies have reported aberrant DNA methylation patterns in genes involved in inflammation, angiogenesis, and oxidative stress in patients with diabetic retinopathy. For example, DNA methylation in the promoter region of the VEGF gene is associated with increased VEGF expression and severity of DR.

Renin–angiotensin system (RAS)

RAS which is activated in DM promotes inflammation, oxidative stress, and vascular dysfunction (Phipps et al., 2019). Inhibition of the RAS pathway has been shown to reduce the incidence and severity of DR in both animal models and clinical trials (Simó and Hernandez, 2015). In addition, the RAS has been shown to contribute to the breakdown of the BRB and the development of retinal edema, which are key features of DR (Rahimi et al., 2014). The efficacy of RAS inhibitors, ACEIs, and angiotensin receptor blockers (ARBs) is reported in preventing and treating DR by improving retinal microvascular function, reducing inflammation and oxidative stress, and preventing the breakdown of the BRB (Sjølie et al., 2008; Chaturvedi et al., 2008).

6.4 TREATMENT

DR is irreversible but can be prevented. After the occurrence, the key focus remains on delaying the progression and remedy for symptoms but removing or suppressing symptom-causing reasons. In Figure 6.1 major treatment options are shown that target pathogenesis in DR.

6.4.1 Standard of care

The preventive standard of care for diabetic retinopathy includes regular eye exams, consistent control of blood sugar levels, and timely treatment when necessary (American Diabetes Association, 2022). For those with no evidence of retinopathy, subsequent exams should occur annually. However, for those with evidence of retinopathy, exams should occur more frequently, as recommended by the eye care professional. Treatment options include laser therapy as a gold standard or injection of medications into the eye (American Diabetes Association, 2022).

6.4.2 As per mode and route

There are several routes and modes of treatment available for DR including one of the most common—laser photocoagulation, which involves using a laser to seal leaking blood vessels

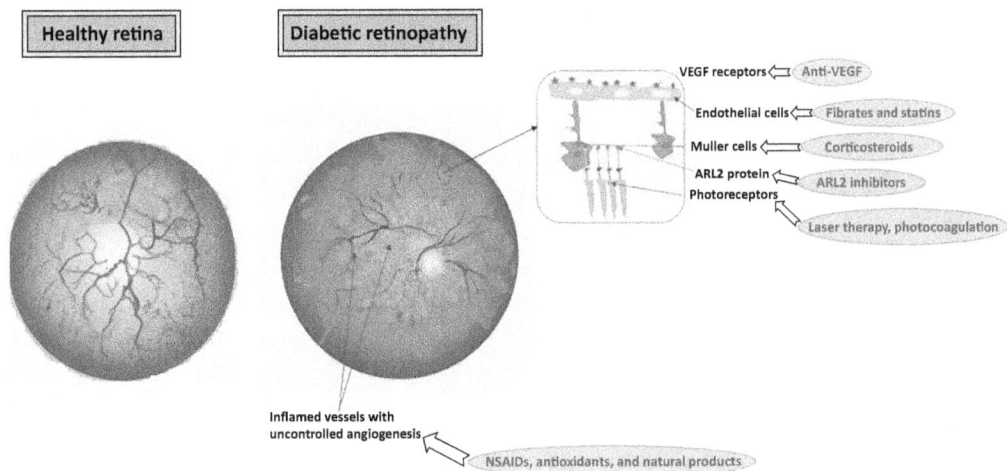

Healthy retina

Diabetic retinopathy

VEGF receptors ⇐ Anti-VEGF

Endothelial cells ⇐ Fibrates and statins

Muller cells ⇐ Corticosteroids

ARL2 protein ⇐ ARL2 inhibitors
Photoreceptors

Laser therapy, photocoagulation

Inflamed vessels with uncontrolled angiogenesis

NSAIDs, antioxidants, and natural products

Figure 6.1 Treatment targeting pathogenesis in diabetic retinopathy.

in the eye (American Diabetes Association, 2022). Another treatment option is intravitreal injections of anti-VEGF medications, which can help reduce swelling and leakage in the eye (National Eye Institute, 2022). A drug that is effective via the intravitreal implant route in treating inflammation in DR is dexamethasone (DI, Ozurdex®). In more severe cases of DR, vitrectomy may be necessary to remove scar tissue and blood from the eye (American Academy of Ophthalmology, 2020). In addition to laser imparting, intravitreal administration, and vitrectomy, other specific routes are periocular for steroids, topical for NSAIDs, and oral for pharmacotherapy with fibrates, ACE inhibitors, statins, etc. medications. It is important to note, however, that oral medication may not be effective for all patients with DR but may prevent progression or delay precipitation in high-risk patients.

6.4.3 Pharmacological agents used for the treatment of DR

Anti-inflammatory

1. Steroids

Dexamethasone is a glucocorticoid that exerts its anti-inflammatory effects by suppressing the production of cytokines and other pro-inflammatory mediators. Its biodegradable implant slowly releases dexamethasone into the vitreous, providing sustained anti-inflammatory effects (Silva et al., 2009).

2. Interleukin inhibitors

Interleukin inhibitors, such as tocilizumab and canakinumab, have been shown to reduce inflammation by inhibiting IL-6 and IL-1β to improve retinal function in patients with DR (Sheemar et al., 2021; Tang et al., 2023).

3. Integrins inhibitors

Integrins are cell surface receptors that play a crucial role in angiogenesis and have been implicated in the pathogenesis of DR. Volociximab and E7820 are integrins inhibitors developed for the treatment of DR, including targeting specific integrins α5β1 and α2 in the retina, which can reduce the progression of DR and show significant improvements in visual acuity and reduction in retinal thickness (Van Hove et al., 2021).

4. NSAIDs

According to the American Diabetes Association (2017), nonsteroidal anti-inflammatory drugs (NSAIDs) can be used to treat DR by reducing inflammation in the retina and improving visual acuity (Tarr et al., 2013). However, the use of NSAIDs in diabetic patients should be carefully monitored due to potential adverse effects on renal function (American Diabetes Association, 2017). Several NSAIDs are currently used for the treatment of diabetic retinopathy. Diclofenac reduces inflammation and improves visual acuity in patients with DR (Seth et al., 2016). Ketorolac improves visual acuity and reduces macular thickness (Bolinger and Antonetti, 2016). Nepafenac reduces inflammation and improves visual acuity in patients with DME (Kern et al., 2007). Aspirin while not typically used specifically for the treatment of DR due to its anti-inflammatory effects may be beneficial in reducing the risk of retinopathy in diabetic patients (Pafundi et al., 2020).

Anti-VEGF agents

Anti-vascular endothelial growth factor (anti-VEGF) agents have emerged as an effective treatment for DR and DME that improves the quality of life too (Simunovic and Maberley, 2015; Sam-Oyerinde et al., 2023). In a study, ranibizumab, aflibercept, and bevacizumab were found to be similarly effective in improving visual acuity and reducing central macular thickness in patients with DME (Wells et al., 2016). Anti-VEGFs are found to be more effective than photocoagulation in PDR (Heier et al., 2016). The use of anti-VEGF for DR and DME should be carefully monitored and managed considering intraocular adverse effects (Cox et al., 2021).

Statins

Statins show promising results in the prevention and treatment of DR (Pranata et al., 2021) by up to 47% delay in progression majorly due to anti-inflammatory and antioxidant properties along with blood lipid lowering (Yau et al., 2012).

Fibrates

Fibrates are effective in the treatment of DR by reducing the progression of the disease (Chew et al., 2014) by 30% in patients with type II diabetes (Keech et al., 2007). Fibrates are a class of medications that activate peroxisome proliferator-activated receptor alpha (PPAR-α), a nuclear receptor that regulates the transcription of genes involved in lipid metabolism, inflammation, and oxidative stress. Activation of PPAR-α by fibrates results in an increase in fatty acid oxidation, leading to a decrease in serum triglyceride levels and an increase in high-density lipoprotein (HDL) cholesterol levels (Tenenbaum et al., 2006). While fibrates are not currently approved by the FDA for the treatment of DR, they may be considered as an adjunct therapy in patients with diabetes and dyslipidemia who are at risk for DR progression (Su et al., 2019).

ARL2 inhibitors

ADP-ribosylation factor-like protein 2 (ARL2)—a small GTPase has been identified as a potential therapeutic target for DR. ARL2 inhibitors have shown promising results in preclinical studies by reducing vascular leakage and inflammation in the retina and improve retinal function by reducing retinal neovascularization in animal models of diabetic retinopathy. These findings suggest that ARL2 inhibitors could be a promising therapeutic option for the treatment of diabetic retinopathy, but safety and efficacy are needed to be considered (Julius et al., 2019).

Antioxidants and polyphenols

Antioxidants such as vitamin C, vitamin E, and carotenoids may have a protective effect against DR (Garcia-Medina et al., 2020). Resveratrol, a polyphenol found in red wine and grapes, may help to prevent or slow the progression of DR (Chen et al., 2019).

6.5 DRAWBACKS OF CONVENTIONAL THERAPY

Conventional therapy for DR has several drawbacks. Laser photocoagulation is effective in only certain stages, can't restore vision, and can cause visual field defects, decreased night vision, and decreased contrast sensitivity too (Flaxel et al., 2019). Conventional therapy often does not address the specific underlying causes and can be invasive and uncomfortable for patients. Anti-VEGF drugs may require frequent injections due to their short duration of action (Stewart, 2016). Another limitation of current formulations is the potential for immune reactions. Some patients may develop antibodies against anti-VEGF drugs, reducing their efficacy (Simó et al., 2014). Moreover, some formulations may not be suitable for all patients due to comorbidities. For instance, intravitreal corticosteroids may not be appropriate for patients with a history of steroid-induced glaucoma or cataracts. To address these limitations, new formulations and delivery methods are being developed. For example, sustained-release drug delivery systems may reduce the need for frequent injections and improve treatment efficacy (Sharma et al., 2021).

6.6 METAL NANOCOMPOSITES

Nanotechnology offers unique merits in the treatment of DR. Currently, the size ranging from 1 to 100 nm permits the drug molecules to smoothly pass through the BRB barrier. The nanoparticles can easily combine with particular receptors present on retinal pigment cells which permits them to get to the epithelial cells in a selected means, and thus the goal of DR treatment is achieved (Muller et al., 2017; Jonas, 2007).

Nanocomposites hold the distinctive properties of nanomaterials like high surface area, controlled release of drugs, improved chemical reactivity along with biodegradability, non-toxicity, and biocompatibility (Borodina et al., 2021; Kaurav et al., 2018). Generally, NPs are 100–10,000 fold smaller than the size of human cells which provides remarkable relation with biomolecules present inside and on the surface of the cells (Alomari et al., 2021). Among the NPs, inorganic NPs offer numerous benefits such as stability in the biological environment and standardized synthesis protocols. As the materials used for the fabrication of inorganic NPs have anti-angiogenic and anti-inflammatory properties, they can be a choice for the treatment of DR.

Metal NPs (MNPs) like magnetic, silver, and gold NPs look to have a favorable ability for the therapy of disorders triggered by higher generation of ROS (Lushchak et al., 2018). MNPs are considered an emerging area with a remarkable impact on drug delivery and imaging. Being small in size, MNPs can easily infiltrate through biological membranes through which macromolecules cannot pass. The surface of the MNPs can be decorated as per the need to modify the pharmacokinetic properties. For example, the coating of MNPs with polyethylene glycol (PEG) is widely employed for enhancing blood circulation time by minimizing the uptake by the mononuclear phagocyte system. The optical properties of MNPs like surface plasmon resonance (SPR) make them capable carriers for use in biomedical areas. The unique features of the MNPs such as tunable size, large surface area, and high pore volume make them suitable candidates for the DR (Chandrakala et al., 2022). They are extensively used for therapeutic agents, such as peptides, anticancer drugs, antibodies, nucleic acid, and so on. MNPs possess tunable optical properties. Moreover, they can enhance the water solubility of lipophilic drugs, increase blood circulation time, and reduce drug elimination.

Nowadays, MNPs are commonly used for early stage treatment, discovery, and diagnosis of various diseases. They are considered exceptional materials having size-dependent physicochemical characteristics which are unobtainable with NPs fabricated using organic material. MNPs-based nanomedicines approved by FDA clinically have been found to augment the bioavailability and effectiveness of drug delivery and reduce side effects owing to improved targeted delivery to active cellular uptake. The surface chemistry and doping techniques enable to design the MNPs, which undergo degradation under physiological conditions and hence can be effectively absorbed by several metabolic pathways without affecting the healthy tissues (Chandrakala et al., 2022; Mody et al., 2010).

MNPs have large surface energies which are crucial to understanding the thermodynamic behavior of the particles. The large surface area-to-volume ratio led to showing governing behavior of the atoms present at the surface as compared to the interior of the particles which eventually increase the overall surface energy. Various techniques such as molecular dynamics simulations, ab initio calculations, and classical thermodynamic calculations are used to determine surface free energies. MNPs possess high surface area which gives additional reactive sites along with high surface energy, and hence they are considered ideal candidates in drug delivery systems (Chandrakala et al., 2022; Zahran et al., 2019). MNPs like magnetic, silver, and gold NPs look to have a favorable ability for treating the disorders caused by excessive generation of ROS (Lushchak et al., 2018; Ahmad et al., 2013). Among MNPs, inorganic NPs like silver, gold, and silica display "self-therapeutic" effects without surface modification. The therapeutic effect is governed by particle size, shape, surface characteristics, and microenvironment of tissue which ultimately regulate the action of NPs in biological systems (Alomari et al., 2021; Ahmad et al., 2013).

Recently, inorganic NPs like noble metal NPs (gold, silver, platinum), magnetic NPs, silicon NPs, and carbon-based NPs are gaining more attention in ophthalmology (Figure 6.2). The MNPs exhibit resonance electron oscillation called localized surface plasmon resonance. They can mix noble metal-based nanoparticles in the biological environment, and their nontoxicity has influenced medicinal research (Yaqoob et al., 2020). Silver- and gold-based NPs display favorable potential in the treatment and prevention of ROS-generated disorders (Lushchak et al., 2018).

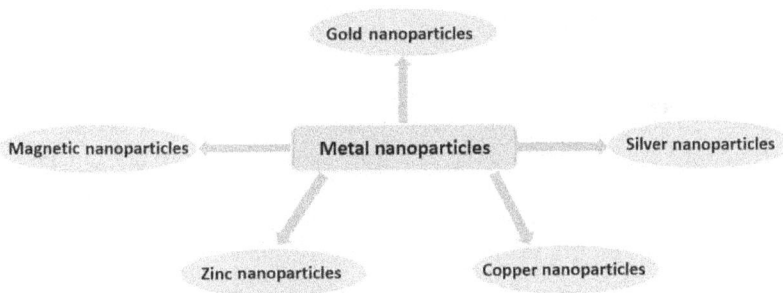

Figure 6.2 Different types of metallic nanoparticles.

6.6.1 Gold nanoparticles (Au NPs)

Au NPs are preferred over other MNPs owing to low toxicity, easy method of preparation, and capability of attachment with biological molecules (Ahmad et al., 2013). They are useful as therapeutic agents for the treatment of diabetes and related problems via their anti-hyperglycemic, antioxidant, antiglycation, anti-angiogenic, anti-inflammatory, and anti-fibrotic effects. Au NPs have a strong affinity to bind to molecules containing –SH and –NH$_2$ groups. Hence, biological molecules especially proteins can be a vital substrate in binding to Au NPs via residues of cysteine and lysine. The attachment of cysteine/lysine-rich proteins to Au NPs may modify their structural and physiological property which allows the Au NPs to be used as a therapeutic agent. It has been reported that uncoated Au NPs accumulated in high amounts in cells through cellular uptake against conjugated ones. It can be ascribed to the surface adsorption of serum proteins, whereas conjugation with materials such as PEG prevents cell surface interactions (Lushchak et al., 2018).

Treatment with Au NPs efficiently disrupts multiple pathogenesis determinants of diabetes and diabetes-related complications in animal models. Au NPs produce a dominant effect against oxidative damage by adding free –SH groups related to the reduced effect of antioxidant enzymes in biomolecules. Moreover, they are used as anti-inflammatory agents owing to their capacity to inhibit IL-6 and TNF-α. They also show antiglycation, anti-angiogenic, anti-hyperglycemic, and antioxidant effects (Alomari et al., 2021; Apaolaza et al., 2020; Masse et al., 2019).

6.6.2 Silver NPs (Ag NPs)

Ag NPs are used to deliver antiviral, antifungal, antibiotics, and anticancer drugs (Lawal et al., 2021). The Ag NPs are proven to be used widely due to their antimicrobial property against lethal viruses, microbes, and other microorganisms (Yaqoob et al., 2020). Silver is known to elicit antiviral, antibacterial, antifungal, and antioxidant properties. They also can enhance optical, thermal, electrical, and catalytic properties. Ag NPs associate with microbes and release the silver ion required for the inactivation of cellular-based enzymes and reduce membrane penetrability. They are found to produce cytotoxicity by apoptosis and necrosis toward different types of cells. Additionally, they display results in opposition to secondary effects of available therapies like DNA damage, generation of ROS, increasing leakage of lactate dehydrogenase (LDH), and reducing stem cell differentiation (Chandrakala et al., 2022; Simos et al., 2021).

6.6.3 Magnetic NPs

Nowadays, functionalized magnetic NPs display a propitious approach for the intraocular delivery of drugs. Magnetic NPs can be metallic, bimetallic, and superparamagnetic iron oxide NPs. As magnetic NPs possess reactive surfaces, they are preferred to functionalize with bioactive molecules and biocompatible coatings which in turn increase the specificity toward targets and avoid damage to the healthy tissues. Also, the magnetic property of the NPs is useful in the biomedical field. For example, as a contrast agent in magnetic resonance imaging or in drug delivery using a magnetic field as an external driving force. Magnetic

NPs are considered safe as they are biodegradable (entering the normal iron metabolism). The various types of magnetic NPs have been approved by FDA for clinical purposes, e.g., MRI contrast agents such as Combidex® by Advanced Magnetic Inc., Cambridge, USA, for the distinction of metastatic and non-metastatic lymph nodes; Endorem® developed by Amag Pharmaceutical Inc., Cambridge, USA, for identification of liver tumors; Resovist® developed by Bayer Schering Pharma AG, Berlin, Germany, for diagnosis of cancer of colon and metastases of liver. Feraheme® by Amag Pharmaceutical Inc., Cambridge, USA, is specified for the iron deficiency anemia treatment in adult patients with chronic kidney disease (Amato et al., 2018).

Magnetic NPs are preferred for the intraocular delivery of drugs. Certainly, intraocularly administered magnetic NPs have exhibited rapid entry into the retina and localization into the retinal pigment epithelium (RPE) of xenopus and zebrafish with no damage to the tissue (Amato et al., 2020).

The ocular applications of the magnetic NPs have not been tested yet in humans; their use has been proposed for the treatment of eye diseases (U.S. Patent 20130225906). The experimental results in the animals revealed that the iron oxide magnetic NPs showed negligible toxicity to the eye tissues. When magnetic NPs were functionalized with recombinant VEGF, the localization in choroid indicates that the magnetic NPs combined with molecules with different bioactivity may bring about tissue-specific targets in the posterior segment of the eye (Amato et al., 2018). Furthermore, magnetic nanoparticles have been shown to enhance the permeability of the BRB, which is often compromised in diabetic retinopathy. This increased permeability allows for better delivery of therapeutic agents to the retina, including anti-VEGF drugs and steroids, which are commonly used to treat DR.

6.6.4 Zinc and copper NPs

Copper (Cu) is pivotal in hyperglycemia, secretion and action of insulin secretion, mechanism, oxidative stress, and immunity. Cu and Zn regulate oxidative stress by controlling the level of superoxide dismutase (SOD), oxidases, and peroxidases. All these enzymes decrease the production of free radicals, preserve homeostasis, and diminish oxidative stress. It is demonstrated that Cu reinstates the pancreatic β cell islets and also facilitates secretion of insulin and the inflammatory response by promoting the secretion of interleukins (Miao et al., 2013). Deficiency of Cu causes neuropathy but also reduces vision and optic nerve involvement. Zn hinders the secretion of glucagon which holds significance in preserving normal ocular function. High levels of it were found in ocular tissue, mainly in the retina and the choroid. Continued Zn reduction causes retinal damage and the involved pathological mechanisms like augmented oxidative stress, lipofuscin buildup in the RPE, and photoreceptor disturbances (Dascalu et al., 2022; Ramasubbu et al., 2023).

6.7 APPLICATIONS OF METAL NANOCOMPOSITES IN DR

Researchers have already reported the positive outcome related to applications of MNPs in the prevention of DR. Some of the examples are shown in Table 6.1.

Table 6.1 Different metal nanoparticles in the treatment of DR

Formulation	Name of the drug	Outcome	References
Gold nanoparticles		The *in vivo* and *in vitro* study of Au NPs performed in the retinal region of C57BL/6 mice pups and microvascular endothelial cells of a retinal region of humans. Au NPs showed to prevent the generation of new vasculature structures by blocking the signals produced through the VEGFR-2 pathway without damaging normal retinal cells.	(Kim et al., 2011)
		Au NPs displayed an anti-inflammatory effect, beneficial for diabetic retinopathy. The effect was proved through the reduced level of IL-1β.	(Paula et al., 2015)
	Resveratrol	Streptozotocin-induced diabetic retinopathy was potentially reduced by Au NPs. The study explained, in brief, the pathways involved in the event and the reduced expression of VEGF-1, tumor necrosis factor (TNFα), monocyte chemotactic proteins-1 (MCP-1), intercellular adhesion molecule-1 (ICAM-1), and Interleukin (IL)-6, IL-1β displayed the effectiveness of Au NPs.	(Dong et al., 2019)
		Gold nanoparticles (Au NPs) were used for the diagnosis of diabetic retinopathy using urine 8-hydroxy-2'-deoxyguanosine (8-OHdG) as a detection marker. The paper-based sensor was able to quantify biomarkers through a colorimetric immunoassay sensor that demonstrated data on a smartphone camera, a feasible and cost-effective tool in the recent era.	(Hainsworth et al., 2020)
	Sorafenib tosylate	Au NPs, prepared using green synthesis, were functionalized with folic acid. FA-modified Au NPs were assessed to check their targetability on VEGF receptors located in the area with angiogenesis related to DR with sustained release of Sorafenib.	(Dave et al., 2020)
Silver nanoparticles		The importance of angiogenesis was highlighted, specifically VEGF since it holds significant importance in the induction of DR. Ag NPs of 50 nm were capable to reduce angiogenesis and PI3K activity associated with VEGF at concentrations as low as 500 nM. This indicated the possible utilization of Ag NPs as a therapeutic molecule for DR.	(Gurunathan et al., 2009)
		Ag NPs demonstrated obstruction of Src signals and anti-vaso permeability effects that prevent the biological events responsible to induce DR.	(Sheikpranbabu et al., 2010)
Magnetic nanoparticles		MNPs of varied magnetic fields were coated with silicon oxide and incorporated into microbubbles. The formulation showed promising results in targeting VEGF receptors of the retina region.	(Heun et al., 2017)
	Somatostatin	Magnetic NPs were modified with somatostatin analog octreotide (OCT) and exhibited efficiency on human retinal endothelial cells (HRECs) and in mouse retinal explants with no observed retinal toxicity. The effect required a very low concentration as compared to free OCT to treat DR. The enhanced activity was attributed to increased retention in the retina.	(Amato et al., 2020)
Zinc nanoparticles	*Cyperus rotundus*	Zinc oxide nanoparticles loaded with *Cyperus rotundus* showed their effectiveness in the treatment of DR through an inhibitory action on high expression of NLRP3 inflammasome, procaspase-1, cleaved-caspase-1, IL-18, ASC, and IL-1β.	(Zhang et al., 2020)

6.8 RECENT ADVANCES AND FUTURE PERSPECTIVES

Gene therapy with metal nanoparticles has shown potential as a treatment option for DR. Additionally, MNPs can be functionalized with targeting molecules, allowing for targeted delivery to specific cells or tissues. In addition to Au NPs, other metals such as silver, copper, and iron have also been explored for their potential use in gene therapy for DR. Further research is needed to determine the optimal MNPs for use in this therapy. Furthermore, the use of metal nanoparticles in gene therapy for diabetic retinopathy also holds promise for reducing the side effects and increasing the efficacy of current treatments, such as the administration of anti-VEGF agents through the intravitreal route (Amadio et al., 2016; Cai et al., 2008). Furthermore, the use of MNPs in gene therapy for DR can also provide a more targeted and precise approach to treatment compared to current therapies. Traditional therapies for DR, such as laser photocoagulation and intravitreal injections of anti-VEGF agents, are nonspecific and can cause damage to healthy retinal tissue. In contrast, metal nanoparticles can be engineered to selectively target specific cells or tissues, allowing for precise delivery of therapeutic genes to the affected areas in the retina. This can potentially minimize off-target effects and reduce the risk of damage to healthy tissues.

6.9 CONCLUSIONS

The usage of MNPs in the medical field has numerous merits like their superior biocompatibility and stability and their ability to hinder and efficiently disturb several proteins responsible to induce disease which is involved in the development of diabetic complications. Different metal nanoparticles could be possibly efficient to treat DR. Though, additional relevant studies regarding the safe effective size and effective dose are essential.

REFERENCES

Abbott, N. Joan, Adjanie A.K. Patabendige, Diana E.M. Dolman, Siti R. Yusof, and David J. Begley. 2010. "Structure and function of the blood: Brain barrier." *Neurobiology of Disease* 37: 13–25. doi: 10.1016/j.nbd.2009.07.030.

Abhary, Sotoodeh, Kathryn P. Burdon, Aanchal Gupta, Stewart Lake, Dinesh Selva, Nikolai Petrovsky, and Jamie E. Craig. 2009. "Common sequence variation in the VEGFA gene predicts risk of diabetic retinopathy." *Investigative Ophthalmology & Visual Science* 50: 5552–5558. doi: 10.1167/iovs.09-3694.

Ahmad, Tokeer, Irshad A. Wani, Nikhat Manzoor, Jahangeer Ahmed, and Abdullah M. Asiri. 2013. "Biosynthesis, structural characterization and antimicrobial activity of gold and silver nanoparticles." *Colloids and Surfaces B: Biointerfaces* 107: 227–234. doi: 10.1016/j.colsurfb.2013.02.004.

Aiello, Lloyd Paul, Robert L. Avery, Paul G. Arrigg, Bruce A. Keyt, Henry D. Jampel, Sabera T. Shah, Louis R. Pasquale et al. 1994. "Vascular endothelial growth factor in ocular fluid of patients with diabetic retinopathy and other retinal disorders." *New England Journal of Medicine* 331: 1480–1487. doi: 10.1056/NEJM199412013312203.

Alomari, Ghada, Salehhuddin Hamdan, and Bahaa Al-Trad. 2021. "Gold nanoparticles as a promising treatment for diabetes and its complications: Current and future potentials." *Brazilian Journal of Pharmaceutical Sciences* 57: e19040. doi: 10.1590/s2175-97902020000419040.

Amadio, Marialaura, Alessia Pascale, Sarha Cupri, Rosario Pignatello, Cecilia Osera, Gian Marco Leggio, Barbara Ruozi, Stefano Govoni, Filippo Drago, and Claudio Bucolo. 2016. "Nanosystems based on siRNA silencing HuR expression counteract diabetic retinopathy in rat." *Pharmacological Research* 111: 713–720. doi: 10.1016/j.phrs.2016.07.042.

Amato, Rosario, Martina Giannaccini, Massimo Dal Monte, Maurizio Cammalleri, Alessandro Pini, Vittoria Raffa, Matteo Lulli, and Giovanni Casini. 2020. "Association of the somatostatin analog octreotide with magnetic nanoparticles for intraocular delivery: A possible approach for the treatment of diabetic retinopathy." *Frontiers in Bioengineering and Biotechnology* 8: 144. doi: 10.3389/fbioe.2020.00144.

Amato, Rosario, Massimo Dal Monte, Matteo Lulli, Vittoria Raffa, and Giovanni Casini. 2018. "Nanoparticle-mediated delivery of neuroprotective substances for the treatment of diabetic retinopathy." *Current Neuropharmacology* 16: 993–1003. doi: 10.2174/1570159X15666170 717115654.

American Academy of Ophthalmology. (2020). *Diabetic Retinopathy*. Retrieved from www.aao.org/ eye-health/diseases/diabetic-retinopathy-cause-symptoms-treatment.

American Diabetes Association. 2017. "Standards of medical care in diabetes-2017." *Diabetes Care* 40: S1–S135.

American Diabetes Association Professional Practice Committee. 2022. "12. Retinopathy, Neuropathy, and Foot Care: Standards of Medical Care in Diabetes—2022." *Diabetes Care* 45: S185–S194. https://doi.org/10.2337/dc22-S012

American Diabetes Association Professional Practice Committee, and American Diabetes Association Professional Practice Committee. 2022. "2. Classification and diagnosis of diabetes: Standards of medical care in diabetes-2022." *Diabetes Care* 45: S17–S38. doi: 10.2337/dc22-S002.

Antonetti, David A., Ronald Klein, and Thomas W. Gardner. 2012. "Diabetic retinopathy." *The New England Journal of Medicine* 366: 1227–1239. doi: 10.1056/NEJMra1005073.

Apaolaza, P.S., M. Busch, E. Asin-Prieto, Karen Peynshaert, R. Rathod, Katrien Remaut, N. Dünker, and A. Göpferich. 2020. "Hyaluronic acid coating of gold nanoparticles for intraocular drug delivery: Evaluation of the surface properties and effect on their distribution." *Experimental Eye Research* 198: 108151. doi: 10.1016/j.exer.2020.108151.

Bolinger, Mark T., and David A. Antonetti. 2016. "Moving past anti-VEGF: Novel therapies for treating diabetic retinopathy." *International Journal of Molecular Sciences* 17: 1498. doi: 10.3390/ ijms17091498.

Borodina, Tatiana, Dmitry Kostyushev, Andrey A. Zamyatnin, Jr., and Alessandro Parodi. 2021. "Nanomedicine for treating diabetic retinopathy vascular degeneration." *International Journal of Translational Medicine* 1: 306–322. doi: 10.3390/ijtm1030018.

Bourne, Rupert R.A., Gretchen A. Stevens, Richard A. White, Jennifer L. Smith, Seth R. Flaxman, Holly Price, Jost B. Jonas et al. 2013. "Causes of vision loss worldwide, 1990–2010: A systematic analysis." *The Lancet Global Health* 1: e339–e349. doi: 10.1016/S2214-109X(13)70113-X.

Bringmann, Andreas, Andreas Reichenbach, and Peter Wiedemann. 2004. "Pathomechanisms of cystoid macular edema." *Ophthalmic Research* 36: 241–249. doi: 10.1159/000081203.

Burns, Marie E., and Denis A. Baylor. 2001. "Activation, deactivation, and adaptation in vertebrate photoreceptor cells." *Annual Review of Neuroscience* 24: 779–805. doi: 10.1146/annurev. neuro.24.1.779.

Cai, Xue, Shannon Conley, and Muna Naash. 2008. "Nanoparticle applications in ocular gene therapy." *Vision Research* 48: 319–324. doi: 10.1016/j.visres.2007.07.012.

Campochiaro, Peter A. 2015. "Molecular pathogenesis of retinal and choroidal vascular diseases." *Progress in Retinal and Eye Research* 49: 67–81. doi: 10.1016/j.preteyeres.2015.06.002.

Chandrakala, V., Valmiki Aruna, and Gangadhara Angajala. 2022. "Review on metal nanoparticles as nanocarriers: Current challenges and perspectives in drug delivery systems." *Emergent Materials* 5: 1593–1615. doi: 10.1007/s42247-021-00335-x.

Chaturvedi, Nish, Massimo Porta, Ronald Klein, Trevor Orchard, John Fuller, Hans Henrik Parving, Rudy Bilous, and Anne Katrin Sjølie. 2008. "Effect of candesartan on prevention (DIRECT-Prevent 1) and progression (DIRECT-Protect 1) of retinopathy in type 1 diabetes: Randomised, placebo-controlled trials." *The Lancet* 372: 1394–1402. doi: 10.1016/S0140-6736(08)61412-9.

Chen, Yuhua, Jiao Meng, Hua Li, Hong Wei, Fangfang Bi, Shi Liu, Kai Tang, Haiyu Guo, and Wei Liu. 2019. "Resveratrol exhibits an effect on attenuating retina inflammatory condition and damage of diabetic retinopathy via PON1." *Experimental Eye Research* 181: 356–366. doi: 10.1016/j.exer.2018.11.023.

Chew, Emily Y., Matthew D. Davis, Ronald P. Danis, James F. Lovato, Letitia H. Perdue, Craig Greven, Saul Genuth et al. 2014. "The effects of medical management on the progression of diabetic retinopathy in persons with type 2 diabetes: The Action to Control Cardiovascular Risk in Diabetes (ACCORD) eye study." *Ophthalmology* 121: 2443–2451. doi: 10.1016/j.ophtha.2014.07.019.

Cho, Heeyoon, and Lucia Sobrin. 2014. "Genetics of diabetic retinopathy." *Current Diabetes reports* 14: 1–7. doi: 10.1007/s11892-014-0515-z.

Cox, Jacob T., Dean Eliott, and Lucia Sobrin. 2021. "Inflammatory complications of intravitreal anti-VEGF injections." *Journal of Clinical Medicine* 10: 981. doi: 10.3390/jcm10050981.

Cunha-Vaz, José G. 2004. "The blood-retinal barriers system: Basic concepts and clinical evaluation." *Experimental Eye Research* 78: 715–721. doi: 10.1016/s0014-4835(03)00213-6.

Daneman, Richard, and Alexandre Prat. 2015. "The blood: Brain barrier." *Cold Spring Harbor Perspectives in Biology* 7: a020412. doi: 10.1101/cshperspect.a020412.

Dascalu, Ana Maria, Anca Anghelache, Daniela Stana, Andreea Cristina Costea, Vanessa Andrada Nicolae, Denisa Tanasescu, Daniel Ovidiu Costea et al. 2022. "Serum levels of copper and zinc in diabetic retinopathy: Potential new therapeutic targets." *Experimental and Therapeutic Medicine* 23: 1–6. doi: 10.3892/etm.2022.11253.

Dave, Vivek, Rekha Sharma, Chavi Gupta, and Srija Sur. 2020. "Folic acid modified gold nanoparticle for targeted delivery of Sorafenib tosylate towards the treatment of diabetic retinopathy." *Colloids and Surfaces B: Biointerfaces* 194: 111151. doi: 10.1016/j.colsurfb.2020.111151.

Dong, Yi, Guangming Wan, Panshi Yan, Cheng Qian, Fuzhen Li, and Guanghua Peng. 2019. "Fabrication of resveratrol coated gold nanoparticles and investigation of their effect on diabetic retinopathy in streptozotocin induced diabetic rats." *Journal of Photochemistry and Photobiology B: Biology* 195: 51–57. doi: 10.1016/j.jphotobiol.2019.04.012.

Dowling[RefCheck65] , John E. 2011. "The retina." In *Neurons and Networks: An Introduction to Neuroscience*, edited by John E. Dowling, 3rd ed., 161–200.

Du, Yunpeng, Alexander Veenstra, Krzysztof Palczewski, and Timothy S. Kern. 2013. "Photoreceptor cells are major contributors to diabetes-induced oxidative stress and local inflammation in the retina." *Proceedings of the National Academy of Sciences* 110: 16586–16591. doi: 10.1073/pnas.1314575110.

Ferrara, Napoleone, and Terri Davis-Smyth. 1997. "The biology of vascular endothelial growth factor." *Endocrine Reviews* 18: 4–25. doi: 10.1210/edrv.18.1.0287.

Flaxel, Christina J., Ron A. Adelman, Steven T. Bailey, Amani Fawzi, Jennifer I. Lim, G. Atma Vemulakonda, and Gui-shuang Ying. 2020. "Diabetic retinopathy preferred practice pattern®." *Ophthalmology* 127: 66–145. doi: 10.1016/j.ophtha.2019.09.025.

Garcia-Medina, Jose Javier, Elena Rubio-Velazquez, Elisa Foulquie-Moreno, Ricardo P. Casaroli-Marano, Maria Dolores Pinazo-Duran, Vicente Zanon-Moreno, and Monica del-Rio-Vellosillo. 2020. "Update on the effects of antioxidants on diabetic retinopathy: In vitro experiments, animal studies and clinical trials." *Antioxidants* 9: 561. doi: 10.3390/antiox9060561.

Gupta, N., S. Mansoor, A. Sharma, A. Sapkal, J. Sheth, P. Falatoonzadeh, B.D. Kuppermann, and M.C. Kenney. 2013. "Diabetic retinopathy and VEGF." *The Open Ophthalmology Journal* 7: 4. doi: 10.2174/1874364101307010004.

Gursoy Ozdemir, Yasemin, Muge Yemisci, and Turgay Dalkara. 2012. "Microvascular protection is essential for successful neuroprotection in stroke." *Journal of Neurochemistry* 123: 2–11. doi: 10.1111/j.1471-4159.2012.07938.x.

Gurunathan, Sangiliyandi, Kyung-Jin Lee, Kalimuthu Kalishwaralal, Sardarpasha Sheikpranbabu, Ramanathan Vaidyanathan, and Soo Hyun Eom. 2009. "Antiangiogenic properties of silver nanoparticles." *Biomaterials* 30: 6341–6350. doi: 10.1016/j.biomaterials.2009.08.008.

Hainsworth, Dean P., Abilash Gangula, Shreya Ghoshdastidar, Raghuraman Kannan, and Anandhi Upendran. 2020. "Diabetic retinopathy screening using a gold nanoparticle: Based paper strip assay for the at-home detection of the urinary biomarker 8-hydroxy-2′-deoxyguanosine." *American Journal of Ophthalmology* 213: 306–319. doi: 10.1016/j.ajo.2020.01.032.

Heier, Jeffrey S., Jean-François Korobelnik, David M. Brown, Ursula Schmidt-Erfurth, Diana V. Do, Edoardo Midena, David S. Boyer et al. 2016. "Intravitreal aflibercept for diabetic macular edema: 148-week results from the VISTA and VIVID studies." *Ophthalmology* 123: 2376–2385. doi: 10.1016/j.ophtha.2016.07.032.

Heun, Yvonn, Staffan Hildebrand, Alexandra Heidsieck, Bernhard Gleich, Martina Anton, Joachim Pircher, Andrea Ribeiro et al. 2017. "Targeting of magnetic nanoparticle-coated microbubbles to the vascular wall empowers site-specific lentiviral gene delivery in vivo." *Theranostics* 7: 295. doi: 10.7150/2Fthno.16192

International Diabetes Federation. (2021). *IDF Diabetes Atlas*, 10th ed. Retrieved from www.diabetesatlas.org/en/.

Jankovic, Milena, Ivana Novakovic, Dejan Nikolic, Jasmina Mitrovic Maksic, Slavko Brankovic, Ivana Petronic, Dragana Cirovic, Sinisa Ducic, Mirko Grajic, and Dragana Bogicevic. 2021. "Genetic and epigenomic modifiers of diabetic neuropathy." *International Journal of Molecular Sciences* 22: 4887. doi: 10.3390/ijms22094887.

Jonas, Jost B. 2007. "Intravitreal triamcinolone acetonide for diabetic retinopathy." *Diabetic Retinopathy* 39: 96–110. doi: 10.1159/000098502.

Julius, Angeline, and Waheeta Hopper. 2019. "A non-invasive, multi-target approach to treat diabetic retinopathy." *Biomedicine & Pharmacotherapy* 109: 708–715. doi: 10.1016/j.biopha.2018.10.185.

Kaur, C., W.S. Foulds, and E.A. Ling. 2008. "Blood-retinal barrier in hypoxic ischaemic conditions: Basic concepts, clinical features and management." *Progress in Retinal and EYE Research* 27: 622–647. doi: 10.1016/j.preteyeres.2008.09.003.

Kaurav, Hemlata, Satish Manchanda, Kamal Dua, and Deepak N. Kapoor. 2018. "Nanocomposites in controlled & targeted drug delivery systems." *Nano Hybrids and Composites* 20: 27–45. doi: 10.4028/www.scientific.net/nhc.20.27.

Keech, Anthony C., Paul Mitchell, P.A. Summanen, Justin O'Day, Timothy M.E. Davis, M.S. Moffitt, Marja-Riitta Taskinen et al. 2007. "Effect of fenofibrate on the need for laser treatment for diabetic retinopathy (FIELD study): A randomised controlled trial." *The Lancet* 370, no. 9600: 1687–1697. doi: 10.1016/S0140-6736(07)61607-9.

Kern, Timothy S., Casey M. Miller, Yunpeng Du, Ling Zheng, Susanne Mohr, Sherry L. Ball, M. Kim, Jeffrey A. Jamison, and David P. Bingaman. 2007. "Topical administration of nepafenac inhibits diabetes-induced retinal microvascular disease and underlying abnormalities of retinal metabolism and physiology." *Diabetes* 56: 373–379. doi: 10.2337/db05-1621.

Kim, Jin Hyoung, Myung Hun Kim, Dong Hyun Jo, Young Suk Yu, Tae Geol Lee, and Jeong Hun Kim. 2011. "The inhibition of retinal neovascularization by gold nanoparticles via suppression of VEGFR-2 activation." *Biomaterials* 32: 1865–1871. doi: 10.1016/j.biomaterials.2010.11.030.

Kolb, Helga, Eduardo Fernandez, and Ralph Nelson. 2014. "Webvision: The organization of the retina and visual system [Internet]." Retrieved from www.ncbi.nlm.nih.gov/books/NBK11535/.

Kurihara, Toshihide, Peter D. Westenskow, Stephen Bravo, Edith Aguilar, and Martin Friedlander. 2012. "Targeted deletion of Vegfa in adult mice induces vision loss." *The Journal of Clinical Investigation* 122: 4213–4217. doi: 10.1172/JCI65157.

Kyari, Fatima, Abubakar Tafida, Selvaraj Sivasubramaniam, Gudlavalleti V.S. Murthy, Tunde Peto, and Clare E. Gilbert. 2014. "Prevalence and risk factors for diabetes and diabetic retinopathy:

Results from the Nigeria national blindness and visual impairment survey." *BMC Public Health* 14: 1–12. doi: 10.1186/1471-2458-14-1299. PMID: 25523434; PMCID: PMC4301086.

Lawal, Sodiq Kolawole, Samuel Oluwaseun Olojede, Ayobami Dare, Oluwaseun Samuel Faborode, Edwin Coleridge S. Naidu, Carmen Olivia Rennie, and Onyemaechi Okpara Azu. 2021. "Silver nanoparticles conjugate attenuates highly active antiretroviral therapy-induced hippocampal NISSL substance and cognitive deficits in diabetic rats." *Journal of Diabetes Research* 2021: 2118538. doi: 10.1155/2021/2118538.

Levin, Leonard A., and Siv F.E. Nilsson. 2011. In Leonard A. Levin, Siv F. E. Nilsson, James Ver Hoeve, Samuel Wu, Paul L. Kaufman, Albert Alm (eds.), *Adler's Physiology of the Eye*, 11th ed., Elsevier, 369–409.

Lushchak, Oleh, Alina Zayachkivska, and Alexander Vaiserman. 2018. "Metallic nanoantioxidants as potential therapeutics for type 2 diabetes: A hypothetical background and translational perspectives." *Oxidative Medicine and Cellular Longevity* 2018: 3407375–3407383. doi: 10.1155%2F2018%2F3407375.

Manchanda, Satish, Kamal Dua, and Deepak Kapoor. 2018. "Guyton and hall textbook of medical physiology." *Nano Hybrids and Composites* 20: 27–45. doi: 10.4028/www.scientific.net/nhc.20.27.

Masland, Richard H. 2012. "The neuronal organization of the retina." *Neuron* 76: 266–280. doi: 10.1016/j.neuron.2012.10.002.

Masse, Florence, Mathieu Ouellette, Guillaume Lamoureux, and Elodie Boisselier. 2019. "Gold nanoparticles in ophthalmology." *Medicinal Research Reviews* 39: 302–327. doi: 10.1002/med.21509.

Miao, Xiao, Weixia Sun, Lining Miao, Yaowen Fu, Yonggang Wang, Guanfang Su, and Quan Liu. 2013. "Zinc and diabetic retinopathy." *Journal of Diabetes Research* 2013: 425854. doi: 10.1155/2013/425854.

Mody, Vicky V., Rodney Siwale, Ajay Singh, and Hardik R. Mody. 2010. "Introduction to metallic nanoparticles." *Journal of Pharmacy and Bioallied Sciences* 2: 282–289. doi: 10.4103/0975-7406.72127.

Muller, Alexandre Pastoris, Gabriela K. Ferreira, Allison Jose Pires, Gustavo de Bem Silveira, Débora Laureano de Souza, Joice de Abreu Brandolfi, Claudio Teodoro de Souza, Marcos M.S. Paula, and Paulo Cesar Lock Silveira. 2017. "Gold nanoparticles prevent cognitive deficits, oxidative stress and inflammation in a rat model of sporadic dementia of Alzheimer's type." *Materials Science and Engineering: C* 77: 476–483. doi: 10.1016/j.msec.2017.03.283.

National Eye Institute. (2022). *Diabetic Eye Disease*. Retrieved from www.nei.nih.gov/learn-about-eye-health/eye-conditions-and-diseases/diabetic-eye-disease.

Pafundi, Pia Clara, Raffaele Galiero, Alfredo Caturano, Carlo Acierno, Chiara de Sio, Erica Vetrano, Riccardo Nevola et al. 2020. "Aspirin in a diabetic retinopathy setting: Insights from NO BLIND study." *Nutrition, Metabolism and Cardiovascular Diseases* 30: 1806–1812. doi: 10.1016/j.numecd.2020.06.021.

Paula, Marcos M.S., Fabricia Petronilho, Francieli Vuolo, Gabriela K. Ferreira, Leandro De Costa, Giulia P. Santos, Pauline S. Effting et al. 2015. "Gold nanoparticles and/or N-acetylcysteine mediate carrageenan-induced inflammation and oxidative stress in a concentration-dependent manner." *Journal of Biomedical Materials Research Part A* 103: 3323–3330. doi: 10.1002/jbm.a.35469.

Phipps, Joanna A., Michael A. Dixon, Andrew I. Jobling, Anna Y. Wang, Ursula Greferath, Kirstan A. Vessey, and Erica L. Fletcher. 2019. "The renin-angiotensin system and the retinal neurovascular unit: A role in vascular regulation and disease." *Experimental Eye Research* 187: 107753. doi: 10.1016/j.exer.2019.107753.

Pranata, Raymond, Rachel Vania, and Andi Arus Victor. 2021. "Statin reduces the incidence of diabetic retinopathy and its need for intervention: A systematic review and meta-analysis." *European Journal of Ophthalmology* 31: 1216–1224. doi: 10.1177/1120672120922444.

Puro, Donald G. 2007. "Physiology and pathobiology of the pericyte-containing retinal microvasculature: New developments." *Microcirculation* 14: 1–10. doi: 10.1080/10739680601072099.

Rahimi, Zohreh, Mahmoudreza Moradi, and Hamid Nasri. 2014. "A systematic review of the role of renin angiotensin aldosterone system genes in diabetes mellitus, diabetic retinopathy and diabetic neuropathy." *Journal of Research in Medical Sciences: The Official Journal of Isfahan University of Medical Sciences* 19: 1090. PMID: 25657757.

Ramasubbu, Kanagavalli, Siddharth Padmanabhan, Khalid A. Al-Ghanim, Marcello Nicoletti, Marimuthu Govindarajan, Nadezhda Sachivkina, and Vijayarangan Devi Rajeswari. 2023. "Green synthesis of copper oxide nanoparticles using sesbania grandiflora leaf extract and their evaluation of anti-diabetic, cytotoxic, anti-microbial, and anti-inflammatory properties in an in-vitro approach." *Fermentation* 9: 332. doi: 10.3390/fermentation9040332.

Reese, Benjamin E. 2011. "Development of the retina and optic pathway." *Vision Research* 51: 613–632. doi: 10.1016%2Fj.visres.2010.07.010.

Riva, Charles E., Eric Logean, and Benedetto Falsini. 2005. "Visually evoked hemodynamical response and assessment of neurovascular coupling in the optic nerve and retina." *Progress in Retinal and Eye Research* 24: 183–215. doi: 10.1016/j.preteyeres.2004.07.002.

Romeo, Giulio, Wei-Hua Liu, Veronica Asnaghi, Timothy S. Kern, and Mara Lorenzi. 2002. "Activation of nuclear factor-κB induced by diabetes and high glucose regulates a proapoptotic program in retinal pericytes." *Diabetes* 51: 2241–2248. doi: 10.2337/diabetes.51.7.2241.

Sam-Oyerinde, Olapeju A., and Praveen J. Patel. 2023. "Real-world outcomes of anti-VEGF therapy in diabetic macular oedema: Barriers to treatment success and implications for low/lower-middle-income countries." *Ophthalmology and Therapy*: 809–826. doi: 10.1007/s40123-023-00672-6.

Sanes, Joshua R., and Richard H. Masland. 2015. "The types of retinal ganglion cells: Current status and implications for neuronal classification." *Annual Review of Neuroscience* 38: 221–246. doi: 10.1146/annurev-neuro-071714-034120.

Seth, Anisha, Basudeb Ghosh, Usha K. Raina, Anika Gupta, and Supriya Arora. 2016. "Intravitreal diclofenac in the treatment of macular edema due to branch retinal vein occlusion." *Ophthalmic Surgery, Lasers and Imaging Retina* 47: 149–155. doi: 10.3928/23258160-20160126-08.

Sharma, Deep Shikha, Sheetu Wadhwa, Monica Gulati, Arya Kadukkattil Ramanunny, Ankit Awasthi, Sachin Kumar Singh, Rubiya Khursheed et al. 2021. "Recent advances in intraocular and novel drug delivery systems for the treatment of diabetic retinopathy." *Expert Opinion on Drug Delivery* 18: 553–576. doi: 10.1080/17425247.2021.1846518.

Sheemar, Abhishek, Deepak Soni, Brijesh Takkar, Soumyava Basu, and Pradeep Venkatesh. 2021. "Inflammatory mediators in diabetic retinopathy: Deriving clinicopathological correlations for potential targeted therapy." *Indian Journal of Ophthalmology* 69: 3035–3049. doi: 10.4103/ijo.IJO_1326_21.

Sheikpranbabu, Sardarpasha, Kalimuthu Kalishwaralal, Kyung-jin Lee, Ramanathan Vaidyanathan, Soo Hyun Eom, and Sangiliyandi Gurunathan. 2010. "The inhibition of advanced glycation end-products-induced retinal vascular permeability by silver nanoparticles." *Biomaterials* 31: 2260–2271. doi: 10.1016/j.biomaterials.2009.11.076.

Silva, Paolo S., Jennifer K. Sun, and Lloyd Paul Aiello. 2009. "Role of steroids in the management of diabetic macular edema and proliferative diabetic retinopathy." *Seminars in Ophthalmology* 24: 93–99. doi: 10.1080/08820530902800355.

Simó, Rafael, and Cristina Hernandez. 2015. "Novel approaches for treating diabetic retinopathy based on recent pathogenic evidence." *Progress in Retinal and Eye Research* 48: 160–180. doi: 10.1016/j.preteyeres.2015.04.003.

Simó, Rafael, Jeffrey M. Sundstrom, and David A. Antonetti. 2014. "Ocular anti-VEGF therapy for diabetic retinopathy: The role of VEGF in the pathogenesis of diabetic retinopathy." *Diabetes care* 37: 893–899. doi: 10.1159/000317909.

Simos, Yannis V., Konstantinos Spyrou, Michaela Patila, Niki Karouta, Haralambos Stamatis, Dimitrios Gournis, Evangelia Dounousi, and Dimitrios Peschos. 2021. "Trends of nanotechnology in type 2 diabetes mellitus treatment." *Asian Journal of Pharmaceutical Sciences* 16: 62–76. doi: 10.1016/j.ajps.2020.05.001.

Simunovic, Matthew P., and David A.L. Maberley. 2015. "Anti-vascular endothelial growth factor therapy for proliferative diabetic retinopathy: A systematic review and meta-analysis." *Retina* 35: 1931–1942. doi: 10.1097/IAE.0000000000000723.

Sivaprasad, Sobha, Bhaskar Gupta, Roxanne Crosby-Nwaobi, and Jennifer Evans. 2012. "Prevalence of diabetic retinopathy in various ethnic groups: A worldwide perspective." *Survey of Ophthalmology* 57: 347–370. doi: 10.1016/j.survophthal.2012.01.004.

Sjølie, Anne Katrin, Ronald Klein, Massimo Porta, Trevor Orchard, John Fuller, Hans Henrik Parving, Rudy Bilous, and Nish Chaturvedi. 2008. "Effect of candesartan on progression and regression of retinopathy in type 2 diabetes (DIRECT-Protect 2): A randomised placebo-controlled trial." *The Lancet* 372: 1385–1393. doi: 10.1016/S0140-6736(08)61411-7.

Snell, Richard S., and Michael A. Lemp. 2013. *Clinical Anatomy of the Eye.* John Wiley & Sons.

Solomon, S.D., E. Chew, E.J. Duh, L. Sobrin, J.K. Sun, B.L. VanderBeek, C.C. Wykoff, and T.W. Gardner. 2017, Mar. "Diabetic retinopathy: A position statement by the American diabetes association." *Diabetes Care* 40, no. 3: 412–418. doi: 10.2337/dc16-2641. Erratum in: *Diabetes Care.* 2017 Jun;40, no. 6: 809. Erratum in: *Diabetes Care* 2017 Jul 13: PMID: 28223445; PMCID: PMC5402875.

Stewart, Michael W. 2016. "Treatment strategies for chorioretinal vascular diseases: Advantages and disadvantages of individualised therapy." *EMJ Diabet* 4: 91–98.

Su, Xing-jie, Lin Han, Yan-Xiu Qi, and Hong-wei Liu. 2019. "Efficacy of fenofibrate for diabetic retinopathy: A systematic review protocol." *Medicine* 98: e14999. doi: 10.1097/MD.0000000000014999.

Tang, Lei, Guo-Tong Xu, and Jing-Fa Zhang. 2023. "Inflammation in diabetic retinopathy: Possible roles in pathogenesis and potential implications for therapy." *Neural Regeneration Research* 18: 976–982. doi: 10.4103/1673-5374.355743.

Tarr, Joanna M., Kirti Kaul, Mohit Chopra, Eva M. Kohner, and Rakesh Chibber. 2013. "Pathophysiology of diabetic retinopathy." *International Scholarly Research Notices*: 343560. doi: 10.1155/2013/343560.

Tenenbaum, Alexander, Enrique Z. Fisman, Valentina Boyko, Michal Benderly, David Tanne, Moti Haim, Zipora Matas, Michael Motro, and Solomon Behar. 2006. "Attenuation of progression of insulin resistance in patients with coronary artery disease by bezafibrate." *Archives of Internal Medicine* 166: 737–741. doi: 10.1001/archinte.166.7.737.

Trott, Mike, Robin Driscoll, and Shahina Pardhan. 2022. "Associations between diabetic retinopathy, mortality, disease, and mental health: An umbrella review of observational meta-analyses." *BMC Endocrine Disorders* 22: 1–10. doi: 10.1186/s12902-022-01236-8.

Van Hove, Inge, Tjing-Tjing Hu, Karen Beets, Tine Van Bergen, Isabelle Etienne, Alan W. Stitt, Elke Vermassen, and Jean H.M. Feyen. 2021. "Targeting RGD-binding integrins as an integrative therapy for diabetic retinopathy and neovascular age-related macular degeneration." *Progress in Retinal and Eye Research* 85: 100966. doi: 10.1016/j.preteyeres.2021.100966.

Wells, John A., Adam R. Glassman, Allison R. Ayala, Lee M. Jampol, Neil M. Bressler, Susan B. Bressler, Alexander J. Brucker et al. 2016. "Aflibercept, bevacizumab, or ranibizumab for diabetic macular edema: Two-year results from a comparative effectiveness randomized clinical trial." *Ophthalmology* 123: 1351–1359. doi: 10.1016/j.ophtha.2016.02.022.

Yaqoob, Asim Ali, Hilal Ahmad, Tabassum Parveen, Akil Ahmad, Mohammad Oves, Iqbal M.I. Ismail, Huda A. Qari, Khalid Umar, and Mohamad Nasir Mohamad Ibrahim. 2020. "Recent advances in metal decorated nanomaterials and their various biological applications: A review." *Frontiers in Chemistry* 8: 341. doi: 10.3389/fchem.2020.00341.

Yau, J.W., S.L. Rogers, R. Kawasaki, E.L. Lamoureux, J.W. Kowalski, T. Bek, S.J. Chen et al., Meta-Analysis for Eye Disease [META-EYE] Study Group. 2012. "Global prevalence and major risk factors of diabetic retinopathy." *Diabetes Care* 35: 556–564. doi: 10.2337/dc11-1909.

Zahran, Moustafa, and Amal H. Marei. 2019. "Innovative natural polymer metal nanocomposites and their antimicrobial activity." *International Journal of Biological Macromolecules* 136: 586–596. doi: 10.1016/j.ijbiomac.2019.06.114.

Zhang, Liwei, Wen Chu, Lei Zheng, Juanjuan Li, Yuling Ren, Liping Xue, Wenhua Duan, Qing Wang, and Hua Li. 2020. "Zinc oxide nanoparticles from Cyperus rotundus attenuates diabetic retinopathy by inhibiting NLRP3 inflammasome activation in STZ induced diabetic rats." *Journal of Biochemical and Molecular Toxicology* 34: e22583. doi: 10.1002/jbt.22583.

Chapter 7

Nanotherapeutics for diabetic nephropathy using metal nanocomposites

Jhuma Samanta, Ashok Behera, and Aman Chaudhary

LIST OF ABBREVIATIONS

ACEIs	Angiotensin-converting enzyme inhibitors
AGEs	Advanced glycation end-products
Ag-NCs	Ag NPs, chitosan, and ascorbic acid nanocomposites
Ag NPs	Silver nanoparticles
AGT	Anti-angiotensinogen
AKI	Acute kidney injury
AKr1B1	Aldo-keto reductase
AMP	Adenosine monophosphate
AQP11	Aquaporin11
ARBs	Angiotensin receptor blockers
ATRA	All-trans-retinoic acid
Au NP	Gold nanoparticle
BWs	Body weights
CKD	Chronic kidney disease
C-Mn$_3$O$_4$	Citrate-functionalized manganese oxide
COX-2	Cyclooxygenase-2
CRP	C-reactive protein
CS/NaLS/Au NPs	Chitosan/sodium lignosulfonate/Au nanoparticles
CS–SeNPs	Chitosan–selenium nanoparticles
CTLHNs	Chitosan/tripolyphosphate lipid hybrid nanoparticles
CT-PLGA-NP	Crocetin-Polylactic-co-glycolic Acid-Nanoparticle
DN	Diabetic nephropathy
DNA	Deoxy ribonucleic acid
FBG	Fasting blood sugar
GFR	Glomerular filtration rate
GLP-1 RA	Glucagon-like peptide-1 receptor agonists
GM-CSF	Granulocyte-macrophage colony-stimulating factor
GPE	Galactose polyethyleneimine glycoprotein urethane
GSH	Glutathione reduced
GSK-3	Glycogen synthase kinase-3
HFD/STZ	High-fat diet/streptozotocin
HMSN	Renoprotective void mesoporous silica nanocomposite
ICAM-1	Intercellular adhesion molecule-1

DOI: 10.1201/9781032621135-7

IL-1	Interleukin-1
IL6	Interleukin-6
LC3	Microtubule-associated protein 1 light chain 3
LOX-1	Low-density lipoprotein receptor 1
MAPK/NF-KB/STAT3/cytokine	Mitogen-activated protein kinase/nuclear factor kappa B/signal transducer and activator of transcription 3
MDA	Malondialdehyde
MET	Metformin
MET-HMSN-CeO2	Multifunctional nanoparticles loaded with metformin
Mg(OH)$_2$NP-Mm	Magnesium-based nanoparticles using *Monodora myristica* seed
miRNA-377	Micro RNA 377
mPTP	Mitochondrial permeability transition pore
mRNA	Messenger ribonucleic acid
mTOR	Mechanistic target of rapamycin
MXene-Au	Titanium carbide (MXene) nanosheets-gold
NADPH	Reduced nicotinamide adenine dinucleotide phosphate
NLRP3	Nucleotide-binding domain, leucine-rich-containing family, pyrin domain-containing-3
NO	Nitric oxide
NOX4/p-47	NADPH oxidase 4
NPs	Nanoparticles
Nrf2	Nuclear factor E2-related factor 2
pAMPK	Protein AMP-activated protein kinase
PEGL	Pegylated liposomes
PI3K/AKT	Phosphoinositide-3-kinase-protein kinase B/Akt
PPE-AuNP	Pomegranate extract-stabilized
Q	Free quercetin
Q-PEGL	Pegylated quercetin liposomes
RH	Rhein
rHI	Renal hypertrophy index
RLN	Rubidian-loaded niosomes
Se NPs	Selenium nanoparticles
SGLT-2i	Sodium–glucose cotransporter 2 inhibitors
SLN	Solid lipid nanoparticles
SOD	Superoxide dismutase
STZ	Streptozotocin
STZ-NA	Streptozotocin-nicotinamide
T2DM	Type 2 diabetes mellitus
TGF-1	Transforming growth factor-1
TNF	Tumor necrosis factor
TP	Total protein levels
TXNIP	Thioredoxin-interacting protein
USPIOs	Ultrasmall nanoparticles of superparamagnetic iron oxide
VEGF-α	Vascular endothelial growth factor alpha
ZnO NPs	Zinc oxide nanoparticles

7.1 INTRODUCTION

In recent years, diabetic nephropathy (DN) has turned into a health risk as type I and type II diabetes prevalence is increasing rapidly (Kakitapalli et al. 2020), which is difficult for the economy and society (Akhtar et al. 2020; Bikbov et al. 2020). DN, a leading cause of kidney disorders in people with diabetes, affects more than 40% of diabetic patients and requires renal replacements. The pathologic underpinning for DN is kidney fibrosis, which is clinically identified as microalbuminuria brought on by an early elevation in glomerular filtration rate (GFR) (Lim and disease 2014; Carlsson et al. 2015; Sharma et al. 2021). Healthy versus diabetic nephropathy kidney and glomerulus have been depicted in Figure 7.1. DN is classified on the basis of urinary albumin excretion: a. microalbuminuria; b. macroalbuminuria. Elevation of blood pressure, blood sugar, and genetic liability are prime risk factors for DN. Smoking, dietary protein and fat, source of dietary protein, and increase in serum lipid level are a few other reasons for DN. Hyperfiltration, microalbuminuria, nephrotic proteinuria, and progressive chronic kidney disease, which renders severe kidney dysfunction, are different clinical stages linked with DN. All parts of the kidney, such as the glomerulus, the tubules, the vasculature, and the interstitium, become affected through structural pathological changes that occur in different clinical stages of DN.

Advanced glycation end products (AGEs), such as proteins or lipids, that become glycated and oxygenated in the presence of aldose sugar affect the intracellular and extracellular structure and function of the cell and are associated with structural pathological changes in different parts of the kidney. Metabolic deviation from its normal pathway is the governing factor for hemodynamic perturbation that causes diabetic kidney diseases at a different level (Kimmelstiel and Clifford, 1936).

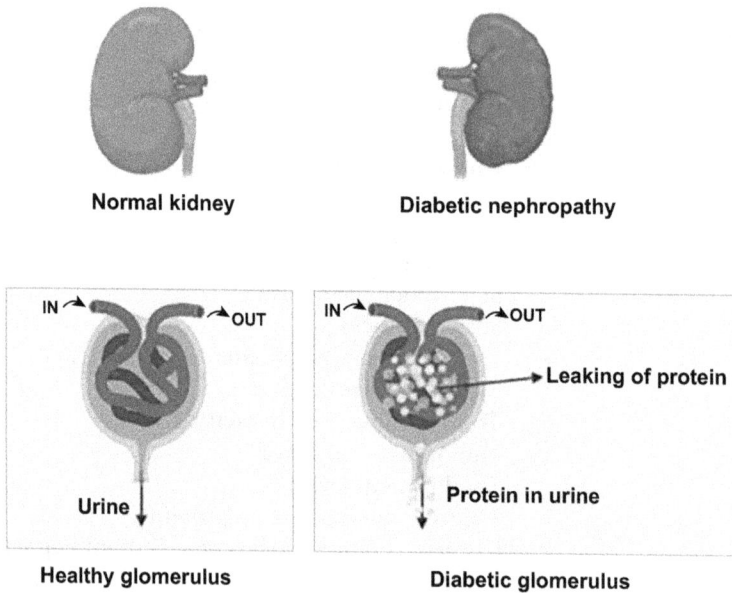

Figure 7.1 Healthy versus diabetic nephropathy kidney and glomerulus.

The risk of cardiovascular morbidity is higher (20–40%) in a patient suffering from DN (Magee et al. 2017). The development of genetics and molecular biology has improved our understanding of the pathophysiology of diabetic nephropathy, but the conventional treatment approach is not sufficient for treating DN as it is associated with multiple factors (Khoury, Chen, and Ziyadeh 2020). By decreasing the production of internal angiotensin II and/or blocking its receptors, numerous studies demonstrated the promise of renin-angiotensin system (RAS) inhibitors as the crucial treatment for DN (Balakumar et al. 2009; Pofi et al. 2016). In search of a new and efficient DN treatment, several investigations have been carried out. RAS blockers are thus the mainstay of DN treatment (Lytvyn et al. 2016). New antidiabetic medications, including SGLT-2I (sodium–glucose cotransporter 2 inhibitors) as well as GLP-1 RA (glucagon-like peptide-1 receptor agonists), are among the most promising treatments for DN (Sawaf et al. 2022; Tong and Adler 2022). Antilipidemic agents, especially statins (atorvastatin), ameliorate DN and considerably reduce cardiovascular diseases (He, Al-Mureish, and Wu 2021). To bring these drugs into the first-line treatment of DN, immense clinical trials are needed (Papademetriou et al. 2020). Even though kidney disorders place a heavy cost on the healthcare system, there are few effective treatments available (Zheng et al. 2020; Yan, Chao, and Lin 2021). Therefore, there is a significant demand for new methods, medications, and technologies that can diagnose and treat renal diseases more effectively, precisely, and easily.

Nanotherapeutics is predicted to provide revolutionary healthcare facilities by targeting specific drug delivery and minimizing the side effects (Murthy 2007). Multifunctional nanotherapeutic agents may close the gap between desired and existing therapeutic strategies (Prasad et al. 2018). Nanocomposites are hybrid materials that contain a component with nanoscale morphology, such as lamellar nanostructures, nanotubes, or nanoparticles. These are multielement substances because they have numerous phases; at the very least, one of these component elements must have a diameter between 10 nm and 100 nm (Pramanik and Das 2019). Today, nanocomposites have emerged as useful alternatives to overcome the constraints of many engineering materials. Scattered matrix and scattered phase elements in nanocomposites can be used to categorize them (Sen and Environment 2020). The matrix allows various materials to join to create novel properties. Due to this, it is split into two sections: the steady phase and the irregular reinforcing phase. Metals, ceramics, and polymers are the three basic building elements that can be used to create nanocomposite materials in any combination. Nanocomposites contain at least one inorganic particle that is a nanometer or smaller in size. The granules could be lamellar, fibrillar, or spherical (metal or ceramic). The concepts of composites and nanoparticles are combined in polymer nanocomposites. Although porous media, colloids, gels, and copolymers can also be included in a nanocomposite, this term is typically used to refer to a solid combination of nano-dimensional phases with different characteristics as a result of structural and chemical differences. Natural structures like the abalone shell, teeth, nacre, and bone contain nanocomposite materials. Nanofillers, which are used in very small quantities, are a key component of nanocomposite technology. The macroscopic characteristics of the polymer nanocomposite can be significantly changed by adding nanofillers. Nanocomposites have the potential for use in a variety of industries, including the biomedical, transportation, aviation, and military sectors. Choices in design and characteristics are available with nanocomposites that are not possible with traditional composites. Nanocomposites meet requirements due to their light weight and versatility without sacrificing the ease of use and elegance of materials. Figure 7.2 provides a schematic illustration demonstrating the general categorization of nanocomposites.

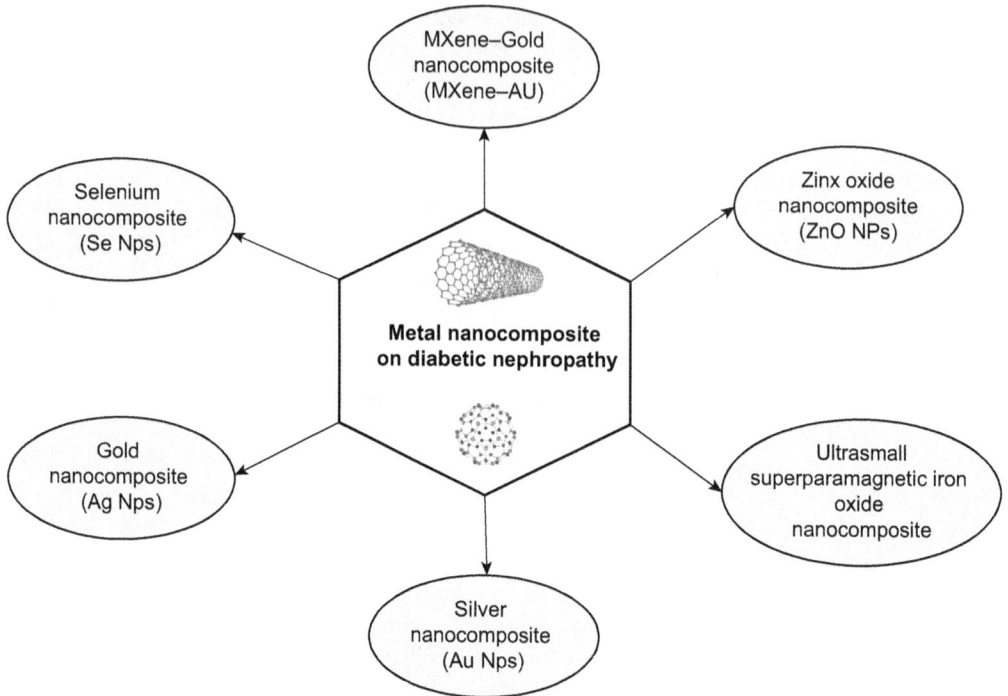

Figure 7.2 Types of metal nanocomposites on diabetic nephropathy.

Metal nanocomposites are non-polymer-based nanocomposites. Due to their enhanced catalytic properties and modifications in the electronic and optical properties linked to individual, separate metals, bimetallic NPs, whether in the form of alloys or core-shell structures, are currently the subject of extensive research. Their intriguing physicochemical characteristics are thought to be the result of the fusion of two different types of metals and their intricate structures. Because of their properties, which depend on their size and dimensions, uniformly sized nanocomposite synthesis is crucial (Rogach et al. 2002). These characteristics include optical, magnetic, electrical, and biological properties. Before they may be extensively assessed in clinical investigations, several obstacles must be successfully overcome. Recently, effective nanomedicines and even nanoparticles (NPs) for detecting, imaging, and treating several types of diseases have appeared (Oroojalian et al. 2020). DN, chronic kidney disease (CKD), and acute kidney injury (AKI) are three kidney conditions for which NPs are beneficial due to their effectiveness, specificity, and variety (Afsharzadeh et al. 2018). Nanomedicines may be successful in tackling the problems related to the treatment of kidney disorders (Liu et al. 2019). However, instead of the remarkable potential of nano-based treatments for the treatment of kidney disorders, their use has been largely constrained by the difficulties of systemic administration and tailored distribution (Kamaly et al. 2016). These barriers are present in the kidney, the blood circulation, and the real tissue once it reaches the target areas. The use of nanotechnology for the therapy of renal illnesses will be covered in this chapter. The implications of the NPs' function as well as their physicochemical attributes on their ability to target certain kidney regions have been the subject of extensive research.

7.2 METAL NANOCOMPOSITES IN DIABETIC NEPHROPATHY

7.2.1 Zinc oxide NPs (ZnO NPs)

In DN caused by STZ, Abd El-Khalik et al. (2022) investigated the molecular renoprotective properties of ZnO NPs. ZnO NPs significantly boosted the biochemical, renal activity, and histological findings in an animal model following six weeks of therapy. ZnO NPs are a promising treatment for DN progression via the interaction of autophagy and Nrf2/TXNIP/NLRP3 inflammasome signaling. As a result, it is important to consider the possibilities of ZnO NPs as an agent of therapy in the cure of DN. In an animal model of DN, ZnO NPs therapy lowered renal function as well as histological alterations. ZnO NPs restored a nearly normal renal structure while maintaining the consistency of the glomerular filter barrier. This was demonstrated by the increased expression of mRNA of the podocyte indicators nephrin and podocin in renal tubules, as well as of the immunological histochemical marker aquaporin 11 (AQP11). The stimulation of autophagy was responsible for the positive effects of ZnO NP caused by inhibition of the mammalian target of rapamycin (mTOR) signaling channel (Abd El-Baset et al. 2022).

7.2.2 Gold nanoparticles (Au NPs)

By modulating the mitogen-activated protein kinase/nuclear factor kappa B/signal transducer and activator of transcription 3 (MAPK/NF-KB/STAT3/cytokine) axis, pomegranate extract-stabilized gold nanoparticle (PPE-AuNP) lowered the proinflammatory load. The administration of PPE-AuNP was also accompanied by phosphoinositide-3-kinase-protein kinase B/Akt (PI3K/AKT)-guided Nrf2 activation, which improved the antioxidant response and preserved hyperglycemic equilibrium. When PPE-AuNP was applied, the hyperglycemia-induced increase in protein glycation and NADPH oxidase 4 (NOX4/p-47) activation were reduced (Manna et al. 2019).

mRNA expression data confirmed that DN frequently displays an increase in the expression of the inflammatory signals tumor necrosis factor-α (TNFα) and vascular endothelial growth factor alpha (VEGF-α). Transforming growth factor-1 (TGF-1), fibronectin, collagen IV, TNF, as well as VEGF-A protein or mRNA expression in the kidneys are shown to be downregulated by Au NPs. Additionally, glomeruli exhibit an increase in the protein expression of the podocyte markers nephrin and podocin (Alomari et al. 2020).

Using G-quadruplex nano-amplification along with MXene–Au nanocomposites, a sensor for miRNA-377 recognition has been constructed. The suggested sensing platform was more convenient, sensitive, specific, and stable than other nanocomposites-based biosensors and previously described miRNA-377 biosensors because it did not require thermal cycling or transcription reversal. Because the biosensor structure also displayed strong selectivity, which was used to precisely identify miRNA-377 in samples of human serum with adequate sensitivity, it has promising potential for biological research and the early clinical diagnosis of DN (Wu et al. 2022).

An antidiabetic and nephroprotective effect of chitosan/sodium lignosulfonate/Au nanoparticles (CS/NaLS/Au NPs) was reported by Gong, Guo, and Zhu (2022). The team created safe, stable, and tiny gold nanoparticles. After receiving a large dose of CS/NaLS/Au NPs, the renal weight, renal volume, and renal structural length all drastically decreased.

CS, NaLS, and Au NPs regulated blood sugar alongside urea levels in STZ-induced diabetes in mice and avoided renal failure.

Edam, Aldokheily, and Al-Yaseen (2022) performed DN experiments using metformin nanoparticles in a microemulsion system and discovered that metformin NPs effectively increased serum albumin while lowering the albumin-to-creatinine ratio, blood urea, creatinine levels, FBS levels, fasting insulin levels, serum urea, and serum creatinine in diabetic nephropathy rats.

Employing a hydrothermal technique, Yu et al. (2021) produced Au NPs to evaluate if they might lower the levels of oxidative stress and have anti-DN effects. They examined the detrimental effects of high hyperglycemia (50 mM) on HK-2 cell lines in addition to the anti-AGE and antioxidant characteristics of Au NPs. By enhancing cell survival and reduction in the production of AGEs and free radicals, Au NPs with a size of 30 nm and colloidal integrity at a pH of 7.4 considerably minimized excess glucose-induced lethality in HK-2 cell lines. Along with mRNA levels, the protein expression of caspase-3, as well as Bax and Bcl-2, was downregulated by Au NPs. The study concluded that using Au NPs might be a successful strategy to stop the spread of DN.

7.2.3 Silver nanoparticles (Ag NPs)

Ag NPs, chitosan, and ascorbic acid nanocomposites (Ag-NCs) were examined for their impact on diabetes in STZ-induced diabetic rats. In the Ag NPs/chitosan/ascorbic acid nanocomposite group, the levels of glucose, nitric oxide (NO), malondialdehyde (MDA), creatinine, urea, and uric acid were reduced. The actions of CAT, superoxide dismutase (SOD), insulin, and reduced glutathione (GSH) increased. The histological analysis showed that the renal architecture had improved (Abu El Qassem Mahmoud et al. 2021).

7.2.4 Selenium nanoparticles (Se NPs)

The bioavailability of selenium and its biological activities are constrained by its functional and toxic limits. With superior biological activity, bioavailability, and low toxicity, selenium nanoparticles (Se NPs) claim to strike a compromise between therapeutic potential and toxicity profiling. Numerous investigations have demonstrated that these Se NPs possess powerful antioxidant properties and 37-fold less toxicity than compounds of selenium or organic selenium (Al-Quraishy, Dkhil, and Abdel 2015; Zhang et al. 2001).

In rats with type 2 diabetes mellitus (T2DM) induced by streptozotocin (STZ), stabilized chitosan selenium nanoparticles (CS–Se NPs) were capable of decreasing inflammation, structural alterations, and renal dysfunction by inhibiting the renal profibrotic protein TGF-β, aldo-keto reductase (AKR1B1), desmin, nestin, as well as vimentin pathways signaling. The antioxidant capacity of the rats receiving both CS–Se NPs and metformin was restored (Khater et al. 2021).

The therapy of metformin and Se NPs increased the levels of the protein AMP-activated protein kinase (pAMPK), demonstrating the direct modulation of the protein kinase B/glycogen synthase kinase-3 (AKT/GSK-3) and insulin receptor substrate 1 pathway that increases insulin sensitivity. Through the reduction of cytokine expression, Se NPs had an anti-inflammatory impact, and the equilibrium between oxidative stress and overall antioxidant activity was re-established. Additionally, after the treatment period, diabetic problems significantly improved (Abdulmalek and Balbaa 2019).

The preventive impact of Se NPs on the development of diabetic nephropathy was studied by Kumar et al. (2014). By reducing oxidative stress, raising the activity of the longevity protein SIRT1, enhancing the heat shock protein (HSP-70), and controlling the expression of apoptotic protein BCL-2 and BAX in the apoptotic kidney, Se NPs generated effective renoprotective effects in STZ-induced DN. Alhazza et al. (2022) reported the protective effect of Se NPs in the kidneys of female pregnant rats with hyperglycemia. Hassan et al. (2021) reported the renal protective mechanism of Se NPs in the offspring of gestational diabetic mother rats. The super antioxidative property of Se NPs protected the kidney and pancreas in the offspring.

7.2.5 Other nano formulations

Fucoidan NPs at a concentration of 300 mg/kg substantially boosted superoxide dismutase and glutathione peroxidase as compared to the STZ-treated group. The STZ treatment worsened renal cell necrosis and the loss of normal kidney cell structure. In contrast, therapy with fucoidan NPs reduced kidney cell death. It was clear that fucoidan NPs have the potential to be preventative medicines against STZ-induced nephropathy by reducing oxidative stress (by lowering malondialdehyde and enhancing superoxide dismutase and glutathione peroxidase) and the inflammatory response (by lowering interleukin-6 and TNF-α) levels (Wardani et al. 2022).

All-trans-retinoic acid (ATRA)-loaded chitosan/tripolyphosphate lipid hybrid nanoparticles (CTLHNs) have been produced to increase the therapeutic effect of ATRA in DN by increasing its solubility and oral distribution. An increased release of ATRA was seen during *in vitro* testing of the formulations to produce chitosan-coated nanoparticles of lipid that are stabilized against acidic pH by sodium tripolyphosphate cross-linking. Higher levels of TNF-α, granulocyte-macrophage colony-stimulating factor (GM-CSF), VEGF, and intercellular adhesion molecule-1 (ICAM-1), serum concentrations of creatinine and urea were observed in the DN rat model. Therapy with free ATRA and the selected formulations greatly reduced the symptoms of DN by raising the amounts of AMPK and LKB1 and by enhancing the activities of AMPK and LKB1 (Asfour, Salama, and Mohsen 2021).

Huang et al. (2021) designed a nanoliposome formulation with poor soluble and low bioavailable polyphenol calycosin available in Radix astragali. Calycosin-loaded nanoliposomes controlled the survival, generation of ROS, lipid peroxidation, and activity of the mitochondria in cells of the kidney of diabetic nephropathic rats. The effects of calycosin-loaded nanoliposomes on the mitochondria of kidney cells were explored and established that calycosin ameliorated the antioxidant mechanism, preventing lipid peroxidation and significantly reduced the oxidative damage to the kidney cells due to diabetes. An STZ-induced DN model was evaluated for the renoprotective effect of PEGylated liposomes of quercetin (Q-PEGL). DN biochemistry and pathological alterations were significantly improved by quercetin and Q-PEGL. The therapeutic effects of Q-PEGL were superior to nonencapsulated quercetin (Tang et al. 2020). Similarly, Tong et al. (2017) reported the renal protective effect of polymeric NP complex of quercetin designed with poly(ethylene glycol)-*block*-(poly(ethylenediamine l-glutamate)-*graft*-poly(ε-benzyloxycarbonyl-l-lysine)) (Q-PEG-*b*-(PELG-*g*-PZLL)). The quercetin polymeric NP complex improved the renal function, renal damage, and renal oxidative injury and downregulated the expression of ICAM-1 and alleviated the symptoms of DN significantly.

Tong et al. (2020) developed a renoprotective void mesoporous silica nanocomposite (HMSN) particles with high drug-loading capacity. The HMSN was loaded with cerium oxide NPs and metformin. The nano-formulation showed renoprotective effects by inhibiting ROS-associated DN pathogenesis. Interestingly, compared to free metformin, the multifunctional nanoparticles (MET-HMSN-CeO$_2$) loaded with metformin (MET) showed considerably higher kidney accumulation. The DN symptoms were lessened by protecting renal damage, suppressing cellular apoptosis, and reduction in oxidative stress both *in vitro* and *in vivo*. Treatment with MET-HMSN-CeO$_2$ effectively improved the renal deficiencies associated with DN.

Rhein (RH) is a pharmacologically diverse anthraquinone derivative that is isolated from herbs and acts in different ways on DN. However, due to its weak ability to dissolve, weak bioavailability, limited distribution into the renal system, and unpleasant side effects, it cannot be employed clinically. Chen et al. (2018) developed a polymeric NPs system for kidney-targeted delivery of rhein using polyethylene glycol, polycaprolactone, and polyethyleneimine (PPP-RH-NPs). A diabetic model of STZ-induced DN was utilized to evaluate the distribution and pharmacodynamics of PPP-RH-NPs, which demonstrated kidney-targeted distribution and boosted the therapeutic advantages of RH on DN.

The efficacy of rubidian-loaded niosomes (RLN) was evaluated in a streptozotocin-nicotinamide (STZ-NA)-induced DN rat model by Tinku et al. (2022). The RLN decreased the blood glucose levels, urine, urea, and creatinine levels significantly on oral administration. The RLN considerably enhanced the levels of TBARS, GSH, SOD, and CAT in DN rats. The RLN formulation improved the lipid profiles of DN rats.

Yang (2019) designed polymeric NPs of crocetin with polylactic-co-glycolic acid (CT-PLGA-NP) and studied the anti-fibrotic and anti-inflammatory effects in the STZ-induced DN model. CT-PLGA-NPs reduced blood sugar and increased plasma insulin and body weight. The NPs downregulated the expression of renal TNF-α, IL-6, IL-1β, and Monocyte Chemoattractant Protein-1 (MCP-1). CT-PLGA-NPs also reduced the activity of protein kinase C and the expression of NF-Kb, p65 activity, and protein formation in renal tissue. CT-PLGA-NPs changed the expression of type IV collagen, fibronectin, and TGF-1β.

Li et al. (2022) designed apigenin-loaded SLN and studied its efficacy in STZ-NA-induced DN rat model. The SLN ameliorated the expression of nuclear factor erythroid 2-related factor 2 and heme oxygenase-1 and suppressed the expression of NF-κB. The apigenin SLN exhibited a renoprotective effect by antioxidant and anti-inflammatory activity. Myricitrin and its solid lipid nanoparticles (SLN) were examined by Ahangarpour et al. (2018) on streptozotocin nicotinamide (STZ-NA)-induced DN rat model. Myricitrin-SLN reduced oxidative stress by increasing antioxidant enzyme levels and reduced the glomerular filtration rate, albumin level in plasma, and blood urea nitrogen (BUN) and creatinine levels in urine, thus ameliorating the symptoms of DN.

Ahad et al. (2018) prepared eprosartan mesylate-containing nano-bilosomes and studied their efficacy in STZ-induced diabetic nephropathy. The renal function of the eprosartan mesylate-loaded nano-bilosomes was preserved by decreasing the serum creatinine, urea, lactate dehydrogenase, total albumin, and malondialdehyde significantly.

The low-density lipoprotein receptor 1 (LOX-1) is expressed in early diabetic nephropathy, which results in extensive inflammation. The effectiveness of anti-LOX-1 ultrasmall nanoparticles of superparamagnetic iron oxide (USPIOs) was examined by Luo et al. (2015)

to identify inflammatory kidney lesions in early DN. The study detected the LOX-1-enriched inflammatory renal injuries in early DN by anti-LOX-1 USPIOs.

7.3 FUTURE PERSPECTIVES

The crucial role that metal nanocomposites play in monotherapy and other biological uses make them appear to have a commanding position in the 21st century. Metal nanocomposites can be produced using different methods, and they can be used successfully in a range of nanomedicines. Future research should concentrate on how to manage the size as well as the form of nanoparticles. Expanding the use of nanoparticles for medicinal purposes and lowering their degree of toxicity represent additional significant challenges. Using noble metal nanoparticles, new approaches are being developed with the promise of nanoscience to surmount the difficulties. Before widespread use, the impact on cultural health variables must be considered. In the future, a complete investigation will be required. Due to their unique atomic and supramolecular characteristics, noble metal nanoparticles can serve as active agents in curative as well as diagnostic procedures. The demand for various kinds of nanomaterials and their compounds is currently very strong in the medical, biological, and healthcare sectors. As a result, focus must be placed on taking safety measures to protect human health. Specific research on safety characteristics is needed for the use of metallic nanoparticles in medical applications. Future research should examine how specific metal nanoparticles can be used for specific purposes. Future noble metal nanocomposite production must be ramped up from laboratory to industrial and medical levels. Metallic nanoparticles have been created and are currently being tested extensively in multiple directions. At the moment, it is employed for the diagnosis and treatment of DN. Metal nanoparticles demonstrate their potency as cutting-edge agents for potential DN treatment approaches. Since many studies are being done on a laboratory basis, modern investigations should concentrate on the possible commercial uses of metal nanocomposites. Commercial exploration has the potential to revolutionize human existence.

7.4 CONCLUSION

Metal-based nanocomposites have drawn a lot of interest as a result of extensive medical research and numerous other biological uses. Due to their special physicochemical characteristics, metal-based nanocomposites have the potential to regulate hostile impacts on various organs, bodily tissues, and subcellular, cellular, and protein ranges. In addition, some metal nanoparticles (Cu, Zn, Ag, and Au) can show a significant toxic effect as their size declines, despite being inert on a bulk scale. It has been proven effective to cure, avoid, and diagnose DN by employing pure metal nanoparticles or metal oxide nanocomposites as a potential substitute. It is quite likely that low-cost treatment medications for DN will be developed using alloys made of metal, metal oxide, or metal oxide/metal-doped metal. It will serve as a substitute for conventional antibiotics. Because of their antibacterial capabilities, nanoparticles of noble metals and their combined form are of the utmost importance. The toxic effects of the metal oxide nanoparticles, which typically occur in high quantities, also restrict their usefulness. One way to lessen the impact of self-toxicity is by using linked polymeric metal oxide nanoparticles, functionalization, and ion doping. Finally, it is possible to speculate about the possibility of future renal cell repair using composite, metal, metal oxide, or less toxic metal nanoparticles.

REFERENCES

Abd El-Baset, Samia A., Nehad F. Mazen, Rehab S. Abdul-Maksoud, and Asmaa A.A. Kattaia. 2022. "The therapeutic prospect of zinc oxide nanoparticles in experimentally induced diabetic nephropathy." *Tissue Barriers*: 2069966.

Abd El-Khalik, Sarah Ragab, Elham Nasif, Heba M. Arakeep, and Hanem Rabah. 2022. "The prospective ameliorative role of zinc oxide nanoparticles in STZ-induced diabetic nephropathy in rats: Mechanistic targeting of autophagy and regulating Nrf2/TXNIP/NLRP3 inflammasome signaling." *Biological Trace Element Research*: 1–11.

Abdulmalek, Shaymaa A., and Mahmoud Balbaa. 2019. "Synergistic effect of nano-selenium and metformin on type 2 diabetic rat model: Diabetic complications alleviation through insulin sensitivity, oxidative mediators and inflammatory markers." *PLoS One*: 14 (8):e0220779.

Abu El Qassem Mahmoud, Esraa A., Ayman S. Mohamed, Sohair R. Fahmy, Amel Mahmoud Soliman, and Khadiga Gaafar. 2021. "Antidiabetic potential of silver/chitosan/ascorbic acid nanocomposites." *Current Nanomedicine*: 11 (4):237–248.

Afsharzadeh, Maryam, Maryam Hashemi, Ahad Mokhtarzadeh, Khalil Abnous, and Mohammad Ramezani. 2018. "Recent advances in co-delivery systems based on polymeric nanoparticle for cancer treatment." *Artificial Cells, Nanomedicine, and Biotechnology*: 46 (6):1095–1110.

Ahad, Abdul, Mohammad Raish, Ajaz Ahmad, Fahad I. Al-Jenoobi, and Abdullah M. Al-Mohizea. 2018. "Eprosartan mesylate loaded bilosomes as potential nano-carriers against diabetic nephropathy in streptozotocin-induced diabetic rats." *European Journal of Pharmaceutical Sciences*: 111:409–417.

Ahangarpour, Akram, Ali Akbar Oroojan, Layasadat Khorsandi, Maryam Kouchak, and Mohammad Badavi. 2018. "Solid lipid nanoparticles of myricitrin have antioxidant and antidiabetic effects on streptozotocin-nicotinamide-induced diabetic model and myotube cell of male mouse." *Oxidative Medicine and Cellular Longevity*:2018.

Akhtar, Mohammed, Noheir M. Taha, Awais Nauman, Imaad B. Mujeeb, and Ajayeb Dakhilalla MH Al-Nabet. 2020. "Diabetic kidney disease: Past and present." *Advances in Anatomic Pathology*: 27 (2):87–97.

Alhazza, Ibrahim M., Hossam Ebaid, Mohamed S. Omar, Iftekhar Hassan, Mohamed A. Habila, Jameel Al-Tamimi, and Mohamed Sheikh. 2022. "Supplementation with selenium nanoparticles alleviates diabetic nephropathy during pregnancy in the diabetic female rats." *Environmental Science and Pollution Research*:1–9.

Alomari, Ghada, Bahaa Al-Trad, Salehhuddin Hamdan, Alaa Aljabali, Mazhar Al-Zoubi, Nesreen Bataineh, Janti Qar, and Murtaza M. Tambuwala. 2020. "Gold nanoparticles attenuate albuminuria by inhibiting podocyte injury in a rat model of diabetic nephropathy." *Drug Delivery and Translational Research*: 10:216–226.

Al-Quraishy, Saleh, Mohamed A. Dkhil, and Ahmed Esmat Abdel. 2015. "Anti-hyperglycemic activity of selenium nanoparticles in streptozotocin-induced diabetic rats." *International Journal of Nanomedicine Moneim*: 10:6741.

Asfour, Marwa Hasanein, Abeer A.A. Salama, and Amira Mohamed Mohsen. 2021. "Fabrication of all-trans retinoic acid loaded chitosan/tripolyphosphate lipid hybrid nanoparticles as a novel oral delivery approach for management of diabetic nephropathy in rats." *Journal of Pharmaceutical Sciences*: 110 (9):3208–3220.

Balakumar, Pitchai, Mandeep Kumar Arora, Subrahmanya S. Ganti, Jayarami Reddy, and Manjeet Singh. 2009. "Recent advances in pharmacotherapy for diabetic nephropathy: Current perspectives and future directions." *Pharmacological Research*: 60 (1):24–32.

Bikbov, Boris, Caroline A. Purcell, Andrew S. Levey, Mari Smith, Amir Abdoli, Molla Abebe, Oladimeji M. Adebayo, Mohsen Afarideh, Sanjay Kumar Agarwal, and Owolabi Marcela. 2020. "Global, regional, and national burden of chronic kidney disease, 1990–2017: A systematic analysis for the Global Burden of Disease Study 2017." *The Lancet Agudelo-Botero*: 395 (10225):709–733.

Carlsson, Axel C., Lina Nordquist, Tobias E. Larsson, Juan-Jesús Carrero, Anders Larsson, Lars Lind, and Johan Ärnlöv. 2015. "Soluble tumor necrosis factor receptor 1 is associated with glomerular filtration rate progression and incidence of chronic kidney disease in two community-based cohorts of elderly individuals." *Cardiorenal Medicine*: 5 (4):278–288.

Chen, Danfei, Shunping Han, Yongqin Zhu, Fang Hu, Yinghui Wei, and Guowei Wang. 2018. "Kidney-targeted drug delivery via rhein-loaded polyethyleneglycol-co-polycaprolactone-co-polyethylenimine nanoparticles for diabetic nephropathy therapy." *International Journal of Nanomedicine*: 13:3507.

Edam, Khalid A., Mohsin E. Aldokheily, and Firas F. Al-Yaseen. 2022. "Evaluation of the renoprotective effects of metformin nanoparticles in rats with diabetic nephropathy." *International Journal of Health Sciences*: 6 (S1):13212–13227. doi.org/10.53730/ijhs.v6nS1.8311

Gong, Yimeng, Xiaoyun Guo, and Qihan Zhu. 2022. "Nephroprotective properties of chitosan/sodium lignosulfonate/Au nanoparticles in streptozotocin-induced nephropathy in mice: Introducing a novel therapeutic drug for the treatment of nephropathy." *Arabian Journal of Chemistry*: 15 (6):103761.

Hassan, Iftekhar, Hossam Ebaid, Jameel Al-Tamimi, Mohamed A. Habila, Ibrahim M. Alhazza, and Ahmed M. Rady. 2021. "Selenium nanoparticles mitigate diabetic nephropathy and pancreatopathy in rat offspring via inhibition of oxidative stress." *Journal of King Saud University-Science*: 33 (1):101265.

He, Yujing, Abdulrahman Al-Mureish, and Na Wu. 2021. "Nanotechnology in the treatment of diabetic complications: A comprehensive narrative review." *Journal of Diabetes Research*: 2021:1–11.

Huang, Chunrong, Lian-Fang Xue, Bo Hu, Huan-Huan Liu, Si-Bo Huang, Suliman Khan, and Yu Meng. 2021. "Calycosin-loaded nanoliposomes as potential nanoplatforms for treatment of diabetic nephropathy through regulation of mitochondrial respiratory function." *Journal of Nanobiotechnology*: 19 (1):1–12.

Kakitapalli, Yesubabu, Janakiram Ampolu, Satya Dinesh Madasu, and M.L.S. Sai Kumar. 2020. "Detailed review of chronic kidney disease." *Kidney Diseases*: 6 (2):85–91.

Kamaly, Nazila, John C. He, Dennis A. Ausiello, and Omid C. Farokhzad. 2016. "Nanomedicines for renal disease: Current status and future applications." *Nature Reviews Nephrology*: 12 (12):738–753.

Khater, Safaa I., Amany Abdel-Rahman Mohamed, Ahmed Hamed Arisha, Lamiaa L.M. Ebraheim, Shefaa A.M. El-Mandrawy, Mohamed A. Nassan, Amany Tharwat Mohammed, and Samar Ahmed Abdo. 2021. "Stabilized-chitosan selenium nanoparticles efficiently reduce renal tissue injury and regulate the expression pattern of aldose reductase in the diabetic-nephropathy rat model." *Life Sciences*: 279:119674.

Khoury, Charbel C., Sheldon Chen, and Fuad N. Ziyadeh. 2020. "Pathophysiology of diabetic nephropathy." In *Chronic Renal Disease*: 279–296. Academic Press.

Kimmelstiel, Paul, and Clifford Wilson. 1936. "Intercapillary lesions in the glomeruli of the kidney." *The American Journal of Pathology*: 12 (1):83.

Kumar, Goru Santosh, Apoorva Kulkarni, Amit Khurana, Jasmine Kaur, and Kulbhushan Tikoo. 2014. "Selenium nanoparticles involve HSP-70 and SIRT1 in preventing the progression of type 1 diabetic nephropathy." *Chemico-Biological Interactions*: 223:125–133.

Li, Pingping, Syed Nasir Abbas Bukhari, Tahseen Khan, Renukaradhya Chitti, Davan B. Bevoor, Anand R. Hiremath, Nagaraja SreeHarsha, Yogendra Singh, and Kumar Shiva Gubbiyappa. 2022. "Apigenin-loaded solid lipid nanoparticle attenuates diabetic nephropathy induced by streptozotocin nicotinamide through Nrf2/HO-1/NF-kB signalling pathway [Retraction]." *International Journal of Nanomedicine*: 17:3457–3458.

Lim, Andy K.H. 2014. "Diabetic nephropathy: Complications and treatment." *International Journal of Nephrology, and Renovascular Disease*:361–381.

Liu, Chun-Ping, You Hu, Ju-Chun Lin, Hua-Lin Fu, Lee Yong Lim, and Zhi-Xiang Yuan. 2019. "Targeting strategies for drug delivery to the kidney: From renal glomeruli to tubules." *Medicinal Research Reviews*: 39 (2):561–578.

Luo, Bing, Song Wen, Yu-Chen Chen, Ying Cui, Fa-Bao Gao, Yu-Yu Yao, Sheng-Hong Ju, and Gao-Jun Teng,. 2015. "LOX-1-targeted iron oxide nanoparticles detect early diabetic nephropathy in db/db mice." *Molecular Imaging and Biology*: 17:652–660.

Lytvyn, Yuliya, Petter Bjornstad, Nicole Pun, David M. Maahs, Bruce Perkins, and David Z.I. Cherney. 2016. "New and old agents in the management of diabetic nephropathy." *Current Opinion in Nephrology and Hypertension*: 25 (3):232.

Magee, Corey, David J. Grieve, Chris J. Watson, and Derek P. Brazil. 2017. "Diabetic nephropathy: A tangled web to unweave." *Cardiovascular Drugs and Therapy*: 31:579–592.

Manna, Krishnendu, Snehasis Mishra, Moumita Saha, Supratim Mahapatra, Chirag Saha, Govind Yenge, Nilesh Gaikwad, Ramkrishna Pal, Dasharath Oulkar, and Kaushik Banerjee. 2019. "Amelioration of diabetic nephropathy using pomegranate peel extract-stabilized gold nanoparticles: Assessment of NF-κB and Nrf2 signaling system." *International Journal of Nanomedicine*: 14:1753.

Murthy, Shashi K. 2007. "Nanoparticles in modern medicine: State of the art and future challenges." *International Journal of Nanomedicine*: 2 (2):129–141.

Oroojalian, Fatemeh, Fahimeh Charbgoo, Maryam Hashemi, Amir Amani, Rezvan Yazdian-Robati, Ahad Mokhtarzadeh, Mohammad Ramezani, and Michael R. Hamblin. 2020. "Recent advances in nanotechnology-based drug delivery systems for the kidney." *Journal of Controlled Release*: 321:442–462.

Papademetriou, Vasilios, Sofia Alataki, Konstantinos Stavropoulos, Christodoulos Papadopoulos, Kostas Bakogiannis, and Kostas Tsioufis. 2020. "Pharmacological management of diabetic nephropathy." *Current Vascular Pharmacology*: 18 (2):139–147.

Pofi, Riccardo, Francesca Di Mario, Antonietta Gigante, Edoardo Rosato, Andrea M. Isidori, Antonio Amoroso, Rosario Cianci, and Biagio Barbano. 2016. "Diabetic nephropathy: Focus on current and future therapeutic strategies." *Current Drug Metabolism*: 17 (5):497–502.

Pramanik, Sujata, and Pankaj Das. 2019. "Metal-based nanomaterials and their polymer nanocomposites." In *Nanomaterials and Polymer Nanocomposites*: 91–121. Elsevier.

Prasad, Minakshi, Upendra P. Lambe, Basanti Brar, Ikbal Shah, J. Manimegalai, Koushlesh Ranjan, Rekha Rao, Sunil Kumar, Sheefali Mahant, and Sandip Kumar Khurana. 2018. "Nanotherapeutics: An insight into healthcare and multi-dimensional applications in medical sector of the modern world." *Biomedicine and Pharmacotherapy*: 97:1521–1537.

Rogach, Andrey L., Dmitri V. Talapin, Elena V. Shevchenko, Andreas Kornowski, Markus Haase, and Horst Weller. 2002. "Organization of matter on different size scales: Monodisperse nanocrystals and their superstructures." *Advanced Functional Materials*: 12 (10):653–664.

Sawaf, Hanny, George Thomas, Jonathan J. Taliercio, Georges Nakhoul, Tushar J. Vachharajani, and Ali Mehdi. 2022. "Therapeutic advances in diabetic nephropathy." *Journal of Clinical Medicine*: 11 (2):378.

Sen, Mousumi. 2020. "Nanocomposite materials." *Nanotechnology, and the Environment*:1–12.

Sharma, Anjul, Raymond E. Bourey, John C. Edwards, David S. Brink, and Stewart G. Albert. 2021. "Nephrotic range proteinuria associated with focal segmental glomerulosclerosis reversed with pioglitazone therapy in a patient with Dunnigan type lipodystrophy." *Diabetes Research and Clinical Practice*: 172:108620.

Tang, Lixia, Ke Li, Yan Zhang, Huifang Li, Ankang Li, Yuancheng Xu, and Bing Wei. 2020. "Quercetin liposomes ameliorate streptozotocin-induced diabetic nephropathy in diabetic rats." *Scientific Reports*: 10 (1):2440.

Tinku, Mohd Mujeeb, Abdul Ahad, Mohd Aqil, Waseem Ahmad Siddiqui, Abul Kalam Najmi, Mymoona Akhtar, Apeksha Shrivastava, Abdul Qadir, and Thasleem Moolakkadath. 2022. "Ameliorative effect of rubiadin-loaded nanocarriers in STZ-NA-induced diabetic nephropathy in rats: Formulation optimization, molecular docking, and in vivo biological evaluation." *Drug Delivery and Translational Research*: 12 (3):615–628.

Tong, Fei, Suhuan Liu, Bing Yan, Xuejun Li, Shiwei Ruan, and Shuyu Yang. 2017. "Quercetin nanoparticle complex attenuated diabetic nephropathy via regulating the expression level of ICAM-1 on endothelium." *International Journal of Nanomedicine*: 12:7799.

Tong, Li-Li, and Sharon G. Adler. 2022. "Diabetic kidney disease treatment: New perspectives." *Kidney Research and Clinical Practice*: 41 (Suppl 2):S63–S73.

Tong, Yuna, Lijuan Zhang, Rong Gong, Jianyou Shi, Lei Zhong, Xingmei Duan, and Yuxuan Zhu. 2020. "A ROS-scavenging multifunctional nanoparticle for combinational therapy of diabetic nephropathy." *Nanoscale*: 12 (46):23607–23619.

Wardani, Giftania, Jusak Nugraha, Mohd Mustafa, and Sri Agus Sudjarwo. 2022. "Antioxidative stress and anti-inflammatory activity of fucoidan nanoparticles against nephropathy of streptozotocin-induced diabetes in rats." *Evidence-Based Complementary and Alternative Medicine*: 2022.

Wu, Qianqing, Zhenhui Li, Qianwei Liang, Rongkai Ye, Shuzhou Guo, Xiaobing Zeng, Jianqiang Hu, and Aiqing J. Li. 2022. "Ultrasensitive electrochemical biosensor for microRNA-377 detection based on MXene-Au nanocomposite and G-quadruplex nano-amplification strategy." *Electrochimica Acta*: 428:140945.

Yan, Ming-Tso, Chia-Ter Chao, and Shih-Hua Lin. 2021. "Chronic kidney disease: Strategies to retard progression." *International Journal of Molecular Sciences*: 22 (18):10084.

Yang, Xiaodong. 2019. "Design and optimization of crocetin loaded PLGA nanoparticles against diabetic nephropathy via suppression of inflammatory biomarkers: A formulation approach to preclinical study." *Drug Delivery*: 26 (1):849–859.

Yu, Y., J. Gao, L. Jiang, and J. Wang. 2021. "Antidiabetic nephropathy effects of synthesized gold nanoparticles through mitigation of oxidative stress." *Arabian Journal of Chemistry*: 14 (3):103007.

Zhang, Jin-Song, Xue-Yun Gao, Li-De Zhang, and Yong-Ping. 2001. "Biological effects of a nano red elemental selenium." *Biofactors Bao*: 15 (1):27–38.

Zheng, Hui Juan, Xueqin Zhang, Jing Guo, Wenting Zhang, Sinan Ai, Fan Zhang, Yaoxian Wang, and Wei Jing Liu. 2020. "Lysosomal dysfunction: Induced autophagic stress in diabetic kidney disease." *Journal of Cellular and Molecular Medicine*: 24 (15):8276–8290.

Chapter 8

Nanotherapeutics for diabetic cardiomyopathy using metal nanocomposites

Jayaraman Rajangam, Narahari N. Palei, G.S.N. Koteswara Rao, R. Prakash, Sagili Varalakshmi, Haja Bava Bakrudeen, and Shruti Srivastava

LIST OF ABBREVIATIONS

a FGF	Acidic fibroblast growth factor
AHA	American Heart Association
AMPK	AMP-activated protein kinase
ARBs	Angiotensin II receptor blockers
CNT	Carbon nanotubes
DCM	Diabetic cardiomyopathy
DD	Diastolic dysfunction
DDP	Dipeptidyl peptidase
ERK	Extracellular signal-regulated protein kinases
FFA	Free fatty acid
GFAT	Glutamine: fructose-6-phosphate aminotransferase
GLP	Glucagon-like peptide
GLUT	Glucose transporter
GO	Graphene oxide
GPCR	G-protein-coupled receptor
HF	Heart failure
JNK	Jun N-terminal kinases
LAA	Left atrial appendage
LVH	Left ventricular hypertrophy
MACE	Major adverse cardiovascular events
MAPK	Mitogen-activated protein kinase
miRNA	micro-RNA
MNC	Metal nanocomposite
PGS	Poly (glycerol sebacate)
PKC	Protein kinase C
PLCL	Poly (L-lactide-co-ε-caprolactone)
PPAR	Peroxisome proliferator-activated receptor
RAAS	Renin–angiotensin–aldosterone system
RAS	Reticular activating system
ROS	Reactive oxygen species
SD	Systolic dysfunction
SGLT	Sodium–glucose co transporter
UDP-GlcNAc	Uridine diphosphate N-acetylglucosamine

DOI: 10.1201/9781032621135-8

8.1 INTRODUCTION

Diabetic cardiomyopathy (DCM) is a clinical disorder defined by aberrant myocardial shape and functioning in patients with diabetes mellitus without additional cardiac risk factors. DCM was initially documented more than 40 years ago, and the key factors in its development were hyperglycemia and a compromised cardiac insulin signaling system (Jia, 2018). Apart from this, oxidative stress, calcium homeostasis imbalance, decreased mitochondrial function, extracellular matrix remodeling, and poor cardiomyocyte contractility are its main root causes. Hyperglycemia, hyperinsulinemia, and insulin resistance are extensive variables contributing to the pathological alteration of the heart, which is preceded by left ventricular hypertrophy (LVH) and perivascular and interstitial fibrosis leading to ventricular dysfunction (Yancy et al., 2013).

Further, DCM is characterized by many morphological and structural myocardial abnormalities that are caused by the activation of diverse changes, with mechanical dysfunction being the most important change, as a result of the lack of antioxidants that favor prooxidative stress. Moreover, the aberrant, anatomical, metabolic, and physiological alterations may disrupt the intracellular signaling of the heart's myocytes, resulting in insufficient energy production and a decline in the contractile efficiency of the cardiac myocytes (Cai and Kang, 2003). These anomalies may lead to extracellular matrix remodeling, heart fibrosis, and other microvascular consequences such as steatosis and fibrosis (Loncarevic et al., 2016). On the other hand, DCM is a prevalent medical illness that predisposes diabetic individuals to ventricular dysfunction in the absence of clinically severe coronary, valvular, or hypertension illness. Moreover, the emergence of new symptoms and the escalation of existing symptoms, as well as the prognosis over the long term, are all significantly impacted by the existence of DCM and will significantly affect the diabetes patients' "quality of life." However, therapy is limited, and the prognosis is bleak.

DCM is a progressive chronic clinical condition that goes through various processes and stages as it develops. The initial stage is distinguished by the gradual emergence and progression of asymptomatic and transitory diastolic dysfunction. Insulin resistance (IR) and hyperglycemia are the primary risk factors for this. As a result of the heart's compensatory response to metabolic abnormalities, the heart shape is nearly normal at this point, and the only alteration is "myocardial and endothelial cell" dysfunction (Adameova et al., 2014). Yet it can be quite challenging to confirm DCM with these abnormalities while going for a diagnosis.

Currently, a differential diagnosis of DCM cannot be made without morphological abnormalities, biochemical markers, or clinical signs. A clear diagnosis is difficult to make since this condition sometimes exhibits no symptoms initially and frequently coexists with comorbidities. But techniques such as MRI and echocardiography, which may provide precise details on the anatomy and physiology of the heart, have advanced tremendously during the last few decades. The functional myocardial anomalies are measured using tissue doppler imaging as well as trans mitral doppler imaging techniques.

8.1.1 Stages of diabetic cardiomyopathy

In diabetic cardiomyopathy, in the initial stage, the heart has a series of characteristic abnormalities such as

1. Raised "left ventricular end-diastolic pressure"
2. High "ventricular stiffness"
3. LVH
4. Left atrial enlargement of left atrium, etc.

All these abnormalities lead to the progression of diastolic dysfunction as the primary root cause of DCM. There is structural damage and little distortion in the internal microvascular structure of the heart's hierarchy during the intermittent stage of DCM. When myocardial cell damage develops, the likelihood of abnormal diastolic cardiac function increases. The effects of metabolic diseases also worsen as the condition worsens, with dysfunction of mitochondria which play a significant role (Duncan, 2011a).

On the other hand, lipotoxicity from hazardous lipid metabolites can activate mitochondrial membrane rearrangements in addition to the excessive generation of reactive oxygen species (ROS). Cardiomyocyte loss and fibrosis are caused by subcellular processes like mitochondrial abnormalities and reduced calcium signaling (Boudina et al., 2009; Boudina and Dale Abel, 2006). The second stage manifests clinical symptoms and signs associated with significant cardiac remodeling and systolic dysfunction. Only during the late phases of the illness (Falcão-Pires et al., 2012), and frequently in conjunction with severe CADs, does systolic dysfunction emerge (Goyal and Mehta, 2013).

Cardiac interstitial fibrosis worsens further and progresses to DCM as the last stage. The difference in myocardial microcirculation is visible. At this stage, severe collagen deposition plays a very vital role where the myocardial basement membrane gets thickened, microaneurysms of blood capillaries gets developed followed by sclerosis of the coronary artery (Marfella et al., 2021; Kumric et al., 2021; Chavali et al., 2013). Since this process may be started by activating neurohumoral processes such as hyperglycemia, IR, renin–angiotensin–aldosterone system (RAAS), and sympathetic nervous system at the cellular level, which results in cardiomyocyte hypertrophy, stiffness, and fibrosis. GLUT-1 and GLUT-4 depletion, calcium homeostasis imbalance, and free fatty acid (FFA) buildup all work together to cause myocyte damage at the molecular level.

8.1.2 Pathogenesis and molecular signaling

Although there is still much to learn about the pathophysiology of DCM in people, several pathways have been linked to the development of this clinical condition. The three primary metabolic anomalies that contribute to cardiac dysfunction are the development of hyperglycemia, compensatory hyperinsulinemia, and insulin resistance. It is critical to keep in mind that the impact of the alterations brought on by diabetes is amplified when these comorbidities exist, especially in those individuals with type II diabetes. This will most certainly hasten the formation of LVH, render the heart more prone to ischemia damage, and upsurge the likelihood of heart failure (Boudina et al., 2010).

DCM patients exhibit several pathophysiological abnormalities in terms of anatomical and physiological differences. The modifications mostly include hyperglycemia, insulin resistance or hyperinsulinemia, oxidative stress caused by ROS production, and poor intracellular calcium handling. Through encouraging myocardial stiffness, hypertrophy, and fibrosis, these pathophysiological aberrations cause cardiac diastolic dysfunction, systolic dysfunction, and heart failure.

Table 8.1 Major pathophysiological abnormalities developing diabetic cardiomyopathy

S. No.	Major pathophysiological abnormalities	Cells/signaling abnormalities
1.	Hyperglycemia	Increased protein kinase C (PKC)
2.	Left ventricular hypertrophy (LVH)	Nuclear factor κ-light-chain-enhancer of activated B cells (NF-κB)
3.	Insulin resistance or hyperinsulinemia	Mitogen-activated protein kinase (MAPK)
4.	Myocardial lipotoxicity	Sodium–glucose cotransporter-2 (SGLT2)
5.	Dysregulation of microRNA (miRNA) and exosomes	Cyclic adenosine 5′-monophosphate responsive element modulator (CREM) signaling
6.	Interstitial fibrosis	Reduction of AMP-activated protein kinase (AMPK)
7.	Mitochondrial dysfunction	Nuclear factor erythroid 2-related factor 2 (Nrf2)
8.	Changes in myocardial metabolism	Peroxisome proliferator-activated receptor (PPAR)-γ
9.	Diastolic dysfunction (DD)	Altered apoptosis signaling
10.	Systolic dysfunction (SD)	Increased oxidative stress due to ROS

As shown in Table 8.1, all the mentioned abnormalities can alter substrate metabolism and cause cardiac lipotoxicity which in turn leads to development of DCM.

8.2 SIGNALING PATHWAYS OF DCM

The development of DCM has been linked to many signaling pathways like HBP pathway, protein kinase-C (PKC), the NFκB pathway, PPAR pathway, PI3K pathway, and MAPK pathway. These pathways and proteins are activated by factors such as high blood sugar, increased reticular activating system (RAS) activity, raised free fatty acid (FFA) levels, ROS production, and release of mediators (pro-inflammatory) (Figure 8.1).

8.2.1 Upregulation of hexosamine biosynthesis pathway (HBP)

Under normal physiological conditions, the HBP pathway diverts a small quantity of fructose-6-phosphate (2–5%) from glycolysis, an alternative route of glucose metabolism (Huynh, 2014). The glutamine: fructose-6-phosphate aminotransferase (GFAT) is considered as the HBP rate-limiting enzyme involved in the conversion of fructose 6-phosphate into glucosamine 6-phosphate. The amino group from glutamine is irreversibly transferred by GFAT, and fructose 6-phosphate is isomerized into glucosamine 6-phosphate (G6P) and glutamate (Broschat, 2002). G6P is converted into uridine diphosphate N-acetylglucosamine (UDP-GlcNAc), which serves as the direct sugar contributor for O-linked beta-N-acetylglucosamine (O-GlcNAc) to proteins (serine/threonine) (Zeidan and Hart, 2010; Kohler, 2010). Since O-GlcNAc undergoes cycling over serine or threonine residues in a similar way of phosphorylation (Hart et al., 2007), the protein's phosphorylation steps may be affected by O-GlcNAc alteration (Brownlee, 2005). Diabetes-related elevated glucose levels are thought to be harmful to the heart. HBP has been identified as a new risk factor for diabetic heart problems that result from poor glycemic control (Cieniewski-Bernard,

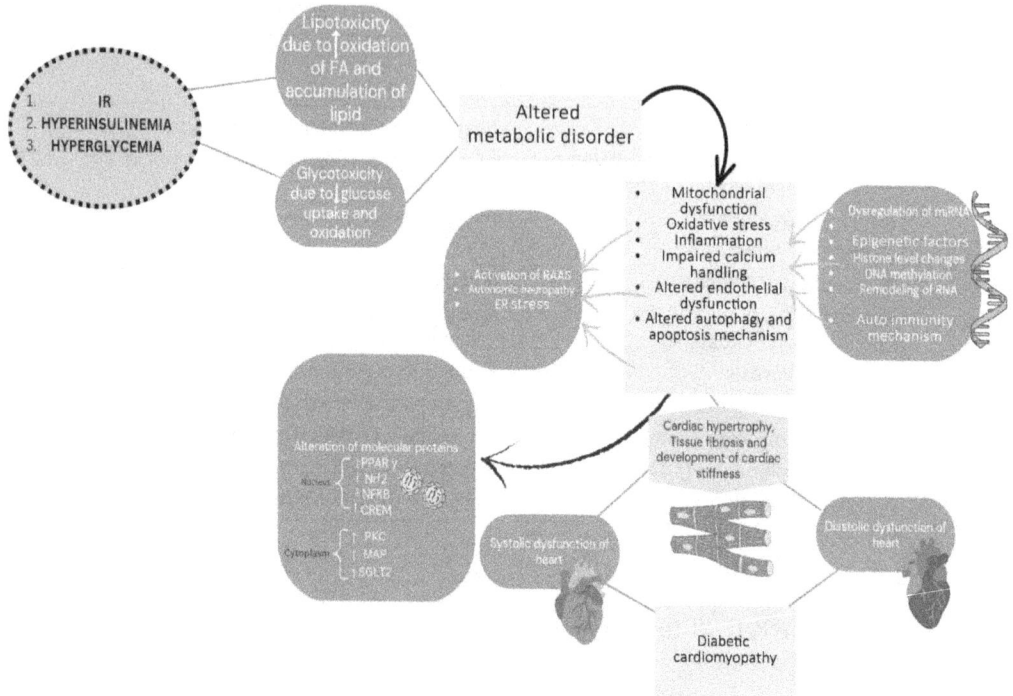

Figure 8.1 Major signaling pathways for developing diabetic cardiomyopathy.

2009). The intact heart and cardiomyocytes, among other cells and organs, experience elevated O-GlcNAc levels as a result of hyperglycemia. Increasing glucose levels has several impacts, including increased oxidative stress, unlike glucosamine, which is predominantly processed by the HBP. Moreover, hyperglycemia was known to increase the production of mitochondrial superoxide, and this rise in ROS causes the HBP to activate and increases the synthesis of OGlcNAc. It leads to a decrease in the activity of the powerhouse of the myocardial cells and results in cell apoptosis (Darley-Usmar et al., 2012; Rajangam et al., 2022).

8.2.2 Activation of protein kinase C (PKC)

PKC is a group of enzymes that phosphorylate proteins at serine/threonine residues that regulate certain hormonal, neural, and growth factor inputs in signal transduction events. PKC is activated by increased levels of calcium, phospholipid, or diacylglycerol (DAG). DAG is produced either by the breakdown of phosphatidylinositol by phospholipase C or by the de novo synthesis from glucose (Way, 2001). There are 12 isoforms in the serine/threonine kinase family PKC. Many tissues are linked to diabetes vascular problems, including renal glomeruli, the heart, aorta, retina, and the kidneys (in both diabetic individuals and animal models), and have been found to have higher total DAG concentrations. Moreover, in diabetic and IR animals, nonvascular organs such as the liver, skeletal muscles, and circulating monocytes have also shown enhanced activation of the DAG-PKC pathway (Noh, 2007).

Hyperglycemia and diabetes-induced PKC, especially PKC-β1 isoform, has shown alterations in the cell functions of glomerulus as well as in gene expressions that are associated with chronic pathological conditions of diabetic nephropathy (Koya, 1997). Around four PKC isoforms (α, β, δ, ε) were known to mediate cardiac hypertrophy and/or pathology (Bernardo, 2010), whereas PKCα and PKCβ isoforms were reported to be upregulated in the diabetic heart (Huynh, 2014). Going further to the sublevel of isoforms, PKCβ2 transgenic overexpression in cardiac myocytes shows significant effects of hypertrophy, fibrosis, and cardiomyocyte necrosis, resulting in reduced performance of the heart. This is in turn associated with the upregulation of TGF-β1, collagen gene expression, and β-myosin heavy chain. Due to the untoward effect of calcium in cardiomyocytes, the risk of heart failure is noticed through PKCα-associated impairment in contractility (Huynh, 2014).

8.2.3 Peroxisome proliferator-activated receptor (PPAR) signaling

PPARα, PPARγ, and PPARβ/δ are the three subtypes of PPARs that come under the nuclear hormone receptor superfamily and are claimed as ligand-activated transcription factors (Tyagi et al., 2011). PPARα has received the most research attention among the available PPAR isoforms in the heart (Finck et al., 2002). The diabetic heart is characterized in large part by metabolic disturbances. In contrast to the healthy heart, the diabetic heart gets almost all of its energy from fatty acid metabolism. In cardiac myocytes, PPARα is expressed at a rather high level. Increased fatty acid levels trigger the PPARα that improves the metabolism of cardiac fatty acids. Activated PPARα increases the gene expression for fatty acid β-oxidation and also causes suppression of glucose utilization. The combined effects of enhanced myocardial fatty acid uptake and oxidation in addition to suppressed glucose uptake and oxidation exhibit cardiac dysfunction leading to diabetic cardiomyopathy (Lee et al., 2017; Duncan, 2011b).

8.2.4 Mitogen-activated protein kinase (MAPK) signaling pathway

In humans, the broad classification of MAPKs can be done under six groups: the extracellular signal-regulated protein kinases (ERK1 and ERK2); c-Jun N-terminal kinases (JNK1, JNK2, and JNK3); p38s (p38α, p38β, p38γ, and p38δ); ERK5 (ERK5); ERK3s (ERK3, p97 MAPK, and ERK4); and ERK7s (ERK7 and ERK8) (Teramoto, 2013). While p38α is assumed to be predominantly expressed in heart muscle, p38α MAPK is present throughout the body (Lemke, 2001; Ge et al., 2002). High glucose conditions, pathogens, heat, osmotic stress, UV radiation, inflammatory cytokines, growth hormones, and other stimuli can all activate p38 (Rose, 2010). p38 mitogen-activated MKK3 or MKK6 transgenic overexpression of signaling pathway in the heart shows increased interstitial fibrosis, thinning of the ventricular wall, and even premature death due to cardiac failure (Rose, 2010).

8.3 CURRENT THERAPEUTIC IMPLICATIONS AND CHALLENGES IN DCM

Raised blood sugar levels and prolonged persistent hyperinsulinemia create microvascular complications followed by structural damage to the heart of diabetic patients. As a result, lowering glucose levels is the prime target for a successful treatment regimen for diabetes-related cardiovascular complications. The following are some categories of medications used commonly for the management of DCM.

8.3.1 GLP-1 receptor agonists

Glucagon-like peptide-1 (GLP-1) is a gut-derived, G-protein-coupled receptor (GPCR) peptide hormone mainly generated after meal intake. When activated, it changes ATP into cAMP, and an increase in cytosolic cAMP in beta-pancreatic cells causes the release of insulin (Holst, 2007). Nevertheless, endogenously released GLP-1 is rapidly degraded, predominantly by dipeptidyl peptidase-4 (DPP-4), after a relatively brief half-life of around 2–3 minutes. To give extended *in vivo* activity and positive benefits for T2DM patients, several synthetic GLP-1 receptor agonists (GLP-1RAs) have been developed (Meier, 2012). But these classes of drugs are mainly employed to enhance the secretion of insulin against hyperglycemic conditions only.

8.3.2 Dipeptidyl peptidase-4 inhibitors (DPP-4 inhibitors)

DPP-4 is expressed widely by a variety of cells and exhibits exopeptidase activity. These DPP-4 inhibitors act on cardiac myocytes and coronary vasculature and influence myocardial metabolism either directly or indirectly. The metabolic effects are brought by DPP-4 inhibitors by increasing insulin and decreasing glucagon levels, as well as by lowering free fatty acid (FFA) levels in the circulation which leads to the development of metabolic abnormalities. This could result in cardiac myocytes absorbing more glucose while using fewer fatty acids. According to the DCM pathophysiological pathways, these metabolic effects of DPP-4 inhibitors result in improved myocardial metabolism and reduce the cause of contractile dysfunction (Nikolaidis et al., 2004).

8.3.3 Oral hypoglycemic agents

Metformin works by improving the insulin resistance of diabetics and thereby helps to improve the complications related to DCM and lowers the risk of morbidity followed by mortality in diabetic patients, according to the data currently available (Liu et al., 2022; Nour et al., 2021). It increases AMPK activity and endothelial nitric oxide synthase while lowering TNF-α and fibroblast growth factor activity (Wang et al., 2011). This causes the LV volume to shrink and undergo LV remodeling, which enhances the systolic and diastolic performances (Raffaele et al., 2019).

However, metformin use was contraindicated in patients with heart failure due to the risk of lactic acidosis, whereas researchers are claiming that it is a misconception and using metformin is to be contraindicated only in those patients of renal or hepatic impairment with a risk of lactic acidosis but need not be avoided in patients with heart failure. So metformin use in heart failure patients with hypoperfusion should be done with caution, whereas with dehydration, hypoxemia, or sepsis, the metformin use should be discontinued (Kinsara, 2018).

On the other hand, drugs like thiazolidinediones (TZDs) that increase the anti-inflammatory effects by activating PPAR and decreasing the activity of inflammatory cells can also be suggested for DCM (Varsha et al., 208). Nevertheless, due to their propensity to cause HF and edema, their use in clinical practice is also constrained. As a result, these medications shouldn't be used in people with severe heart failure and should be administered only with caution to patients who already exhibit HF signs or symptoms (Kshitij et al., 2021). Sulfonylureas (SUs), on the other hand, are often not the recommended class of medications for DCM since they have a considerable impact on weight, elevate BMI, and increase the risk of HF.

8.3.4 Sodium–glucose co-transporter 2 (SGLT2) inhibitors

The use of sodium–glucose co-transporter 2 (SGLT2) inhibitors is one of the best treatments for diabetic cum heart failure. It shows excellent cardiorenal benefits in clinical trials of people with type II diabetes. Heart failure can be treated with SGLT2 inhibitors in patients with and without type II diabetes (T2D). In T2D patients, all three medications improve heart failure, but dapagliflozin also does so in patients without diabetes. Dapagliflozin does not affect major adverse cardiovascular events (MACE); however, both canagliflozin and empagliflozin assist T2D patients in having fewer of these events. Moreover, safety issues have been raised, either during clinical trials or through post-marketing surveillance, such as a potential link between SGLT2i exposure and left atrial appendage (LAA), diabetic ketoacidosis, bone fracture, and glucose management indicator. In addition to the generally established beneficial effects, SGLT2-Is have certain mild and substantially adverse effects that have been observed in clinical trials (Pelletier, 2021).

According to these SGLT2i categories, each class has their individual adverse effects compared to placebos. A greater risk of lower limb amputations and genital infections such as soreness and pain were also linked to canagliflozin medication when compared to placebo and active comparators (Xiong et al., 2016). Compared to placebo and active comparators, using dapagliflozin was significantly linked to an increase in genital tract infections and urinary tract infections. However, the use of dapagliflozin with metformin and sulfonylureas caused an increased risk of hypoglycemia (Zhang et al., 2014). Empagliflozin while compared to other classes of SGLT 2 inhibitors, empagliflozin showed much potency. In comparison to placebo and active comparators, there was a considerably greater frequency of genital tract infections, according to nine reviews from empagliflozin. The incidence of urinary tract infections was shown to be considerably higher with empagliflozin compared to dapagliflozin. Empagliflozin was inconsistently connected to both a significant reduction in the risk of hypoglycemia against a placebo and a significant increase in hypoglycemia over canagliflozin (Zhang et al., 2014).

As per the reports, after treatment of 81 HF patients for a median of 7 days (34 individuals received canagliflozin, 24 patients received dapagliflozin, and 23 patients received empagliflozin), lower limb edema and dyspnea were significantly reduced. The only difference existing in the background of patients treated with these three SGLT2i drugs was the usage of renin–angiotensin-system medications ($p<0.011$). The mean values for HbA1c and fasting blood sugar were 7.3% and 134 mg/dL, respectively. About 42% of cases of heart failure had an ischemic origin, while 58 % did not. About 40 individuals (or 50%) had a reduced left ventricular ejection fraction, 14 patients (17%) had a mid-range ejection fraction (40–47%), and a preserved ejection fraction (50%) (Nakagaito et al., 2019).

8.3.5 Other therapeutic implications

A metabolic abnormality is caused by cardiac fibrosis, microvascular dysfunction, and IR in diabetic cardiomyopathy, a variety of treatment modalities may be effective in slowing the disease's progression and its effects. Enhancing diabetes management, using calcium blockers, ACE inhibitors, or comparable drugs, engaging in physical activity, receiving lipid-lowering therapy, and taking antioxidant and insulin-sensitizing medications are a few examples.

Controlling diabetes and associated conditions, such as increased FFA, oxidative stress, altered calcium homeostasis, and abnormal lipid metabolism, maybe the main strategy for preventing the onset of DCM. Microvascular disease, heart fibrosis, and IR may be due to the distinct pathophysiology of type I and type II diabetes. Replacement for C-peptide improves the baseline cardiac performance and perfusion of type I diabetic patients, according to Hansen et al. (2002) and Norby (2002). The dearth of evidence demonstrating the effectiveness of strict glycemic control can prevent the development of DCM.

8.4 NANOMATERIALS AS DRUG THERAPY FOR DIABETIC CARDIOMYOPATHY

Nanomaterials are engineered structures that range in size from 1 nm to 100 nm. They possess unique physicochemical properties that allow them to be used as drug-delivery vehicles. The small size of nanomaterials allows for efficient penetration into tissues, including the heart, while their surface chemistry can be modified to increase drug loading and reduce toxicity. Additionally, nanomaterials can be functionalized by targeting ligands, which can selectively bind to receptors on the surface of cells in the heart, improving drug delivery to the affected area (Palei et al., 2016). Several types of nanomaterials have been investigated for their potential use as drug carriers for DCM. These include liposomes, dendrimers, polymeric nanoparticles, and mesoporous silica nanoparticles, and all have shown their potential as drug carriers for DCM therapy, and further studies are needed to explore their safety and efficacy (Figure 8.2).

Nanotechnology-based drug delivery systems have the potential to revolutionize the treatment of various diseases, and their usage in DCM treatment may lead to improved outcomes for patients (He et al., 2021). Studies have shown that nanomaterials can improve drug solubility, prolong drug release, and target the affected site in the heart. For example, a study has shown that poly(lactide-co-glycolide) (PLGA) nanoparticles loaded with the anti-inflammatory drug curcumin significantly reduced cardiac fibrosis in a rat model of DCM (Danhier, 2012; Tong et al., 2018). One of the most promising biomaterials for drug delivery is graphene oxide (GO) due to its large surface area and high biocompatibility. Researchers have successfully loaded GO with various drugs, such as metformin, and have demonstrated its efficacy in reducing inflammation and oxidative stress in diabetic models (Zhang et al., 2021; Liang et al., 2022). However, Zhang et al. (2021) have reported that GO has shown low cardiotoxicity which might be attributed to oxidative stress, lipid peroxidation, and mitochondrial dysfunction.

Another nanomaterial that shows promise in DCM therapy is chitosan-conjugated nanoformulations. Chitosan is a biodegradable and biocompatible polymer that can be modified to improve drug delivery. Researchers have used chitosan nanoparticles to deliver insulin to diabetic rats and observed improved glucose control and reduced oxidative stress in the heart (Sing et al., 2018; Heidarisasan et al., 2018; Mohanta et al., 2019).

Carbon nanotubes (CNTs) have also been investigated as drug carriers for DCM. CNTs have a high surface area and can be functionalized to target specific tissues. Researchers have demonstrated that CNTs can be used to deliver cardioprotective drugs, such as

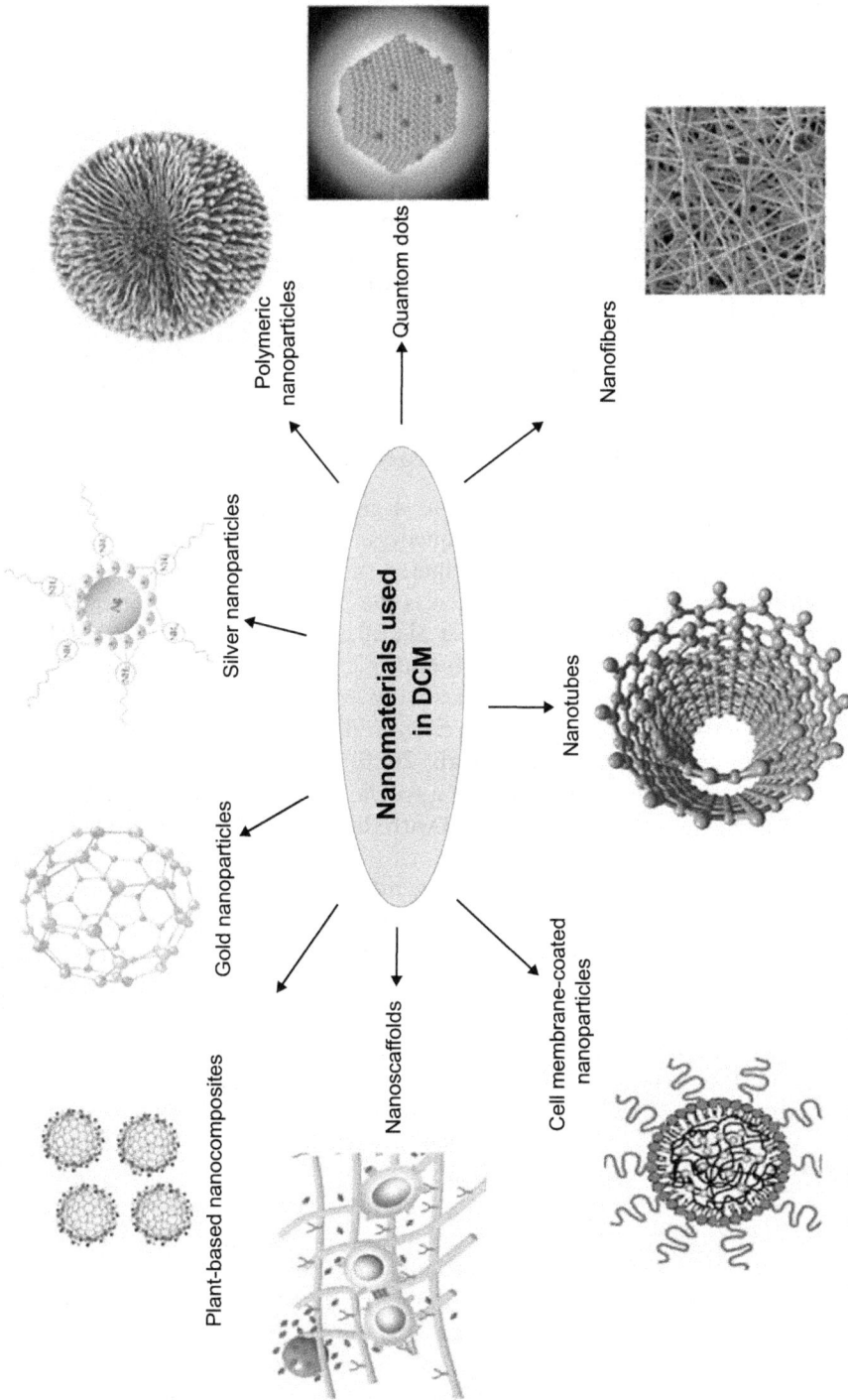

Figure 8.2 Different types of nanomaterials used for DCM.

erythropoietin, to diabetic animal models, resulting in reduced inflammation and improved cardiac function (Hosseinpour et al., 2016). Many studies demonstrated that functionalized CNTs effectively delivered cardioprotective drugs to diabetic animals, resulting in improved cardiac function and reduced inflammation. The use of CNTs as drug carriers for cardiovascular disease treatment is an active area of research, and further studies are needed to fully understand their potential and limitations (Saleemi et al., 2020; Jashandeep et al., 2019).

Traditional drug delivery systems have shown limited success in the treatment of DCM due to their inability to target the affected site and efficiently deliver the drug. Nanomaterials have emerged as a promising option for drug delivery in DCM, owing to their success rate in targeting the diseased tissues. These particles possess unique physicochemical properties that enable efficient drug delivery to the affected site in the heart while reducing toxicity. The development of targeted nanomaterials for DCM treatment may provide new opportunities for improved clinical outcomes in people with diabetes mellitus.

8.4.1 Why nanomaterials for diabetic cardiomyopathy?

Diabetes is a silent metabolic disorder that has a greater impact on public health globally and causes several potentially fatal consequences. The incidence rates are dramatically increasing each day, and this number is anticipated to reach 693 million by 2045. Diabetes is the seventh major cause of mortality and is the main reason for kidney failure, loss of vision, and cardiac problems (Abubakar et al., 2015; WHO, 2019; Cho et al., 2018). Persistent diabetes causes a variety of macro and microvascular complications including retinopathy, neuropathy, nephropathy, and cardiomyopathy, among others (Harding et al., 2019). The risk of getting a myocardial infarction in diabetes people is 6–10 times higher than in nondiabetic persons (Ferrannini et al., 2020). Additionally, the American Heart Association (AHA) states that there should be a special concern while dealing the coronary artery disease patients with diabetes compared with those who don't have diabetes (Arnold et al., 2020).

Researchers currently confront a tough management dilemma when it comes to treating diabetic patients because of the comorbidities that are connected with the disease as epidemiological rates soar. Diabetic cardiomyopathy (DCM) is becoming more frequent in diabetic individuals as a result of hyperglycemia and hyperinsulinemia, which lead to cardiac dysfunction along with myocardial fibrosis and apoptosis. A drug delivery system should be able to allow the utilization of the full potential of a potent drug and preferably should be target specific to avoid damage to other healthy tissues. To fulfill such needs, the emerging field of nanotechnology is being tailored and several researchers have reported their promising applications in the literature. Nano-formulations are offering effective therapy for both acute and chronic ailments (Kaurav et al., 2018).

Nanocomposites made of zinc, silver, and gold offer diversified applications due to their biocompatible nature, increased surface area, and cell permeability. With their attractive features, nanomaterials and their composites act as novel nanocarriers for several diseases including DCM (Ali and Shakeel, 2018). Nanomaterials play a key role in the targeted delivery of drugs/nucleic acid/protein in addition to the diagnosis and biomedical applications. The same applies to the diagnosis and treatment of diabetes along with associated complications (Luo et al., 2021). Nanoparticles (NPs) are prepared from natural, semisynthetic, or synthetic substances either organic or inorganic in nature, and hence they have a wide variety of physicochemical properties. In the recent past, metal NPs have dragged the attention of researchers owing to their potential role in diagnostic and therapeutic applications for several diseases.

Gold NPs were found to show antihyperglycemic effects in streptozotocin-induced animal models of diabetes (Aljabali et al., 2021; Pradhan et al., 2022, 2023). Gold nanoclusters were reported to have promising applications in the management of cardiomyopathy by inhibiting mitophagy in diabetic patients. Two modified ligands namely 3-mercaptopropionic acid and N-acetyl-l-cysteine were used for stabilization of these gold nanoclusters along with positively, relatively positive, and negatively charged polymers, namely polyetherimide, polyethylene glycol, and polyvinyl pyrrolidone, respectively. The results indicated their applicability in the treatment of DCM (Shen et al., 2022; Li et al., 2019; Dzulkharnien et al., 2022).

8.4.2 Metal nanocomposites

Metal nanocomposites are nanostructures of biphasic nature comprising metals as well as polymers creating hybrid materials with dual characteristics. Their special features include distinctive electronic, magnetic, and optical properties, low coordination numbers, and strongly reactive edge atoms (Pramanik et al., 2019). The metallic polymer-based nanocomposites show varied properties according to their particle size, and hence they can be tailored as per the requirement. Hence, it is evident that the combination of metals and polymers in nanocomposites can provide higher and more synergistic efficacy than the components used separately for enhanced specific applications. Nanomaterials made of metals provide improved qualities and capabilities, and as a result, they offer an alternative to many current and upcoming technological requirements. The field of metal nanocomposites (MNCs) represents material prosperity in the contemporary environment of sophisticated technologies. The robustness of developing metal-based polymeric nanocomposites with a wide variety of combinations and processing parameters helps the researchers to develop a unique class of materials that can suit the needs of drug delivery to complicated treatments like DCM (Faupel et al., 2010). By varying the composition of various metals, which will give the polymer specified qualities, these nanocomposites can be fitted to their desired use.

8.5 RECENT TRENDS IN THE TREATMENT OF DCM WITH NANOCOMPOSITES

Advances in nanomaterial research have applications in the diagnosis and treatment of diabetes, as well as its consequences. Researching innovative inorganic nanoparticulate drug delivery methods lead to the creation of diverse nanomaterials with varied functionalities and features that can be used to treat diabetes and related problems. Recent advances in inorganic nanocomposites have greatly expanded their use in pharmacological and biological applications. Because of their numerous therapeutic uses, inorganic nanocomposites are now growing. These are chemical compounds that lack carbon–hydrogen bonds in their structure. Quantum dots are ultrasmall carbon-containing nanostructures that are also known as nanoceramics. Inorganic nanomaterials are metallic substances that are inorganic nanosized and generated from natural sources or through biosynthetic or synthetic routes. These generated metallic nanoparticles and nanocomposites are then employed to fabricate a wide spectrum of nanomaterials for biological purposes. Manufactured nanoparticles come in a variety of shapes and sizes, including micelles, vesicles, and spheres. Despite differences in the size and structure of nanoparticles, long clearance times boost the therapeutic potential of drug delivery systems (Liu et al., 2022).

Nanomaterials containing various plant extracts as well as inorganic elements such as Au, Pt, Ag, and ZnO are being developed for use in the treatment of diabetic cardiomyopathy. These nanoparticles provide therapeutic effectiveness while causing few adverse effects and significantly improving biocompatibility. The use of natural derivatives to treat diabetic cardiomyopathy is primarily based on the antioxidant mechanism of phytoconstituents and metals (Rajangam, 2013; Mounika et al., 2021; Krishnan et al., 2021; Christina et al., 2018).

8.5.1 Metal nanocomposites for DCM

Nanotechnology-based drug delivery systems, such as MNCs, have emerged as promising therapeutic options for the treatment of DCM. Targeted nanotherapy strategies using metallic nanocomposites offer a hopeful approach to improve drug delivery to the heart and enhance therapeutic efficacy. MNCs, such as gold, silver, and iron oxide NPs, have unique physical and chemical properties that make them attractive candidates for drug delivery. These nanocomposites can be functionalized with ligands that target specific tissues or cells, allowing for targeted drug delivery to the heart. In the case of DCM, targeted nanotherapy can deliver drugs directly to the heart, reducing the need for high doses and minimizing off-target effects. For instance, gold nanoparticles functionalized with angiotensin II receptor blockers (ARBs) have been shown to selectively target cardiac fibroblasts and reduce fibrosis in a mouse model of diabetic cardiomyopathy (Enzan et al., 2022). Similarly, iron oxide nanoparticles conjugated with a glucagon-like peptide-1 (GLP-1) receptor agonist have been used to target GLP-1 receptors on cardiomyocytes and improve cardiac function in a rat model of type II diabetes (Nour et al., 2021).

Tin oxide nanoparticles, in addition to other forms of nanoparticles like silver, zinc, and gold, have a wide range of pharmacological properties (Roopan et al., 2015). Due to their surface characteristics, nanocomposites are lighter in weight and used as innovative medicine nanocarriers (Ali and Shakeel, 2018). Rongrong et al. (2022) designed a nanocomposite of tin oxide–chitosan–polyethylene glycol–D pinitol (SCP-D-P) and studied the myocardial ischemia-related p62/Keap1/Nrf2 signaling pathway. The SCP-D-P nanocomposites significantly reduced the diabetic variables and improved heart function. The histopathological studies revealed the reduction of myocardial tissue damage with the application of SCP-D-P nanocomposites.

Liraglutide is a GLP-1 receptor agonist that has shown a dual role in antidiabetic activity and improved cardiovascular functions in patients. To overcome its problem of short half-life, He et al. (2020) developed a ternary nanosystem comprising liraglutide, tannic acid, and Al^{3+}. This ternary nanosystem offered long-term glycemic control along with improved cardiovascular functions. The ternary nanosystem showed significant improvements in reducing DCM, with inhibition in lipotoxicity by a 40% reduction of triglyceride, 30% reduction of diacylglycerol, and 50% reduction of PKC level in the heart. It also ameliorated oxidative stress and cell apoptosis activities via positive regulation of superoxidase, malondialdehyde, caspase-3, and BAX (He et al., 2020).

ZnO NPs at a moderate dose of 3 mg/kg body weight have shown protection against cardiac damage in streptozotocin-induced cardiomyopathy in the rat model. At this dose several other abnormal values of lipoproteins, TNF-α, cardiac malondialdehyde, caspase-3, and atherogenic index were also significantly controlled. However, an elevated dose of 10 mg/kg has shown toxic effects (Asri-Rezaei et al., 2017). Zinc oxide nanoparticles (ZnO NPs) were used to deliver the antioxidant quercetin to diabetic rat hearts (Varsha Rani et al., 2018). The study found that the nanocomposites significantly reduced oxidative stress and improved heart function.

Similarly, Ahmad et al. (2020) synthesized a chitosan-coated copper sulfide nanocomposite to deliver the anti-inflammatory drug, resveratrol, to the heart of diabetic rats. The study found that the nanocomposites significantly reduced inflammation and oxidative stress, leading to improved heart function. In a recent study, manganese dioxide nanocomposites were synthesized to deliver the anti-inflammatory drug, curcumin, to the heart of diabetic rats (Kshitij et al., 2021). The study found that the nanocomposites significantly reduced inflammation and oxidative stress in the heart, improving cardiac function. Correspondingly, multiwalled carbon nanotube-based nanocarriers were developed for the delivery of the anti-inflammatory drug, dexamethasone, to diabetic cardiomyocytes. The study found that the nanocarriers improved drug uptake and reduced inflammation in the cells (Sandeep Kumar et al., 2011).

Curcumin, a chemical compound produced from natural sources, is used to treat cardiovascular diseases. The aqueous dispersion of nano curcumin has demonstrated potential reductions in myocardial fibrosis, cardiac inflammation, and programmed myocardial cell death caused by oxidative stress in cardiac tissues in rat models (Shome et al., 2016). On the other hand, there are reports which claim that "Acidic fibroblast growth factor" (aFGF), a peptide with a molecular weight of 15.8 kDa, is effective in treating DCM (Zhang et al., 2013). aFGF protects cardiovascular functioning by reducing oxidative stress and cardiac damage. The therapeutic efficacy and side effects of aFGF-loaded heparin-based nanocarrier systems demonstrated increased *in vitro* and *in vivo* stability, reduced cardiomyocyte fibrosis and apoptosis, as well as improved microvasculature (Zhou et al., 2021).

Liu et al. (2022) reported a dosage of 3 mg/kg body weight ZnO nanoparticles showed a reversal impact on heart damage. But this work warrants additional research on dose-response relationship. miR155-conjugated Au NPs exhibited therapeutic efficacy against DCM. miR155-Au NPs increased anti-inflammatory type 2 macrophages, which in turn reduced the rate of inflammation, led to a reduction in cell death, which dramatically improves cardiac function (Jia et al., 2017). Adenosine-loaded silica nanoparticles, a cardioprotective medication that is delivered into cardiac tissues, are also being developed as a therapy for DCM (Galagudza et al., 2012).

Myriad investigators have developed drug-embedded nanofibers for the treatment of DCM through tissue engineering. These nanofibers, whose sizes range from 1 nm to 1,000 nm, have been used for tissue engineering and cardiac tissue regeneration. Moreover, PLGA nanofiber scaffolds were being developed for the regeneration of heart tissue. By using the electrospinning technology, nanofibrous scaffolds with embedded poly (2-ethyl-2-oxazoline) (PEOz) and poly (L-lactide-co-ε-caprolactone) (PLCL) was created, and hydrophobic or hydrophilic drugs were incorporated. Gelatin and poly (glycerol sebacate) (PGS), and PGS-incorporated CNTs had better mechanical toughness and electrical conductivity, and they also demonstrated a significant improvement in cardiomyocyte function. The inclusion of PLGA in carbon nanofibers improved cytocompatibility and conductivity, leading to better cardiomyocyte function (Norouzi et al., 2015).

For cardiac tissue engineering, new scaffolds that can be delivered parenterally have also been designed. Injectable carbon nanofibers (CNF) in conjunction with self-assembled rosette nanotubes in a poly (2-hydroxyethyl methacrylate) (pHEMA) hydrogel exhibited improved conductivity. Both horizontal and vertical conductivities were improved by PLGA/CNF composites (Chopra et al., 2022). Recently, due to widespread interest, the use of herbal extracts for the creation of metallic bio-nanocomposites, owing to several advantages such as eco-friendliness, cost-effectiveness, medicinal efficacy, and stability have been developed (Zeinab et al., 2020).

8.6 CHALLENGES, LIMITATIONS, AND FUTURE PERSPECTIVES

In the pharmaceutical area, many metallic nanocomposites have been created for various therapeutic purposes including for the management of DCM. These nanocomposites may be used for diagnostic research to therapeutic intervention. Despite the potential benefits of targeted nanotherapy using MNCs for DCM treatment, several challenges must be addressed. One of the major challenges is the potential toxicity of metallic nanoparticles. Metal NPs can induce oxidative stress, inflammation, and DNA damage, which can lead to cell death and tissue damage. Therefore, the safety and toxicity of MNCs must be carefully evaluated in animal models and clinical trials. Recent studies have shown that surface modifications, such as PEGylation, can improve the biocompatibility and safety of MNCs for cardiovascular applications. Another challenge is the difficulty of targeting the heart *in vivo*. The heart is a highly dynamic organ that undergoes constant changes in size, shape, and blood flow, making it difficult to target using conventional drug delivery systems.

Targeted nanotherapy strategies using MNCs have shown promise in the treatment of DCM. MNCs can be engineered to target the heart and improve drug delivery, resulting in reduced inflammation and improved cardiac function. However, further research is needed to address the challenges associated with MNCs and to translate these therapies into clinical practice. Targeted nanotherapy using MNCs has the potential to revolutionize the treatment of DCM and the quality of life of diabetic individuals. Several metallic nanoparticles have been created for general application based on their site-specific activity and therapeutic efficacy (Rajasekharreddy et al., 2020; Rajangam et al., 2022; Palei et al., 2022).

Material stability, processability, and solubility should all be addressed during the manufacture of metallic nanocomposite with structural alteration. The fundamental constraint of MNCs is connected with unfavorable consequences such as cytotoxicity, which may be caused by the interaction of the NPs and cells. As a result, while designing MNCs for DCM, the preparation procedure must be optimized critically for biocompatibility with the heart cells and should not interfere with their functions (Rajasekharreddy et al., 2021). MNCs have recently been investigated for *in vitro* and *in vivo* tests, although clinical trials are still in the works. Institutional and industrial collaboration, inadequate infrastructure, competent persons for patient care, and registered individuals for clinical trials are the key barriers to the development of MNCs (Dzulkharnien, and Rosiah, 2022).

8.7 CONCLUSION

The prevalence of DCM, which affects around 12% of diabetics and causes overt heart failure and mortality, is becoming more well-acknowledged. Despite important advances in the treatment of diabetes, both the incidence and death rates are raising progressively. The hunt for innovative and enhanced treatment alternatives that are more effective is thus at the forefront of current research. Nanocomposite materials have lately piqued the interest of scientists due to their better characteristics and numerous alternatives for carrying out a wide range of biological activities. These nanocomposites have a wide range of uses from diagnosis to disease management. On the other hand, therapy with an SGLT2 inhibitor is now one of the most effective therapies for diabetes-related heart failure. Unfortunately, no effective therapy for DCM has been identified, and the establishment of an effective treatment for DCM is vital. MNCs have been developed for a variety of therapeutic applications, including the treatment of DCM. Still, there are significant obstacles to overcome, including the potential toxicity of metal NPs and the difficulty of targeting the heart *in vivo*. Future studies should investigate the safety and efficacy of these NPs in animal models and clinical trials, and further research is needed to address the challenges and translate these therapies into clinical practice.

REFERENCES

Abubakar, I. I., Taavi Tillmann, and Amitava Banerjee. "Global, regional, and national age-sex specific all-cause and cause-specific mortality for 240 causes of death, 1990–2013: A systematic analysis for the Global Burden of Disease Study 2013." *Lancet* 385, no. 9963 (2015): 117–171.

Adameova, Adriana, and Naranjan S. Dhalla. "Role of microangiopathy in diabetic cardiomyopathy." *Heart Failure Reviews* 19 (2014): 25–33.

Ahmad, Awais, N. M. Mubarak, Khalida Naseem, Hina Tabassum, Muhammad Rizwan, Agnieszka Najda, M. Kashif, May Bin-Jumah, Afzal Hussain, Asma Shaheen, Mohamed M. Abdel-Daim, Shafaqat Ali, and Shahid Hussain. "Recent advancement and development of chitin and chitosan-based nanocomposite for drug delivery: Critical approach to clinical research." *Arabian Journal of Chemistry* 13, no. 12 (2020): 8935–8964.

Ali, Akbar, and Shakeel Ahmed. "A review on chitosan and its nanocomposites in drug delivery." *International Journal of Biological Macromolecules* 109 (2018): 273–286. doi: 10.1016/j.ijbiomac.2017.12.078

Aljabali, Alaa A. A., Bahaa Al-Trad, Lina Al Gazo, Ghada Alomari, Mazhar Al Zoubi, Walhan Alshaer, Khalid Al-Batayneh, Bahja Kanan, Kaushik Pal, and Murtaza M. Tambuwala. "Gold nanoparticles ameliorate diabetic cardiomyopathy in streptozotocin-induced diabetic rats." *Journal of Molecular Structure* 1231 (2021): 130009.

Arnold, Suzanne V., Deepak L. Bhatt, Gregory W. Barsness, Alexis L. Beatty, Prakash C. Deedwania, Silvio E. Inzucchi, Mikhail Kosiborod et al. "Clinical management of stable coronary artery disease in patients with type 2 diabetes mellitus: A scientific statement from the American Heart Association." *Circulation* 141, no. 19 (2020): e779–e806.

Asri-Rezaei, S., B. Dalir-Naghadeh, A. Nazarizadeh, and Z. Noori-Sabzikar. "Comparative study of cardio-protective effects of zinc oxide nanoparticles and zinc sulfate in streptozotocin-induced diabetic rats." *J Trace Elem Med Biol* 42 (2017 July): 129–141.

Bernardo, B. C., K. L. Weeks, L. Pretorius, and J. R. McMullen. "Molecular distinction between physiological and pathological cardiac hypertrophy: Experimental findings and therapeutic strategies." *Pharmacology & Therapeutics* 128, no. 1 (2010): 191–227.

Boudina, Sihem, H. Bugger, S. Sena, B. T. O'Neill, V. G. Zaha, O. Ilkun, et al. "Contribution of impaired myocardial insulin signaling to mitochondrial dysfunction and oxidative stress in the heart." *Circulation* 119, no. 9 (2009): 1272–1283.

Boudina, Sihem, and Evan Dale Abel. "Diabetic cardiomyopathy, causes and effects." *Reviews in Endocrine and Metabolic Disorders* 11 (2010): 31–39.

Boudina, Sihem, and Evan Dale Abel. "Mitochondrial uncoupling: A key contributor to reduced cardiac efficiency in diabetes." *Physiology* 21, no. 4 (2006): 250–258.

Broschat, Kay O., Christine Gorka, Jimmy D. Page, Cynthia L. Martin-Berger, Michael S. Davies, Horng-chih Huang, Eric A. Gulve, William J. Salsgiver, and Thomas P. Kasten. "Kinetic characterization of human glutamine-fructose-6-phosphate amidotransferase I: Potent feedback inhibition by glucosamine 6-phosphate." *Journal of Biological Chemistry* 277, no. 17 (2002): 14764–14770.

Brownlee, M. "The pathobiology of diabetic complications: A unifying mechanism." *Diabetes* 54, no. 6 (2005): 1615–1625.

Cai, L., and Y. J. Kang. "Cell death and diabetic cardiomyopathy." *Cardiovasc Toxicol* 3 (2003): 219–228.

Chavali, Vishalakshi, Suresh C. Tyagi, and Paras K. Mishra. "Predictors and prevention of diabetic cardiomyopathy." *Diabetes, Metabolic Syndrome and Obesity: Targets and Therapy* (2013): 151–160.

Cho, Nam H., J. E. Shaw, Suvi Karuranga, Yafang Huang, J. D. da Rocha Fernandes, A. W. Ohlrogge, and B. I. D. F. Malanda. "IDF diabetes atlas: Global estimates of diabetes prevalence for 2017 and projections for 2045." *Diabetes Research and Clinical Practice* 138 (2018): 271–281.

Chopra, Hitesh, Shabana Bibi, Awdhesh Kumar Mishra et al. "Nanomaterials: A promising therapeutic approach for cardiovascular diseases." *Journal of Nanomaterials* 1–25 (2022): 4155729.

Christina, A. J. M., S. Pushpa Latha, and Narahari N. Palei. "Antihyperlipidemic and antio citrullus colocynthis fruits on hyperlipidemic and STZ-induc." *Int J Pharma Res Health Sci* 6, no. 5 (2018): 2808–2813.

Cieniewski-Bernard, C., V. Montel, L. Stevens, and B. Bastide. "O-GlcNAcylation, an original modulator of contractile activity in striated muscle." *Journal of Muscle Research and Cell Motility* 30 (2009): 281–287.

Danhier, F., E. Ansorena, J. M. Silva, R. Coco, A. Le Breton, and V. Préat. "PLGA-based nanoparticles: An overview of biomedical applications." *J. Control. Release* 161 (2012): 505–522.

Darley-Usmar, V. M., L. E. Ball, and J. C. Chatham. "Protein O-linked β-N-acetylglucosamine: A novel effector of cardiomyocyte metabolism and function." *Journal of Molecular and Cellular Cardiology* 52, no. 3 (2012): 538–549.

Duncan, Jennifer G. "Mitochondrial dysfunction in diabetic cardiomyopathy." *Biochimica et Biophysica Acta (BBA)-Molecular Cell Research* 1813, no. 7 (2011a): 1351–1359.

Duncan, Jennifer G. "Peroxisome proliferator activated receptor-alpha (PPARα) and PPAR gamma coactivator-1alpha (PGC-1α) regulation of cardiac metabolism in diabetes." *Pediatric Cardiology* 32 (2011b): 323–328.

Dzulkharnien, Nur Syafiqah Farhanah, and Rosiah Rohani. "A review on current designation of metallic nanocomposite hydrogel in biomedical applications." *Nanomaterials* 12, no. 10 (2022): 1629.

Enzan, N., S. Matsushima, T. Ide, T. Tohyama, K. Funakoshi, T. Higo, and H. Tsutsui. "The use of angiotensin II-receptor blocker is associated with greater recovery of cardiac function than angiotensin-converting enzyme inhibitor in dilated cardiomyopathy." *ESC Heart Fail* 9, no. 2 (2022 Apr): 1175–1185.

Falcão-Pires, Inês, and Adelino F. Leite-Moreira. "Diabetic cardiomyopathy: Understanding the molecular and cellular basis to progress in diagnosis and treatment." *Heart Failure Reviews* 17 (2012): 325–344.

Faupel, Franz, Vladimir Zaporojtchenko, Thomas Strunskus, and Mady Elbahri. "Metal-polymer nanocomposites for functional applications." *Advanced Engineering Materials* 12, no. 12 (2010): 1177–1190.

Ferrannini, Giulia, Maria Laura Manca, Marco Magnoni, Felicita Andreotti, Daniele Andreini, Roberto Latini, Attilio Maseri et al. "Coronary artery disease and type 2 diabetes: A proteomic study." *Diabetes Care* 43, no. 4 (2020): 843–851.

Finck, B. N., J. J. Lehman, T. C. Leone, M. J. Welch, M. J. Bennett, A. Kovacs, and D. P. Kelly. "The cardiac phenotype induced by PPARα overexpression mimics that caused by diabetes mellitus." *The Journal of Clinical Investigation* 109, no. 1 (2002): 121–130.

Galagudza, Michael., Dmitry Korolev, Viktor Postnov, Elena Naumisheva, Yulia Grigorova, Ivan Uskov, and Eugene Shlyakhto. "Passive targeting of ischemic-reperfused myocardium with adenosine-loaded silica nanoparticles." *International Journal of Nanomedicine* (2012): 1671. doi: 10.2147/ijn.s29511.

Ge, B., H. Gram, F. Di Padova, B. Huang, L. New, R. J. Ulevitch, and J. Han. "MAPKK-independent activation of p38α mediated by TAB1-dependent autophosphorylation of p38α." *Science* 295, no. 5558 (2002): 1291–1294.

Goyal, B. R., and A. A. Mehta. "Diabetic cardiomyopathy: Pathophysiological mechanisms and cardiac dysfunction." *Human & Experimental Toxicology* 32, no. 6 (2013): 571–590.

Hansen, A., B. L. Johansson, J. Wahren, and H. von Bibra. "C-peptide exerts beneficial effects on myocardial blood flow and function in patients with type 1 diabetes." *Diabetes* 51 (2002): 3077–3082.

Harding, Jessica L., Meda E. Pavkov, Dianna J. Magliano, Jonathan E. Shaw, and Edward W. Gregg. "Global trends in diabetes complications: A review of current evidence." *Diabetologia* 62 (2019): 3–16.

Hart, G. W., M. P. Housley, and C. Slawson. "Cycling of O-linked beta-N-acetylglucosamine on nucleocytoplasmic proteins." *Nature* 446 (2007): 1017–1022.

He, Yujing, Abdulrahman Al-Mureish, and Na Wu. "Nanotechnology in the treatment of diabetic complications: A comprehensive narrative review." *Journal of Diabetes Research* 2021 (2021): 1–11.

He, Zhiyu, Tianqi Nie, Yizong Hu, Yang Zhou, Jinchang Zhu, Zhijia Liu, Lixin Liu, Kam W. Leong, Yongming Chen, and Hai-Quan Mao. "A polyphenol-metal nanoparticle platform for tunable release of liraglutide to improve blood glycemic control and reduce cardiovascular complications in a mouse model of type II diabetes." *Journal of Controlled Release* 318 (2020): 86–97.

Heidarisasan, S., N. Ziamajidi, J. Karimi, and R. Abbasalipourkabir. "Effects of insulin-loaded chitosan-alginate nanoparticles on RAGE expression and oxidative stress status in the kidney tissue of rats with type 1 diabetes." *Iran J Basic Med Sci* 21, no. 10 (2018 Oct): 1035–1042.

Holst, J. J. "The physiology of glucagon-like peptide 1." *Physiol. Rev.* 87 (2007): 1409–1439.

Hosseinpour, M., V. Azimirad, M. Alimohammadi, P. Shahabi, M. Sadighi, and G. G. Nejad. "The cardiac effects of carbon nanotubes in rat." *BioImpacts: BI* 6, no. 2 (2016): 79.

Huynh, K., B. C. Bernardo, J. R. McMullen, and R. H. Ritchie. "Diabetic cardiomyopathy: Mechanisms and new treatment strategies targeting antioxidant signaling pathways." *Pharmacology & therapeutics* 142, no. 3 (2014): 375–415.

Jashandeep, Kaur, Gurlal Singh Gill, Kiran Jeet. "Chapter 5: Applications of carbon nanotubes in drug delivery: A comprehensive review." In Shyam S. Mohapatra, Shivendu Ranjan, Nandita Dasgupta, Raghvendra Kumar Mishra, and Sabu Thomas (eds.), *In Micro and Nano Technologies, Characterization and Biology of Nanomaterials for Drug Delivery*, pp. 113–135. Elsevier, 2019.

Jia, Chengming, Hui Chen, Mengying Wei, Xiangjie Chen, Yajun Zhang, Liang Cao, Ping Yuan, Fangyuan Wang, Guodong Yang, and Jing Ma. "Gold nanoparticle-based MIR155 antagonist macrophage delivery restores the cardiac function in ovariectomized diabetic mouse model." *International Journal of Nanomedicine* 12 (2017): 4963–4979.

Jia, G. M. A. "Hill, and JR Sowers, diabetic cardiomyopathy: An update of mechanisms contributing to this clinical entity." *Circ Res* 122, no. 4 (2018): 624–638.

Kaurav, Hemlata, Satish Manchanda, Kamal Dua, and Deepak N. Kapoor. "Nanocomposites in controlled & targeted drug delivery systems." In *Nano Hybrids and Composites*, vol. 20, pp. 27–45. Trans Tech Publications Ltd, 2018.

Kinsara, A. J., and Y. M. Ismail. "Metformin in heart failure patients." *Indian Heart J* 70, no. 1 (2018 Jan–Feb): 175–176.

Kohler, J. J. "A shift for the O-Glcnac paradigm." *Nature Chemical Biology* 6, no. 9 (2010): 634–635.

Koya, D., M. R. Jirousek, Y. W. Lin, H. Ishii, K. Kuboki, and G. L. King. "Characterization of protein kinase C beta isoform activation on the gene expression of transforming growth factor-beta, extracellular matrix components, and prostanoids in the glomeruli of diabetic rats." *The Journal of Clinical Investigation* 100, no. 1 (1997): 115–126.

Krishnan, Sadayan N., Divya Talari, Jayaraman Rajangam, Latha Pujari, Narahari N. Palei, Anna Balaji, and Vijayaraj Surendran. "Protective effect of vernonia cinerea against sunset yellow induced anxiogenic behaviour in mice model: Counteract oxidative damage based approach." *The Natural Products Journal* 11, no. 4 (2021): 537–545.

Kshitij, R. B. Singh, Vanya Nayak, Jay Singh, Ajaya Kumar Singh, and Ravindra Pratap Singh. "Potentialities of bioinspired metal and metal oxide nanoparticles in biomedical sciences." *RSC Advances* 11 (2021): 24722–24746.

Kumric, Marko, Tina Ticinovic Kurir, Josip A. Borovac, and Josko Bozic. "Role of novel biomarkers in diabetic cardiomyopathy." *World Journal of Diabetes* 12, no. 6 (2021): 685.

Lee, T. W., K. J. Bai, T. I. Lee, T. F. Chao, Y. H. Kao, and Y. J. Chen. "PPARs modulate cardiac metabolism and mitochondrial function in diabetes." *J Biomed Sci* 24, no. 1 (2017 Jan 10): 5. doi: 10.1186/s12929-016-0309-5.

Lemke, L. E., L. J. Bloem, R. Fouts, M. Esterman, G. Sandusky, and C. J. Vlahos. "Decreased p38 MAPK activity in end-stage failing human myocardium: p38 MAPK α is the predominant isoform expressed in human heart." *Journal of Molecular and Cellular Cardiology* 33, no. 8 (2001): 1527–1540.

Li, Mingyuan, Chunyang Du, Na Guo, Yuou Teng, Xin Meng, Hua Sun, Shuangshuang Li, Peng Yu, and Hervé Galons. "Composition design and medical application of liposomes." *European Journal of Medicinal Chemistry* 164 (2019): 640–653.

Liang, Y., M. Li, Y. Yang, L. Qiao, H. Xu, and B. Guo. "pH/glucose dual responsive metformin release hydrogel dressings with adhesion and self-healing via dual-dynamic bonding for athletic diabetic foot wound healing." *ACS Nano* 16, no. 2 (2022): 3194–3207.

Liu, Yuntao, Siqi Zeng, Wei Ji, Huan Yao, Lin Lin, Haiying Cui, Hélder A. Santos, and Guoqing Pan. "Emerging theranostic nanomaterials in diabetes and its complications." *Advanced Science* 9, no. 3 (2022): 2102466, 1–54. doi: 10.1002/advs.202102466.

Loncarevic, Brane, Danijela Trifunovic, Ivan Soldatovic, and Bosiljka Vujisic-Tesic. "Silent diabetic cardiomyopathy in everyday practice: A clinical and echocardiographic study." *BMC Cardiovascular Disorders* 16 (2016): 1–11.

Luo, Xiao-Min, Cen Yan, and Ying-Mei Feng. "Nanomedicine for the treatment of diabetes-associated cardiovascular diseases and fibrosis." *Advanced Drug Delivery Reviews* 172 (2021): 234–248.

Marfella, Raffaele, Celestino Sardu, Gelsomina Mansueto, Claudio Napoli, and Giuseppe Paolisso. "Evidence for human diabetic cardiomyopathy." *Acta Diabetologica* 58 (2021): 983–988.

Meier, J. J. "GLP-1 receptor agonists for individualized treatment of type 2 diabetes mellitus." *Nat. Rev. Endocrinol.* 8 (2012): 728–742. doi: 10.1038/nrendo.2012.140.

Mohanta, Bibhash C., Narahari N. Palei, Vijayaraj Surendran, Subas C. Dinda, Jayaraman Rajangam, Jyotirmoy Deb, and Biswa M. Sahoo. "Lipid based nanoparticles: Current strategies for brain tumor targeting." *Current Nanomaterials* 4, no. 2 (2019): 84–100.

Mounika, S., R. Jayaraman, D. Jayashree, K. Hanna Pravalika, Anna Balaji, Moghal Sadiya Banu, and Madavaneri Prathyusha. "A comprehensive review of medicinal plants for cardioprotective potential." *International Journal of Advances in Pharmacy and Biotechnology* 7, no. 1 (2021): 24–29.

Nakagaito, M., S. Joho, R. Ushijima, M. Nakamura, and K. Kinugawa. "Comparison of canagliflozin, dapagliflozin and empagliflozin added to heart failure treatment in decompensated heart failure patients with type 2 diabetes mellitus." *Circulation Reports* 1, no. 10 (2019): 405–413.

Nikolaidis, L. A., S. Mankad, G. G. Sokos, G. Miske, A. Shah, D. Elahi, and R. P. Shannon. "Effects of glucagon-like peptide-1 in patients with acute myocardial infarction and left ventricular dysfunction after successful reperfusion." *Circulation* 109 (2004): 962–965.

Noh, H., and G. L. King. "The role of protein kinase C activation in diabetic nephropathy." *Kidney International* 72 (2007): S49–S53.

Norby, F. L., L. E. Wold, J. Duan, K. K. Hintz, and J. Ren. "IGF-I attenuates diabetes-induced cardiac contractile dysfunction in ventricular myocytes." *Am J Physiol Endocrinol Metab* 283 (2002): E658–E666.

Norouzi, M., I. Shabani, H. H. Ahvaz, and M. Soleimani. "PLGA/gelatin hybrid nanofibrous scaffolds encapsulating EGF for skin regeneration." *J Biomed Mater Res A* 103, no. 7 (2015 Jul): 2225–2235.

Nour, K. Younis, Joseph A. Ghoubaira, Emmanuel P. Bassil, Houda N. Tantawi, Ali H. Eid. "Metal-based nanoparticles: Promising tools for the management of cardiovascular diseases." *Nanomedicine: Nanotechnology, Biology and Medicine* 36 (2021): 102433.

Palei, Narahari N., Santhosh K. Mamidi, and Jayaraman Rajangam. "Formulation and evaluation of lamivudine sustained release tablet using okra mucilage." *Journal of Applied Pharmaceutical Science* 6, no. 9 (2016): 69–75.

Pelletier, R., K. Ng, W. Alkabbani, Y. Labib, N. Mourad, and J. M. Gamble. "Adverse events associated with sodium glucose co-transporter 2 inhibitors: An overview of quantitative systematic reviews." *Ther. Adv. Drug Saf.* 12 (2021): 2042098621989134.

Pradhan, Sweta Priyadarshini, Sonali Sahoo, Anindita Behera, Rajesh Sahoo, and Pratap Kumar Sahu. "Memory amelioration by hesperidin conjugated gold nanoparticles in diabetes induced cognitive impaired rats." *Journal of Drug Delivery Science and Technology* 69 (2022): 103145.

Pradhan, Sweta Priyadarshini, P. Tejaswani, Nishigandha Sa, Anindita Behera, Rajesh Kumar Sahoo, and Pratap Kumar Sahu. "Mechanistic study of gold nanoparticles of vildagliptin and vitamin E in diabetic cognitive impairment." *Journal of Drug Delivery Science and Technology* 4 (2023): 104508.

Pramanik, Sujata, and Pankaj Das. "Metal-based nanomaterials and their polymer nanocomposites." In *Nanomaterials and Polymer Nanocomposites*, pp. 91–121. Elsevier, 2019.

Raffaele, M., V. Pittalà, V. Zingales, I. Barbagallo, L. Salerno, G. Li Volti, ... and L. Vanella. "Heme oxygenase-1 inhibition sensitizes human prostate cancer cells towards glucose deprivation and metformin-mediated cell death." *International Journal of Molecular Sciences* 20 no. 10 (2019): 2593.

Rajangam, Jayaraman. "Citrullus colocynthis attenuates hyperlipidemia and hyperglycemia through its anti-oxidant property against hyperlipidemic and diabetic animal models." *Der pharmacia sinica* 4 (2013): 60–66.

Rajangam, Jayaraman, Narahari N. Palei, Shvetank Bhatt, Manas Kumar Das, D. A. S. Saumya, and Krishnapillai Mathusoothanan. "Ameliorative potential of rosuvastatin on doxorubicin-induced cardiotoxicity by modulating oxidative damage in rats." *Turkish Journal of Pharmaceutical Sciences* 19, no. 1 (2022): 28.

Rajasekharreddy, Pala, V. T. Anju, Madhu Dyavaiah, Siddhardha Busi, Surya M. Naul. "Nanoparticle-mediated drug delivery for the treatment of cardiovascular diseases." *International Journal of Nanomedicine* 15 (2020): 3741–3769.

Rajasekharreddy, Pala, Subhaswaraj Pattnaik, Siddhardha Busi, and Surya M. Nauli. "Nanomaterials as novel cardiovascular theranostics." *Pharmaceutics* 13 (2021): 348. https://doi.org/10.3390.

Rongrong, Hou, Yin Tao, Kong Ying, Jia Fang, Jiang Wei, Yang Qiang, and Xu Jing. "Tin oxide-chitosan-polyethylene glycol-d-pinitol nanocomposite ameliorates cardiac ischemia in diabetic rats via activating p62/Keap1/Nrf2 signaling." *Journal of King Saud University-Science* 34, no. 3 (2022): 101827. doi: 10.1016/j.jksus.2022.101827.

Roopan, Selvaraj Mohana, Subramanian Hari Subbish Kumar, Gunabalan Madhumitha, and Krishnamurthy Suthindhiran. "Biogenic-production of SnO2 nanoparticles and its cytotoxic effect against hepatocellular carcinoma cell line (HepG2)." *Applied Biochemistry and Biotechnology* 175 (2015): 1567–1575. doi: 10.1007/s12010-014-1381-5.

Rose, B. A., T. Force, and Y. Wang. "Mitogen-activated protein kinase signaling in the heart: Angels versus demons in a heart-breaking tale." *Physiological Reviews* 90, no. 4 (2010): 1507–1546.

Saleemi, M. A., Y. L. Kong, P. V. C. Yong, and E. H. Wong. "An overview of recent development in therapeutic drug carrier system using carbon nanotubes." *Journal of Drug Delivery Science and Technology* 59 (2020): 101855.

Sandeep Kumar, Vashist, Dan Zheng, Giorgia Pastorin, Khalid Al-Rubeaan, John H. T. Luong, and Fwu-Shan Sheu. "Delivery of drugs and biomolecules using carbon nanotubes." *Carbon* 49, no. 13 (2011): 4077–4097.

Shen, Xiaolei, Dan Li, Pengfei Zhuang, Yang Yu, Zuqiang Shi, Xifan Mei, and Chang Liu. "Negatively charged gold nanoclusters protect against diabetic cardiomyopathy by inhibiting mitophagy." *New Journal of Chemistry* 46, no. 22 (2022): 10878–10886.

Shome, S., A. D. Talukdar, M. D. Choudhury, M. K. Bhattacharya, and H. Upadhyaya. "Curcumin as potential therapeutic natural product: A nanobiotechnological perspective." *J Pharm Pharmacol* 68, no. 12 (2016 Dec): 1481–1500. doi: 10.1111/jphp.12611. Epub 2016 Oct 17. PMID: 27747859.

Singh, D. P., C. E. Herrera, B. Singh, S. Singh, R. K. Singh, and R. Kumar. "Graphene oxide: An efficient material and recent approach for biotechnological and biomedical applications." *Materials Science and Engineering: C* 86 (2018): 173–197.

Teramoto, H., and J. S. Gutkind. "Mitogen-activated protein kinase family." In *Encyclopedia of Biological Chemistry*, Second Edition, pp. 176–180. Academic Press, 2013. https://doi.org/10.1016/B978-0-12-378630-2.00362-5.

Tong, F., R. Chai, H. Jiang, and B. Dong. "In vitro/vivo drug release and anti-diabetic cardiomyopathy properties of curcumin/PBLG-PEG-PBLG nanoparticles." *Int. J. Nanomedicine* 13 (2018): 1945–1962.

Tyagi, S., P. Gupta, A. S. Saini, C. Kaushal, and S. Sharma. "The peroxisome proliferator-activated receptor: A family of nuclear receptors role in various diseases." *J Adv Pharm Technol Res* 2, no. 4 (2011 Oct): 236–240. doi: 10.4103/2231-4040.90879.

Varsha Rani, Yeshvandra Verma, Kavita Rana, and Suresh Vir Singh Rana. "Zinc oxide nanoparticles inhibit dimethylnitrosamine induced liver injury in rat." *Chemico-Biological Interactions* 295 (2018): 84–92.

Wang, X. F., J. Y. Zhang, L. Li, X. Y. Zhao, H. L. Tao, and L. Zhang. "Metformin improves cardiac function in rats via activation of AMP-activated protein kinase." *Clinical and Experimental Pharmacology and Physiology* 38, no. 2 (2011 Feb 1): 94–101.

Way, K. J., N. Katai, and G. L. King. "Protein kinase C and the development of diabetic vascular complications." *Diabetic Medicine* 18, no. 12 (2001): 945–959.

World Health Organization. *WHO Global Report on Traditional and Complementary Medicine 2019*. World Health Organization, 2019.

Xiong, Wei, Ming Yue Xiao, Mei Zhang, and Fei Chang. "Efficacy and safety of canagliflozin in patients with type 2 diabetes: A meta-analysis of randomized controlled trials." *Medicine* 95, no. 48 (2016).

Yancy, Clyde W., Mariell Jessup, Biykem Bozkurt, Javed Butler, Donald E. Casey, Mark H. Drazner, Gregg C. Fonarow et al. "2013 ACCF/AHA guideline for the management of heart failure: A report of the American College of Cardiology Foundation/American Heart Association Task Force on Practice Guidelines." *Journal of the American College of Cardiology* 62, no. 16 (2013): e147–e239.

Zeidan, Quira, and Gerald W. Hart. "The intersections between O-GlcNAcylation and phosphorylation: Implications for multiple signaling pathways." *Journal of Cell Science* 123, no. 1 (2010): 13–22.

Zeinab, Nouri, Marziyeh Hajialyani, Zhila Izadi, Roodabeh Bahramsoltani, Mohammad Hosein Far-
zae, and Mohammad Abdollahi. "Nanophytomedicines for the prevention of metabolic syndrome: A pharmacological and biopharmaceutical review." *Frontiers in Bioengineering and
Biotechnology* 8 (2020): 425.

Zhang, Chi, Linbo Zhang, Shali Chen, Biao Feng, Xuemian Lu, Yang Bai, Guang Liang et al. "The prevention of diabetic cardiomyopathy by non-mitogenic acidic fibroblast growth factor is probably mediated by the suppression of oxidative stress and damage." *PLoS One* 8, no. 12 (2013):
e82287.

Zhang, J., H.-Y. Cao, J.-Q. Wang, G.-D. Wu, and L. Wang. "Graphene oxide and reduced graphene
oxide exhibit cardiotoxicity through the regulation of lipid peroxidation, oxidative stress,
and mitochondrial dysfunction." *Front. Cell Dev. Biol.* 9 (2021): 616888. doi: 10.3389/
fcell.2021.616888.

Zhang, Mei, Lin Zhang, Bin Wu, Haolan Song, Zhenmei An, and Shuangqing Li. "Dapagliflozin treatment for type 2 diabetes: A systematic review and meta-analysis of randomized controlled trials." *Diabetes/Metabolism Research and Reviews* 30, no. 3 (2014): 204–221.

Zhou, N. Q., Z. X. Fang, N. Huang, Y. Zuo, Y. Qiu, L. J. Guo, P. Song, J. Xu, G. R. Wan, X. Q. Tian,
Y. L. Yin, and P. Li. "aFGF targeted mediated by novel nanoparticles-microbubble complex combined with ultrasound-targeted microbubble destruction attenuates doxorubicin-induced heart
failure via anti-apoptosis and promoting cardiac angiogenesis." *Front Pharmacol* 12 (2021):
607785.

Chapter 9

Nanotherapeutics for diabetic foot ulcer and wound healing using metal nanocomposites

Sopan Nangare, Jidnyasa Pantwalawalkar, Namdeo Jadhav, Zamir Khan, Ganesh Patil, Mahendra Mahajan, Rutuja Chougale, and Pravin Patil

LIST OF ABBREVIATIONS

$AgNO_3$	Silver nitrate
Ag NPs	Silver nanoparticles
AgO	Silver oxide
ALG	Alginate
Au NPs	Gold nanoparticles
CeO_2 NPs	Cerium dioxide NPs
CIP	Ciprofloxacin
CMC	Carboxymethyl cellulose
CoHA	Cobalt-substituted hydroxyapatite
CT-CS-LPs	Citicoline–chitosan-coated liposomes
Cu/TiO_2-SiO_2	Copper-titanium dioxide and silicon dioxide
Cu^{2+}	Copper
CuNPs	Copper nanoparticles
DFUs	Diabetic foot ulcers
DM	Diabetes mellitus
DNA	Deoxyribonucleic acid
DsiRNA	Dicer substrate small interfering RNA
E. coli	*Escherichia coli*
FA	Folic acid
Fe^{3+}	Iron
HA	Hyaluronic acid
$HAuCl_4$	Chloroauric acid
HDFs	Human dermal fibroblasts
hEGF	Human epidermal growth factor
HEKas	Human epithelial keratinocytes
HUVEC	Human umbilical vein endothelial cells
MMT	3-[4,5-dimethylthiazol-2-yl]-2,5 diphenyl tetrazolium bromide
MOF	Metal–organic framework
nHDF	Normal human dermal fibroblast
Nic	Nicotinamide
NPs	Nanoparticles
PGE2	Prostaglandin E2
PPG	Propylene glycol

DOI: 10.1201/9781032621135-9

PVA	Polyvinyl alcohol
PVDF	Polyvinylidene fluoride
PVP	Polyvinyl pyrrolidone
S. aureus	*Staphylococcus aureus*
SA	Sodium alginate
SF	Silk fibroin
SLNs	Solid lipid nanoparticles
TiO_2 NPs	Titanium dioxide
WHO	World Health Organization
Y_2O_3	Yttrium oxide
Zn^{2+}	Zinc
ZnO NPs	Zinc oxide NPs

9.1 DIABETIC FOOT ULCER AND WOUND HEALING

9.1.1 Introduction

Diabetes mellitus (DM) is a chronic endocrine disorder that affects the population the world over. It is distinguished by diverse metabolic abnormalities resulting in anomalously high blood sugar levels caused by either deficient insulin secretion or deficient insulin action or a combination of both (Forouhi and Wareham 2019). Every year, the prevalence of DM increases globally. Reportedly, 171 million individuals worldwide had DM in 2000, which is expected to escalate to 366 million by 2030 (Saraswat et al. 2021). Because of this growth, WHO declared it a global fast-growing epidemic (Afonso et al. 2021). According to PubMed, the scientific community has published about 17,535 research publications about diabetes and diabetic foot ulcers (DFUs) and wound healing. Figure 9.1 is a bar chart representing the number of papers with the term diabetic foot ulcer published from 2001 to March 2023 (PubMed, 2023). The primary causes of this health threat include altered dietary patterns, physical inactivity, and consequential accumulation of body fat. Moreover, the secondary complications associated with this disease can be listed as retinopathy, neuropathy, atherosclerosis, nephropathy, and DFUs upsurging its severity.

Among them, DFUs depict a higher prevalence (15–25%) (Afonso et al. 2021; Saraswat et al. 2021). Briefly, it is characterized by the destruction of the foot's deep tissues. The key risk factors for the concerned disorder include old age, smoking, hypertension, longer diabetic duration, among others (Zhang et al. 2017). The pathophysiology of DFUs involves distortion of neuropathic, vascular, and immune system components due to hyperglycemia. This state builds oxidative stress stemming from neuropathy (Singh, Young, and McNaught 2017). Additionally, glycosylation of nerve cell proteins produces ischemia. These changes alter motor, autonomic, and sensory components. Impairment of motor neurons may lead to anatomic deformities while damaged autonomic nerves impair sweat gland function leading to skin breakdown (Sumpio 2012). Regarding vascular changes, hyperglycemia induces alteration in the peripheral arteries of the foot, and thus endothelial cell dysfunction decreases vasodilators along with elevation of plasma thromboxane A2. Overall, it results in vasoconstriction and plasma hypercoagulation which further increases the risk of ulceration (Rask-Madsen and King 2013). Lastly, immune changes comprised decreased healing response in DFUs. This can be attributed to elevated T lymphocyte apoptosis, which hinders healing (Kumar et al. 2013). Further, an open ulcer can be colonized by various aerobic and anaerobic species of pathogens, such as *Pseudomonas aeruginosa*, *Staphylococcus* spp., coliform bacteria, and so on. Different microorganisms in DFUs can exist in either a sessile or a planktonic state leading to infection. Additionally, biofilm formation facilitates antibiotic resistance. If not treated efficiently, DFUs can result in gangrene, amputation, and even death (Noor, Zubair, and Ahmad 2015).

In addition, impaired wound healing is significantly observed in diabetes patients. The wound healing process involves several phases including hemostasis, inflammation, re-epithelization, tissue granulation, and tissue remodeling. All these phases overlap in certain stages resulting in a complex physiological process (Singh, Young, and McNaught 2017). However, this systematic process is disrupted in diabetic patients, which eventually deviates from the standard time course of wound healing. This can be attributed to several factors. Initially, epidermal hyper-proliferation is observed at the injury site, which promotes ulcer base formation. Further, the absence of granulation, neutrophil infiltration, keratinocyte proliferation, and reduced transforming growth factor-beta level contribute to extending diabetic wound healing. Moreover, the presence of microbes results in chronic wounds, and their healing is significantly reduced due to a compromised immune system (Wang et al. 2022). Thus, this serious complication endorses a huge economic burden on the patient. To prevent this, appropriate therapy should be provided to the patients. Briefly, a thorough patient history evaluation is the first step to offering appropriate medication. This includes considering the duration of diabetes, insulin dependence, current comorbidities, surgical history, family history, social history (tobacco or alcohol intake), and current medications. Additionally, symptoms of neuropathy should be noted. Also, the categorization of DFUs based on the most commonly employed Wagner and University of Texas Wound Classification Systems would help to assess the severity of the wound and to provide appropriate treatment (Kruse and Edelman 2006). Several treatments have been developed to treat

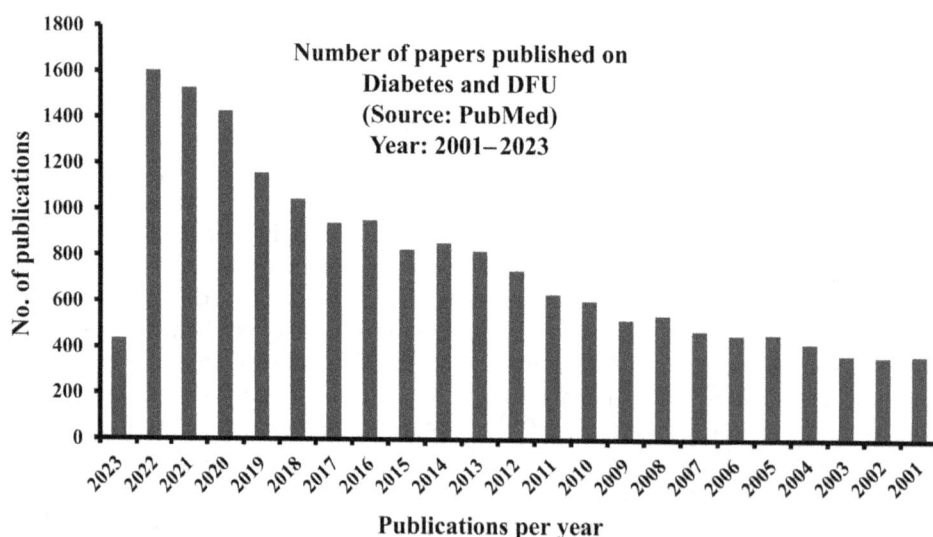

Figure 9.1 Number of papers published on diabetes and DFU from 2001 to 2023.

diabetic wounds. In some cases, oral medication is prescribed which imposes a multitude of adverse effects (Ogurtsova et al. 2017). Alternatively, Indigenous people utilize herbal and massage remedies for topical application to reduce adverse effects. However, considering the time consumption and delayed response, these methods are insufficient to heal diabetic wounds. In line with the recent literature reports, diabetic wounds can be treated with bioengineered grafts, growth factor therapy, artificial hydrophobic polymer dressings, and natural polymer therapy. Nevertheless, these therapies are unable to effectively heal diabetic wounds without causing any side effects (Dabiri, Damstetter, and Phillips 2016). Thus, the exploration of novel approaches to treat DFUs and wound healing is of utmost importance.

9.1.2 Traditional nanotherapeutics for DFUs and wound healing

As we know, DFUs are frequently observed complications in diabetic patients due to repetitive trauma in an insensate foot. These diabetic injuries have a substantial impact on the healing process and may prolong it. Undeniably, the capacity to withstand a long-term wound is a crucial concern. Thus, these problems frequently result in morbidity and entail a significant burden on the patient as well as society (Mariadoss et al. 2022). Because of rapidly increasing cases of DM and subsequently DFUs, the world requires speedy recovery from diabetic wounds to avoid further diabetic complications. Considering the potential of nanotechnology to efficiently address the complexity of diabetic wounds, it is now widely explored (Sethuram et al. 2022).

Moreover, the advancement of nanotechnology has made it possible to increase the bioavailability of target molecules at the site of the wound, hence accelerating healing, preventing complications, and enhancing patient compliance. Specifically, nanoemulsions, nanoliposomes, nanofibers, solid lipid nanoparticles (SLNs), dendrimers, films, carbon nanotubes, and nanoparticles (NPs) have been widely explored. Nanoemulsions offer promising drug delivery topically. Because of its superior physicochemical characteristics and high patient compliance, it has been widely investigated in diabetic wound healing approaches (Alhakamy et al. 2022). Figure 9.2 shows the different types of traditional therapies reported for DFUs and wound healing. Reportedly, the application of insulin nanoemulsion-based aloe vera gel, tocotrienol-naringenin loaded nanoemulgel, and levofloxacin nanoemulgel have been assessed for wound healing. Another innovative technique i.e., nanoliposomes has gained much attention owing to biodegradability and biocompatibility (Tereshkina et al. 2022).

In this line, Umar and co-researchers have designed human epidermal growth factor (hEGF)-based liposomes for wound dressing (Umar et al. 2021). In another study, Eid et al. (2022) constructed citicoline–chitosan-coated liposomes (CT-CS-LPs) for wound healing in a diabetic animal model. One more widely investigated novel technique for the treatment of diabetic wounds is drug-loaded nanofibers. Nanofibers facilitate nutrient exchange between injured tissue and the environment, thus promoting the exudates' absorption from the wound site (Gao et al. 2021).

Considering the immense applicability of nanofibers in the biomedical field, Meamar et al. (2021) developed nanofibers to deliver doxycycline and venlafaxine to treat neuropathy and inflammation in patients suffering from DFUs. Researchers also utilized nanofibers to deliver curcumin (Mitra et al. 2022). Bahadoran and co-investigators developed asiaticoside-incorporated polyvinyl alcohol (PVA)–silk fibroin (SF)–sodium alginate (SA) for the diabetic wound healing process (Bahadoran, Shamloo, and Nokoorani 2020).

Moreover, polymeric nanoparticles depicting biocompatibility have been widely investigated in DFU treatment. Researchers have also adopted SLNs to deliver activities considering biodegradability and modified release (Ezhilarasu et al. 2020). Recently, dendrimers have demonstrated noteworthy potential in DFU treatment owing to their flexibility for drug entrapment (Tetteh-Quarshie, Blough, and Jones 2021). Investigators have also explored the application of biomaterials like chitosan gel/film for DFU treatment considering its antimicrobial property (Escárcega-Galaz et al. 2018). In some cases, carbon nanotubes have also shown promising effects for DFU treatment (Sethuram et al. 2022) with special reference to the Indian scenario. Despite the optimum outcome of all the aforementioned nanocomposites, certain inherent characteristics may compromise their potential viz. unsatisfactory

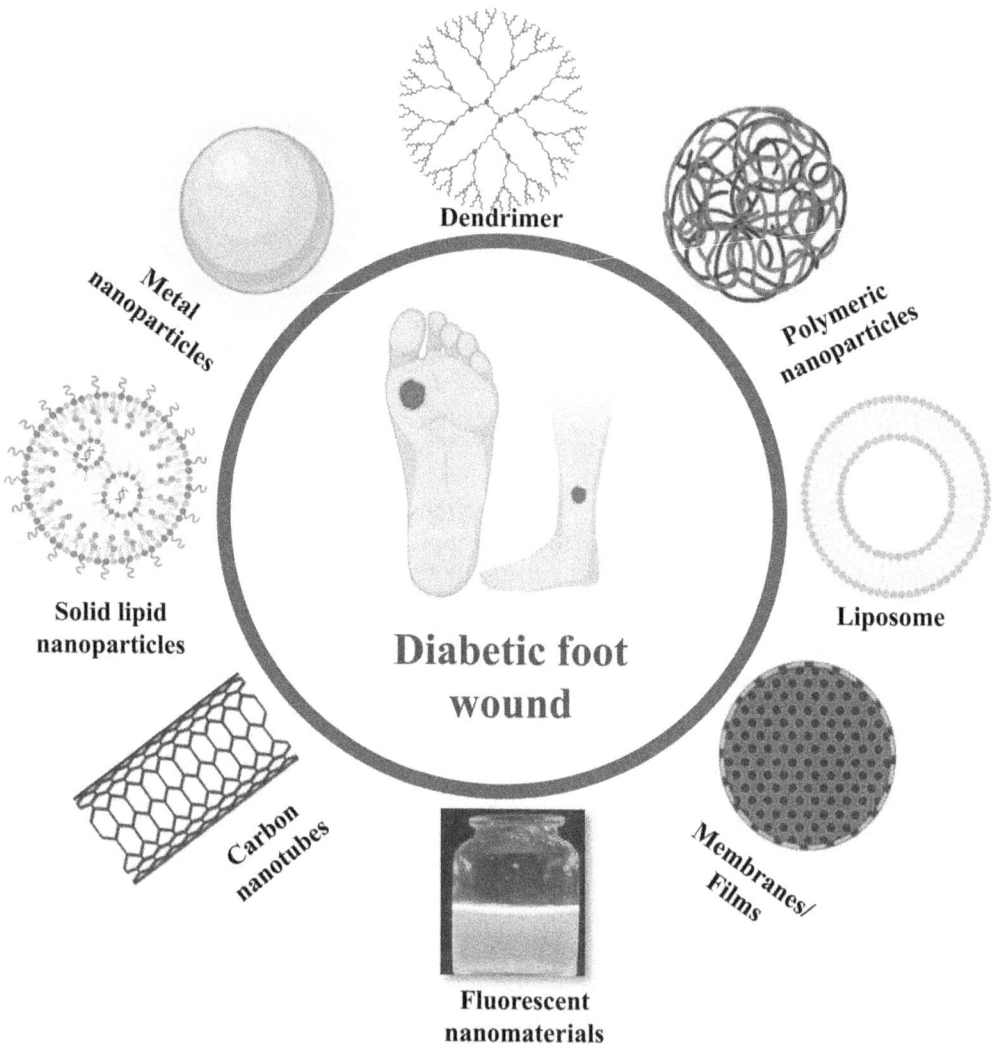

Figure 9.2 Different types of traditional therapies used for DFUs and wound healing.

mechanical properties, low solubility, stability, leakage, and so on. Comparatively, the metal NPs-based approach demonstrated a great potential for topical drug delivery applications owing to its small size, high surface area, and inherent antibacterial activity which improves biological interaction at the wound site (Kushwaha, Goswami, and Kim 2022).

9.2 DESIGN OF METAL NANOCOMPOSITES AND THEIR APPLICATIONS IN WOUND HEALING

The applicability of metallic nanocarriers as drug cargoes for biomedical applications has been comprehensively evaluated. As depicted in Figure 9.3, owing to distinctive features they have gained an exceptional position in the fields of diagnosis and medication delivery. Relatively, silver nanoparticles (Ag NPs), gold nanoparticles (Au NPs), and copper nanoparticles (Cu NPs) are employed frequently as therapeutic agents because of their anti-infective and anti-inflammatory properties (Kushwaha, Goswami, and Kim 2022). Relatively, Ag NPs are the majorly investigated NPs in diabetic wound care management due to their widely known antibacterial effect through the interaction of Ag^+ with deoxyribonucleic acid (DNA), cell walls, proteins, and enzymes of bacteria (Almonaci Hernández et al. 2017). Moreover, it offers anti-inflammatory activity by reducing cytokine release and promoting collagen deposition, which accelerates wound healing. Overall, it also provides novel anti-biofilm characteristics which are of utmost importance (Pandey et al. 2022). Au NPs are another class of widely researched NPs for wound healing applications. In brief, these NPs

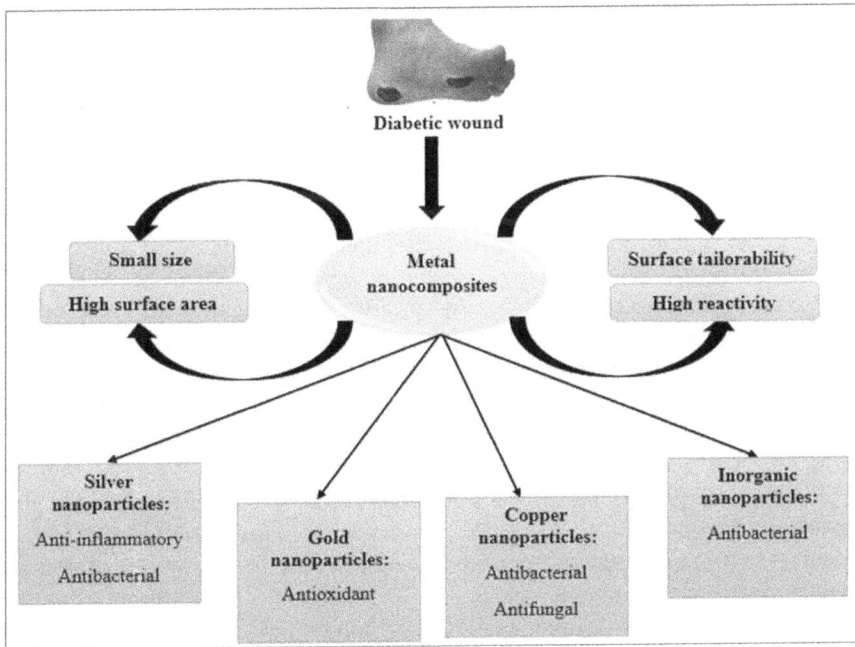

Figure 9.3 Different distinctive features of silver, gold, copper, and inorganic NPs.

inhibit lipid peroxidation and promote wound healing through antioxidant activity (Pandey et al. 2022).

Recently, Cu NPs have also received interest for their use in treating DFU infections corresponding to their antifungal and antibacterial activity. These NPs accelerate wound healing by improving angiogenesis and collagen stabilization (López-Goerne et al. 2020). The recent advancement has highlighted the importance of inorganic NPs viz. zinc oxide NPs (ZnO NPs), titanium dioxide NPs (TiO$_2$ NPs), cerium dioxide NPs (CeO$_2$ NPs), and yttrium oxide NPs (Y$_2$O$_3$ NPs) in the treatment of DFUs. Being inorganic, they are stable for a longer duration. Fundamentally, they are composed of essential mineral elements of the human body. Their antibacterial property promotes wound healing (Gupta et al. 2013). Considering these properties, metal NPs are superior for wound healing compared to most of the reported approaches. Recently, several types of nanocomposites have been found for the treatment of DFUs and other types of wound healing. It mostly consists of metal-based NPs combined with polymeric components, both natural and synthetic. Several dosage forms based on this combination have been described, including nanofibers, hydrogels, fibers, films, and so on (Renuka et al. 2022). For the synthesis of metal-based nanomaterials, different methods have been included such as co-precipitation, chemical reduction (Das et al. 2019), green synthesis (Alavi and Karimi 2020), and so on. Figure 9.4 depicts the advancement of nanomaterials-based therapeutics for DFU therapy and wound healing.

Nanosilver has been widely documented for broad-spectrum microbial prevention since its beginnings. As a result, Anisha and colleagues developed freeze-dried Ag NPs loaded chitosan (CS) and hyaluronic acid (HA) based sponges for treating diabetic wounds. To create the powder sponge, Ag NPs have been combined with CS–HA composite. Using this fabricated composite, the lyophilized sponges have been designed. As a result, it demonstrated antibacterial action against several bacterial species due to the presence of nanosized Ag NPs in combination with other polymeric components. The mechanism involved in antibacterial activity is still unknown. The human dermal fibroblast cells confirmed Ag NPs' concentration-based toxicity, with a greater concentration of Ag NPs resulting in lower cell viability. Overall, it shows good antimicrobial potential and biocompatibility. In the future, these nanosized Ag NPs-loaded sponges might be employed as an effective option for wound healing in diabetes patients (Anisha et al. 2013).

Wound healing is a difficult procedure in the biomedical sector. In this regard, *in vitro* and *in vivo* assessments of natural and synthetic polymeric composites with appropriate metal ions yield outstanding results. Dutta and colleagues developed a silver oxide (AgO)-loaded CS and polyvinyl pyrrolidone (PVP) premised film for the mitigation of bacterial infection during the healing process of wounds. Because of the antibacterial capability of AgO and CS, this film has increased antibacterial activity against selected bacteria strains. Furthermore, the cell compatibility research on L929 cell lines confirmed the biocompatibility of the film. In this case, the AgO can interact with intracellular proteins that comprise sulfur. Moreover, it interacts with enzymes found in bacterium cells, resulting in the demise of the bacterial cell. Ultimately, *in vivo* testing confirmed that the developed film had a high wound healing rate due to the silver, CS, and PVP conjugation. This expected AgO-loaded CS-PVP film will give a novel option to presently used dressing for wound healing in diabetes patients in the future (Archana et al. 2015).

In another work, Ag NPs have been glued with alginate (ALG) due to their ability to stabilize and reduce the potential for silver ions. Moreover, nicotinamide (Nic) has been combined with Ag NPs-ALG to create an anti-inflammatory and antibacterial nanocomposite

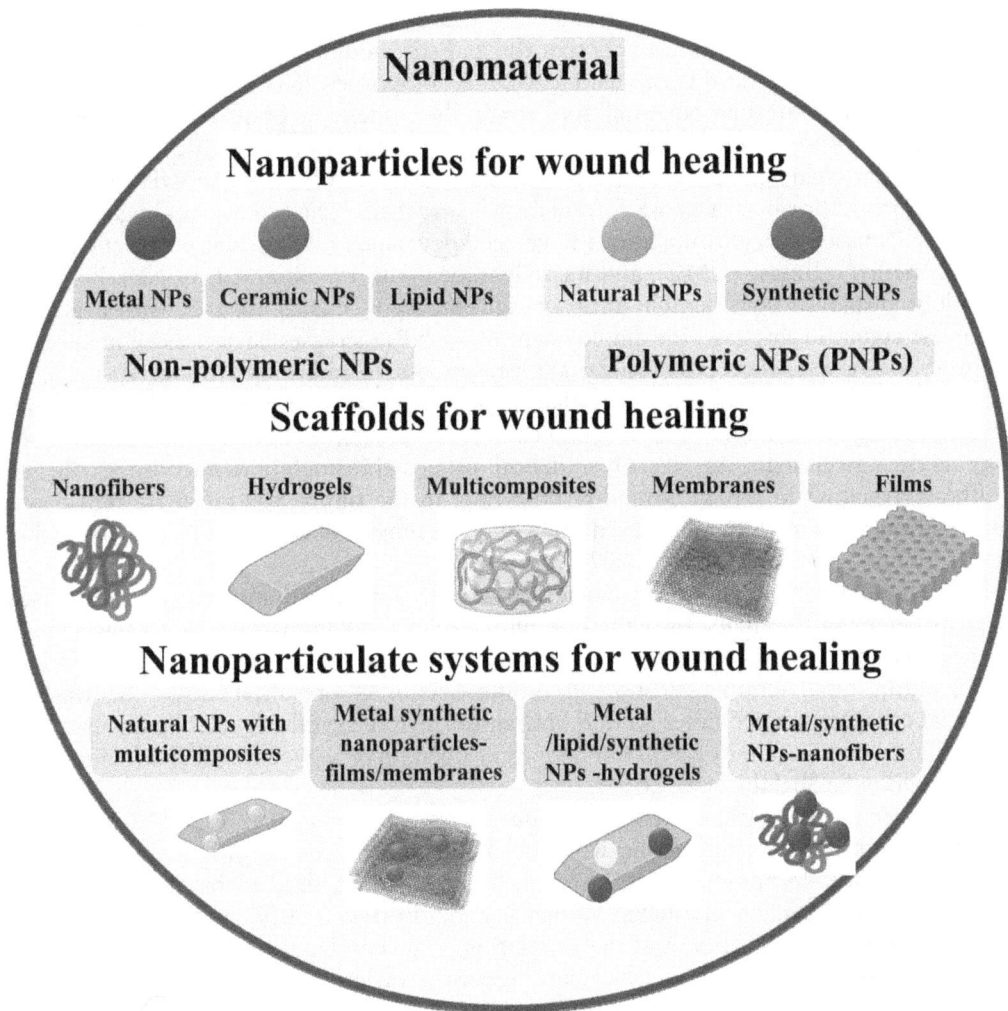

Figure 9.4 Advancement of nanomaterials-based therapeutics for DFU therapy and wound healing.

for the therapy of diabetic wounds. Following that, this nanocomposite has been patterned into nonwoven viscous textiles. As a result, this fabric is effective against gram-negative and gram-positive bacteria species including *Escherichia coli (E. coli)* and *Staphylococcus aureus (S. aureus)*. In this case, bacterial cell death occurred because of the direct bonding between silver ions and bacteria. The wound healing activity demonstrated that the existence of Ag NPs promotes wound healing in the burn diabetic rat model. As a result, this wound dressing model has ensured its preclinical application for the treatment of diabetic wounds. Overall, the anticipated metal-Nic-based nanocomposites showed good antimicrobial and wound-healing potential. It will pave the way for the appropriate care of diabetic wounds in patients in the future (Montaser et al. 2016).

Badhwar and co-authors designed the quercetin and Ag NPs co-loaded carbopol 934—aloe vera-based hydrogel for the treatment of diabetic wounds. In brief, this developed Ag

NPs hydrogel has a high antibacterial capability against *S. aureus* and *E. coli*. In addition, a preclinical investigation demonstrates that the diabetic wound gap was reduced following the application of produced hydrogel. Hence, this composite offers superior antimicrobial activity and wound healing potential. As a result, the co-delivery of quercetin with Ag NPs may open up a new avenue for diabetic wound therapy (Badhwar et al. 2021).

In a similar vein, green synthesized and chemically synthesized Ag NPs have been employed in the form of a spray formulation to treat diabetic wound healing. In short, two Ag NPs-based spray formulations have been developed by blending propylene glycol, carboxymethyl cellulose (CMC), and methylparaben with the obtained Ag NPs. Both formulations displayed considerable antibacterial efficacy against *S. aureus*, with a greater fibroblast count. In this case, the employment of Ag NPs increases bacterial cell membrane permeability. As a result, it offers bacterial nutrition loss. In addition to this, green-made Ag NPs had a greater wound healing rate than pomegranate peel extract, whereas chemically synthesized Ag NPs have a higher wound healing rate than extract but a lower wound healing rate than green-made Ag NPs. Overall, both proposed formulations encourage collagen synthesis, resulting in a high wound-healing rate. In the future, the green Ag NPs-based spray formulation might be employed for wound healing and antibacterial activity in diabetic foot ulcers (Scappaticci et al. 2021).

In 2015, Shalaby and colleagues developed Ag NPs-loaded cellulose acetate (CA)-based electrospun nanofibers mats for diabetic wound therapy and antibacterial activities. Owing to the denaturation and oxidization of bacterial organelles, the release of Ag^+ ions provides strong antibacterial action against *S. aureus* and *E. coli*. As a result, it caused the death of bacterial cells. Also, it promotes wound healing through the synthesis of collagen fibers. As a result, the combination of Ag NPs and CA may open up new avenues for the treatment of diabetic wounds (Shalaby et al. 2015).

In another study, Azlan and colleagues developed a Pluronic F-127 (PF127) gel based on dicer substrate small interfering RNA (DsiRNA)-loaded Au NPs for the treatment of diabetic wounds. In summary, the preclinical investigation demonstrated that utilizing the proposed gel resulted in quicker wound healing. In this, DsiRNA co-delivery with Au NPs enhances vascularization and the generation of prostaglandin E2 (PGE2). Moreover, it exhibited significant antibacterial action against a variety of gram-positive and gram-negative microorganisms. Overall, it has been determined that the developed Au NPs and DsiRNA-based gel may be a good solution for diabetic wound dressing. In the upcoming days, it can be an excellent preference for DFU and wound healing (Azlan et al. 2021).

Martnez et al. disclosed Au NPs functionalized calreticulin nanostructures for diabetic wound healing in 2019. These designed nanocomposites have been utilized for wound healing management. Briefly, the use of tailored nanocomposite increases the clonogenicity of various cells such as fibroblasts, endothelial cells, and keratinocytes. Additionally, it boosted fibroblast migration in a diabetic mouse model. Furthermore, the increase in collagen deposition has been confirmed utilizing histological evaluation due to Au NPs decorated calreticulin nanocomposite. As a result, it demonstrates the potential to be employed as a therapy for diabetic ulcers (Martínez et al. 2019).

Das and colleagues created an iron (Fe^{3+}) and copper (Cu^{2+}) bimetallic nanomaterial-infused cotton swap-based reinforced dressing for the treatment of diabetic wounds in 2019. Concisely, the resulting bimetallic nanocomposite has been combined into an absorbent cotton swab, and then the dressing materials thus produced have been employed for antimicrobial research. As a result, it has high antibacterial and antifungal action against

S. aureus and several fungus species. Furthermore, a preclinical investigation on diabetic wound-injected Wister albino rats confirmed the predicted dressing material's usefulness for the treatment of diabetic wounds (Das et al. 2019).

López-Goerne *et al.* published a case study of a copper–TiO_2 and silicon dioxide (Cu/TiO_2-SiO_2)-based nanocomposite-integrated polymeric gel for the treatment of diabetic foot ulcer on an elderly woman (age: 62 years). In summary, Cu/TiO_2-SiO_2 nanoparticles have been placed in a polymeric system, and the resulting nano gel has been employed to cure ulcers. As a result, there has been considerable improvement in wound healing with less infection. Also, it promoted epithelialization. As a result, the generated metal-based Cu/TiO_2-SiO_2 nanoparticles nanogel can be utilized to treat diabetic foot ulcers in patients (López-Goerne et al. 2020).

To improve wound healing, Xiao *et al.* created folic acid (FA)–Cu-based metal–organic framework NPs (FA–Cu–MOF NPs, or FA–HKUST-1). In this case, the functionalization of HKUST-1 with FA allows for the gradual release of 'Cu' ions, which helps to reduce toxicity to normal cells. Furthermore, it increased cell migration. The wound healing investigation demonstrates the increase in collagen deposition rate, resulting in an improvement in wound healing. Hence, the designed FA–HKUST-1 shows good biocompatibility and wound-healing potential. Overall, FA–HKUST-1 may one day be deemed a potential option for the treatment of diabetic wounds (Xiao et al. 2018).

Kumar and colleagues created CS hydrogel-mediated lyophilized microporous and flexible bandages with ZnO NPs for wound dressing applications. In summary, ZnO NPs have been effectively integrated into the CS hydrogel by intermolecular hydrogen bonding. The freeze-dried composite has been tested for antibacterial activity against *S. aureus* and *E. coli*. The interaction of zinc (Zn^{2+}) ions with the negatively charged cell wall of bacteria ensured good antibacterial activity. As a result, it promotes bacterial cell wall rupture and, thereby, death. The wound healing investigation confirmed that the combination of ZnO and CS has synergistic advantages, with the presence of ZnO promoting faster wound healing (Kumar et al. 2012).

The Cu–Zn–manganese ternary metal oxide nanocomposite has been created for glucose sensing, antimicrobial activity, and other applications. Because of the physical contact between NPs and cell walls, it has significant antibacterial activity against *E. coli*. As a result, the cell structure ruptures, resulting in bacterial mortality. Moreover, this nanocomposite demonstrated remarkable antioxidant activity. As a result of its high antibacterial and antioxidant action, this ternary oxide-based metal nanocomposite can be employed in the therapy of diabetes such as DFU and wound healing management (Alam et al. 2022).

Ansarizadeh et al. published ciprofloxacin (CIP)-incorporated TiO_2 nanoparticles-based polyvinylidene fluoride (PVDF) and starch composite-based electrospun fibers-based mats as a wound dressing material in 2020. This mat demonstrates high biocompatibility and increased L292 cell proliferation. Furthermore, the antibacterial's potential efficacy against *S. aureus* and *E. coli* demonstrated a boost in the growth suppression of both bacteria. As a result, the predicted electrospun mat as an antibacterial wound dressing material will give a new therapy option for diabetics (Ansarizadeh et al. 2020).

Lin and Tang presented wound dressing materials for the treatment of DFUs based on PVA–cobalt-substituted hydroxyapatite (CoHA) nanocomposite. The incorporation of CoHA into the PVA matrix improves mechanical properties while also exhibiting good antibacterial capabilities due to the release of cobalt (Co) ions from the nanocomposite. Moreover, it has high biocompatibility with L929 fibroblast cells. In addition, the release of 'Co' ions into the cell culture promotes cell proliferation. Based on this, this 'Co' nanocomposite will open up new avenues for the treatment of diabetic wounds in the future (Lin and Tang 2020).

Steffy and colleagues created the ZnO–*Strychnos nuxvomica* nanocomposite to treat DFU infections. In brief, *Strychnos nuxvomica* leaf extract has been employed to create this nanocomposite. The antibacterial ability of the ZnO–*Strychnos nuxvomica* nanocomposite has been tested against *S. aureus, E. coli, Pseudomonas aeruginosa (P. aeruginosa), Acinetobacter baumannii*, and other bacteria. As a result, it demonstrates strong antibacterial action since the release of Zn^{2+} ions causes oxidative stress and so bacterium death. Moreover, the interaction of Zn^{2+} ions with negatively charged bacterium cell membranes results in leakage and, eventually, cell death. It also shows much more wound-healing activity than the control group. Overall, the biosynthesized nanocomposite exhibits enhanced biological effects for the management of DFU infections (Steffy et al. 2018).

Ag NPs combined with metal oxide have a higher antibacterial capability than bare Ag NPs. Akavi and Karimi created the Ag–MgO nanocomposites for antibacterial activity in 2020. In a nutshell, it has been created by green synthesis utilizing extracts of *Artemisia haussknechtii* and *Protoparmeliopsis muralis*. In summary, the antibacterial activity of Ag–MgO nanocomposite has been investigated utilizing gram-positive and gram-negative microorganisms. As a result, it has a higher antibacterial capability against *E. coli* than the others reported. Moreover, because of the thiol and imidazole functionalities present in the nanocomposite, this biosynthesized nanoconjugate exhibits hemoglobin aggregation. As a result, it can be employed in wound healing treatments. It may be employed in the future to manage wound healing in various illness states such as DFU infection (Alavi and Karimi 2020).

Overall, the application of Ag NPs, Au NPs, Cu NPs, MgO, ZnO, and other metal-based nanocomposite materials shows great promise in terms of antibacterial, antifungal, and wound healing properties. As a result, over the coming days, it may show a new path for properly managing DFUs and other sorts of wounds in various illness situations. Table 9.1 represents the summary of the application of metal nanocomposites in DFUs management.

9.3 BIOCOMPATIBILITY OF METAL NANOCOMPOSITES

Biocompatibility and cytotoxicity are important considerations for nanomaterials intended for biological applications (Kyriakides et al. 2021). Biocompatibility refers to the ability of produced nanoparticles to perform their intended purpose in biomedical therapy without causing any adverse effects (Li et al. 2012). In the case of cytotoxicity, cell viability and function must be evaluated (Kyriakides et al. 2021). According to a review of the literature, the use of metal-based nanomaterials for wound healing applications can be hazardous to normal cells.

The usage of metal nanoparticles, in particular, has the potential to be hazardous to normal cells. The toxicity of metal-based nanocomposites is widely established to be related to surface coating, solubility, functionalization, growth conditions, exposure period, cell type, size, surface charge, particle concentration, and other factors. As a result, before undergoing clinical testing, the metal nanocomposite's biocompatibility must be confirmed (Nor Azlan et al. 2021). Till now, the precise mechanism behind the interaction between nanomaterials with cells is unclear. As a result, it is necessary to emphasize this interaction process, which might be useful in determining toxicity and biocompatibility (Li et al. 2012).

Table 9.1 Summary of application of metal nanocomposites in DFUs management

S. No.	Name of metal particles	Nanocomposite component(s)	Metal nanocomposites synthesis mechanism	Metal nanocomposites assembly	Activity performed	Mechanism in DFUs treatment	References
1.	Ag NPs	CS HA	Physical adsorption of Ag NPs on the surface of CS–SA	Sponge	Antibacterial, cell viability	The release of silver ions offers the antibacterial potential	(Anisha et al. 2013)
2.	AgO	CS, PVP	Hydrogen bonding between CS–PVP and metal oxide	Film	Antibacterial	Ag inactivates the enzymes of bacteria that leads to cell death	(Archana et al. 2015)
3.	Ag NPs	ALG, Nic	Interaction of COOH/OH functionality of ALG to Ag NPs	Nonwoven fabrics	Antibacterial, wound healing	Interaction between bacterial and Ag NPs	(Montaser et al. 2016)
4.	Ag NPs	Quercetin, Carbapol 934	---	Hydrogel	Antibacterial	---	(Badhwar et al. 2021)
5.	Ag NPs	CMC, propylene glycol (PPG)	---	Spray	Wound healing	Improves the synthesis of collagen and kills the bacteria	(Scappaticci et al. 2021)
6.	Ag NPs	CA	Adsorption of Ag NPs on the surface of CA via reordering of CA	Nanofiber mats	Antibacterial, wound healing	Silver offers the oxidization and denaturation of cell organelles of bacteria	(Shalaby et al. 2015)
7.	Au NPs	DsiRNA, PF127	---	Gel	Antibacterial	---	(Azlan et al. 2021)
8.	Au NPs	Calreticulin	Adsorption of Au NPs with amide functionality of calreticulin (chemical interaction)	Nanocomposite	Wound healing	Promotes clonogenicity and boosts the fibroblast migration	(Martinez et al. 2019)
9.	Fe–Cu–nanocomposite	Cotton swab	Hydrogen bonding between iron oxide and cellulose (–OH), physical attachment or chemical absorption of Cu NPs on cotton	Dressing bed	Antibacterial, antifungal, wound healing	The use of Cu NPs offers the cell wall damage followed by cell death	(Das et al. 2019)
10.	Cu NPs	TiO$_2$, SiO$_2$, CMC, Polyacrylic acid	---	Nanogel	Wound healing	---	(López-Goerne et al. 2020)

(Continued)

Table 9.1 (Continued)

S. No.	Name of metal particles	Nanocomposite component(s)	Metal nanocomposites synthesis mechanism	Metal nanocomposites assembly	Activity performed	Mechanism in DFUs treatment	References
11.	Cu²⁺	FA, HKUST-1	Metal ions-ligand interaction	F-HKUST-1	Wound healing	The customized release of Cu²⁺ ions promotes dermal tissue regeneration	(Xiao et al. 2018)
12.	ZnO	CS	Intermolecular hydrogen bonding among ZnO nanoparticles–chitosan	Dressing bandage	Antibacterial, wound healing	Zinc ions attack negatively charged bacterial cell walls resulting in the bacteria's death.	(Kumar et al. 2012)
13.	Zinc, Copper, Manganese	---	Chemical interaction and oxygen–metal bonding	Nanoparticles	Antibacterial, antioxidant	Physical interaction with cells and nanoparticles	(Alam et al. 2022)
14.	TiO₂	Polyvinylidene fluoride (PVDF), starch	Interaction of carboxylic group with CIP	Nanocomposite mats	Antibacterial	The antibacterial activity of TiO₂ with CIP helps to provide a synergistic effect	(Ansarizadeh et al. 2020)
15.	Co	PVA, CoHA	---	Nanocomposite membrane	Antibacterial	Release of cobalt ions from composite confers the antibacterial activity	(Lin and Tang 2020)
16.	ZnO	Leaf extract of Strychnos nux-vomica	ZnO stabilized by biomolecules present in the extract	Nanocomposite	Antibacterial, Wound healing	Release of Zn²⁺ ions causing oxidative stress and therefore cell death	(Steffy et al. 2018)
17.	Ag–MgO	A. haussknechtii plant and P. muralis lichen extracts	Bonding between macromolecules in extract and metal ions	Nanocomposite	Antibacterial	---	(Alavi and Karimi 2020)

Cell survival tests on HEK 293 cells were used to assess the cytotoxicity of iron–copper nanocomposite (human embryonic kidney cell lines). In this situation, the higher the concentration of copper ions, greater cell viability (84%) has been obtained. Furthermore, the MTT test (MMT: 3-[4,5-dimethylthiazol-2-yl]-2,5 diphenyl tetrazolium bromide) demonstrated that the suggested dressing materials were not harmful to HEK 292 cells even after 48 hours. As a result, it suggests that the designed iron–copper nanocomposite-based dressing materials are safe for further testing, such as in preclinical and clinical studies (Das et al. 2019).

The Au NPs–calreticulin nanocomposites did not affect the cell survival of HaCaT cells (Human keratinocyte cells) and other cell types such as HUVEC (human umbilical vein endothelial cells: HUVEC) and mouse fibroblast (NIH/3T3) cells. Furthermore, it reveals an increase in HaCaT cell viability following Au NPs treatment. In addition, the concentration of prepared Au NPs–calreticulin nanocomposites plays an imperative part in wound healing (Martínez et al. 2019).

F-HKUST-1 has been designed to be less harmful to HEKas (Human epithelial keratinocytes) and HDFs at high concentrations of 0.5 mM (HDFs: human dermal fibroblasts). Moreover, it provides a high cell migration rate due to copper ions at a sub-cytotoxic level (Xiao et al. 2018). After 24 hours, the proposed conjugate with nanosized ZnO resulted in 30–60% cell viability following the addition of 0.01% and 0.005% nanocomposite in normal human dermal fibroblast (HDF) cells. The interaction of ZnO with cells resulted in a decrease in cell viability after 24 hours (Kumar et al. 2012).

The TiO_2 electrospun biocompatibility was recently reported utilizing L292 cells (Mouse fibroblast cell line) and the MTT test. In this case, the increased concentration of tailored TiO_2 demonstrates cell death caused by DNA synthesis inhibition and cell cycle halt (Ansarizadeh et al. 2020). The nano-silver-based CS and HA-based nanocomposite exhibit concentration-dependent cytotoxicity for HDF cells, as demonstrated by the Alamar blue experiment. Moreover, nano-silver at 0.01% concentration in nano-sponge provides 46% cell viability, whereas lower concentrations of silver ions provide 75% cell viability (Anisha et al. 2013). Overall, these investigations show that toxicity and cell viability are concentration-dependent and depend on electrostatic interactions. There is no longer any emphasis on other elements of metal nanocomposite biocompatibility. As a result, in the future, the concentration of metal ions in developed nanocomposites must be considered before usage in preclinical investigations and clinical applications.

9.4 PATENTS ON METAL NANOCOMPOSITES FOR DFU AND WOUND HEALING

Poly (polyethyleneglycol citrate-co-N-isopropyl acrylamide) and Cu^{2+} MOFs nanocomposites have been designed by Xiao and colleagues for wound healing applications in DFU. Herein, the polymer-coated nanoparticles containing Cu^{2+}-based MOFs have been utilized to enhance wound healing and tissue repair while providing antibacterial, antiseptic, and analgesic properties (Xiao and Ameer 2016).

Researchers reported nanocomposite based on silver and eugenol nanoemulsion for gentle application on foot ulcers in diabetic conditions. These metal nanocomposite features promoted wound healing and effectively controlled the growth of microorganisms at the wound site. As well, it played an important role in the treatment of chronic wounds. A biomaterial composed of Ag NPs and biopolymer material reported was highly biocompatible and biodegradable in physiological conditions, making it a promising material for treating diabetic wounds as a wound dressing. This innovative creation is anticipated to represent a significant advancement in nanomedical investigation toward addressing DFU (Chandrasekaran et al. 2019).

Similarly, another invention published a wound healing system using dual metallic ferric/Cu^{2+} nanocomposite powder and a wound bed consisting of an absorbent cotton swab infused with dual metallic ferric/copper nanocomposites. When tested *in vivo* on Wistar albino rats with infected diabetic wounds, the reported nanocomposite displayed significant antibacterial activity against gram-positive and gram-negative bacteria as well as it shows the potential against fungi. Therefore, these novel biocompatible nanocomposites assisted in the healing of infected diabetic wounds and encouraged prospects in the management of other infectious wounds (Chattopadhyay et al. 2020).

Additionally, another invention pertained to a nanocomposite biofilm that contained gentamicin sulfate and silver nanoparticles. The biofilm is made up of HA and pullulan, which encases the Ag NPs. Additionally, the invention reported the nanocomposite biofilm as a wound dressing with an antimicrobial agent to prevent or treat bacterial and fungal infections (Mahajan et al. 2021).

Researchers have created a new method for photodynamic therapy that involves using Au–Ag core-shell nanoparticles conjugated with TBO (toluidine blue). This approach has been designed to address DFUs that are caused by bacteria that are resistant to multiple drugs, whether the infection is present in a single microbe or multiple microbes (Khan 2022).

In another invention, the Ag and Au nanocomposites have been created through "biosynthesis," which involved preparing a first aqueous solution of a noble metal salt or a noble metal oxide, a second solution of an aqueous plant extract, and combining the first and second solutions to create a solution containing noble metal nanoparticles. A plant product derived from *Trigonella foenumgraecum*, *Solenostemma argel*, and *Cinnamomum cassia* was used in the second solution. Silver nitrate ($AgNO_3$) and chloroauric acid ($HAuCl_4$) are examples of metal salts. Separate Au NPs and Ag NPs were prepared and then mixed to create a composition comprising an Au- and Ag-based bimetallic nanocomposite. These nanocomposites could be applied directly in the treatment of diabetic wounds (Manal et al. 2017).

Furthermore, platelet-like particles encapsulating antimicrobial metallic nanoparticles have been reported for diabetic wound dressing application. Herein, an ultra-low crosslinked polymeric microgel and a fibrin-targeting component are included in the platelet-like particles. The antimicrobial metallic nanoparticles could be integrated into the platelet-like particles covalently or noncovalently. The particles helped to halt bleeding and promoted wound healing while also suppressing bacterial infections that could occur because of tissue injury (Brown and Sproul 2019). Overall, it assured the patentability involved in metal-based nanocomposites for DFU and wound healing applications.

9.5 CURRENT CHALLENGES AND PROSPECTS

Given the severity of DFUs and wound healing-related issues such as bacterial and fungal infections, there is a desire to find a new acceptable option. Several types of conventional dosage forms and dressing materials are available for wound healing in diabetes patients, but their uses are limited owing to numerous difficulties. As a result, there is an urgent need to develop an appropriate substitute for DFUs and wound healing. Metal nanoparticles have been widely researched for therapeutic systems during the last two decades due to their unique and varied features, which include antibacterial capabilities. Many studies have demonstrated the use of metal-based nanomaterials such as Ag NPs, Cu NPs, Au NPs, ZnO NPs, MgO NPs, and AgO NPs as a component for DFUs and wound healing treatment. Despite this advancement, only a few types of literature do preclinical investigations on diabetes-indicated models. Furthermore, the design of metal nanocomposite-loaded dressing materials does not adequately describe the role of the employed metal nanoparticles in the composite. As a result, a preclinical experiment is required to investigate the specific mechanism involved in the therapy of DFUs and wound healing utilizing metal-based nanocomposite. Furthermore, the usage of transition metals in composites has been recorded, and cytotoxicity studies are needed to ensure the biocompatible concentration of metallic nanomaterials in dressing materials and dosage forms.

Furthermore, the utilization of appropriate and antibacterial polymeric components for nanotherapeutics is in great demand since it can provide synergistic effects in the treatment of DFUs and wound healing rate. Metal nanoparticle toxicity is a big hurdle in their biological uses. Few publications have provided biocompatibility and toxicity data in the publications about their usage in DFUs and wound healing. As a result, while creating metal nanocomposite-based nanotherapeutics for DFUs and wound healing applications, it is necessary to keep this in mind to reduce toxicity. There is an urgent necessity to emphasize a green approach to metal nanomaterial production. This can assist to reduce the toxicity of metal particles to cells. Several bimetallic and ternary composites have been recorded in which the metal percentage in the final formulation must be controlled. Using diverse biomaterials with metal nanoparticles may also assist to minimize toxicity in this scenario. Despite advances in the biomedical realm, developing multifunctional dressing materials and dosage forms remains difficult. As a result, controlling resistant bacterial strains is extremely difficult. There have been few publications on the delivery of antibacterial and anti-inflammatory drugs using metal nanoconjugate-based nanotherapeutics in DFUs and wound healing, which might be a new research topic for aspiring researchers. In future, there will be a greater emphasis on stable and repeatable dose forms and dressing materials. In the coming days, clinical trials will be necessary to ensure the efficacy of suggested metal nanocomposite-based nanotherapeutics. Overall, it might be a realistic and cost-effective therapy option for DFUs and other wounds.

9.6 CONCLUSION

Numerous research teams created sophisticated metal-based nanocomposites for DFUs and wound healing. Metal NPs such as Ag NPs, Cu NPs, Au NPs, and metal oxides such as ZnO, MgO, and AgO have primarily been used. Other metals, such as manganese, Co, and titanium, have also been found to have antibacterial and wound-healing properties. Design nanocomposites have been documented using various natural and synthetic polymers to create a metal-based gel, film, dressing material, nanocomposite, sponge, and so on. As a result, these produced nanocomposites exhibit increased antimicrobial activity against several gram-negative and gram-positive bacteria, which are the primary causes of DFUs and wounds. In the case of metal-based nanocomposites, the mechanism behind antibacterial action is yet unclear. Metal ion release interacts with negatively charged cell membranes, causing oxidative stress and the generation of reactive oxygen species in the bacterial cell. It also interacts with enzymes and cell organelles, as well as denatures protein production. Ultimately, bacterial cell death occurs. Furthermore, preclinical research on diabetes-induced animal models indicated that the wound healing rate was faster than in the control group. As a result, it concludes that the suggested nanotherapeutics have clinical applications for the treatment of DFUs and wound healing. Moreover, these nanocomposites have good stability and biocompatibility with normal cells. Finally, metal-based nanocomposites provide a nontoxic and efficient treatment solution for DFUs and wound healing. As a result, in the next years, the employment of metal nanocomposite in various forms such as gel, film, powder, and dressing will usher in a new era and eliminate the drawbacks of the currently available choices for wound healing in diabetics and other health problems.

CONFLICT OF INTEREST

The authors state that they have no conflicts of interest.

ACKNOWLEDGMENTS

The Indian Council of Medical Research (ICMR), New Delhi, has awarded Sopan Nangare with a Research Associate Fellowship. The authors are grateful to Principal Dr. S. B. Bari for providing facilities to complete the book chapter.

REFERENCES

Afonso, Ana C., Diana Oliveira, Maria José Saavedra, Anabela Borges, and Manuel Simões. 2021. "Biofilms in diabetic foot ulcers: Impact, risk factors and control strategies." *International Journal of Molecular Sciences* 22 (15):8278.

Alam, Mir Waqas, Hassan S. Al Qahtani, Basma Souayeh, Waqar Ahmed, Hind Albalawi, Mohd Farhan, Alaaedeen Abuzir, and Sumaira Naeem. 2022. "Novel copper-zinc-manganese ternary metal oxide nanocomposite as heterogeneous catalyst for glucose sensor and antibacterial activity." *Antioxidants* 11 (6):1064.

Alavi, Mehran, and Naser Karimi. 2020. "Hemoglobin self-assembly and antibacterial activities of bio-modified Ag-MgO nanocomposites by different concentrations of *Artemisia haussknechtii* and *Protoparmeliopsis muralis* extracts." *International Journal of Biological Macromolecules* 152:1174–1185.

Alhakamy, Nabil A., Giuseppe Caruso, Anna Privitera, Osama A.A. Ahmed, Usama A. Fahmy, Shadab Md, Gamal A. Mohamed, Sabrin R.M. Ibrahim, Basma G. Eid, and Ashraf B. Abdel-Naim. 2022. "Fluoxetine ecofriendly nanoemulsion enhances wound healing in diabetic rats: *In vivo* efficacy assessment." *Pharmaceutics* 14 (6):1133.

Almonaci Hernández, C.A., K. Juarez-Moreno, M.E. Castañeda-Juarez, H. Almanza-Reyes, A. Pestryakov, and N. Bogdanchikova. 2017. "Silver nanoparticles for the rapid healing of diabetic foot ulcers." *Int. J. Med. Nano Res* 4 (01910.23937):2378–3664.

Anisha, B.S., Raja Biswas, K.P. Chennazhi, and R. Jayakumar. 2013. "Chitosan: Hyaluronic acid/nano silver composite sponges for drug resistant bacteria infected diabetic wounds." *International Journal of Biological Macromolecules* 62:310–320.

Ansarizadeh, Mohamadhasan, Seyyed Arash Haddadi, Majed Amini, Masoud Hasany, and Ahmad Ramazani Saadat Abadi. 2020. "Sustained release of CIP from TiO_2-PVDF/starch nanocomposite mats with potential application in wound dressing." *Journal of Applied Polymer Science* 137 (30):48916.

Archana, D., Brijesh K. Singh, Joydeep Dutta, and P.K. Dutta. 2015. "Chitosan-PVP-nano silver oxide wound dressing: *In vitro* and *in vivo* evaluation." *International Journal of Biological Macromolecules* 73:49–57.

Azlan, Ahmad Yasser Hamdi Nor, Haliza Katas, Noraziah Mohamad Zin, and Mh Busra Fauzi. 2021. "Dual action gels containing DsiRNA loaded gold nanoparticles: Augmenting diabetic wound healing by promoting angiogenesis and inhibiting infection." *European Journal of Pharmaceutics and Biopharmaceutics* 169:78–90.

Badhwar, Reena, Bharti Mangla, Yub Raj Neupane, Kushagra Khanna, and Harvinder Popli. 2021. "Quercetin loaded silver nanoparticles in hydrogel matrices for diabetic wound healing." *Nanotechnology* 32 (50):505102.

Bahadoran, Maedeh, Amir Shamloo, and Yeganeh Dorri Nokoorani. 2020. "Development of a polyvinyl alcohol/sodium alginate hydrogel-based scaffold incorporating bFGF-encapsulated microspheres for accelerated wound healing." *Scientific Reports* 10 (1):1–18.

Brown, Ashley, and Erin Sproul. 2019. North Carolina State University. "Antimicrobial platelet-like particles." US Patent PCT/US2019/016232, filed February 01, 2019, and issued August 8.

Chandrasekaran, N., A. Mukherjee, J. Thomas, and S. Lakshimipriya. 2019. "Eugenol embedded with silver nanocomposites on polymer matrix for the treatment of diabetic foot ulcer." Indian patent 201941046272, filed November 14, 2019, and issued November 29.

Chattopadhyay, A., S. Ghosh, S. Das, M. Goswami, K. Upashi, K. Raghuram, and Kalita Sanjeeb. 2020. Indian Institute of Technology Guwahati. "Bimetallic nanocomposite based wound healing system and method of manufacture thereof." Indian patent 201931014175, filed April 8, 2019, and issued October 9.

Dabiri, Ganary, Elizabeth Damstetter, and Tania Phillips. 2016. "Choosing a wound dressing based on common wound characteristics." *Advances in Wound Care* 5 (1):32–41.

Das, Madhumita, Upashi Goswami, Raghuram Kandimalla, Sanjeeb Kalita, Siddhartha Sankar Ghosh, and Arun Chattopadhyay. 2019. "Iron: Copper bimetallic nanocomposite reinforced dressing materials for infection control and healing of diabetic wound." *ACS Applied Bio Materials* 2 (12):5434–5445.

Eid, Hussein M., Adel A. Ali, Ahmed M. Abdelhaleem Ali, Essam M. Eissa, Randa M. Hassan, Fatma I. Abo El-Ela, and Amira H. Hassan. 2022. "Potential use of tailored citicoline chitosan-coated liposomes for effective wound healing in diabetic rat model." *International Journal of Nanomedicine*:555–575.

Escárcega-Galaz, Ana Aglahe, José Luis De La Cruz-Mercado, Jaime López-Cervantes, Dalia Isabel Sánchez-Machado, Olga Rosa Brito-Zurita, and José Manuel Ornelas-Aguirre. 2018. "Chitosan treatment for skin ulcers associated with diabetes." *Saudi Journal of Biological Sciences* 25 (1):130–135.

Ezhilarasu, Hariharan, Dinesh Vishalli, S. Thameem Dheen, Boon-Huat Bay, and Dinesh Kumar Srinivasan. 2020. "Nanoparticle-based therapeutic approach for diabetic wound healing." *Nanomaterials* 10 (6):1234.

Forouhi, Nita Gandhi, and Nicholas J. Wareham. 2019. "Epidemiology of diabetes." *Medicine* 47 (1):22–27.

Gao, Zhaoju, Qiuxiang Wang, Qingqiang Yao, and Pingping Zhang. 2021. "Application of electrospun nanofiber membrane in the treatment of diabetic wounds." *Pharmaceutics* 14 (1):6.

Gupta, Sanjeev, Radhika Bansal, Sunita Gupta, Nidhi Jindal, and Abhinav Jindal. 2013. "Nanocarriers and nanoparticles for skin care and dermatological treatments." *Indian Dermatology Online Journal* 4 (4):267.

Khan, Farheen Akhtar A. 2022. "TBO conjugated CHIT-AU-AgNPS mediated photodynamic therapy against diabetic foot ulcer caused by multi-drug resistant bacteria." Indian patent 202011046527, filed October 26, 2020, and issued May 31.

Kruse, Ingrid, and Steven Edelman. 2006. "Evaluation and treatment of diabetic foot ulcers." *Clinical Diabetes* 24 (2):91–93.

Kumar, Arya Awadhesh, Sunny Garg, Santosh Kumar, Lalit P. Meena, and Kamlakar Tripathi. 2013. "Estimation of lymphocyte apoptosis in patients with chronic, non healing diabetic foot ulcer." *International Journal of Medical Science and Public Health* 2 (4):811–813.

Kumar, P.T., Vinoth-Kumar Lakshmanan, T.V. Anilkumar, C. Ramya, P. Reshmi, A.G. Unnikrishnan, Shantikumar V. Nair, and R. Jayakumar. 2012. "Flexible and microporous chitosan hydrogel/ nano ZnO composite bandages for wound dressing: *In vitro* and *in vivo* evaluation." *ACS Applied Materials & Interfaces* 4 (5):2618–2629.

Kushwaha, Anamika, Lalit Goswami, and Beom Soo Kim. 2022. "Nanomaterial-based therapy for wound healing." *Nanomaterials* 12 (4):618.

Kyriakides, Themis R., Arindam Raj, Tiffany H. Tseng, Hugh Xiao, Ryan Nguyen, Farrah S. Mohammed, Saiti Halder, Mengqing Xu, Michelle J. Wu, and Shuozhen Bao. 2021. "Biocompatibility of nanomaterials and their immunological properties." *Biomedical Materials* 16 (4):042005.

Li, Xiaoming, Sang Cheon Lee, Shuming Zhang, and Tsukasa Akasaka. 2012. "Biocompatibility and toxicity of nanobiomaterials." *Journal of Nanomaterials* 2012:17–17.

Lin, Wei-Chun, and Cheng-Ming Tang. 2020. "Evaluation of polyvinyl alcohol/cobalt substituted hydroxyapatite nanocomposite as a potential wound dressing for diabetic foot ulcers." *International Journal of Molecular Sciences* 21 (22):8831.

López-Goerne, Tessy, Paola Ramírez-Olivares, Luis A. Pérez-Dávalos, Javier Alejandro Velázquez-Muñoz, and Jesús Reyes-González. 2020. "Catalytic nanomedicine: Cu/TiO$_2$-SiO$_2$ nanoparticles as treatment of diabetic foot ulcer: A case report." *Current Nanomedicine (Formerly: Recent Patents on Nanomedicine)* 10 (3):290–295.

Mahajan, A., P. Pandey, P. Patel, and M. Jadav. 2021. "Silver nanocomposite biofilm loaded with gentamicin sulphate for the wound healing application." Indian patent 2021106682, filed August 24, 2021, and issued November 18.

Manal, Ahmed Gasmelseed A., H. Awatif Ahmed, E. W. Mai Abdelrahman, V. Promy, O. Khalid Mustafa. 2017. King Saud University. "Method of treating diabetic wounds using biosynthesized nanoparticles." US Patent 14877916. filed October 7, 2015, and issued April 13.

Mariadoss, Arokia Vijaya Anand, Allur Subramaniyan Sivakumar, Chang-Hun Lee, and Sung Jae Kim. 2022. "Diabetes mellitus and diabetic foot ulcer: Etiology, biochemical and molecular based treatment strategies via gene and nanotherapy." *Biomedicine & Pharmacotherapy* 151:113134.

Martínez, Sara Paola Hernández, Teodoro Iván Rivera González, Moisés Armides Franco Molina, Juan José Bollain y Goytia, Juan José Martínez Sanmiguel, Diana Ginette Zárate Triviño, and Cristina Rodríguez Padilla. 2019. "A novel gold calreticulin nanocomposite based on chitosan for wound healing in a diabetic mice model." *Nanomaterials* 9 (1):75.

Meamar, Rokhsareh, Sana Chegini, Jaleh Varshosaz, Ashraf Aminorroaya, Masoud Amini, and Mansour Siavosh. 2021. "Alleviating neuropathy of diabetic foot ulcer by co-delivery of venlafaxine and matrix metalloproteinase drug-loaded cellulose nanofiber sheets: Production, in vitro characterization and clinical trial." *Pharmacological Reports* 73 (3): 806–819.

Mitra, Souradeep, Tarun Mateti, Seeram Ramakrishna, and Anindita Laha. 2022. "A review on curcumin-loaded electrospun nanofibers and their application in modern medicine." *JOM* 74 (9):3392–3407.

Montaser, A.S., A.M. Abdel-Mohsen, M.A. Ramadan, A.A. Sleem, N.M. Sahffie, J. Jancar, and A. Hebeish. 2016. "Preparation and characterization of alginate/silver/nicotinamide nanocomposites for treating diabetic wounds." *International Journal of Biological Macromolecules* 92:739–747.

Noor, Saba, Mohammad Zubair, and Jamal Ahmad. 2015. "Diabetic foot ulcer: A review on patho-physiology, classification and microbial etiology." *Diabetes & Metabolic Syndrome: Clinical Research & Reviews* 9 (3):192–199.

Nor Azlan, Ahmad Yasser Hamdi, Haliza Katas, Mohd Fauzi Mh Busra, Nur Atiqah Mohamad Salleh, and Ali Smandri. 2021. "Metal nanoparticles and biomaterials: The multipronged approach for potential diabetic wound therapy." *Nanotechnology Reviews* 10 (1):653–670.

Ogurtsova, Katherine, J.D. da Rocha Fernandes, Y. Huang, Ute Linnenkamp, L. Guariguata, Nam H. Cho, David Cavan, J.E. Shaw, and L.E. Makaroff. 2017. "IDF diabetes atlas: Global estimates for the prevalence of diabetes for 2015 and 2040." *Diabetes Research and Clinical Practice* 128:40–50.

Pandey, Supriya, Mohammad Shaif, Tarique M. Ansari, Arshiya Shamim, and Poonam Kushwaha. 2022. "Leveraging potential of nanotherapeutics in management of diabetic foot ulcer." *Experimental and Clinical Endocrinology & Diabetes* 130 (10):678–686.

Rask-Madsen, Christian, and George L. King. 2013. "Vascular complications of diabetes: Mechanisms of injury and protective factors." *Cell Metabolism* 17 (1):20–33.

Renuka, Remya Rajan, Angeline Julius, Suman Thodhal Yoganandham, Dhamodharan Umapathy, Ramya Ramadoss, Antony V. Samrot, and Danis D. Vijay. 2022. "Diverse nanocomposites as a potential dressing for diabetic wound healing." *Frontiers in Endocrinology* 13.

Saraswat, Bharti, Kapil Kumar Gill, Ashok Yadav, and Krishan Kumar. 2021. "A prospective observation study on diabetic foot ulcer using diabetic ulcer severity score at tertiary care hospital." *International Surgery Journal* 8 (12):3553.

Scappaticci, Renan Aparecido Fernandes, Andresa Aparecida Berretta, Elina Cassia Torres, Andrei Felipe Moreira Buszinski, Gabriela Lopes Fernandes, Thaila Fernanda Dos Reis, Francisco Nunes de Souza-Neto, Luiz Fernando Gorup, Emerson Rodrigues de Camargo, and Debora Barros Barbosa. 2021. "Green and chemical silver nanoparticles and pomegranate formulations to heal infected wounds in diabetic rats." *Antibiotics* 10 (11):1343.

Sethuram, Lakshimipriya, Thomas John, Amitava Mukherjee, and Chandrasekaran Natarajan. 2022. "A review on contemporary nanomaterial based therapeutics for the treatment of Diabetic Foot Ulcer (DFU) with special reference to Indian Scenario." *Nanoscale Advances* 4:2367–2398.

Shalaby, Thanaa Ibrahim, Nivan Mahmoud Fekry, Amal Sobhy El Sodfy, Amel Gaber El Sheredy, and Maisa El Sayed Sayed Ahmed Moustafa. 2015. "Preparation and characterization of antibacterial silver-containing nanofibres for wound healing in diabetic mice." *International Journal of Nanoparticles* 8 (1):82–98.

Singh, Shailendra, Alistair Young, and Clare-Ellen McNaught. 2017. "The physiology of wound healing." *Surgery (Oxford)* 35 (9):473–477.

Steffy, Katherin, G. Shanthi, Anson S, Maroky, and S. Selvakumar. 2018. "Potential bactericidal activity of S. nux-vomica: ZnO nanocomposite against multidrug-resistant bacterial pathogens and wound-healing properties." *Journal of Trace Elements in Medicine and Biology* 50:229–239.

Sumpio, Bauer E. 2012. "Contemporary evaluation and management of the diabetic foot." *Scientifica* 2012:435487.

Tereshkina, Yu A., T.I. Torkhovskaya, E.G. Tikhonova, L.V. Kostryukova, M.A. Sanzhakov, E.I. Korotkevich, Yu Yu Khudoklinova, N.A. Orlova, and E.F. Kolesanova. 2022. "Nanoliposomes as drug delivery systems: Safety concerns." *Journal of Drug Targeting* 30 (3):313–325.

Tetteh-Quarshie, Samuel, Eric R. Blough, and Cynthia B. Jones. 2021. "Exploring dendrimer nanoparticles for chronic wound healing." *Frontiers in Medical Technology* 3:661421.

Umar, Abd Kakhar, Sriwidodo Sriwidodo, Iman Permana Maksum, and Nasrul Wathoni. 2021. "Film-forming spray of water-soluble chitosan containing liposome-coated human epidermal growth factor for wound healing." *Molecules* 26 (17):5326.

Wang, Feng, Wenyao Zhang, Hao Li, Xiaonan Chen, Sining Feng, and Ziqing Mei. 2022. "How effective are nano-based dressings in diabetic wound healing? A comprehensive review of literature." *International Journal of Nanomedicine* 17:2097–2119.

Xiao, Jisheng, and G. A. Ameer. 2016. Northwestern University, US. "Polymer metal-organic framework composites." US Patent PCT/US2015/066731, filed December 18, 2015, and issued June 23.

Xiao, Jisheng, Yunxiao Zhu, Samantha Huddleston, Peng Li, Baixue Xiao, Omar K. Farha, and Guillermo A. Ameer. 2018. "Copper metal: Organic framework nanoparticles stabilized with folic
acid improve wound healing in diabetes." *ACS Nano* 12 (2):1023–1032.

Zhang, Pengzi, Jing Lu, Yali Jing, Sunyinyan Tang, Dalong Zhu, and Yan Bi. 2017. "Global epidemiology of diabetic foot ulceration: A systematic review and meta-analysis." *Annals of Medicine* 49
(2):106–116.

Chapter 10

Nanotherapeutics for management of dyslipidemia

P. K. Sahu, P. S. Kumar, and S. K. Prusty

LIST OF ABBREVIATIONS

ACE	Angiotensin-converting enzyme
AMPK	AMP-activated protein kinase
CE	Cholesterol esters
CETP	Cholesterol ester transfer protein
CM	Chylomicrons
CoQ-10	Co-enzyme Q10
DNA	Deoxyribonucleic acid
GIT	Gastrointestinal tract
HDL	High-density lipoprotein
HMG-CoA	3-Hydroxy 3-methylglutaryl-coenzyme A
ICAM	Intercellular adhesion molecule
IDL	Intermediate density lipoprotein
IL-1	Interleukin 1
IL-6	Interleukin 6
JAK/STAT	Janus kinase/signal transducers and activators of transcription
LDL	Low-density lipoprotein
LOX	Lipoxygenase
MAPK	Mitogen-activated protein kinase
NF-κB	Nuclear factor κ-B
NLC	Nanostructured lipid carriers
NPC1L1	Niemann-Pick C1-Like 1
NPs	Nanoparticles
PCSK9	Proprotein convertase subtilisin/kexin type 9
PEG	Polyethylene glycol
PL	Phospholipids
PLA	Poly lactic acid
PLGA	Poly lactic-co-glycolic acid
PPAR-α	Peroxisome proliferation activated receptor-α
ROS	Reactive oxygen species
SLN	Solid lipid nanoparticles
SNEDDS	Self-nanoemulsifying drug delivery system
SOD	Superoxide dismutase
TG	Triglycerides

DOI: 10.1201/9781032621135-10

TNF-α   Tumor necrosis factor-α
VEGF   Vascular endothelial growth factor
VLDL   Very low-density lipoprotein

10.1 INTRODUCTION

Lipids are fatty acid-derived compounds. They are of three types: simple lipids, compound lipids, and neutral lipids. Esters of fatty acid and alcohol are called simple lipids. Sulfur lipids and phospholipids (PL) are examples of compound lipids. Cholesterol, cholesterol esters (CE), and triglycerides (TG) are examples of neutral lipids (Turkish and Sturley 2009). They are transported in blood with the help of lipoproteins. Lipoproteins are spherical particles of water-soluble proteins. Based on the density, lipoproteins can be very low-density lipoprotein (VLDL), low-density lipoprotein (LDL), intermediate-density lipoprotein (IDL), and high-density lipoproteins (HDL). On their surface, they have a protein called apoprotein. Apoproteins provide structural stability to lipoproteins 9 (Su and Peng 2020).

Apoproteins are of different types: Apo-A, Apo-C, Apo-E, Apo-B48, and Apo-B100. Apo-B48 is synthesized in the GIT and found in chylomicrons (CM) and chylomicron remnants to help in their assembly and secretion. Apo-B100 is synthesized in the liver and is found in VLDL, LDL, and IDL. They deliver lipids to the artery wall. Apo-A/C/E is found in HDL. HDL reverses cholesterol transport from peripheral cells or artery walls to the liver (Liu et al. 2021). The source, core lipids, and apoproteins present in different types of lipoproteins are given in Table 10.1.

The triglycerides and cholesterol from the diet are transported from the gastrointestinal tract (GIT) to the tissues by chylomicrons. Chylomicrons are converted to free fatty acids and chylomicron remnants by lipoprotein lipase in the tissues. Chylomicron remnants are uptaken and stored in the liver as very VLDL. VLDL transports cholesterol and newly synthesized triglyceride to the tissues and releases free fatty acids to become LDL. A part of LDL is taken by tissues and arteries, whereas the other part is taken by the liver (by endocytosis via LDL receptor). In muscle and adipose tissue, VLDL is converted to IDL. Some of the IDL is rapidly uptaken by the liver and converted to LDL by hepatic lipase. Lipoprotein (a) is LDL with additional Apo-a. HDLs adsorb cholesterol derived from cell breakdown in tissues (including arteries) and transport it to VLDL and LDL. An increase in HDL is beneficial, whereas an increase in other lipoproteins is harmful (Hernáez et al. 2019).

Table 10.1 Source, core lipids, and apoproteins present in different types of lipoproteins

Source	Lipoprotein	Core lipid	Apoproteins
Diet	CM	TG > CE	B-48, C, E, A
	CM remnant	TG < CE	B-48, C, E, A
Liver (hepatocytes)	VLDL	TG > CE	C, B100, E
	IDL	TG = CE	C, B100, E
	LDL	CE	B100
	HDL	CE + PL	A, C, E

Note: CM= Chylomicron, CE = cholesterol esters, TG = triglycerides, VLDL = very low-density lipoprotein, LDL = low-density lipoprotein, IDL = intermediate-density lipoprotein, HDL = high-density lipoproteins.

Dyslipidemia is a metabolic disorder characterized by abnormal levels of lipids and lipo-proteins. Four types of lipoprotein abnormalities are frequently found in dyslipidemia. They are:

a. Increased LDL cholesterol levels
b. Increased CM remnants and IDL level
c. Increased levels of lipoprotein (a)
d. Decreased HDL cholesterol levels

In addition, there is an increased level of plasma cholesterol and triglycerides. Dyslipid-emia may occur alone or together with a cluster of medical conditions collectively known as metabolic syndrome. Dyslipidemia is usually associated with atherosclerosis, but it may also be associated with conditions like hypertension, obesity, diabetes, and so on (Durrington 2003).

10.2 ETIOPATHOGENESIS OF DYSLIPIDEMIA

Dyslipidemia itself does not produce symptoms. Because of dyslipidemia-associated ath-erosclerosis, lipids such as cholesterol, triglycerides, and LDL are deposited in blood ves-sel walls and restrict blood flow. So dyslipidemia is a major risk factor for cardiovascular diseases such as angina pectoris, heart attack, and stroke. The factors causing dyslipidemia may be categorized as:

- Extrinsic: It can result from external factors like
  - Diet: A diet rich in saturated fat, cholesterol, and trans-fats increases the risk of dyslipidemia.
  - Sedentary lifestyle: Sedentary lifestyle and lack of exercise aggravates dyslipidemia.
  - Tobacco exposure: Tobacco exposure is harmful. It increases oxidative stress and promotes dyslipidemia.
  - Exposure to harmful radiation: Harmful radiations induce oxidative stress and hence may cause dyslipidemia.
- Intrinsic: Dyslipidemia is more frequently associated with
  - Oxidative stress: Dyslipidemia is associated closely with increased endothelial pro-duction of reactive oxygen species (ROS).
  - Cardiovascular diseases: There is strong clinical evidence that dyslipidemia is asso-ciated with cardiovascular diseases.
  - Diabetes mellitus: Insulin level is increased in diabetes. A prolonged increase in insulin causes dyslipidemia.
- Genetic: The causes of dyslipidemia also include genetic (familial) abnormalities of lipid metabolism.

All these factors causing dyslipidemia accumulate lipids. These lipids on oxidation pro-duce intercellular adhesion molecules (ICAM) and endothelial selectin. They cause monocyte adhesion and an influx of inflammatory cells like monocytes, T-lymphocytes, and mast cells. Monocytes are differentiated into macrophages. They release cytokines like interleukin 1 (IL-1), interleukin 6 (IL-6), and tumor necrosis factor-α (TNF-α). These cytokines promote

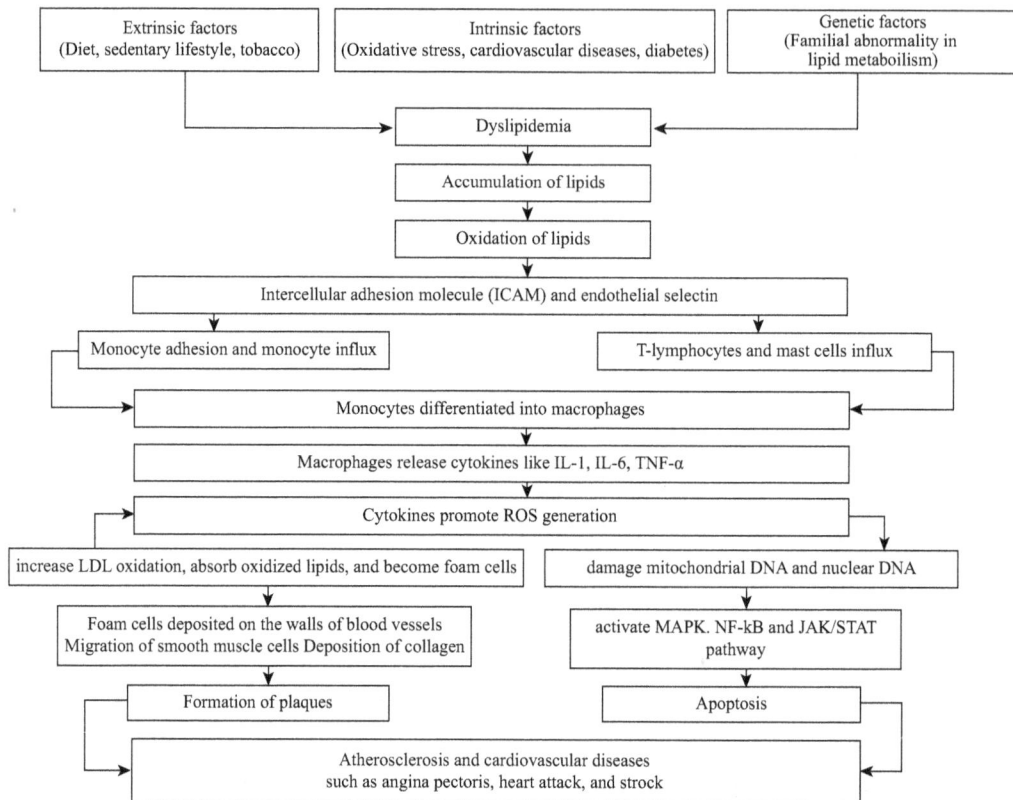

```
┌─────────────────────────┐   ┌─────────────────────────────────┐   ┌─────────────────────────┐
│     Extrinsic factors    │   │        Intrinsic factors         │   │      Genetic factors     │
│ (Diet, sedentary         │   │ (Oxidative stress, cardiovascular│   │ (Familial abnormality in │
│  lifestyle, tobacco)     │   │  diseases, diabetes)             │   │  lipid metaboilism)      │
└─────────────────────────┘   └─────────────────────────────────┘   └─────────────────────────┘
                          │                    │                                    │
                          └──────────► ┌──────────────────┐ ◄────────────────────┘
                                       │   Dyslipidemia    │
                                       └──────────────────┘
                                                │
                                       ┌──────────────────────┐
                                       │ Accumulation of lipids│
                                       └──────────────────────┘
                                                │
                                       ┌──────────────────────┐
                                       │  Oxidation of lipids  │
                                       └──────────────────────┘
```

Intercellular adhesion molecule (ICAM) and endothelial selectin

Monocyte adhesion and monocyte influx | T-lymphocytes and mast cells influx

Monocytes differentiated into macrophages

Macrophages release cytokines like IL-1, IL-6, TNF-α

Cytokines promote ROS generation

increase LDL oxidation, absorb oxidized lipids, and become foam cells | damage mitochondrial DNA and nuclear DNA

Foam cells deposited on the walls of blood vessels Migration of smooth muscle cells Deposition of collagen | activate MAPK. NF-kB and JAK/STAT pathway

Formation of plaques | Apoptosis

Atherosclerosis and cardiovascular diseases such as angina pectoris, heart attack, and strock

Figure 10.1 Etiopathogenesis of dyslipidemia.

ROS generation. ROS increases LDL oxidation. Macrophages absorb oxidized lipids and become foam cells. Foam cells are deposited on the walls of blood vessels and form plaques. ROS also damages mitochondrial DNA and nuclear DNA and activates mitogen-activated protein kinase (MAPK), nuclear factor κ-B (NF-κB), and Janus kinase/signal transducers and activators of transcription (JAK/STAT) pathway to induce apoptosis (Bereda 2022; Vergès 2015; Kolovou et al. 2005). In this way, dyslipidemia leads to atherosclerosis and cardiovascular diseases (Figure 10.1).

10.3 MANAGEMENT OF DYSLIPIDEMIA

Lifestyle modification, especially diet control and exercise, is the first line of treatment for hyperlipidemia. A diet with low calorie intake, high fiber content, and natural antioxidants is preferred in dyslipidemia. Eating marine fish containing omega-3 fatty acids is also beneficial. Avoiding active or passive tobacco smoke exposure and sound sleep every day is also beneficial. Daily exercise or physical activity of 35–40 minutes can burn fats and is beneficial for dyslipidemia (Bereda 2022).

Pharmacotherapy (Thongtang et al. 2022; Bays and Stein 2003) is reserved for use in patients who do not respond to lifestyle modifications or are at high risk of cardiovascular diseases. Among lipid-lowering drugs, statins are the drugs of first choice. Statins inhibit the enzyme HMG-CoA reductase thereby inhibiting cholesterol synthesis. Other LDL-lowering drugs include NPC1L1 (Niemann-Pick C1-Like 1) inhibitor, bile acid binding resins, PCSK9 (Proprotein convertase subtilisin/kexin type 9) inhibitor, and CETP (cholesterol ester transfer protein) inhibitor. Ezetimibe inhibits NPC1L1 and inhibits intestinal absorption of cholesterol. Cholestyramine sequesters the bile acid in the gastrointestinal tract (GIT) and inhibits its reabsorption. This increases the catabolism of plasma cholesterol to bile acid in the liver thereby decreasing cholesterol levels. CETP transfers cholesterol from HDL to VLDL or LDL. Anacetrapib increases HDL cholesterol and decreases LDL cholesterol by inhibiting CETP. Triglyceride-lowering drugs are fibrates and niacin; they have a mild LDL-lowering action. These drugs are not effective when used as add-on drugs to statin therapy. PCSK9 is a protein found in the liver, intestine, and kidney. PCSK9 binds and degrades LDL receptors thereby inhibiting the uptake of LDL. PCSK9 inhibitor inhibits the degradation of LDL receptors, increases LDL uptake, and hence decreases plasma LDL.

- First-line lipid-lowering drugs: They are the drugs of the first choice. Statins are preferred over any other drugs.
  a. HMG-CoA reductase inhibitors (Statins): Simvastatin, Rosuvastatin, Atorvastatin, Lovastatin, Pravastatin, Mevastatin
  b. Bile acid binding resins: Cholestyramine, Colestipol, Colesevelam
  c. NPC1L1 (Niemann-Pick C1-Like 1) inhibitors: Ezetimibe
  d. CETP (cholesterol ester transfer protein) inhibitors: Anacetrapib, Torcetrapib
- Second-line lipid-lowering drugs: They are used when the first-line drugs are ineffective or contraindicated.
  a. Activate lipoprotein lipase (PPAR-α agonist/fibrates): Gemfibrozil, Fenofibrate, Bezafibrate
  b. Inhibit VLDL secretion and lipolysis: Niacin (nicotinic acid)
  c. PCSK9 inhibitors: Alirocumab, Evolocumab
  d. Miscellaneous: Fish oil, guggulipid, and antioxidants like Probucol

10.4 PLANT-BASED DRUGS FOR DYSLIPIDEMIA

Many medicinal plants and their phytochemicals (Table 10.2) can be used as alternatives to synthetic drugs in the treatment of dyslipidemia. It is their safety and efficacy which increased interest in the development of plant products (Scicchitano et al. 2014). Resveratrol is a polyphenolic stilbenoid phytochemical present in grapes and berries. Resveratrol-based supplements improve sugar control, decrease LDL and triglycerides levels significantly in animals but concurrently increase HDL and also prevent damage to blood vessels (Kan et al. 2018). Gingerol is a phytochemical with phenolic content with β-hydroxy ketone derivative, found in the ginger rhizome. It has a role in inhibiting adipogenesis (Haris et al. 2018). Curcumin is a natural food dyestuff that is rich in the plant *Curcuma longa* (turmeric) and is chemically a phyto-polyphenolic pigment. It is an anti-inflammatory agent, anticancer agent, hepatoprotective agent, a flavoring component

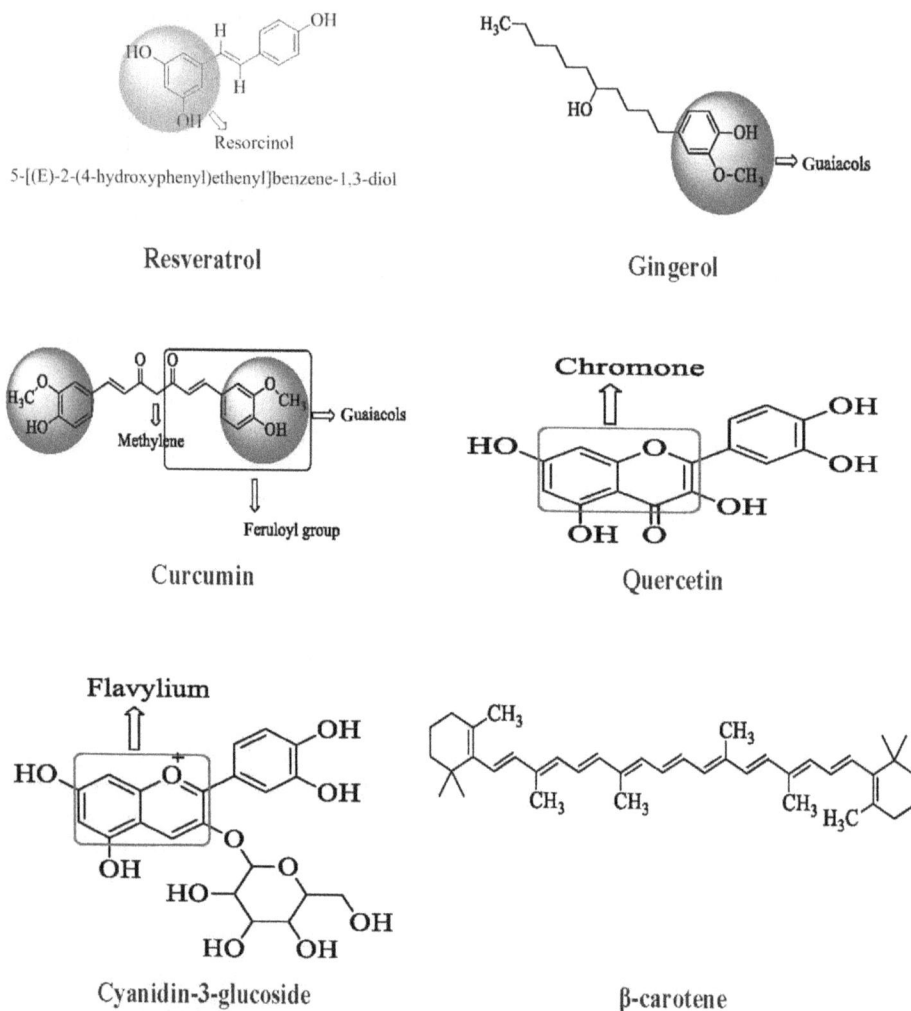

Figure 10.2 Structures of different natural antioxidants.

in pharmaceuticals, a biological indicator, a lipoxygenase (LOX) inhibitor, and a free radical scavenger (Labban et al. 2021). Quercetin is a flavonol-derived polyphenolic phytochemical, chemically 3,3′,4′,5,7-pentahydroxy flavone. It is a free radical scavenger that has been used to treat or prevent diverse conditions including cardiovascular diseases and hypercholesterolemia (Zhuang et al. 2018; Lee et al. 2016). Anthocyanin (Cyanidin-3-glucoside) is the major anthocyanin and ethereal-linked glycoside. It is obtained from colored legumes. It is an anti-inflammatory drug that inhibits LDL oxidation and may treat or prevent hypercholesterolemia (Wu et al. 2018). As an antioxidant, β-carotene inhibits free-radical damage to DNA (Edge and Truscott 2018). The structure of these natural compounds is given in Figure 10.2.

Table 10.2 Plants and their specific phytochemicals possessing efficacy against dyslipidemia

Source	Active constituents	Chemical class
Curcuma longa	Curcumin	Phyto-polyphenolic derivative
Capsicum annuum	Oleoresin, β-carotene	Oleoresin polyterpenoid
Coptis chinensis	Berberine	Benzylisoquinoline-derived alkaloid
Citrus sinensis (orange)	Naringin (glycoside) Naringenin (algycone)	flavanone-7-O-glycoside
Citrus grandis (grapefruit)	Resveratrol	Polyphenolic stilbenoid
Malus domestica (Apple) Allium cepa (onion) Thea sinesis (tea)	Quercetin	Flavonol-derived polyphenol phytochemical, chemically 3,3',4',5,7-pentahydroxy flavone
Rheum emodi (rhubarb)	Emodin	Anthraquinone glycoside
Gymnemasylvestre	Gymnemic acid, gymnemasaponins	Triterpenoid
Scutellaria radix	baicalein, baicalin, and wogonin	Flavonoid,5,6,7-trihydroxyflavone
Silybum marianum	Silibinin	flavonolignans – composed of two parts taxifolin and coniferyl alcohol
Myrica cerifera Myrica rubra Manikara zapota, Eugenia uniflora Pouteria gender	Myricitrin	Flavonol glucoside, 3,3',4',5,5' and 7 hexahydroxy flavone
Stevia rebaudiana	Stevioside	Diterpene glycoside
Momordica charantia	α-eleostearic acid, stearic acid	Octadeca-9,11,13-trienoic acid Unsaturated fatty acid
Rubus armeniacus	Cyanidin-3-glucoside	Anthocyanin and etheral-linked glycoside
Zingiber officinale	Gingerol	Phenolic compound

10.5 NANOTHERAPEUTICS FOR DYSLIPIDEMIA

Lipid-lowering drugs have the following limitations (Dayar and Pechanova 2022):

- Occurrence of long-term adverse events
- Intolerance to drugs like statins in several patients
- Requires to be used in combination to reduce the dose
- Low water solubility

Nanotherapeutics can be employed to address the above issues. Nanotherapeutics have the following advantages (Dayar and Pechanova 2022):

- Enhanced bioavailability
- Enhanced absorption
- Enhanced efficacy
- Enhanced solubility
- Enhanced dissolution

- Reduced adverse effects
- Improved stability

The commonly studied nano-based drug delivery systems are as follows.

10.5.1 Polymeric nanoparticles

Drugs like atorvastatin, rosuvastatin, and pitavastatin have been loaded in polymeric nanoparticles like poly lactic-co-glycolic acid (PLGA) and poly L-lactide-co-caprolactone. Atorvastatin-loaded PLGA nanoparticles were able to reduce the daily dose by 66% (Meena et al. 2008). Similarly, the PLGA nanoparticles loaded with pitavastatin showed anti-inflammatory (Chen et al. 2011) and cardioprotective (Yokoyama et al. 2019; Ichimura et al. 2016) effects. Rosuvastatin-loaded polycaprolactone nanoparticles reduced the thrombotic potential by improving vascular endothelial growth factor (VEGF) signaling (Liu et al. 2018). Atorvastatin-loaded polycaprolactone nanoparticles have improved efficacy and bioavailability and reduced adverse effects (Kumar et al. 2016). Berberine-encapsulated PLGA–PEG nanoparticles modulate PCSK-9 (Proprotein convertase subtilisin/kexin) and has the potential to reduce LDL cholesterol (Ochin and Garelnabi 2018).

10.5.2 Lipid-based nanoparticles

Solid lipid nanoparticles (SLN) and nanostructured lipid carriers (NLC) are the extensively studied lipid-based nanoparticles. SLNs of statins like atorvastatin, simvastatin, lovastatin, and so on have increased their oral bioavailability (Zhang et al. 2010; Sathali et al. 2013). Atorvastatin-loaded SLN not only improved its hypolipidemic action but also reduced its adverse effects like myopathy and hepatotoxicity when supplemented with CoQ-10 and vitamin E (Farrag et al. 2018). Simvastatin SLN significantly improved antihyperlipidemic activity of simvastatin in hyperlipidemic rat model (Rizvi et al. 2019). Similarly, ezetimibe-loaded SLNs have improved their stability as well as bioavailability (ud Din et al. 2019).

Statin-loaded NLCs have shown better-sustained release than SLNs (Fathi et al. 2018). Simvastatin-loaded NLC improved its bioavailability and significantly reduced blood lipid levels and atherogenic risk in hypolipidemic rats (Harisa et al. 2017). Lovastatin-loaded myverol (a lipophilic emulsifier) containing nanostructured lipid carriers (NLCs) increased its stability in the gastric environment (Chen et al. 2010).

Atorvastatin-loaded nanotransferosomal carriers increased the lipid-lowering efficacy of atorvastatin in hyperlipidemic rats (Mahmoud et al. 2017).

10.5.3 Chitosan-based nanoparticles

Statins like atorvastatin and rosuvastatin have been successfully loaded in chitosan nanoparticles. Statin–chitosan nanoparticles have shown sustained release (Bathool et al. 2012) of the drug and a better improvement in the lipid profile (Chen et al. 2020).

Thymoquinone-loaded chitosan nanoparticles improved the bioavailability of thymoquinone. It ameliorated dyslipidemia, inflammation, and oxidative stress in streptozotocin/nicotinamide-induced diabetic rats (Hosni et al. 2022).

10.5.4 Nanoliposomes

Nanoliposomes are useful in the delivery of both hydrophilic and lipophilic statins (Askarizadeh et al. 2019). Simvastatin-loaded nanoliposomes exhibited higher plasma concentration (Tuerdi et al. 2016). Anionic nanoliposomes showed anti-dyslipidemic effect in tyloxapol-induced hyperlipidemic mouse model (Sahebkar et al. 2015, 2014). Nanoliposomal PCSK9 vaccine has the potential therapeutic and preventive effect against dyslipidemia and atherosclerosis in a C57BL/6 mouse model (Momtazi-Borojeni et al. 2021, 2019).

10.5.5 Nanoemulsions

Nanoemulsion of simvastatin improved its oral bioavailability (Reddy et al. 2013). Self-nanoemulsifying drug delivery system of statins increased the dissolution rate, oral bioavailability, and drug release (Kassem et al. 2017; Kulkarni et al. 2015). Nanoemulsion of garlic oil improved the efficacy of garlic oil in high-fat diet-induced dyslipidemia in rats. The lipid-lowering effect of garlic oil nanoemulsion is better than that of atorvastatin (Ragavan et al. 2017). Nanoemulsion of Curcumin enhances antihypertensive and lipid lowering effect of Curcumin by inhibiting angiotensin-converting enzyme (ACE) and HMG CoA reductase respectively (Rachmawati et al. 2016).

Alginate double layer nanoemulsion of capsicum oleoresin is more effective than its single-layer nanoemulsion in inducing lipolytic gene expression and suppressing adipogenic gene expression in 3T3-L1 adipocytes (Lee et al. 2017). Nanoemulsion of capsicum oleoresin showed anti-obesity effect by stimulating AMP-activated protein kinase (AMPK) activity and decreasing glycerol-3-phosphate dehydrogenase activity (Kim et al. 2014).

Solidified self-nano-emulsifying drug delivery systems (SNEDDS) of rosuvastatin calcium significantly reduced diet-induced hyperlipidemia and atherogenic index in rats. It also improved its dissolution and hence oral absorption (Ahsan and Verma 2017).

Nanoencapsulation significantly reduced the dose of volatile oils of fennel and rosemary to produce significant improvement in lipid profile in diet-induced dyslipidemia in rats (Al-Okbi et al. 2018).

10.5.6 Nanosuspensions

The dissolution and hence the bioavailability of rosuvastatin were improved when given in the nanosuspension formulation (Sharannavar et al. 2021). The lipid-lowering effect of simvastatin was significantly improved when given as nanosuspension (Amsa et al. 2021). Nanosuspensions of ezetimibe improved the dissolution and hence absorption of ezetimibe (Thadkala et al. 2014).

10.5.7 Nanocrystals

Nanocrystals of ezetimibe improved its solubility, dissolution, and lipid-lowering efficiency (Srivalli and Mishra 2015). The nanocrystals of fenofibrate are clinically used in adult patients with primary hypercholesterolemia or mixed dyslipidemia (Ansquer et al. 2009). Surface-stabilized atorvastatin nanocrystals have improved the antihyperlipidemic activity, safety, and bioavailability of atorvastatin (Sharma and Mehta 2019).

10.5.8 Cerium oxide nanoparticles

Cerium oxide nanoparticles showed insulin sensitizing effect in mouse 3T3L1 adipocytes and C2C12 myotubes (Lopez-Pascual et al. 2019). They also lowered the plasma levels of insulin, leptin, glucose, and triglycerides (Dhall and Self 2018). Atorvastatin-loaded cerium oxide nanoparticles have shown antioxidant, anti-inflammatory, and anti-apoptotic activity (Yu et al. 2020).

10.5.9 Metal-based nanocomposites

Poria coccos gold nanoparticles significantly reduced total cholesterol, triglycerides, LDL, VLDL in high-fat diet rat model. It also significantly reduced the atherogenic index and coronary index. In addition, it showed antioxidant, anti-inflammatory, and anti-obesity effect (Li et al. 2020). Gold nanoparticles of *Ziziphus jujuba* fruit extract improved the effects of *Ziziphus jujuba* on oxidative stress, dyslipidemia, atherogenicity index, and insulin resistance in streptozotocin-induced diabetic rats (Javanshir et al. 2020).

Silver nanoparticles of methanolic extract of *Taverniera counefolia* significantly improved the lipid profile in alloxan-induced diabetic rats (Ul Haq et al. 2022). *Solanum nigrum* silver nanoparticles significantly improved the antihyperlipidemic effects of *Solanum nigrum* leaf extract in alloxan-induced diabetic rats (Sengottaiyan et al. 2016).

10.6 ROS-BASED NANOTHERAPEUTICS AGAINST DYSLIPIDEMIA

ROS are the products of aerobic metabolism and play a role in a variety of physiological as well as pathological processes in the body. The ROS family comprises both free and non-free radicals (Yao et al. 2019). Nonetheless, excessive ROS can trigger a cascade of inflammations that aggravate oxidative stress. More importantly, ROS accumulation initiates DNA oxidation, lipid peroxidation, glycoxidation, and protein aggregation, thereby inducing apoptosis. ROS act as a key factor in the onset and progression of dyslipidemia. Protective mechanisms in our body include enzymes like superoxide dismutase (SOD), catalase, glutathione peroxidase, and so on. Clinical use of these natural enzymes is limited because of stability and mass production issues (Yahya et al. 2019).

Since ROS are produced constantly in living systems, protecting antioxidants and delivering them specifically to the area experiencing oxidative stress can significantly increase their effectiveness. In this regard, the potential for conjugating nanoparticle surfaces to enable local environment responsiveness and thereby targeting the drug offers great promise and progressive advantage in the application of such monotherapies. Again, ROS-based nanoparticles (NPs) can facilitate the penetration of drugs along with the enhancement of therapeutic value. It also promotes targeted controlled drug release with minimum side effects. ROS-responsive NPs are frequently used as sophisticated drug delivery systems. ROS-responsive polymeric scaffolds and hydrogels being porous in structure can support cell adhesion and growth and are regarded as potential biomedical materials in the future (Yao et al. 2019).

Target specificity, increased cell membrane permeability, and the use of catalytic scavengers are just a few of the strategies for scavenging free radicals under oxidative stress that have been significantly improved by the application of nanomedicine to ROS-mediated

Table 10.3 ROS-based nanoparticles of different drugs effective against atherosclerosis

Drug	NP composition	Model	Efficacy	Reference
Atorvastatin	Poly lactic-co-glycolic acid (PLGA)	ApoE−/− mice fed with a high-fat diet	Atherosclerosis	Zhao et al. 2018
Propofol	Indole	C57BL/6 mice fed with a high-fat diet	Atherosclerosis	Chen et al. 2021
Prednisolone	β-cyclodextrin	ApoE−/− (male) mice fed with a high-fat diet	Anti-inflammatory	Ma et al. 2020
Dexamethasone	Hyaluronic acid	High-fat-fed rats	Anti-inflammatory	Hou et al. 2018
Tempol	β-cyclodextrin	ApoE−/− (male) mice fed with high-fat diet	Atherosclerosis	Wang et al. 2018
-	Zinc oxide nanoparticles	Cyclophosphamide-induced dyslipidemia	Anti-dyslipidemic	Karema et al. 2020
Curcumin	Nanomicelles-based on oligomeric hyaluronic acid	High-fat-fed rats	Atherosclerosis	Hou et al. 2020
Ferulic acid	Ferulic acid-based polymers linked by diglycolic acid	Human monocyte-derived macrophages	Anti-inflammatory	Chmielowski et al. 2017
Vanillin	PVAX	Human umbilical vein endothelial cells	Antioxidant	Kwon et al. 2016

pathologies (Nash and Ahmed 2015). ROS-based nanoparticles have been successfully tried against atherosclerosis (Table 10.3). Zinc oxide nanoparticles are effective against cyclophosphamide-induced dyslipidemia (Karema et al. 2020). Atorvastatin as PLGA nanoparticles has shown efficacy in atherosclerosis in ApoE−/− mice fed with a high-fat diet. Similar results were shown by propofol in C57BL/6 mice fed with a high-fat diet model (Zhao et al. 2018). In addition to this, prednisolone and dexamethasone nanoparticles possess anti-inflammatory properties for which they may be used against atherosclerosis (Ma et al. 2020; Hou et al. 2018). Tempol being a superoxide dismutase mimetic agent has been conjugated with β-cyclodextrin to form nanoparticles against high-fat diet-induced atherosclerosis in ApoE−/− (male) mice (Wang et al. 2018). Polymeric nanoparticles of natural molecules like curcumin, vanillin, and ferulic acid have been reported as efficacious against atherosclerosis, oxidative stress, and inflammation, respectively.

10.7 CONCLUSION

Dyslipidemia is a major risk factor for atherosclerosis development. Atherogenic lipoproteins induce the reactive oxygen species to produce atherosclerosis and other cardiovascular diseases. Nanoparticles are small in size and have large surfaces with a high possibility of biological modification. They can effectively target lesions. Different lipid-lowering drugs including statins and ezetimbe have been loaded with polymeric nanoparticles, lipid-based nanoparticles, cerium oxide nanoparticles, nanoemulsions, nanoliposomes, nanocapsules, nanosuspensions, nanorods, nanostructured lipid carriers, solid lipid nanoparticles (SLN), etc. to improve their solubility, dissolution, bioavailability, and stability and to reduce their dose and adverse effects. ROS-based nanoparticles have also been carried out successfully in recent times for dyslipidemia treatment for drugs like atorvastatin, propofol, prednisolone, dexamethasone, tempol, curcumin, vanillin, and ferulic acid. These ROS-based nanoparticles possess anti-inflammatory and antioxidant properties for which they may be used against atherosclerosis. In future, statins being the first-line drugs may be supplemented with natural antioxidants and delivered as nanoparticles. Metal-based nanocomposites can also be developed against dyslipidemia for their potential antioxidant activity.

REFERENCES

Ahsan, Mohd Neyaz, and Priya Ranjan Prasad Verma. "Solidified self nano-emulsifying drug delivery system of rosuvastatin calcium to treat diet-induced hyperlipidemia in rat: In vitro and in vivo evaluations." *Therapeutic Delivery* 8, no. 3 (2017): 125–136.

Al-Okbi, Sahar Y., Ahmed M. S. Hussein, Hagar F. H. Elbakry, Karem Aly Fouda, Khaled F. Mahmoud, and Mohamed E. Hassan. "Health benefits of fennel, rosemary volatile oils and their nano-forms in dyslipidemic rat model." *Pak. J. Biol. Sci* 21, no. 7 (2018): 348–358.

Amsa, P., S. Punitha, V. Lalitha, S. Tamizharasi, and T. Sivakumar. "In-vivo pharmacokinetic evaluation of simvastatin nanosuspension and its antihyperlipidemic activity against high-fat diet-induced hyperlipidemia in animal model." *Efflatounia* 5, no. 2 (2021): 3106–3116.

Ansquer, Jean-Claude, Ivan Bekaert, Martine Guy, Markolf Hanefeld, Alain Simon, and Study Investigators. "Efficacy and safety of coadministration of fenofibrate and ezetimibe compared with each as monotherapy in patients with type IIb dyslipidemia and features of the metabolic syndrome: A prospective, randomized, double-blind, three-parallel arm, multicenter, comparative study." *American Journal of Cardiovascular Drugs* 9 (2009): 91–101.

Askarizadeh, Anis, Alexandra E. Butler, Ali Badiee, and Amirhossein Sahebkar. "Liposomal nano-carriers for statins: A pharmacokinetic and pharmacodynamics appraisal." *Journal of Cellular Physiology* 234, no. 2 (2019): 1219–1229.

Bathool, Afifa, Gowda D. Vishakante, Mohammed S. Khan, and H. G. Shivakumar. "Development and characterization of atorvastatin calcium loaded chitosan nanoparticles for sustain drug delivery." *Advanced Materials Letters* 3, no. 6 (2012): 466–470.

Bays, Harold, and Evan A. Stein. "Pharmacotherapy for dyslipidemia: Current therapies and future agents." *Expert Opinion on Pharmacotherapy* 4, no. 11 (2003): 1901–1938.

Bereda, Gudisa. "Pathophysiology and management of dyslipidemia." *Biomed J Sci & Tech Res* 43, no. 2 (2022).

Chen, Chih-Chieh, Tung-Hu Tsai, Zih-Rou Huang, and Jia-You Fang. "Effects of lipophilic emulsifiers on the oral administration of lovastatin from nanostructured lipid carriers: Physicochemical characterization and pharmacokinetics." *European Journal of Pharmaceutics and Biopharmaceutics* 74, no. 3 (2010): 474–482.

Chen, Feng, Jun Chen, Chuyi Han, Zhangyou Yang, Tao Deng, Yunfei Zhao, Tianye Zheng, Xuelan Gan, and Chao Yu. "Theranostics of atherosclerosis by the indole molecule-templated self-assembly of probucol nanoparticles." *Journal of Materials Chemistry B* 9, no. 20 (2021): 4134–4142.

Chen, Lin, Caihong Wang, and Yuanchu Wu. "Cholesterol (Blood lipid) lowering potential of rosuvastatin chitosan nanoparticles for atherosclerosis: Preclinical study in rabbit model." *Acta Biochimica Polonica* 67, no. 4 (2020): 495–499.

Chen, Ling, Kaku Nakano, Satoshi Kimura, Tetsuya Matoba, Eiko Iwata, Miho Miyagawa, Hiroyuki Tsujimoto et al. "Nanoparticle-mediated delivery of pitavastatin into lungs ameliorates the development and induces regression of monocrotaline-induced pulmonary artery hypertension." *Hypertension* 57, no. 2 (2011): 343–350.

Chmielowski, Rebecca A., Dalia S. Abdelhamid, Jonathan J. Faig, Latrisha K. Petersen, Carol R. Gardner, Kathryn E. Uhrich, Laurie B. Joseph, and Prabhas V. Moghe. "Athero-inflammatory nanotherapeutics: Ferulic acid-based poly (anhydride-ester) nanoparticles attenuate foam cell formation by regulating macrophage lipogenesis and reactive oxygen species generation." *Acta Biomaterialia* 57 (2017): 85–94.

Dayar, Ezgi, and Olga Pechanova. "Targeted strategy in lipid-lowering therapy." *Biomedicines* 10, no. 5 (2022): 1090.

Dhall, Atul, and William Self. "Cerium oxide nanoparticles: A brief review of their synthesis methods and biomedical applications." *Antioxidants* 7, no. 8 (2018): 97.

Durrington, Paul. "Dyslipidemia." *The Lancet* 362, no. 9385 (2003): 717–731.

Edge, Ruth, and T. George Truscott. "Singlet oxygen and free radical reactions of retinoids and carotenoids: A review." *Antioxidants* 7, no. 1 (2018): 5.

Farrag, S. M., M. A. Hamzawy, M. F. El-Yamany, M. A. Saad, and N. N. Nassar. "Atorvastatin in nanoparticulate formulation abates muscle and liver affliction when coalesced with coenzyme Q10 and/or vitamin E in hyperlipidemic rats." *Life Sciences* 203 (2018): 129–140.

Fathi, Heba A., Ayat Allam, Mahmoud Elsabahy, Gihan Fetih, and Mahmoud El-Badry. "Nanostructured lipid carriers for improved oral delivery and prolonged antihyperlipidemic effect of simvastatin." *Colloids and Surfaces B: Biointerfaces* 162 (2018): 236–245.

Haris, Poovvathingal, Varughese Mary, and Chellappanpillai Sudarsanakumar. "Probing the interaction of the phytochemical 6-gingerol from the spice ginger with DNA." *International Journal of Biological Macromolecules* 113 (2018): 124–131.

Harisa, Gamaleldin I., Abdullah H. Alomrani, and Mohamed M. Badran. "Simvastatin-loaded nanostructured lipid carriers attenuate the atherogenic risk of erythrocytes in hyperlipidemic rats." *European Journal of Pharmaceutical Sciences* 96 (2017): 62–71.

Hernáez, Álvaro, María Trinidad Soria-Florido, Helmut Schröder, Emilio Ros, Xavier Pinto, Ramón Estruch, Jordi Salas-Salvadó et al. "Role of HDL function and LDL atherogenicity on cardiovascular risk: A comprehensive examination." *PLoS One* 14, no. 6 (2019): e0218533.

Hosni, Ahmed, Adel Abdel-Moneim, Mohammed Hussien, Mohamed I. Zanaty, Zienab E. Eldin, and Ahmed A. G. El-Shahawy. "Therapeutic significance of thymoquinone-loaded chitosan nanoparticles on streptozotocin/nicotinamide-induced diabetic rats: In vitro and in vivo functional analysis." *International Journal of Biological Macromolecules* 221 (2022): 1415–1427.

Hou, Chunyan, Hu Bai, Zhaojie Wang, Yuanhao Qiu, Li-Li Kong, Feifei Sun, Dongdong Wang et al. "A hyaluronan-based nanosystem enables combined anti-inflammation of mTOR gene silencing and pharmacotherapy." *Carbohydrate Polymers* 195 (2018): 339–348.

Hou, Xiaoya, Hua Lin, Xiudi Zhou, Ziting Cheng, Yi Li, Xue Liu, Feng Zhao, Yanping Zhu, Peng Zhang, and Daquan Chen. "Novel dual ROS-sensitive and CD44 receptor targeting nanomicelles based on oligomeric hyaluronic acid for the efficient therapy of atherosclerosis." *Carbohydrate Polymers* 232 (2020): 115787.

Ichimura, Kenzo, Tetsuya Matoba, Kaku Nakano, Masaki Tokutome, Katsuya Honda, Jun-ichiro Koga, and Kensuke Egashira. "A translational study of a new therapeutic approach for acute myocardial infarction: Nanoparticle-mediated delivery of pitavastatin into reperfused myocardium reduces ischemia-reperfusion injury in a preclinical porcine model." *PLoS One* 11, no. 9 (2016): e0162425.

Javanshir, Reyhane, Moones Honarmand, Mehran Hosseini, and Mina Hemmati. "Anti-dyslipidemic properties of green gold nanoparticle: Improvement in oxidative antioxidative balance and associated atherogenicity and insulin resistance." *Clinical Phytoscience* 6, no. 1 (2020): 1–10.

Kan, Nai-Wen, Mon-Chien Lee, Yu-Tang Tung, Chien-Chao Chiu, Chi-Chang Huang, and Wen-Ching Huang. "The synergistic effects of resveratrol combined with resistant training on exercise performance and physiological adaption." *Nutrients* 10, no. 10 (2018): 1360.

Karema, El M. Shkal, Ahmed M. Attia, Sabah G. El-Banna, Azab Elsayed Azab, and Rabia A. M. Yahya. "Anti-dyslipidemic effect of zinc oxide nanoparticles against cyclophosphamide induced dyslipidemia in male albino rats." *The Gazette of Medical Science* 1, no. 1 (2020): 055–063.

Kassem, Abdulsalam M., Hany M. Ibrahim, and Ahmed M. Samy. "Development and optimisation of atorvastatin calcium loaded self-nanoemulsifying drug delivery system (SNEDDS) for enhancing oral bioavailability: In vitro and in vivo evaluation." *Journal of Microencapsulation* 34, no. 3 (2017): 319–333.

Kim, Joo-Yeon, Mak-Soon Lee, Sunyoon Jung, Hyunjin Joo, Chong-Tai Kim, In-Hwan Kim, Sangjin Seo, Soojung Oh, and Yangha Kim. "Anti-obesity efficacy of nanoemulsion oleoresin capsicum in obese rats fed a high-fat diet." *International Journal of Nanomedicine* 9 (2014): 301.

Kolovou, G. D., K. K. Anagnostopoulou, and D. V. Cokkinos. "Pathophysiology of dyslipidemia in the metabolic syndrome." *Postgraduate Medical Journal* 81, no. 956 (2005): 358–366.

Kulkarni, Nilesh S., Nisharani S. Ranpise, and Govind Mohan. "Development and evaluation of solid self-nano-emulsifying formulation of rosuvastatin calcium for improved bioavailability." *Tropical Journal of Pharmaceutical Research* 14, no. 4 (2015): 575–582.

Kumar, Nagendra, Sundeep Chaurasia, Ravi R. Patel, Gayasuddin Khan, Vikas Kumar, and Brahmeshwar Mishra. "Atorvastatin calcium loaded PCL nanoparticles: Development, optimization, in vitro and in vivo assessments." *RSC Advances* 6, no. 20 (2016): 16520–16532.

Kwon, Byeongsu, Changsun Kang, Jinsub Kim, Donghyuck Yoo, Byung-Ryul Cho, Peter M. Kang, and Dongwon Lee. "H$_2$O$_2$-responsive antioxidant polymeric nanoparticles as therapeutic agents for peripheral arterial disease." *International Journal of Pharmaceutics* 511, no. 2 (2016): 1022–1032.

Labban, Ranyah Shaker M., Hanan A. Alfawaz, Ahmed T. Almnaizel, May N. Al-Muammar, Ramesa Shafi Bhat, and Afaf El-Ansary. "Garcinia mangostana extract and curcumin ameliorate oxidative stress, dyslipidemia, and hyperglycemia in high fat diet-induced obese Wistar albino rats." *Scientific Reports* 11, no. 1 (2021): 7278.

Lee, Ji-Sook, Yong-Jun Cha, Kyung-Hea Lee, and Jung-Eun Yim. "Onion peel extract reduces the percentage of body fat in overweight and obese subjects: A 12-week, randomized, double-blind, placebo-controlled study." *Nutrition Research and Practice* 10, no. 2 (2016): 175–181.

Lee, Mak-Soon, Sunyoon Jung, Yoonjin Shin, Seohyun Lee, Chong-Tai Kim, In-Hwan Kim, and Yangha Kim. "Lipolytic efficacy of alginate double-layer nanoemulsion containing oleoresin capsicum in differentiated 3T3-L1 adipocytes." *Food & Nutrition Research* 61, no. 1 (2017): 1339553.

Li, Wansen, Hong Wan, Shuxun Yan, Zhao Yan, Yalin Chen, Panpan Guo, Thiyagarajan Ramesh, Ying Cui, and Lei Ning. "Gold nanoparticles synthesized with Poria cocos modulates the anti-obesity parameters in high-fat diet and streptozotocin induced obese diabetes rat model." *Arabian Journal of Chemistry* 13, no. 7 (2020): 5966–5977.

Liu, Peixi, Yingjun Liu, Peiliang Li, Yingjie Zhou, Yaying Song, Yuan Shi, Wenhao Feng et al. "Rosuvastatin-and heparin-loaded poly (l-lactide-co-caprolactone) nanofiber aneurysm stent promotes endothelialization via vascular endothelial growth factor type A modulation." *ACS Applied Materials & Interfaces* 10, no. 48 (2018): 41012–41018.

Liu, Ting, Jia-Mao Chen, Dan Zhang, Qian Zhang, Bowen Peng, Lei Xu, and Hua Tang. "ApoPred: Identification of apolipoproteins and their subfamilies with multifarious features." *Frontiers in Cell and Developmental Biology* 8 (2021): 621144.

Lopez-Pascual, Amaya, Andoni Urrutia-Sarratea, Silvia Lorente-Cebrián, J. Alfredo Martinez, and Pedro González-Muniesa. "Cerium oxide nanoparticles regulate insulin sensitivity and oxidative markers in 3T3-L1 adipocytes and C2C12 myotubes." *Oxidative Medicine and Cellular Longevity* 2019 (2019).

Ma, Boxuan, Hong Xu, Weihua Zhuang, Yanan Wang, Gaocan Li, and Yunbing Wang. "Reactive oxygen species responsive theranostic nanoplatform for two-photon aggregation-induced emission imaging and therapy of acute and chronic inflammation." *ACS Nano* 14, no. 5 (2020): 5862–5873.

Mahmoud, Mohamed O., Heba M. Aboud, Amira H. Hassan, Adel A. Ali, and Thomas P. Johnston. "Transdermal delivery of atorvastatin calcium from novel nanovesicular systems using polyethylene glycol fatty acid esters: Ameliorated effect without liver toxicity in poloxamer 407-induced hyperlipidemic rats." *Journal of Controlled Release* 254 (2017): 10–22.

Meena, A. K., D. Venkat Ratnam, G. Chandraiah, D. D. Ankola, P. Rama Rao, and M. N. V. Ravi Kumar. "Oral nanoparticulate atorvastatin calcium is more efficient and safe in comparison to Lipicure® in treating hyperlipidemia." *Lipids* 43 (2008): 231–241.

Momtazi-Borojeni, Amir Abbas, Mahmoud Reza Jaafari, Mohammad Afshar, Maciej Banach, and Amirhossein Sahebkar. "PCSK9 immunization using nanoliposomes: Preventive efficacy against hypercholesterolemia and atherosclerosis." *Archives of Medical Science: AMS* 17, no. 5 (2021): 1365.

Momtazi-Borojeni, Amir Abbas, Mahmoud Reza Jaafari, Ali Badiee, Maciej Banach, and Amirhossein Sahebkar. "Therapeutic effect of nanoliposomal PCSK9 vaccine in a mouse model of atherosclerosis." *BMC Medicine* 17 (2019): 1–15.

Nash, Kevin M., and Salahuddin Ahmed. "Nanomedicine in the ROS-mediated pathophysiology: Applications and clinical advances." *Nanomedicine: Nanotechnology, Biology and Medicine* 11, no. 8 (2015): 2033–2040.

Ochin, Chinedu C., and Mahdi Garelnabi. "Berberine encapsulated PLGA-PEG nanoparticles modulate PCSK-9 in HepG2 cells." *Cardiovascular & Haematological Disorders-Drug Targets (Formerly Current Drug Targets-Cardiovascular & Hematological Disorders)* 18, no. 1 (2018): 61–70.

Rachmawati, Heni, Irene Surya Soraya, Neng Fisheri Kurniati, and Annisa Rahma. "In vitro study on antihypertensive and antihypercholesterolemic effects of a curcumin nanoemulsion." *Scientia Pharmaceutica* 84, no. 1 (2016): 131–140.

Ragavan, Gokulakannan, Yuvashree Muralidaran, Badrinathan Sridharan, Rajesh Nachiappa Ganesh, and Pragasam Viswanathan. "Evaluation of garlic oil in nano-emulsified form: Optimization and its efficacy in high-fat diet induced dyslipidemia in Wistar rats." *Food and Chemical Toxicology* 105 (2017): 203–213.

Reddy, A. K. B., S. Debnath, and N. M. Babu. "Design development and evaluation of novel nanoemulsion of simvastatin." *Int. J. Adv. Pharm* 3, no. 2 (2013): 94–101.

Rizvi, Syed Zaki Husain, Fawad Ali Shah, Namrah Khan, Iftikhar Muhammad, Khan Hashim Ali, Muhammad Mohsin Ansari, Fakhar ud Din et al. "Simvastatin-loaded solid lipid nanoparticles for enhanced anti-hyperlipidemic activity in hyperlipidemia animal model." *International Journal of Pharmaceutics* 560 (2019): 136–143.

Sahebkar, Amirhossein, Ali Badiee, Majid Ghayour-Mobarhan, Seyed Reza Goldouzian, and Mahmoud Reza Jaafari. "A simple and effective approach for the treatment of dyslipidemia using anionic nanoliposomes." *Colloids and Surfaces B: Biointerfaces* 122 (2014): 645–652.

Sahebkar, Amirhossein, Ali Badiee, Mahdi Hatamipour, Majid Ghayour-Mobarhan, and Mahmoud Reza Jaafari. "Apolipoprotein B-100-targeted negatively charged nanoliposomes for the treatment of dyslipidemia." *Colloids and Surfaces B: Biointerfaces* 129 (2015): 71–78.

Sathali, H., A. Abdul, and N. Nisha. "Development of solid lipid nanoparticles of rosuvastatin calcium." *J Pharm Res* 1 (2013): 536–548.

Scicchitano, Pietro, Matteo Cameli, Maria Maiello, Pietro Amedeo Modesti, Maria Lorenza Muiesan, Salvatore Novo, Pasquale Palmiero et al. "Nutraceuticals and dyslipidemia: Beyond the common therapeutics." *Journal of Functional Foods* 6 (2014): 11–32.

Sengottaiyan, Arumugam, Adithan Aravinthan, Chinnapan Sudhakar, Kandasamy Selvam, Palanisamy Srinivasan, Muthusamy Govarthanan, Koildhasan Manoharan, and Thangaswamy Selvankumar. "Synthesis and characterization of solanum nigrum-mediated silver nanoparticles and its protective effect on alloxan-induced diabetic rats." *Journal of Nanostructure in Chemistry* 6 (2016): 41–48.

Sharannavar, B., and S. Sawant. "Formulation and evaluation of nanosuspension of rosuvastatin for solubility enhancement by quality by design approach." *Int J Pharm Sci Res* 12, no. 11 (2021): 5949–5958.

Sharma, Manu, and Isha Mehta. "Surface stabilized atorvastatin nanocrystals with improved bioavailability, safety and antihyperlipidemic potential." *Scientific Reports* 9, no. 1 (2019): 16105.

Srivalli, Kale Mohana Raghava, and Brahmeshwar Mishra. "Preparation and pharmacodynamic assessment of ezetimibe nanocrystals: Effect of P-gp inhibitory stabilizer on particle size and oral absorption." *Colloids and Surfaces B: Biointerfaces* 135 (2015): 756–764.

Su, Xin, and Daoquan Peng. "The exchangeable apolipoproteins in lipid metabolism and obesity." *Clinica Chimica Acta* 503 (2020): 128–135.

Thadkala, Kiran, Prema Kumari Nanam, Bathini Rambabu, Chinta Sailu, and Jithan Aukunuru. "Preparation and characterization of amorphous ezetimibe nanosuspensions intended for enhancement of oral bioavailability." *International Journal of Pharmaceutical Investigation* 4, no. 3 (2014): 131.

Thongtang, Nuntakorn, Renan Sukmawan, Elmer Jasper B. Llanes, and Zhen-Vin Lee. "Dyslipidemia management for primary prevention of cardiovascular events: Best in-clinic practices." *Preventive Medicine Reports* (2022): 101819.

Tuerdi, Nuerbiye, Lu Xu, Baoling Zhu, Cong Chen, Yini Cao, Yunan Wang, Qiang Zhang, Zijian Li, and Rong Qi. "Preventive effects of simvastatin nanoliposome on isoproterenol-induced cardiac remodeling in mice." *Nanomedicine: Nanotechnology, Biology and Medicine* 12, no. 7 (2016): 1899–1907.

Turkish, Aaron R., and Stephen L. Sturley. "The genetics of neutral lipid biosynthesis: An evolutionary perspective." *American Journal of Physiology-Endocrinology and Metabolism* 297, no. 1 (2009): E19–E27.

ud Din, Fakhar, Alam Zeb, and KifayatUllah Shah. "Development, in-vitro and in-vivo evaluation of ezetimibe-loaded solid lipid nanoparticles and their comparison with marketed product." *Journal of Drug Delivery Science and Technology* 51 (2019): 583–590.

Ul Haq, Muhammad Nisar, Ghulam Mujtaba Shah, Farid Menaa, Rahmat Ali Khan, Norah A. Althobaiti, Aishah E. Albalawi, and Huda Mohammed Alkreathy. "Green silver nanoparticles synthesized from *Taverniera couneifolia* elicits effective anti-diabetic effect in alloxan-induced diabetic wistar rats." *Nanomaterials* 12, no. 7 (2022): 1035.

Vergès, Bruno. "Pathophysiology of diabetic dyslipidemia: Where are we?" *Diabetologia* 58, no. 5 (2015): 886–899.

Wang, Yuquan, Lanlan Li, Weibo Zhao, Yin Dou, Huijie An, Hui Tao, Xiaoqiu Xu et al. "Targeted therapy of atherosclerosis by a broad-spectrum reactive oxygen species scavenging nanoparticle with intrinsic anti-inflammatory activity." *ACS Nano* 12, no. 9 (2018): 8943–8960.

Wu, Tao, Yufang Gao, Xueqi Guo, Min Zhang, and Lingxiao Gong. "Blackberry and blueberry anthocyanin supplementation counteract high-fat-diet-induced obesity by alleviating oxidative stress and inflammation and accelerating energy expenditure." *Oxidative Medicine and Cellular Longevity* 2018 (2018).

Yahya, Rabia A. M., Azab Elsayed Azab, and Shkal Karema El M. "Effects of copper oxide and/or zinc oxide nanoparticles on oxidative damage and antioxidant defense system in male albino rats." *EASJ Pharm Pharmacol* 1, no. 6 (2019): 135–144.

Yao, Yuejun, Haolan Zhang, Zhaoyi Wang, Jie Ding, Shuqin Wang, Baiqiang Huang, Shifeng Ke, and Changyou Gao. "Reactive Oxygen Species (ROS)-responsive biomaterials mediate tissue microenvironments and tissue regeneration." *Journal of Materials Chemistry B* 7, no. 33 (2019): 5019–5037.

Yokoyama, Ryo, Masaaki Ii, Yasuhiko Tabata, Masaaki Hoshiga, Nobukazu Ishizaka, and Michio Asahi. "Cardiac regeneration by statin-polymer nanoparticle-loaded adipose-derived stem cell therapy in myocardial infarction." *Stem Cells Translational Medicine* 8, no. 10 (2019): 1055–1067.

Yu, Hui, Feiyang Jin, Di Liu, Gaofeng Shu, Xiaojuan Wang, Jing Qi, Mingchen Sun et al. "ROS-responsive nano-drug delivery system combining mitochondria-targeting ceria nanoparticles with atorvastatin for acute kidney injury." *Theranostics* 10, no. 5 (2020): 2342.

Zhang, Zhiwen, Huihui Bu, Zhiwei Gao, Yan Huang, Fang Gao, and Yaping Li. "The characteristics and mechanism of simvastatin loaded lipid nanoparticles to increase oral bioavailability in rats." *International Journal of Pharmaceutics* 394, no. 1–2 (2010): 147–153.

Zhao, Yi, Hai Gao, Jianhua He, Cuiping Jiang, Jing Lu, Wenli Zhang, Hu Yang, and Jianping Liu. "Co-delivery of LOX-1 siRNA and statin to endothelial cells and macrophages in the atherosclerotic lesions by a dual-targeting core-shell nanoplatform: A dual cell therapy to regress plaques." *Journal of Controlled Release* 283 (2018): 241–260.

Zhuang, Manjiao, Honghong Qiu, Ping Li, Lihua Hu, Yayu Wang, and Lei Rao. "Islet protection and amelioration of type 2 diabetes mellitus by treatment with quercetin from the flowers of edgeworthiagardneri." *Drug Design, Development and Therapy* (2018): 955–966.

Chapter 11

Nanotherapeutics for endometrial cancer using metal nanocomposites

Shraddha M. Gupta, Dinesh D. Rishipathak, and Siddharth Singh

LIST OF ABBREVIATIONS

AUB	abnormal uterine bleeding
AuNP/PEI/rGO	gold nanoparticle/polyethyleneimine/reduced graphene oxide
BJOE	*Brucea javanica* oil emulsion
BMI	body mass index
CCK-8	cell counting kit-8
CPs	coordination polymers
CMB	cationic microbubbles
DCs	dendritic cells
DUB	dysfunctional anovulatory uterine bleeding
EC	endometrial carcinoma
EECs	endometrioid endometrial cancers
Erk	extracellular signal-regulated kinase
FIGO	International Federation of Obstetrics and Gynecology
LDL	low-density lipoprotein
MA	megestrol acetate
MAPK	mitogen-activated protein kinase
MMR	mismatch repair
MNPs	magnetic nanoparticles
MPA	medroxyprogesterone acetate
MSI	microsatellite instability assay
MVD	microvascular density
NHDF	normal human dermal fibroblast
PDA	polydopamine
PTEN	phosphatase and tensin homolog
ROS	reactive oxygen species
SAHA	suberoylanilide hydroxamic acid
SPE	screen-printed electrode
VEGF	vascular endothelial growth factor
LNG-IUS	levonorgestrel intrauterine device
PNP	polymeric nanoparticle
HDAC	histone deacetylase
POLE	polymerase epsilon
MSS	microsatellite stable

DOI: 10.1201/9781032621135-11

CTNNB1	catenin beta 1
KRAS	Kirsten rat sarcoma viral oncogene homologue
TIMP-2	tissue inhibitor of metalloproteinase-2
AEH	atypical endometrial hyperplasia
IGF-1	insulin-like growth factor-1
mTOR	mammalian/mechanistic target of rapamycin
AMPK	adenosine mono phosphate-activated protein kinase
GLUT	glucose transporter
ADAM	a disintegrin and metalloproteinases
PTX	paclitaxel
LHRH	luteinizing hormone-releasing hormone
LDE	lipid nanoemulsion
BIBF	triple angiokinase inhibitor developed by Boehringer Ingelheim
PLGA	poly (lactic-co-glycolic acid)

11.1 INTRODUCTION

One of the most frequent gynecological cancers is endometrial carcinoma (EC). Obesity, diabetes, and high blood pressure are frequently linked to EC. The metabolic trio of endometrial cancer refers to these three factors. Epidemiology research found that the likelihood of developing endometrial cancer increased by 2.12 times in individuals with diabetes compared to normal patients, and 2.45 times more common among people who have a body mass index (BMI) of 25 or more. Having both excess weight and high blood pressure elevated the risk of endometrial cancer by a factor of 3.5. Rising rates of endometrial cancer have been seen, especially among younger populations, as a result of the epidemic of metabolic illnesses (such as obesity, diabetes, and hypertension). Treatment of early lesions, as well as the necessity to preserve fertility in late and recurrent patients, is still constrained, despite recent advances in diagnostics, surgery, radiation, and chemotherapy. Several EC histological subtypes have different frequencies, clinical presentations, prognoses, and related epidemiological risk factors. Among these are endometrioid cysts, serous endometrioid cysts, clear-cell endometrioid cysts, mixed endometrioid cysts, and uterine carcinosarcoma. Endometrioid cancer is the most prevalent and usual form of adenocarcinoma. Cancers of the endometrium's glandular cells, known as endometrioid, mimic the structure of healthy endometrium. Endometrioid malignancies come in numerous forms, such as adenocarcinoma of the squamous cell type, adenoacanthoma, adenosquamous or mixed cell malignancy, secretory carcinoma, ciliated carcinoma, and villoglandular adenocarcinoma.

The extent of glandular differentiation as well as cellular architecture was presented as criteria for a grading system for EC via the International Federation of Obstetrics and Gynecology (FIGO). Endometrioid Endometrial Cancers (EECs) and mucinous adenocarcinomas with similar architecture and differentiation are the only cancers for which this method is appropriate. The other ECs fall into the "high grade" category. Grade 1 EECs have a glandular squamous development pattern of 95% or more as well as close to 5% solid growth tissue; the glandular differentiation may be seen in just under half of the tissue in grade 2 EECs, whereas it can be found in between 6% and 50% of grade 3 EECs. Grade 2 EECs have between 6% and 50% solid growth tissue, as defined by FIGO. The

Table 11.1 The relationship between several categorization schemes and the molecular classification of endometrial cancer

Genetic aberration	Epidemiology	Histology	FIGO rating	Phase	Risk cluster
Microsatellite instability, somatic or germline mutation in MMR genes as well as epigenetic alterations	24.7% of G1–2 tumors	Endometrioid Endometrial Carcinoma (EEC)	Low (G1–2)	IA	Low
				IB	Intermediate
			High (G3)	IA	Intermediate
				IB	High intermediate
			Regardless of the grade	I	High intermediate
				II	
			High	III-IVA	High
	39.7% of G3 tumors	Non-EEC *	High	I-IV A	High
Stability of the microsatellites and a low number of mutations	63.5% of G1–2 tumors	Endometrioid Endometrial Carcinoma (EEC)	Low (G1–2)	IA	Low
				IB	Intermediate
			High (G3)	IA	Intermediate
				IB	High intermediate
			Regardless of the grade	I	High intermediate
				II	
			High	III-IVA	High
	28% of G3 tumors	Non-EEC*	High	I-IV A	High
TP53 somatic mutation in the majority of instances (91%)	4.7% of G1–2 tumors	Non-EEC*	Not Applicable	IA	Intermediate
	5% of G3 EEC	EEC or non-EEC*		I-IVA	High
Somatic mutation of POLE, TP53 mutation in 35% of cases	6.2% of G1–2 tumors	EEC or non-EEC*	Low (G1–2)	I-II	Low
	12.1% of G3 tumors		High (G3)		

* *Note:* Non-endometrioid: clear cell, serous, undifferentiated, carcinosarcoma, mixed.

"binary FIGO scheme" is presently the preferred method of classification for ECs due to the fact the epidemiological and molecular indicators of Grades 1 and 2 ECs are quite similar to one another. Grades 1 and 2 of a tumor are referred to as being "low-grade;" however, grade 3 of a tumor is referred to as being "high-grade." In 2013, the "The Cancer Genome Atlas" project incorporated genomic data from wide-genome studies, proteomic analyses, and Microsatellite instability (MSI) assays to better characterize and stratify endometrial malignancies. Molecular classification may reallocate a significant proportion of patients to a different risk group, prompting the development of algorithms to perform more specific tests and minimize the number of tests without impacting risk categorization (Kalampokas et al., 2022; Conlon et al., 2014; Benda et al., 1994) as per Table 11.1.

The sequencing of the whole genome, as well as the sequencing of the exome, microsatellite instability assay (MSI), as well as single gene analysis, were all a part of The Cancer Genome Atlas (TCGA) project, the most comprehensive molecular study of ECs that has been done to this moment. A total of 232 cases of endometrioid and serous endometrial cancer were divided into four subtypes based on molecular data: those with POLE ultra-mutations, MSI hypermutations, copy-number (CN) low or high, and intermediate or poor progression-free survival. The TCGA's discovery of the ultra-mutated POLE subgroup piqued curiosity because of its remarkably good outcomes, especially among high-grade cancers (Talhouk and McAlpine 2016).

Insights into disease biology and diagnostic classification are gained from this combined genomic and proteomic investigation of 373 endometrial malignancies, which may have direct therapeutic applicability. With the help of integrated genomic data, we were able to identify four distinct types of tumors, one of which was a novel POLE subtype present in 10% of endometrioid tumors. This subtype is characterized by a large number of mutations that occur inside the somatic cells, MSS, and frequent, recently discovered hotspots when there are mutations in the exonuclease domain of POLE (Levine et al., 2013).

11.2 MOLECULAR MECHANISM OF ENDOMETRIAL CANCER

Long-term overstimulation of the endometrium with non-progesterone estrogen is widely held to have a significant role in the progression of endometrial hyperplasia and malignancy. According to the conventional theory of endometrial cancer's etiology via its interaction with the nuclear estrogen receptor, estrogen can control the expression of certain genes (a process known as "genotype regulation"). By interaction with the G protein-coupled estrogen receptor present on the surface of the cell, estrogen can induce Ca^{2+} influx, which in turn immediately stimulates the signal transduction pathway known as mitogen-activated protein kinase/extracellular signal-regulated kinase (MAPK/Erk) and promotes the growth of endometrial cancer. This procedure is referred to as the "non-gene-transcription effect" since it does not involve the transcription of genes or the synthesis of proteins. There are several potential mechanisms at play here that may explain why obesity is linked to an increased risk of endometrial cancer including the presence of insulin resistance (hyperinsulinemia), aberrant fat metabolism (leptin, adiponectin abnormalities), diabetes, high cholesterol, and persistent inflammation in obese patients. These factors may aid in tumor growth and progression.

EC is still a risk for women beyond menopause. Reportedly, elevated serum estradiol levels result from aromatase is produced in fat cells and converts androstenedione in the bloodstream to estradiol, which binds to estrogen receptors (ER), ultimately resulting in the activation of transcription factor binding and possible activation or repression of gene transcription. Hence, adipose tissue is the primary site of estrogen production in postmenopausal women. Type 1 (endometrioid) EC accounts for around 80% of cases while type 2 (non-endometrioid) accounts for about 20% based on epidemiological studies, histology, diagnosis, as well as course of therapy. Atypical glandular hyperplasia is the precursor to type 1 cancer. Endometrial hyperplasia usually occurs first and is linked to prolonged unopposed estrogen stimulation. There is a correlation between type 1 carcinomas and near-diploid karyotypes, deficiencies in deoxyribonucleic acid mismatch repair, CTNNB1, and other oncogene mutations, phosphatase and tensin homolog (PTEN) tumor suppressor gene mutations. ECs more closely resemble proliferative endometrium than secretory endometrium when viewed from a molecular perspective. The upregulation of such tumor-suppressing gene PTEN in an estrogen-rich environment is a possible cause of the beginning of the disease (Martini et al., 2002; Zondervan et al., 2007; Painter et al., 2014; Parasar et al., 2017; Zondervan et al., 2020; Terzic et al., 2021).

11.2.1 Variables of uncertainty

The incidence of EC is highest in North America and northern Europe, whereas it is lower in Eastern Europe and South America, and it is at its lowest in Asia and Africa. This points to the fact that racial background plays a significant role in its development. Another crucial factor is age. It is well established that women who are beyond menopause, above 60 years old contribute to the primary demography for EC. Most instances (85%) of occurrence are in people aged 50 and over, with just 5% recorded in those less than 40 years old. The peak age range is 75–79 years (Lee et al., 2017; Singh et al., 2022; Colombo et al., 2016).

11.2.2 Signs and biological markers

Ninety percent of women with EC experience abnormal uterine bleeding (AUB). Many other female genital illnesses also appear to have this characteristic. The treatment of a woman having irregular bleeding from the uterus will vary depending on whether she is in her reproductive or postmenopausal years. AUB is also an indicator of the presence of EC in premenopausal women (20% of EC cases). All postmenopausal women suffering from AUB should get an endometrial biopsy, particularly if any of the aforementioned risk factors are present. Up to 10% of postmenopausal women who experience uterine bleeding have EC. Thickening of the endometrium may be discovered incidentally during imaging studies, or women may come with nonspecific complaints of increased vaginal discharge. Individuals with the disease in its later stages may experience symptoms such as dyspnea due to pleural effusion, pain during sexual activity, early satiation, alterations in the bowel or bladder function, and pelvic pain and abdominal distension. Remember, though, that as many as 5% of people with EC experience no symptoms at all. In individuals with AUB, transvaginal ultrasound is a common method for additional examination. Most women will have endometrial samples performed after a comprehensive sonographic examination. Endometrial sampling followed by histological evaluation serves as the industry benchmark for diagnosis and confirmation of EC.

Matrix metalloproteinases (MMPs) are zinc-dependent proteases that break down collagen and elastin and other components of the extracellular matrix. They contribute to the breakdown and repair of the endometrium throughout the menstrual cycle. The lack of progesterone can lead to elevated estrogen levels, which in turn can increase MMP-2 and MMP-9 production through both direct and indirect mechanisms. Among these pathways is the activation of vascular endothelial growth factor (VEGF), interleukin-8, cytokine-stimulated gene expression, monocyte chemoattractant protein-1, and cyclooxygenase-2 (Michalczyk et al., 2021; Critchley et al., 2006; Curry and Osteen, 2001).

Women suffering from anovulatory dysfunctional uterine bleeding (DUB) have been found to have higher expression of endometrial MMP-2 and MMP-9, as well as VEGF and endometrial MVD (microvascular density), in comparison to the control group. This finding suggests MMPs may have a significant impact on the incidence of irregular uterine bleeding in women with anovulatory DUB and endometrial hyperplasia. The histological grade of endometrial carcinoma is correlated with a rise in the level of expression of MMP-2 and a decrease in tissue inhibitors of metalloproteinase-2 (TIMP-2) expression in endometrial carcinoma tissue. TIMP-2 expression was also related to the extent of invasion of the lymphovascular space, involvement of the lymph nodes, and invasion of the myometrium (Graesslin et al., 2006; Zhang et al., 2010).

mRNA and protein expression of MMP-11, -23, -24, and -28 has also been shown by EC cell lines, which may be linked to the progression of this cancer. MMP-1, or matrix metalloproteinase-1, has been linked to numerous malignancies. Metal NPs have been reported to have antitumor properties. The mechanisms of action involved in cancer therapeutics have been shown (Figure 11.1). Metal NPs can act as vectors for targeted drug delivery. Both hydrophobic drugs, e.g., paclitaxel-loaded selenium nanoparticles and hydrophilic drugs, e.g., doxorubicin-loaded iron oxide nanoparticulate drug delivery system can be loaded into metal nanoparticles. The small size and stability of the metal nanoparticles enhance the bioavailability of the drugs (Xu et al., 2022).

Gold nanoparticle/polyethyleneimine/reduced graphene oxide (Au NPs/PEI/rGO)-modified disposable screen-printed electrode (SPE) for matrix metalloproteinase-1 immunosensor manufacturing was described. To prevent Au NPs agglomeration in a PEI environment, the concomitant reduction of gold and graphene oxide was accomplished using a microwave-assisted single-step technique. Researchers looked into the interior morphology, optical properties, chemical composition, and crystal structure of the materials. Because of its high sulfur–gold bond strength, 3-mercaptopropionic acid was used to build layer by layer to make a label-free MMP-1 immunosensor; its carboxyl group at the end allowed using the standard cross-linking method, the MMP-1 antibody (anti-MMP-1) will be linked to N-(3-dimethylaminopropyl) and N-(3-ethylcarbodiimide) hydrochloride technique. Results from using this immunosensor were equivalent to those obtained using the trusted ELISA technique for the detection. MMP-1 was found in the urine, saliva, bovine serum, and cell culture medium of individuals with oral and brain cancers (HSC-3 and C6) (Liu et al., 2021; Xu et al., 2022).

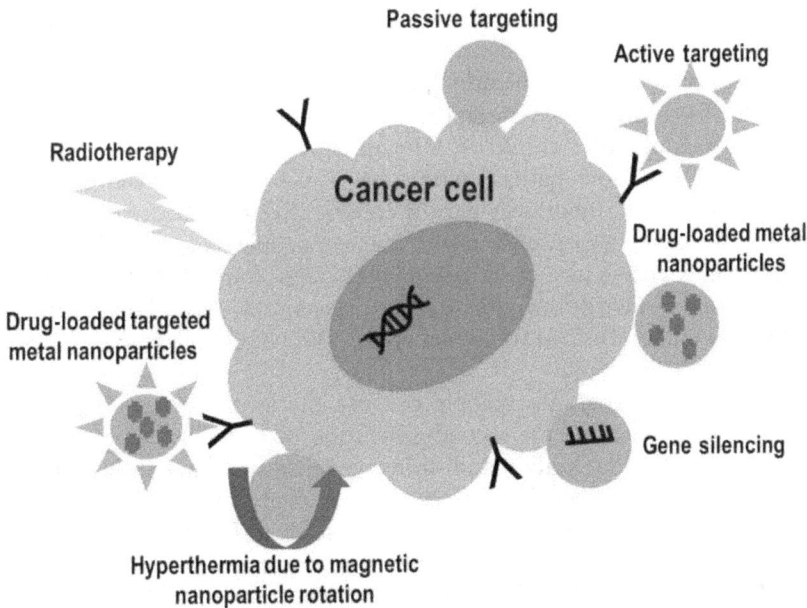

Figure 11.1 Mechanisms involved in cancer targeting.

DJ-1 protein is suggested as an identified biomarker to enhance differential diagnosis between EC subtypes because of its differential expression in endometrioid endometrial cancers (EEC) and serous endometrial cancers tissues in comparison to healthy controls. DJ-1 is primarily responsible for defending cells against oxidative stress. In addition to this role, DJ-1 also regulates a slew of other cellular activities, including migration, adhesion, chemotaxis, proliferation, and apoptosis. The expression of DJ-1 in SEC tissues was found to be substantially higher than in EEC-G1 and EEC-G2 tissues. More than one-third of EEC-G3 showed DJ-1 expression more similar to SEC than the corresponding endometrioid histotype, which was a surprising finding. Using western blotting with sera from ten patients with EEC, five patients with SEC, and 20 healthy women, we also discovered that DJ-1 overexpression in tissues was reflected at the serum level. The data presented here support the possibility of proposing DJ-1 as a novel and accurate serum biomarker for EC (Di Cello et al., 2017).

11.3 RECENT DIFFICULTIES IN ENDOMETRIAL CANCER TREATMENT

Conservative EC treatment typically entails the use of medroxyprogesterone acetate (MPA) and megestrol acetate (MA), and they are effective and safe in several trials. The typical daily dose of MPA is between 20 mg and 1,500 mg, while the typical daily dose of MA is between 40 mg and 480 mg. The most commonly reported dosages of MPA and MA, respectively, are 400–600 mg per day and 160–320 mg per day (Eftekhar et al., 2009). Twenty-eight women with stage IA EC and 17 with atypical endometrial hyperplasia participated in a Japanese multicenter prospective study (AEH). For 26 weeks, they took 600 mg of MPA every day in addition to a low dose of aspirin. After eight weeks and 16 weeks of treatment, endometrial samples were taken. Overall, 67% of women who were diagnosed with EC and 82% of women who were diagnosed with AEH achieved a complete response (CR). After therapy, 12 pregnancies and seven healthy births were observed during follow-up. Between 7 and 36 months, 14 recurrences were reported (47%) (Hahn et al., 2009; Park et al., 2012; Park et al., 2013b; Kudesia et al., 2014; Ohyagi-Hara et al., 2015; Leone Roberti Maggiore et al., 2021). Twenty-one patients with stage IA G1 EC were given 160 mg MA once daily (starting dose) for six months as part of conservative treatment, and the results were analyzed in a single-center prospective trial. Nineteen people had a CR (85.7%), and three women underwent extensive surgery. Five women (27.2%) found CR at 160 mg/day, and 13 (72.2%) at 320 mg/day. Five of the patients were pregnant (27.8%). There was a recurrence in three of 18 individuals (16.7%) (Ushijima et al., 2007; Signorelli et al., 2009; Perri et al., 2011; Shirali et al., 2012).

Instead of using oral progestins, people can use the levonorgestrel intrauterine device (LNG-IUS). By locally releasing the second-generation progestin levonorgestrel, this device produced endometrial decidualization and shrinkage while simultaneously reducing systemic adverse effects and increasing local effectiveness (Park et al., 2013a). The treatment efficacy of LNG-IUS for AEH and EC in female patients was studied. The study included 48 patients, 13/16 (81.3%) of the women with G1 EC and 3/4 of the patients with G2 EC achieved a CR, with a mean ± (SD) time to CR of 5.0 ± 2.9 months and 4.0 ± 0 months, respectively. Forty-four women with endometrial-confined G1 EC were enrolled in a Korean prospective multicenter study in 2019 and treated with oral MPA (500 mg/day) and LNG-IUS. There was a CR rate of 37.1% (13/35) at six months. Nine out of 35 patients (25.7%) experienced a partial response. There was no mention of the condition worsening or any consequences of the treatment. Seventy women under the age of 42 who had been diagnosed with AEH or G1 EC participated in a prospective trial. Monotherapy using a levonorgestrel-releasing intrauterine device was given to patients suffering from AEH. LNG-IUS in conjunction with a gonadotropin-releasing hormone agonist was used to treat patients with G1 EC (Goserelin acetate, every 28 days, 3.6 mg). For a minimum of six months, each woman underwent hormone therapy. Twenty-three (72%) EC women and 35 (92%) AEH women reported CR. Two of these with EC and one with AEH had recurrences during follow-up. Disease persisted in nine women (7 with EC and 2 with AEH). There were ten successful pregnancies for a total of eight mothers (Wang et al., 2014; Pronin et al., 2015).

The treatment of malignant tumors with anti-tumor immunotherapy has shown great promise, but numerous issues remain in its clinical use. The reason for this is that additional study is required to identify new cellular and molecular pathways involved in the immune response to tumors. To solve the issues that arise during immunotherapy, it is vital to develop patient-specific treatment plans. Second, it's important to create fair evaluation criteria that take immunotherapy's effectiveness into account. Finally, immunological tolerance to tumors must be broken. Reducing the immunosuppressive environment in patients is essential when employing the activated immune system to treat malignancies, particularly the immunosuppressive microenvironment within tumor tissues (Cao et al., 2021).

11.3.1 Metformin: as an endometrial cancer therapy

Systemic hyperglycemia creates an environment where cancer cells can efficiently metabolize energy. The glucose transporter is the primary glucose carrier for cellular glucose absorption. The high rate of glycolysis is maintained by the allosteric action of the glucose transporter upon glucose entry into the cell. It has been shown that excessive glucose can stimulate the proliferation and invasion of endometrial cancer cells by influencing the expression of ER/GLUT4, which in turn promotes the production of vascular endothelial growth factor (VEGF)/VEGFR. Patients with endometrial cancer are benefited from typical doses of Metformin, which reduces plasma hyperglycemia and elevated IGF-1 levels. By activating AMPK signaling and blocking the mTOR signaling pathway, metformin drastically reduced the growth of endometrial cancer cells (Cantrell et al., 2010; Gu et al., 2018; Kheirandish et al., 2018; Huang et al., 2023).

11.4 FUTURE TREATMENT APPROACHES IN ENDOMETRIAL CANCER: METALLOPROTEINASE INHIBITORS

TIMPs, or tissue inhibitors of metalloproteinases, are natural inhibitors that interact with matrix metalloproteinases to regulate their activity (TIMP-1, TIMP-2, TIMP-3 etc.). The specificity is regulated by the C-terminus of TIMP, while the N-terminus facilitates complex formation with the zinc-bonded site of active forms. They impede the activity of MMPs by forming noncovalent complexes with active enzymes.

Through controlling matrix alignment, structure, and immune cell entry, TIMPs and MMPs affect the tumor microenvironment and facilitate tumor survival and dissemination. By inhibiting a disintegrin and metalloproteinase (ADAM) protein, TIMPs affect pericellular communication and the local tumor microenvironment by controlling the availability of protease-reliant growth factors or cytokine signals. More advanced tumor stages and shorter relapse durations in EC patients were associated with higher TIMP-1 levels.

TIMPs, as endogenous inhibitors, garnered a lot of interest and sparked research into creating the next generation of MMP inhibitors that are more selective and have other side groups. TIMP engineering for targeted MMP suppression has proven successful through alterations and adjustments to TIMP that connect with the MMP active site, specifically the AB loop.

Human epidermal growth factor interacts with the transmembrane tyrosine kinase receptor C-erbB2s. Overexpression of C-erbB2 has been observed in many malignant tumors, including EC, breast cancer, ovarian cancer, stomach cancer, and lung cancer. Therapeutic targeting of C-erbB2 in EC has shown promise. Thus, knocking down C-erbB2 directly to investigate its effect on the endometrium is a practical approach.

It is widely agreed that the most effective approach for editing genes is the CRISPR/ CRISPR-associated protein 9 (Cas9) system. Experiments on cell function have shown that the proliferation rate, wound healing, cloning, migration, and invasion ability of endometrial cancer cells (HEC1A cells) were all reduced after treatment with CRISPR/Cas9 cationic microbubbles (CMB) or CRISPR/Cas9 cationic microbubbles (PTX) knockout C-erbB2. These results have shown that the EC cell line HEC-1A could proliferate *in vitro* even after C-erbB2 knockdown. CRISPR/Cas9, PTX, and CMB all worked together to significantly reduce HEC-1A cell proliferation and invasion. This research proved that CMB-targeted destruction could be used to transport genes or drugs to their intended targets. The anti-tumor impact of CMBs after PTX release was enhanced at the same medication concentration as compared to the PTX group. Traditional treatments that rely on a single medication have drawbacks, such as the risk of side effects and toxicity from increasing drug doses that don't make a difference, the unintended consequences of gene editing, and the collateral damage done to healthy tissues and organs. The results indicated that P27 and P21 expression regulation may play a role in the process of C-erbB-2/EC gene regulation. Tumor suppressor genes like p21 and p27, also known as negative cell cycle regulators, substantially restrict tumor cell growth and division. The results demonstrated that P21 may downregulate C-erbB2 and that the elevated expression of P21 and P27 in EC largely serves a role in tumor suppression. The proliferation of tumor cells is aided by the expression of P27. Drugs and genes were transferred by CMBs. The use of an ultrasound-based CMB-assisted transfection method for the treatment of EC may enable the discovery of new therapies (Rodrigues et al., 2002; Pires et al., 2009; Erickson et al., 2020; Peng et al., 2021).

11.5 NANOMEDICINES AND CANCER TREATMENTS BASED ON GLYCANS

Glycosyltransferase inhibitors, antibody-based immunotherapies, and vaccinations against glycosylated antigens are the three primary types of immunotherapies based on modifications to glycosylation. Some monoclonal antibodies (mAbs) that target glycosylation-associated tumor-linked epitopes are specific for glycolipids, like gangliosides (GM2, GM3, GD2, and GD3) while others bind to the carbohydrate haptens present on both glycolipids and glycoproteins, like the Lex/Ley and SLex/SLea glycan hapten structures. Proteins targeted with a specific glycoside showed increased anti-tumor specificity and decreased off-target effects. Therapeutic anticancer vaccines have long focused on tumor-associated carbohydrate antigens or sugars attached to proteins with immunogenic properties. Certain cancers can be prevented or treated using vaccines; they include breast cancer (by targeting mucin-related antigens), melanoma (through targeting gangliosides GM2 and GD3), and prostate cancer (via targeting the glycosphingolipid, GloboH). Dendritic cells are pivotal in anti-tumor immunity because they provide real-time updates on tumor dynamics by cross-presenting antigens to T cells, which can kill cancer cells. Defective antigen cross-presentation by tumor-associated dendritic cells severely hinders the efficacy of cancer vaccines and immunotherapies. To maintain antigen cross-presentation and boost the anticancer immune response, activating dendritic cells is crucial. Phase I clinical trials demonstrated the efficacy of the Mannan-Mucin-1 fusion protein in stimulating dendritic cells (DCs). Breast cancer immunotherapy, for instance, makes use of MUC1 lysate-pulsed DCs to increase the production of Mannan-Mucin-1-specific CD8+T cells. Mannan-Mucin-1 is an O-glycosylated protein that is highly expressed in breast carcinoma. The anti-tumor immunotherapeutic impact of chimeric antigen receptor T cell immunotherapy (CAR-T) is improved through the modification of T cells to increase their recognition of tumor-specific glycosylated antigens. Clinical trials are being conducted on several of the aforementioned studies at present. Unfortunately, most studies on cancer immunotherapy have not specifically addressed EC; therefore, research on glycosylation-targeted immunotherapy in EC has had to rely on studies of other malignancies as a point of reference (Cadena et al., 2019; Pu et al., 2022).

Nanotherapies are being developed at a rapid pace as a response to the limitations of conventional therapies for cancers. Therapeutic techniques like chemotherapy, heat, radiation, gene or RNA interference, and immunotherapy have been combined with multiple nanotechnology platforms like liposomes, albumin nanoparticles, and polymeric micelles (Shi et al., 2017; Huang et al., 2023).

Micellar nanoparticles have been reported to generate reactive oxygen species (ROS) and to decrease antioxidants simultaneously, which could render cancer cells more sensitive to ROS, leading to enhanced apoptotic cell death and prevention of tumor development. This would represent a unique anticancer method (Park et al., 2014).

Since HA receptors on the surface of cells, such as CD44, are upregulated in many types of cancer cells, HA-based hydrogels have demonstrated promising results in drug delivery and targeted cancer therapy. Cancer cells that overexpress CD44 may be particularly receptive to HA-based nanogels loaded with anticancer medicines, leading to significant suppression of tumor growth and spread. There was a correlation between high CD44 expression and poor overall survival in 35.4% of EC patients. Suberoylanilide hydroxamic acid (SAHA), one inhibitor of Histone Deacetylase (HDAC), was shown to be delivered to and decrease the proliferation of EC cells expressing CD44 in a previously conducted *in vitro* investigation (Edwards et al., 2021).

Endometrial and ovarian cancers express LHRH receptors heavily. Using these receptors, cytotoxic luteinizing hormone-releasing hormone analogs like AN-152 can bind doxorubicin to [D-Lys (6)]. After one week of treatment with 700 or 300 nmol/20 g AN-152, luteinizing hormone-releasing hormone receptor-positive HEC-1B and NIH: OVCAR-3 tumor volumes were significantly ($p < 0.001$) reduced. There were no harmful impacts noticed. Epithelial cells called OVCAR-3 were taken from the abdominal fluid of a patient with advanced ovarian cancer. Doxorubicin treatment slowed the growth of the tumor, but it did not lessen the size of the tumor. The lethal dose of doxorubicin was 700 nmol/20 g, while the nonlethal dose was 300 nmol/20 g (8.7 mg/kg). SKOV-3 tumors, which lacked luteinizing hormone-releasing hormone receptors, were unaffected by AN-152. The ovary, fallopian tube, myometrium, cervix, and endometrium have luteinizing hormone-releasing hormone receptors but not normal human nonreproductive tissues, hematopoietic stem cells, or vaginal tissue. LHRH receptor-positive tumors benefited more efficiently from AN-152 than doxorubicin. AN-152 helped cure ovarian or endometrial cancers (Gründker et al., 2002).

A bio-nanocomposites-based electrochemical immunosensor for endometriosis diagnosis is sensitive and label-free. Multiwalled carbon nanotube and magnetite nanoparticle ($MWCNT-Fe_3O_4$) were disseminated in chitosan to create a bio-nanocomposite for electrochemical immuno-sensing of carbohydrate antigen 19–9 (CA19–9), a possible endometriosis biomarker (CS). This was done by cross-linking the CS with glutaraldehyde. Researchers detected CA 19–9 with a detection limit of 0.163 pg/mL, a detection range of 1.0–100 ng/mL, and a sensitivity of 2.55 A pg^{-1}/cm using a glassy carbon electrode and a well-characterized anti-AbsCA19–9/CS-MWCNT-Fe_3O_4 immune-electrode. The electrochemical AbsCA19–9/CS-MWCNT-Fe3O4 immunosensor can detect CA19–9 at physiological levels and in early stage diagnoses, sickness monitoring, and therapy optimization. Enzyme-linked immune-sorbent assay (ELISA) and real-world samples proved the sensor's sensing performance and clinical viability. The research suggested that an electrochemical CA19–9 immunosensor based on AbsCA19–9/CS-MWCNT-Fe_3O_4 could be a cost-effective replacement for more established methods like ELISA (Thangapandi et al., 2021).

Endometriosis affects 3% of women, making it the third most common gynecological condition. Late-stage endometriosis is typically diagnosed and treated using the serum marker cancer antigen 125 (CA 125). A label-free immunosensor was created using electrochemical detection of CA 125 in endometriosis blood serum samples. Gold nanoparticles (Au NPs) and reduced graphene oxide (rGO) nanocomposite were deposited in a single electrochemical step to create the sensor. By reducing $HAuCl_4$ and graphene oxide on-site, electrocatalytic activity was improved. The appearance and structure of the Au NP/RGO nanocomposite were validated by several analytical methods. Through the self-assembly monolayer, the antibody was also present on the surface of the customized electrode. The binding of antibodies to antigens has been quantified using square wave voltammetry (Ag). For detecting CA125, the as-fabricated sensor shows a dynamic linear range of 0.0001 300 U/mL and a lower limit of detection of 0.000042 U/mL. The created sensor is sufficiently stable, selective, and repeatable to be useful. Blood samples from patients with endometriosis have been tested with the suggested immunosensor, and positive findings for CA 125 detection have confirmed the validity of the as-made-up sensor, which is linked to the gold-standard ELISA method (Singh et al., 2018; Sangili et al., 2020).

11.6 NANOCOMPOSITES OF POLYMERS AND METALS IN THE TREATMENT OF ENDOMETRIAL CANCER

Antimicrobial wound dressings, anti-infective topical lotions, and cancer treatments are just a few examples of the many ways silver nanoparticles (Ag NPs) are finding widespread use in biomedicine (Sondi and Salopek-Sondi 2004).

ROS, oxidative tension, and DNA break are the key ways by which Ag NPs exert their effects. The significance of ROS in keeping homeostasis within cells make them vital to their continued existence. ROS, a metabolic by-product, plays an important role in cellular signaling networks. Nevertheless, Ag NP-induced toxicity is mediated by an increase in intracellular ROS, which ultimately destroys DNA, lipids, and proteins. The overall anticancer activity on human cancer cells is enhanced by Ag NPs functionalized paclitaxel nanocrystals.

The paclitaxel was conjugated with Ag NPs to create nanocrystals. The paclitaxel nanocrystals served as a mold for the polydopamine (PDA) coating. The PDA layer served as a connecting bridge for the in-situ production and placement of Ag NPs and the tumor-targeting peptide NR1 (RGDARF). The cellular transport efficiency, *in vitro* anticancer activity, and antimigration impact of drug nanocrystals coated using NR1/Ag NPs were all considerably enhanced. In addition to their individual effects on the connection between NR1 receptors, drug release that is pH-responsive, and minuscule size, the data suggested that Ag NPs plus paclitaxel had a synergistic effect when used together. Paclitaxel nanocrystals coated with NR1-AgNP showed excellent selectivity and biocompatibility. Apoptosis was effectively induced by these nanocrystals, with consequences involving the lysis of membranes, damage to nuclear material, dysfunction of mitochondria, excess ROS, and breakage in DNA strands. The activity may involve activating P53 and caspase 3, along with implementing changes to the ratio of BAX to BCL-2 (Umapathi et al., 2020; Muhammad et al., 2021).

After undergoing the standard treatment for endometrial cancer, which involves the removal of the uterus and adnexa along with the collection of lymph node samples, the patient is often left with long-term complications such as diabetes, cardiovascular disease, neuropathy, and immunological decline. Inducing angiogenetic processes and epidermal remodeling as a means of mitigating damage can effectively result in the formation of new connective tissue. Most modern wound dressings were developed to shield an injury from further infection rather than to speed up the healing process. Intensive wound dressings and integrated growth factor-organic molecule dressings to stimulate the enhancement of remodeling of cells are only some of the more than 3,000 items that have been developed recently. By triggering the production of collagen and a variety of growth factors, using nanocomposites immersed in silver molecules accelerates the healing process and promotes re-epithelialization, vasculogenesis (neovascularization), and collagen fiber deposits. Furthermore, Ag NPs can stimulate keratinocytes, causing them to multiply and migrate to the wound site, and they can also cause fibroblasts to differentiate into myofibroblasts, the cells responsible for contracting the wound and hastening the healing process. The impact of Ag NPs on the recovery process has been investigated. It was discovered that the right dose of nanoparticles can aid in the relocation of fibroblasts from the wound's periphery, increasing the amount of alpha-smooth muscle actin, and eventually leading to the differentiation of fibroblasts into myofibroblasts and the contraction of the wound, both of which hasten the remodeling process. When applied to large surgical wounds following a complicated process, such as the one required for the ovarian or endometrial cancer protocol, biopolymers synthesized in conjunction with bioactive antimicrobials, antibacterials, and anti-inflammatory nanoparticles demonstrate significant enhancement of the wound healing perspective (Toczek et al., 2022).

Chitosan, a natural, hydrophilic, and biodegradable polymer linked onto Fe_3O_4 magnetic nanoparticles, carried the poorly water-soluble anticancer drug telmisartan (MNPs). MNPs and chitosan-coated MNPs were examined using X-ray diffraction, field emission scanning electron microscopy, transmission electron microscopy, vibrating sample magnetometer, and BET surface area analyzer (MNP-CS). FTIR and TGA showed MNPs have chitosan coatings. MNP–CS-loaded telmisartan was synthesized by an amide connection between amino groups of chitosan and carboxylic groups of telmisartan. Mesoporous MNP–CS has a 50% drug-loading capacity at 134 m^2/g BET surface area. Telmisartan-loaded MNP–CS showed pH-responsive regulated release. MNP–CS–telmisartan boosted PC-3 prostate cancer cell cytotoxicity with dosage (Dhavale et al., 2021).

The body's main cholesterol transporter, low-density lipoprotein (LDL), is taken up more readily by tumor cells than by normal ones. This condition happens as a result of tumor cells' increased requirement for cholesterol for membrane production during cell growth and division. This is the primary driver behind research into LDL as a particular delivery system for antitumor agents; the transport system is created by incorporating substitute molecules in place of the original molecule and changing the internal lipids to other hydrophobic substances to reduce adverse effects and preventing the emergence of drug resistance. Today, chemotherapy drugs like paclitaxel, carmustine, and etoposide may be delivered in an LDE, a synthetic lipid emulsion that is comparable to the LDL and has been found to have good cellular destroying potential in experimental animals and individuals with advanced malignancies and atherosclerosis. LDE enables the use of LDL receptor-mediated endocytosis in clinical settings because, in contrast to native LDL, LDE preparations of medicines strongly associated with LDE may be produced at an industrial production level (Vitols et al., 1992; Dorlhiac-Llacer et al., 2001; Maranhão et al., 2002; Gabitova et al., 2014; Simko and Ginter, 2014; Graziani et al., 2017; Lima et al., 2017; Bedin et al., 2019).

Green synthetic metallic nanoparticles found in medicinal plants have been found to offer potent anticancer effects, according to recent scientific studies. In recent years, metallic nanoparticles have garnered a lot of attention in the medical community. Recent research has proven that certain nanoparticles have therapeutic characteristics, making them a promising substitute for conventional antimicrobial, and most importantly cancer, medications (Saad et al., 2008; Taylor and Webster, 2011; Zangeneh et al., 2019, 2020; Zhaleh, 2019).

In an *in vitro* MTT experiment, human endometrium's cytotoxicity and anticancer potential were tested against gold (Au) salt, *Spinacia oleracea* leaf extract, and Au nanoparticles. Normal (HUVEC) and common endometrial cancer cell lines (Ishikawa, KLE, HEC-1-A, and HEC-1-B) were evaluated for cytotoxicity using the MTT assay 48 hours after treatment with various doses of $HAuCl_4$, *Spinacia oleracea* leaf aqueous extract, and Au nanoparticles. Even at 1000 µg/mL, $HAuCl_4$, *Spinacia oleracea* leaf aqueous extract, and Au nanoparticles demonstrated extremely strong cytotoxicity against a normal cell line (HUVEC). Endometrial cancer cell lines were dose-dependently killed by $HAuCl_4$, *Spinacia oleracea* leaf aqueous extract, and Au NPs. *Spinacia oleracea* and Au NPs had IC_{50} values of 484 and 324 mg/mL against the HEC-1-B cell line, 458 and 316 against the Human HEC-1-A cell line, 460 and 335 against the KLE cell line, and 468 and 341 against the Ishikawa cell line. The HEC-1-A cell line had the greatest anti-endometrial cancer potential. In the anticancer test employing conventional cancer cell lines, the size effect is more prominent among metallic nanoparticle parameters like size, texture, and surface functions. Due to cell line penetration, smaller particles are more anticancerous. In relevant cancer cell types, particles under 50 nm show increased activity (Namvar et al., 2014). *S. oleracea* leaf aqueous extract can be used to create gold nanoparticles with an average size of 16.7 nm (Zhu et al., 2022).

Green synthetic Au NPs are a subset of these well-known metallic nanoparticles that have found new medical applications in the treatment of various diseases and tumors (Zangeneh et al., 2019; Zhaleh et al., 2019; Shahriari et al., 2019).

Research into the therapeutic effect of $H-MNO_2$-polyethylene glycol (PEG) on endometrial cancer has been conducted because nano-MnO_2 acts as a novel nanocarrier responsive to tumor microenvironment for cancer therapy. Endometrial cancer cells (RL95–2) were exposed to varying concentrations of H-MnO$_2$-PEG (0, 25, and 50 micrograms/mL) to investigate its potential as a monotherapy. As the concentration of H-MnO$_2$-PEG was raised, the intensity reduced, indicating that it acted as an inhibitor of endometrial cancer cell proliferation. These findings establish a link between the H-MnO$_2$-PEG nano platform and endometrial cancer for the first time. *Brucea javanica* oil emulsion (BJOE) has the potential to directly destroy cancer cells by controlling their proliferation and proliferation genes. In addition, it was shown to greatly boost immunity and suppress tumor cells' resistant reaction to chemotherapy. Cell counting Kit-8 test and 4′,6-Diamidine-2′-phenylindole dihydrochloride assay were used to investigate the cellular activity of RL95–2 for 48 hours following the single BJOE (50 g/mL), the single H-MnO$_2$-PEG, or the combination of H-MnO$_2$-PEG/BJOE. Cellular activity and cell number were shown to decrease in the H-MnO$_2$-PEG and H-MnO$_2$-PEG/BJOE treatment groups as compared to the BJOE treatment group. A more potent medicinal impact of H-MnO$_2$-PEG/BJOE on killing endometrial cancer cells was also demonstrated by the combination's superior efficacy in preventing cancer cell proliferation. H-MnO$_2$-PEG/BJOE is synergistic in inducing apoptosis in endometrial cancer. Apoptosis in endometrial cancer cells was investigated after exposure to HMnO$_2$-PEG/BJOE. Using flow cytometry, researchers were able to determine that H-MnO$_2$-PEG considerably boosted endometrial cancer cell death, whereas H-MnO$_2$-PEG/BJOE further accelerated endometrial carcinoma cell apoptosis. Furthermore, the western blot experiment demonstrated that the expression levels of Caspase-3, KRas, and Raf1 were considerably increased in the H-MnO$_2$-PEG/BJOE group compared to the just H-MnO$_2$-PEG group. Overall, the data suggested that H-MnO$_2$-PEG/BJOE may efficiently inhibit tumor growth and promote cell death by regulating the expression of relevant genes (Hu et al., 2021).

Azadirachta indica leaf extract, *Hibiscus rosa-sinensis*, *Murraya koenigii*, *Moringa oleifera*, and *Tamarindus indica* were used in a green chemistry method to produce copper oxide (CuO) nanoparticles. Human breast (MCF-7), cervical (HeLa), epithelioma (Hep-2), and lung (A549) cancer cell lines, together with a normal human dermal fibroblast (NHDF) cell line, were tested to determine the cytotoxicity of CuO nanoparticles. Green synthesis of CuO nanoparticles resulted in significantly higher antioxidant and cytotoxicity properties than their chemical counterparts (Rehana et al., 2017).

To treat cancer with a combination of photodynamic, photothermal, and chemotherapeutic approaches, a drug delivery system based on black phosphorus (BP) is built. BP is a 2D nanosheet with an extremely substantial drug-loading capacity in addition to pH- and light-sensitive drug release. BP's anticancer actions are boosted by its inherent photothermal and photodynamic properties. Multifunctional nanomedicine platform exhibits synergetic photodynamic/photothermal/chemotherapeutic BP-based drug delivery system (Chen et al., 2017).

The molecular inhibitor of triple angiokinases, nintedanib, which blocks three different tyrosine kinase receptors (VEGFRs, PDGFRs, and FGF receptors), has shown promising results (Hilberg et al., 2008) and when coupled with paclitaxel, causes cell death in uterine serous carcinoma cells (Meng et al., 2013). Nonetheless, the incidence of gastrointestinal adverse events was higher in the nintedanib-treated groups, suggesting the need for further efforts to enhance the combinatorial strategy's safety. To boost the efficiency and security of the combinatorial technique, a polymeric nanoparticle (PNP) delivery device was created. It is well known that NPs do the following: (1) increase dissolution, which compensates for Paclitaxel's and BIBF's reported low water solubility; (2) enhance pharmacokinetics; (3)

reduce side effects by reducing off-target effects; and (4) passively target tumors by increasing permeability and retention (Jain et al., 2010). This happens when there are abnormally high levels of angiogenic signals within a solid tumor, leading to the development of defective, "leaky," tumor vasculature via which NPs can escape (Acharya et al., 2011).

BIBF-loaded PLGA (75/T) NPs consist of a biodegradable poly (lactic-co-glycolic acid) NP. As compared to the other therapies, such as PTXp (75/T) + BIBFs, the combination therapy induced a dramatic drop in cell viability. Hence, the synthetic lethality of BIBF in NPs when coupled with PTXp (75/T) is preserved. Loading BIBF in NPs that include D-tocopherol polyethylene glycol 1000 succinate may improve intracellular accumulation and therapeutic efficacy since, like PTX, BIBF is a documented substrate of P-gp. Hec50co cells were used to create a xenograft model of USC that was studied *in vivo*. Compared to PTXs, PTXp (75/T) inhibited tumor development independently, suggesting that delivering PTX in an NP formulation enhances efficacy. Synergistic lethality between BIBF and PTX was induced *in vivo*. Overall, PTXp (75/T) + BIBFp (75/T) therapy performed better than treatment with PTXp (75/T) alone. When comparing PTXs and the naive control group, only when PTXp (75/T) was combined with BIBFp (75/T) was tumor growth considerably suppressed (Ebeid et al., 2018).

11.7 CONCLUSION

Endometrial carcinoma ranks high among gynecological cancers in terms of prevalence. ECs can be categorized according to their cellular architecture and their degree of glandular differentiation, as described by the International Federation of Obstetrics and Gynecology (FIGO). The overstimulation of the endometrium by estrogens other than progesterone for an extended period is the primary factor in endometrial hyperplasia and endometrial cancer. Abnormal uterine bleeding is the most common indication of endometrial cancer, affecting 90% of women. Polymer–metal nanocomposites are currently well-investigated for the treatment of endometrial cancers. The antitumor action that paclitaxel nanocrystals have on human cancer cells is increased overall. The Au NPs/rGO-based sensor for the electrochemical sensing of CA 125 in blood serum samples taken from endometriosis patients has been developed. Chitosan is a naturally occurring polymer that is hydrophilic and biodegradable and was grafted onto Fe_3O_4 magnetic nanoparticles to serve as a carrier for anticancer medications with low water solubility. Thus, there are great opportunities to explore metal nanocomposites for the theranostics of endometrial cancers.

REFERENCES

Acharya, S., Sahoo, S. "PLGA nanoparticles containing various anticancer agents and tumour delivery by EPR effect". Advanced Drug Delivery Reviews. 63, no. 3 (2011): 170–183. doi: 10.1016/j. addr.2010.10.008

Bedin, A., Maranhão, R.C., Tavares, E.R., Carvalho, P.O., Baracat, E.C., Podgaec, S. "Nanotechnology for the treatment of deep endometriosis: Uptake of lipid core nanoparticles by LDL receptors in endometriotic foci". Clinics (Sao Paulo). 74 (2019): e989. doi: 10.6061/clinics/2019/e989

Benda, J.A., Zaino, R. Gynecologic Oncology Group Pathology Manual. Gynecologic Oncology Group, Buffalo, New York, 1994.

Cadena, A.P., Cushman, T.R., Welsh, J.W. "Glycosylation and antitumor immunity". In International Reviews in Cell and Molecular Biology, Ed. Lorenzo Galluzzi, 111–127. Academic Press, New York, 2019. doi: 10.1016/bs.ircmb.2018.05.014

Cantrell, L.A., Zhou, C., Mendivil, A., Malloy, K.M., Gehrig, P.A., Bae-Jump, V.L. "Metformin is a potent inhibitor of endometrial cancer cell proliferation: Implications for a novel treatment strategy". Gynecologic Oncology. 116, no.1 (2010): 92–98. doi: 10.1016/j.ygyno.2009.09.024

Cao, W., Ma, X., Fischer, J.V., Sun, C., Kong, B., Zhang, Q. "Immunotherapy in endometrial cancer: Rationale, practice and perspectives". Biomarker Research. 9, no.49 (2021): 1–30. doi: 10.1186/s40364-021-00301-z

Chen, W., Ouyang, J., Liu, H., Chen, M., Zeng, K., Sheng, J., Liu, Z., et al. "Black phosphorus nanosheet-based drug delivery system for synergistic photodynamic/photothermal/chemotherapy of cancer". Advanced Materials. 29, no.5 (2017): 1603864. doi: 10.1002/adma.201603864

Colombo, N., Creutzberg, C., Amant, F., Bosse, T., González-Martín, J., Ledermann, C., Marth, R., Nout, D. et al. "ESMO-ESGO-ESTRO consensus conference on endometrial cancer: Diagnosis, treatment and follow-up". Annals of Oncology. 27 (2016): 16–41.

Conlon, N., MB, Leitao, M.M., Jr. MD, Abu-Rustum, N.R., MD, Soslow, R.A., MD. "Grading uterine endometrioid carcinoma: A proposal that binary is best". The American Journal of Surgical Pathology. 38, no.12 (2014): 1583–1587. doi: 10.1097/PAS.0000000000000327

Critchley, H.O., Kelly, R.W., Baird, D.T., Brenner, R.M. "Regulation of human endometrial function: Mechanisms relevant to uterine bleeding". Reproductive Biology and Endocrinology. 4, Suppl 1 (2006): S5. doi: 10.1186/1477-7827-4-S1-S5

Curry, T.E., Jr., Osteen, K.G. "Cyclic changes in the matrix metalloproteinase system in the ovary and uterus". Biology of Reproduction. 64, no.5 (2001): 1285–1296. doi: 10.1095/biolreprod64.5.1285

Dhavale, R.P., Sahoo, S.C., Kollu, P., Jadhav, S.U., Patil, P.S., Dongale, T.D., Chougale, A.D., Patil, P.B. "Chitosan coated magnetic nanoparticles as carriers of anticancer drug Telmisartan: pH-responsive controlled drug release and cytotoxicity studies". Journal of Physics and Chemistry of Solids. 148 (2021): 109749. doi: 10.1016/j.jpcs.2020.109749

Di Cello, A., Di Sanzo, M., Perrone, F.M., Santamaria, G., Rania, E., Angotti, E., Venturella, R. et al. "DJ-1 is a reliable serum biomarker for discriminating high-risk endometrial cancer". Tumour Biology. 39, no.6 (2017): 1010428317705746. doi: 10.1177/1010428317705746

Dorlhiac-Llacer, P.E., Marquezini, M.V., Toffoletto, O., Carneiro, R.C., Maranhão, R.C., Chamone, D.A. "In vitro cytotoxicity of the LDE: Daunorubicin complex in acute myelogenous leukemia blast cells." Brazilian Journal of Medical Biology Research. 34, no.10 (2001): 1257–1263. doi: 10.1590/S0100-879X2001001000004

Ebeid, K., Meng, X., Thiel, K.W., Do, A.V., Geary, S.M. et al. "Synthetically lethal nanoparticles for treatment of endometrial cancer". Nature Nanotechnology. 13, no.1 (2018): 72–81. doi: 10.1038/s41565-017-0009-7

Edwards, K., Yao, S., Pisano, S., Feltracco, V., Brusehafer, K., Samanta, S. et al. "Hyaluronic acid-functionalized nanomicelles enhance SAHA efficacy in 3D endometrial cancer models". Cancers. 13, no.16 (2021): 4032. doi: 10.3390/cancers13164032

Eftekhar, Z., Izadi-Mood, N., Yarandi, F., Shojaei, H., Rezaei, Z., Mohagheghi, S. "Efficacy of megestrol acetate (megace) in the treatment of patients with early endometrial adenocarcinoma: Our experiences with 21 patients". International Journal of Gynecological Cancer. 19, no.2 (2009): 249–252. doi: 10.1111/IGC.0b013e31819c5372

Erickson, B.K., Zeybek, B., Santin, A.D., Fader, A.N. "Targeting human epidermal growth factor receptor 2 (HER2) in gynecologic malignancies". Current Opinion in Obstetrics and Gynecology. 32, no.1 (2020): 57–64. doi: 10.1097/GCO.0000000000000599

Gabitova, L., Gorin, A., Astsaturov, I. "Molecular pathways: Sterols and receptor signaling in cancer". Clinical Cancer Research. 20, no.1 (2014): 28–34. doi: 10.1158/1078-0432.CCR-13-0122

Graesslin, O., Cortez, A., Fauvet, R., Lorenzato, M., Birembaut, P., Daraï, E. "Metalloproteinase-2, -7 and -9 and tissue inhibitor of metalloproteinase-1 and -2 expression in normal, hyperplastic and neoplastic endometrium: A clinical-pathological correlation study". Annals of Oncology. 17, no.4 (2006): 637–645. doi: 10.1093/annonc/mdj129

Graziani, S.R., Vital, C.G., Morikawa, A.T., Van Eyll, B.M., Fernandes, H.J., Jr., Kalil Filho, R. et al. "Phase II study of paclitaxel associated with lipid core nanoparticles (LDE) as third-line treatment of patients with epithelial ovarian carcinoma". Medical Oncology. 34, no.9 (2017): 151. doi: 10.1007/s12032-017-1009-z

Gründker, C., Völker, P., Griesinger, F., Ramaswamy, A., Nagy, A., Schally, A.V., Emons, G. "Antitumor effects of the cytotoxic luteinizing hormone-releasing hormone analog AN-152 on human endometrial and ovarian cancers xenografted into nude mice". American Journal of Obstetrics and Gynecology. 187, no.3 (2002): 528–537. doi: 10.1067/mob.2002.124278

Gu, C.J., Xie, F., Zhang, B., Yang, H.L., Cheng, J., He, Y.Y. et al. "High glucose promotes epithelial-mesenchymal transition of uterus endometrial cancer cells by increasing ER/GLUT4-mediated VEGF secretion". Cellular Physiology and Biochemistry. 50 (2018): 706–720. doi: 10.1159/000494237

Hahn, H.S., Yoon, S.G., Hong, J.S., Hong, S.R., Park, S.J., Lim, J.Y., Kwon, Y.S., Lee, I.H., Lim, K.T., Lee, K.H. et al. "Conservative treatment with progestin and pregnancy outcomes in endometrial cancer". International Journal of Gynecological Cancer. 19 (2009): 1068–1073. doi: 10.1111/IGC.0b013e3181aae1fb

Hilberg, F., Roth, G.J., Krssak, M., Kautschitsch, S., Sommergruber, W. et al. "BIBF 1120: Triple angiokinase inhibitor with sustained receptor blockade and good antitumor efficacy". Cancer Research. 68, no.12 (2008): 4774–4782. doi: 10.1158/0008-5472

Hu, Q., Zhang, S., Zhu, J., Yin, L. et al. "The promotional effect of hollow MnO2 with Brucea Javanica Oil Emulsion (BJOE) on endometrial cancer apoptosis". BioMed Research International. 2021 (2021): Article ID 6631533. https://doi.org/10.1155/2021/6631533

Huang, P., Fan, X., Yu, H. et al. "Glucose metabolic reprogramming and its therapeutic potential in obesity-associated endometrial cancer". Journal of Translational Medicine. 21 (2023): 94. doi: 10.1186/s12967-022-03851-4

Jain, R., Stylianopoulos, T. "Delivering nanomedicine to solid tumors". Nature Reviews Clinical Oncology. 7 (2010): 653–664. doi: 10.1038/nrclinonc.2010.139

Kalampokas, E., Georgios, G., Kalampokas, T., Papathanasiou, A., Dimitra Mitsopoulou, D., Tsironi, E. et al. "Current approaches to the management of patients with endometrial cancer". Cancers. 14, no. 18 (2022): 4500. https://doi.org/10.3390/cancers14184500

Kheirandish, M., Mahboobi, H., Yazdanparast, M., Kamal, W., Kamal, M.A. "Anti-cancer effects of metformin: Recent evidences for its role in prevention and treatment of cancer". Current Drug Metabolism. 19, no.9 (2018): 793–797. doi: 10.2174/1389200219666180416161846

Kudesia, R., Singer, T., Caputo, T.A., Holcomb, K.M., Kligman, I., Rosenwaks, Z., Gupta, D. "Reproductive and oncologic outcomes after progestin therapy for endometrial complex atypical hyperplasia or carcinoma". American Journal of Obstetrics and Gynecology. 210 (2014): 255. e1–255.e4. doi: 10.1016/j.ajog.2013.11.001

Lee, Y.C., Lheureux, S., Oza, A.M. "Treatment strategies for endometrial cancer: Current practice and perspective". Current Opinion in Obstetrics and Gynecology. 29 (2017): 47–58. doi: 10.1097/GCO.0000000000000338

Leone Roberti Maggiore, U., Khamisy-Farah, R., Bragazzi, N.L., Bogani, G., Martinelli, F., Lopez, S., Chiappa, V., Signorelli, M., Ditto, A., Raspagliesi, F. "Fertility-sparing treatment of patients with endometrial cancer: A review of the literature". Journal of Clinical Medicine. 10, no.20 (2021): 4784. doi: 10.3390/jcm10204784

Levine, D., The Cancer Genome Atlas Research Network. "Integrated genomic characterization of endometrial carcinoma". Nature. 497 (2013): 67–73. doi: 10.1038/nature12113

Lima, A.D., Hua, N., Maranhão, R.C., Hamilton, J.A. "Evaluation of atherosclerotic lesions in cholesterol-fed mice during treatment with paclitaxel in lipid nanoparticles: A magnetic resonance imaging study". Journal of Biomedical Research. 31, no.2 (2017):116–121. doi: 10.7555/JBR.31.20160123

Liu, X., Lin, L.-Y., Tseng, F.-Y., Tan, Y.-C., Li, J., Feng, L., Song, L. et al. "Label-free electrochemical immunosensor based on gold nanoparticle/polyethyleneimine/reduced graphene oxide nanocomposites for the ultrasensitive detection of cancer biomarker matrix metalloproteinase-1". Analyst. 12, no.146 (2021): 4066–4079. doi: 10.1039/D1AN00537E

Maranhão, R.C., Graziani, S.R., Yamaguchi, N., Melo, R.F., Latrilha, M.C., Rodrigues, D.G. et al. "Association of carmustine with a lipid emulsion: In vitro, in vivo and preliminary studies in cancer patients". Cancer Chemotherapeutics and Pharmacology. 49, no.6 (2002): 487–498. doi: 10.1007/s00280-002-0437-3

Martini, M., Ciccarone, M., Garganese, G., Maggiore, C., Evangelista, A., Rahimi, S., Zannoni, G., Vittori, G., Larocca, L.M. "Possible involvement of hMLH1, p16INK4a and PTEN in the malignant transformation of endometriosis". International Journal of Cancer. 102 (2002): 398–406. doi: 10.1002/ijc.10715

Meng, X., Laidler, L.L., Kosmacek, E.A., Yang, S., Xiong, Z. et al. "Induction of mitotic cell death by overriding G2/M checkpoint in endometrial cancer cells with non-functional p53". Gynecologic Oncology. 128, no.3 (2013): 461–469. doi: 10.1016/j.ygyno.2012.11.004

Michalczyk, Kaja, Cymbaluk-Płoska, A. "Metalloproteinases in endometrial cancer: Are they worth measuring?" International Journal of Molecular Sciences. 22, no.22 (2021): 12472. doi: 10.3390/ijms222212472

Muhammad, N., Zhao, H., Song, W. et al. "Silver nanoparticles functionalized paclitaxel nanocrystals enhance overall anti-cancer effect on human cancer cells". Nanotechnology 32 (2021): 85105. doi: 10.1088/1361-6528/abcacb

Namvar, F. et al. "Cytotoxic effect of magnetic iron oxide nanoparticles synthesized via seaweed aqueous extract". International Journal of Nanomedicine. 9, no.1 (2014): 2479–2488. doi: 10.2147/IJN.S59661

Ohyagi-Hara, C., Sawada, K., Aki, I., Mabuchi, S., Kobayashi, E., Ueda, Y., Yoshino, K., Fujita, M., Tsutsui, T., Kimura, T. "Efficacies and pregnant outcomes of fertility-sparing treatment with medroxyprogesterone acetate for endometrioid adenocarcinoma and complex atypical hyperplasia: Our experience and a review of the literature". Archives in Gynecology and Obstetrics. 291, no.1 (2015): 151–157. doi: 10.1007/s00404-014-3417-z

Painter, J.N., Nyholt, D.R., Krause, L., Zhao, Z.Z., Chapman, B., Zhang, C., Medland, S., Martin, N.G., Kennedy, S., Treloar, S. et al. "Common variants in the CYP2C19 gene are associated with susceptibility to endometriosis". Fertility and Sterility. 102 (2014): 496–502. doi: 10.1016/j.fertnstert.2014.04.015

Parasar, P., Ozcan, P., Terry, K.L. "Endometriosis: Epidemiology, diagnosis and clinical management". Current Obstetrics and Gynecology Reports. 6 (2017): 34–41. doi: 10.1007/s13669-017-0187-1

Park, H., Seok, J.M., Yoon, B.S., Seong, S.J., Kim, J.Y., Shim, J.Y., Park, C.T. "Effectiveness of high-dose progestin and long-term outcomes in young women with early-stage, well-differentiated endometrioid adenocarcinoma of uterine endometrium". Archives in Gynecology and Obstetics. 285, no.2 (2012): 473–478. doi: 10.1007/s00404-011-1959-x

Park, J.Y., Kim, D.Y., Kim, J.H., Kim, Y.M., Kim, K.R., Kim, Y.T., Seong, S.J., Kim, T.J., Kim, J.W., Kim, S.M. et al. Long-term oncologic outcomes after fertility-sparing management using oral progestin for young women with endometrial cancer (KGOG 2002). European Journal of Cancer. 49, no.4 (2013a): 868–874. doi: 10.1016/j.ejca.2012.09.017

Park, J.Y., Kim, D.Y., Kim, T.J., Kim, J.W., Kim, J.H., Kim, Y.M., Kim, Y.T., Bae, D.S., Nam, J.H. "Hormonal therapy for women with stage IA endometrial cancer of all grades". Obstetrics and Gynecology. 122, no.1 (2013b): 7–14. doi: 10.1097/AOG.0b013e3182964ce3

Park, S., Kwon, B., Yang, W., Han, E., Yoo, W., Kwon, B.M. et al. "Dual pH-sensitive oxidative stress generating micellar nanoparticles as a novel anticancer therapeutic agent". Journal of Controlled Release. 196 (2014): 19–27. doi: 10.1016/j.jconrel.2014.09.017

Peng, S., Cai, J., Bao, S. "CMBs carrying PTX and CRISPR/Cas9 targeting C erbB 2 plasmids interfere with endometrial cancer cells". Molecular Medicine Reports. 24, no.6 (2021): 830. doi: 10.3892/mmr.2021.12470

Perri, T., Korach, J., Gotlieb, W.H., Beiner, M., Meirow, D., Friedman, E., Ferenczy, A., Ben-Baruch, G. "Prolonged conservative treatment of endometrial cancer patients: More than 1 pregnancy can be achieved". International Journal of Gynecological Cancer. 21, no.1 (2011): 72–78. doi: 10.1097/IGC.0b013e31820003de

Pires, L.A., Hegg, R., Valduga, C.J., Graziani, S.R., Rodrigues, D.G., Maranhão, R.C. "Use of cholesterol-rich nanoparticles that bind to lipoprotein receptors as a vehicle to paclitaxel in the treatment of breast cancer: Pharmacokinetics, tumor uptake and a pilot clinical study". Cancer Chemotherapeutics and Pharmacology. 63, no.2 (2009): 281–287. doi: 10.1007/s00280-008-0738-2

Pronin, S.M., Novikova, O.V., Andreeva, J.Y., Novikova, E.G. "Fertility-sparing treatment of early endometrial cancer and complex atypical hyperplasia in young women of childbearing potential". International Journal of Gynecological Cancer. 25, no.6 (2015): 1010–1014. doi: 10.1097/IGC.0000000000000467

Pu, C., Biyuan, Xu, K. et al. "Glycosylation and its research progress in endometrial cancer". Clinical and Translational Oncology. 24, no.10 (2022): 1865–1880. doi: 10.1007/s12094-022-02858-z

Rehana, D. et al. "Evaluation of antioxidant and anticancer activity of copper oxide nanoparticles synthesized using medicinally important plant extracts". Biomedicine and Pharmacotherapy. 89 (2017): 1067–1077. doi: 10.1016/j.biopha.2017.02.101

Rodrigues, D.G., Covolan, C.C., Coradi, S.T., Barboza, R., Maranhão, R.C. "Use of a cholesterol-rich emulsion that binds to low-density lipoprotein receptors as a vehicle for paclitaxel". Journal of Pharma Pharmacology. 54, no.6 (2002): 765–772. doi: 10.1211/0022357021779104

Saad, M., Garbuzenko, O.B., Ber, E., Chandna, P., Khandare, J.J., Pozharov, V.P., et al. "Receptor targeted polymers, dendrimers, liposomes: Which nanocarrier is the most efficient for tumor-specific treatment and imaging?" Journal of Controlled Release. 130, no.2 (2008):107–114. doi: 10.1016/j.jconrel. 2008.05.024

Sangili, A., Thangapandi, K., Shen-Ming, C., Nanda, A., Jana, S.K. "Label-free electrochemical immunosensor based on one-step electrochemical deposition of AuNP-RGO nanocomposites for detection of endometriosis marker CA 125". ACS Applied Bio Materials. 3, no.11 (2020): 7620–7630. doi: 10.1021/acsabm.0c00821

Shahriari, M. et al. "Biosynthesis of gold nanoparticles using *Allium noeanum* Reutex Regel leaves aqueous extract: Characterization and analysis of their cytotoxicity, antioxidant, and antibacterial properties". Applied Organometallic Chemistry. 33 (2019): e5189. doi: 10.1002/aoc.5189 e5189

Shi, J., Kantof, P.W., Wooster, R., Farokhzad, O.C. "Cancer nanomedicine: Progress, challenges and opportunities". Nature Reviews Cancer. 17, no.1 (2017): 20–37. doi: 10.1038/nrc.2016.108

Shirali, E., Yarandi, F., Eftekhar, Z., Shojaei, H., Khazaeipour, Z. "Pregnancy outcome in patients with stage 1a endometrial adenocarcinoma, who conservatively treated with megestrol acetate". Archives of Gynecology and Obstetics. 285, no.3 (2012): 791–795. doi: 10.1007/s00404-011-2021-8

Signorelli, M., Caspani, G., Bonazzi, C., Chiappa, V., Perego, P., Mangioni, C. "Fertility-sparing treatment in young women with endometrial cancer or atypical complex hyperplasia: A prospective single-institution experience of 21 cases". BJOG an International Journal of Obstetrics & Gynaecology. 116, no.1 (2009): 114–118. doi: 10.1111/j.1471-0528.2008.02024.x

Simko, V., Ginter, E. "Understanding cholesterol: High is bad but too low may also be risky: Is low cholesterol associated with cancer?" Bratislava Medical Journal. 115, no.2 (2014): 59–65. doi: 10.4149/bll_2014_013

Singh, N., Jamieson, A., Morrison, J., Taylor, A., Ganesan, R. "BAGP POLE NGS testing guidance". British Gynaecological Cancer Society: London, UK, v1.1, 8 April 2022.

Singh, P., Pandit, S., Mokkapati, V.R.S.S., et al. "Gold nanoparticles in diagnostics and therapeutics for human cancer." International Journal of Molecular Sciences. 19, no.7 (2018): 1979. doi: 10.3390/ijms19071979

Sondi, I., Salopek-Sondi, B. "Silver nanoparticles as antimicrobial agent: A case study on E. coli as a model for Gram-negative bacteria". Journal of Colloid and Interface Science. 275, no.1 (2004): 177–182. doi: 10.1016/j.jcis.2004.02.012

Talhouk, A., McAlpine, J.N. "New classification of endometrial cancers: The development and potential applications of genomic based classification in research". Gynecologic Oncology Research and Practice. 3 (2016): 14. doi: 10.1186/s40661-016-0035-4

Taylor, E., Webster, T.J. "Reducing infections through nanotechnology." Paper presented at the Twenty-First International Offshore and Polar Engineering Conference, Maui, Hawaii, USA, June 2011.

Terzic, M., Aimagambetova, G., Kunz, J., Bapayeva, G., Aitbayeva, B., Terzic, S., Laganà, A.S. "Molecular basis of endometriosis and endometrial cancer: Current knowledge and future perspectives". International Journal of Molecular Sciences. 22, no. 17 (2021): 9274. doi: 10.3390/ijms22179274

Thangapandi, K., Sangili, A., Nanda, A., Sengodu, P., Kaushik, A., Jana, S.K. "Bio-nanocomposite based highly sensitive and label-free electrochemical immunosensor for endometriosis diagnostics application". Bioelectrochemistry. 139 (2021): 107740. doi: 10.1016/j.bioelechem.2021.107740

Toczek, J., Sadłocha, M., Major, K., Stojko, R. "Benefit of silver and gold nanoparticles in wound healing process after endometrial cancer protocol". Biomedicines. 10, no.3 (2022): 679. doi: 10.3390/biomedicines10030679

Umapathi, A., Navya, P.N., Madhyastha, H. et al. "Curcumin and isonicotinic acid hydrazide functionalized gold nanoparticles for selective anticancer action". Colloids Surfaces A: Physicochemical Engineering Aspects. 607 (2020): 125484. doi: 10.1016/j.colsurfa.2020.125484

Ushijima, K., Yahata, H., Yoshikawa, H., Konishi, I., Yasugi, T., Saito, T., Nakanishi, T., Sasaki, H., Saji, F., Iwasaka, T. et al. "Multicenter phase II study of fertility-sparing treatment with medroxyprogesterone acetate for endometrial carcinoma and atypical hyperplasia in young women". Journal of Clinical Oncology. 25, no.19 (2007): 2798–2803. doi: 10.1200/JCO.2006.08.8344

Vitols, S., Peterson, C., Larsson, O., Holm, P., Aberg, B. "Elevated uptake of low-density lipoproteins by human lung cancer tissue in vivo". Cancer Research. 52, no.22 (1992): 6244–6247.

Wang, C.J., Chao, A., Yang, L.Y., Hsueh, S., Huang, Y.T., Chou, H.H., Chang, T.C., Lai, C.H. "Fertility-preserving treatment in young women with endometrial adenocarcinoma: A long-term cohort study". International Journal of Gynecological Cancer. 24 (2014): 718–728. doi: 10.1097/IGC.0000000000000098

Xu, J., Zhang, W.C., Guo, Y.W., Chen, X.Y., Zhang, Y.N. "Metal nanoparticles as a promising technology in targeted cancer treatment". Drug Delivery. 29, no.1 (2022): 664–678. doi: 10.1080/10717544.2022.2039804

Zangeneh, M.M. "Green synthesis and formulation a modern chemotherapeutic drug of Spinacia oleracea L. leaf aqueous extract conjugated silver nanoparticles: Chemical characterization and analysis of their cytotoxicity, antioxidant, and anti-acute myeloid leukemia properties in comparison to doxorubicin in a leukemic mouse model". Applied Organometallic Chemistry. 34 (2020). doi: 10.1002/aoc.5295

Zangeneh, M.M., Saneei, S., Zangeneh, A. et al. "Preparation, characterization, and evaluation of cytotoxicity, antioxidant, cutaneous wound healing, antibacterial, and antifungal effects of gold nanoparticles using the aqueous extract of Falcaria vulgaris leaves". Applied Organometalllic Chemistry. 33 (2019): e5216. doi: 10.1002/aoc.5216

Zhaleh, M. et al. "In vitro and in vivo evaluation of cytotoxicity, antioxidant, antibacterial, antifungal, and cutaneous wound healing properties of gold nanoparticles produced via a green chemistry synthesis using *Gundelia tournefortii* L. as a capping and reducing agent". Applied Organometallic Chemistry. 33 (2019): e5015. doi: 10.1002/aoc.5015

Zhang, X., Qi, C., Lin, J. "Enhanced expressions of matrix metalloproteinase (MMP)-2 and -9 and vascular endothelial growth factors (VEGF) and increased microvascular density in the endometrial hyperplasia of women with anovulatory dysfunctional uterine bleeding". Fertility and Sterility. 93, no.7 (2010):2362–2367. doi: 10.1016/j.fertnstert.2008.12.142

Zhu, B., Xie, N., Yue, L., Wang, K., Bani-Fwaz, M.Z. et al. "Formulation and characterization of a novel anti-human endometrial cancer supplement by gold nanoparticles green-synthesized using Spinacia oleracea L. Leaf aqueous extract". Arabian Journal of Chemistry. 15 (2022): 103576. doi: 10.1016/j.arabjc.2021.103576

Zondervan, K.T., Becker, C.M., Missmer, S.A. "Endometriosis". New England Journal of Medicine. 382 (2020): 1244–1256. doi: 10.1056/NEJMra1810764

Zondervan, K.T., Treloar, S.A., Lin, J., Weeks, D., Nyholt, D., Mangion, J., Mackay, I.J., Cardon, L.R., Martin, N., Kennedy, S.H. et al. "Significant evidence of one or more susceptibility loci for endometriosis with near-Mendelian inheritance on chromosome 7p13–15". Human Reproduction. 22 (2007): 717–728. doi: 10.1093/humrep/del446

Chapter 12

Nanotherapeutics for colorectal cancer using metal nanocomposites

Yuvraj Patil and Shvetank Bhatt

LIST OF ABBREVIATIONS

5FU	5-fluorouracil
AE	Adverse events
ATP	Adenosine triphosphate
CBS	Cystathionine beta-synthase
CCK8	Cell Counting Kit 8
CD	Cluster of differentiation protein
CEACAM	Carcinoembryonic antigen cell adhesion molecule
CNT	Carbon nanotube
CRC	Colorectal cancer
DMEM	Dubelco's minimum essential media
DOX	Doxorubicin
EDC	N-(3-dimethyl aminopropyl)-N'-ethyl carbodiimide hydrochloride
EGFR	Endothelial growth factor receptor
EMT	Epithelial-mesenchymal transition
EpCAM	Epithelial cellular adhesion molecule
FAT1	FAT atypical cadherin 1
FDA	Food and Drug Administration, see USFDA
G	Graphene
GPC	Glypcian
HPAM	Hydrolyzed polyacrylamide
IBD	Inflammatory bowel disease
IC_{50}	Inhibitory concentration-50%
L-o-C	Lab on a chip
LV	Leucovorin
MCAM	Melanoma cell adhesion molecule
MFH	Magnetic fluid hyperthermia
MNB	Micro-nanobot
MNC	Metal nanocomposite
MNP	Magnetic nanoparticle
MPS	Mononuclear phagocyte system
MSN	Mesoporous silica nanoparticles
MTB	Magnetically tagged bacteria
MTDH	Metadherin

MTT	(3-[4,5-Dimethylthiazol-2-yl]-2,5-diphenyltetrazolium bromide)
MWCNT	Multiwalled carbon nanotube
NC	Nanocomposite
NHS	N-Hydroxysuccinimide
NP	Nanoparticle
PBS	Phosphate buffered saline
PD-1	Programmed cell death protein 1
PD-L1	Programmed death-ligand 1
PEG	Polyethylene glycol
PSBMA	Polydopamine/poly (sulfobetaine methacrylate)
PTX	Paclitaxel
QL	Quality of life
RES	Reticuloendothelial system
TEAE	Treatment-emergent adverse events
Tf	Transferrin
TfR	Transferrin receptor
TGF-β	Transforming growth factor β
TME	Tumor microenvironment
UC	Ulcerative colitis
USFDA	United States Food and Drug Administration, see FDA
VEGF	Vascular endothelial growth factor
XTT	(2,3-Bis-[2-Methoxy-4-Nitro-5-Sulfophenyl]-2H-Tetrazolium-5-Carboxanilide)

12.1 INTRODUCTION

Colorectal cancer (CRC) has remained a leading healthcare challenge over the past two decades. With worldwide mortality as a result of CRC, accounting for roughly 9% of all cancer-related deaths and an incidence rate that has more than doubled over the last two decades, CRC is the fourth leading type of cancer in the world (Parkin et al. 2001; Sung et al. 2021). While the world population has increased roughly 30% since the year 2000, the incidence and mortality due to CRC have increased by 116% and 56%, respectively. The high rate of incidence and deaths suggests an active player in the risk factors that are currently attributed to CRC. Interestingly, recent healthcare data shows a 1–2% decrease in incidence rate in Western countries, especially in older adults (>50 years) while this number is compensated by an increasing incidence rate in younger people under 50 years of age. While this is attributed to greater awareness and screening in older individuals, the rising numbers of CRC in younger people is concerning. Among the risk factors for CRC, alcohol consumption and consumption of red meats, fried and grilled foods are likely culturally linked; however, the rising standard of living for population groups around the globe may likely be responsible for the steady increase in incidences of CRC. According to the Centers for Disease Control and Prevention, US Department of Health and Human Services, CRC patients have a five-year survival rate of almost 88% if the cancer is localized, while the survival rate drops to 16% if distant cancer foci are identified. This is consistent with metastatic growth and a strongly limited probability of treatment efficacy and subsequent survival of the patient (Division of Cancer Prevention and Control 2022).

In light of the poor outcome of metastatic CRC, it is critical to developing therapeutic strategies that combat the disease and help in reversing the disease progression to improve the quality of life and in extending the life of a patient suffering from CRC. To present a clinical perspective on treating colorectal cancer, especially in advanced cancer stages, consider the impact of conventional anticancer strategies which involve combinations of conventional drugs such as irinotecan, 5-fluorouracil (5FU), and leucovorin (LV). In a recent trial conducted in France, the quality of life (QL) metric monitored at baseline (day 0) and then throughout the treatment period of about two years showed an immediate drop from a median value of about 70 to negative indices. The predetermined range of 0 (worst imaginable health outcome) to 100 (best imaginable outcome) indicated the physical condition and self-perception of patients of CRC. The near-absolute decrease in QL values within the first treatment cycle is evidence of the side effects which severely affect the lives of treated patients (Abdelghani et al. 2015). Irinotecan, which is a topoisomer inhibitor, and 5FU, an antimetabolite are complementary drugging molecules to arrest cellular proliferation (Rustum 1990; Saltz 2001; Vanhoefer et al. 2001). LV serves as a stabilizing agent to sustain and enhance the activity of 5FU. Several clinical studies have established the efficacy of 5FU+LV in managing advanced and metastatic CRC, which is to say the survival outcome is modestly improved making the combination a first line of treatment until recently. Multiple trials have demonstrated and established the inclusion of anticancer agents such as irinotecan, raltitrexed, and oxaliplatin in combinations that significantly improved clinical outcomes, especially in advanced CRC (Delaunoit et al. 2004; Hind et al. 2008; Saltz 2001).

Adverse events (AEs) as a direct result of the treatment are unavoidable due to the nature of the anticancer agents, which actively interfere with the DNA and RNA synthesis mechanism (Delaunoit et al. 2004). Treatment-emergent AEs have driven the exploration of alternatives to conventional pharmaceuticals resulting in the development of immunotherapeutic agents such as pembrolizumab, atezolizumab, bevacizumab, and decoy receptors such as aflibercept (Abdelghani et al. 2015; Antoniotti et al. 2022; Diaz et al. 2022; Elez and Baraibar 2022; Ferrara et al. 2004; Tian et al. 2023). Pembrolizumab is modestly effective for the treatment of microsatellite instability or mismatch repair-deficient CRCs; however, immune-checkpoint inhibitors are gradually gaining traction in CRC therapy due to their ability to disable the PD-1 and PD-L1/2-mediated immune-avoidance. Strategies to treat CRC with cytokine blockades such as galunisertib (TGF-β) and bevacizumab (VEGF-A) are gaining traction with the USFDA on account of lowered or at least unchanged treatment-emergent adverse events (TEAEs), with better survival outcomes (Ferrara et al. 2004; Yamazaki et al. 2022). With modest gains over conventional therapeutics, the new class of anticancer agents is being established as a viable first-line avenue in advanced CRC.

Given the relatively poor efficacy of the drugs, their prohibitive price, and demonstrable risk of TEAE mortality, there persists an unmet need for advanced therapeutic systems which can focus on selectivity and target-specific drugging (Elez and Baraibar 2022) and possess cargo-flexibility to deliver not just conventional chemical agents but also protein-based therapeutic agents as well as nucleic acid payloads (Wu et al. 2019) to make inroads into the translation of currently available technologies in medicine. While there has been tremendous headway in nanoparticle/nanocomposite/nanobot-based experimental anticancer therapeutics, the growth has been largely divergent with an emphasis on material research. The independent exploration has led to an explosive diversity of nanotherapeutics with capabilities of highly selective site-directed drug delivery, with specializations in cell organelle targeting and active-site retention. The ultra-localization of nanotherapeutics is precise (Nandi et al. 2020) and may well lead to the abrogation of current clinical standards of dosing based on subject weight or surface area normalization. With advances in precision of diagnosis and personalized medicine, it would be possible to minimize the anticancer payload and in concert with tunable attributes of nanocomposites (NCs) such as self-regulated cytotoxicity, thermal and sonic excitability, still exert maximum anticancer efficacy while significantly downplaying the inherent adverse effects of cancer medicine. There are, however, several considerations that warrant critical attention such as the carrying capacity (Wang 2021) of the NCs, its intended target site, the viscosity (Apmann et al. 2021) of the media at the target site, the chemical milieu, and its motive capacity. Metal nanocomposites (MNCs) have been reported with sensitivity to chemical environment, to enable site-specific activity, while others can self-propel in conditions found specifically in tumor microenvironment, discussed in this chapter (Andhari et al. 2020).

Systemic administration of MNCs, for instance, blends of iron oxide nanoparticles (NPs) with graphene sheets or carbon nanotubes, will lead to differential activity and organ localization. The mononuclear phagocytic system is known to target NPs and depending on their size, charge, hardness, and shape have a rapid or slow circulatory clearance. Furthermore, NCs can extravasate into tissues, especially in the liver, spleen, and kidney where they accumulate. Very interestingly, some MNC researchers believe that while the accumulation of metallic particles may be assumed to be toxic, this may not be the case in reality (Lachowicz et al. 2021; Locci et al. 2019; Mitchell et al. 2021). The European clinical study explored the presence of MNCs in organs such as the liver and kidney in cadavers from healthy individuals and found MNCs accumulated in organs of virtually all cadavers suggesting the long-term and routine presence of nanoparticles in humans. This may not appear surprising given the FDA approval of iron oxide NPs as biocompatible nanomaterial; however, the precise impact of NP accumulation in tissues warrants further investigation.

The chapter explores cancer biology and the various features of note that are currently exploited to synthesize effective targeting strategies. Locomotory aspects such as self-propulsion using multiple methods are discussed along with practical considerations such as limitation in cargo-carrying capacity in progressively smaller MNCs. Validation of nanotherapeutics is carried out using multiple platforms such as 2D, 3D cell culture, tumor spheroids, and *in vivo* using rodent models, which will be briefly covered in terms of their relevance and utility. Analytical considerations for executing MNC-based studies are discussed, and finally, the clinical perspective of MNCs in colorectal cancer is presented with a brief discussion of the path ahead.

12.2 CANCER CELL BIOLOGY: EXPLOITING DEVIANT PHYSIOLOGICAL FEATURES

The incidence of CRC is closely tied to inflammation of the colon and rectum due to pro-carcinogenic events such as chronic inflammatory bowel disease (IBD) and ulcerative colitis (UC) (Shahnazari et al. 2023). The colon inflammation originates from the rectum and left unaddressed over time, envelopes the large intestine. Subjects with IBD and UC experience a CRC risk of 1.5-fold and higher, compared to the general population. The change in physiology in chronically inflamed individuals is key in CRC biology.

12.2.1 Altered metabolic poise

As first described by Warburg, cancer cells are fundamentally altered in metabolic poise compared to healthy cells. A majority of cancer types rely on glycolysis to (Liberti and Locasale 2016; Warburg 1956)

a. Achieve a greater turnover of ATP;
b. Avoid dependence on oxygen, especially in low-oxygen environments on the deep-seated tumor; and
c. Avoid of cell death by apoptotic programs.

However, cancer cells also demonstrate metabolic flexibility and frequently show oxidative phosphorylation equivalent to healthy (aerobic) cells (Bose, Zhang, and Le 2021). The phenomenon is postulated to be exploited by deep-tumor/hypoxic cancer cells for survival and proliferation. The hallmark of the Warburg metabolism is hyperglycolytic poise with the underutilized pyruvate by-product being shunted to lactic acid production for recharging the spent nicotinamide adenine dinucleotide equivalents. A buildup of lactic acid in healthy cells and tissues can negatively impact normal physiological roles; however, in cancerous cells, expression of lactic acid transporters assists with the evacuation of lactic acid from the intracellular compartment to the interstitial space, which now turns acidic (Cassim et al. 2020; Wang et al. 2021).

The acidity serves multiple roles in the tumor microenvironment (TME) including decreasing the integrity of tight-junctions in surrounding tissue to enable movement of shed cancer cells, which also aids shed cells to enter the circulatory system. This is aided by the fact that cancers, especially tumors have an "angiogenetic poise" implying new vascularization, at least in some cancer types, of the neoplastic growth (Ferrara et al. 2004). Localized lactic acidosis is thus a metabolic feature occurring in the majority of tumors, especially in hypoxic cancer types. Glycolytic flux also generates a carbon–electron potential in cancer cells, regardless of oxygen availability. The ready electron availability results in the rapid generation of reactive oxygen species, including hydrogen peroxide. Hydrogen peroxide transported out to the extracellular medium enriches over time and becomes an appreciable metabolite in the TME (Andhari et al. 2020).

12.2.2 Endogenous fuel for propulsion and TME-selective chemical switch

One of several innovations designed with the TME in mind was the utilization of iron oxide MNCs in targeting tumors. Given its superior biocompatibility and ability to be subject to surface chemistry tuning, iron oxide nanoparticles can undergo the Fenton reaction (Scheme 12.1) in the presence of hydrogen peroxide (Andhari et al. 2020; Shin, Yoon, and Jang 2008). The reaction is characterized by the formation of molecular hydrogen, which can generate bubbles creating either minor forward propulsion or more likely a partial buoyancy which allows upwards movement. The upward trajectory is balanced by gravity-driven return after the gas bubble disengages upon coalescence.

As demonstrated by the Khandare group, iron oxide MNCs, even in Janus configuration can sustain the reaction for significant periods to afford mobility and turbulent mixing in various fluids such as cell culture media, serum, and even whole blood (Andhari et al. 2020; Wavhale et al. 2022, 2021).

$$Fe^{2+} + H_2O_2 \longrightarrow Fe^{3+} + HO \cdot + OH^-$$

$$Fe^{3+} + H_2O_2 \longrightarrow Fe^{2+} + HOO \cdot + H^+$$

$$2H_2O \longrightarrow HO \cdot + HOO \cdot + H_2O$$

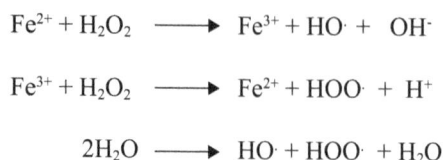

Scheme 12.1 Fenton reaction catalyzed by iron oxide MNCs.

Further, the group has also demonstrated an innovative approach to drug release from MNCs by utilizing amide linkage chemistry. The iron oxide NPs, which serve as caps on carbon nanotubes (CNTs) loaded with doxorubicin, are conjugated to the CNTs using amide linkage. Amide bonds are acid-labile linkages that separate in acidic media, such as that observed in lysosomes as well as in acidic TMEs, as a result of lactate accumulation. Lysosomal processing also involves cathepsin-like enzymes that can cleave amide bonds. Thus, MNCs loaded with cytotoxic drugs, regulated by acid-labile gatekeepers, are enabled for effective site-directed drug release within the exclusive TME zone (Figure 12.1). The drug cannot be released in neutral media unless taken up within the cell and subject to endosomal/lysosomal digestion.

12.2.3 Overexpression of cell surface proteins in cancer cells

Another hallmark of cancerous cells is the overexpression of survival-centric surface proteins. As cancer cells selectively emphasize survival and proliferative mechanisms, several proteins are overexpressed. These molecules are involved in signaling pathways, cellular uptake pathways, metabolic pathways, drug-efflux programs, modified cell adhesion and attachment mechanisms, etc. Notable instances are folate receptors 1, 2, and α, transferrin receptor, EpCAM, and so on. Extensive work in the area has established the overexpression of FR1 and FR2 in multiple cancers (Nawaz and Kipreos 2022; Scaranti et al. 2020; Ledermann, Canevari, and Thigpen 2015). Simple strategies for site-directed targeting include MNCs designed with folate ligands, which prove to be quite selective for cancerous cells as compared to healthy, non-cancerous cells (Angelopoulou et al. 2019; Jurczyk et al. 2021; Zwicke, Mansoori, and Jeffery 2012). Advanced and more specific approaches include using antibodies directed against folate receptors (Daniels et al. 2006). The choice of targeting moiety would be largely dependent on pharma-economics and to an extent on the desired degree of selectivity.

Recently, significant advances have been made in elaborating on the cancer cell surface proteome, cross-referenced with cancer type, and critical involvement in the disease process. Surface biomarkers such as CD151, CD19, CD22, CD30, MCAM, CEACAM5, MTDH (metadherin), FXYD3, GPC-1, and GPC-2 are being critically examined for their role as bona fide cancer cell surface biomarkers (Castillo et al. 2018; Orentas et al. 2012; Lee et al. 2018). While glypican (GPC-1 & -2) remains potentially controversial due to the presence of these surface proteins in healthy cells, immune validation for the potential candidates appears promising in roles as biomarkers and targets for site-directed therapy. Careful assessment of biomarkers is critical here to avoid damage to healthy cells bearing similar signatures.

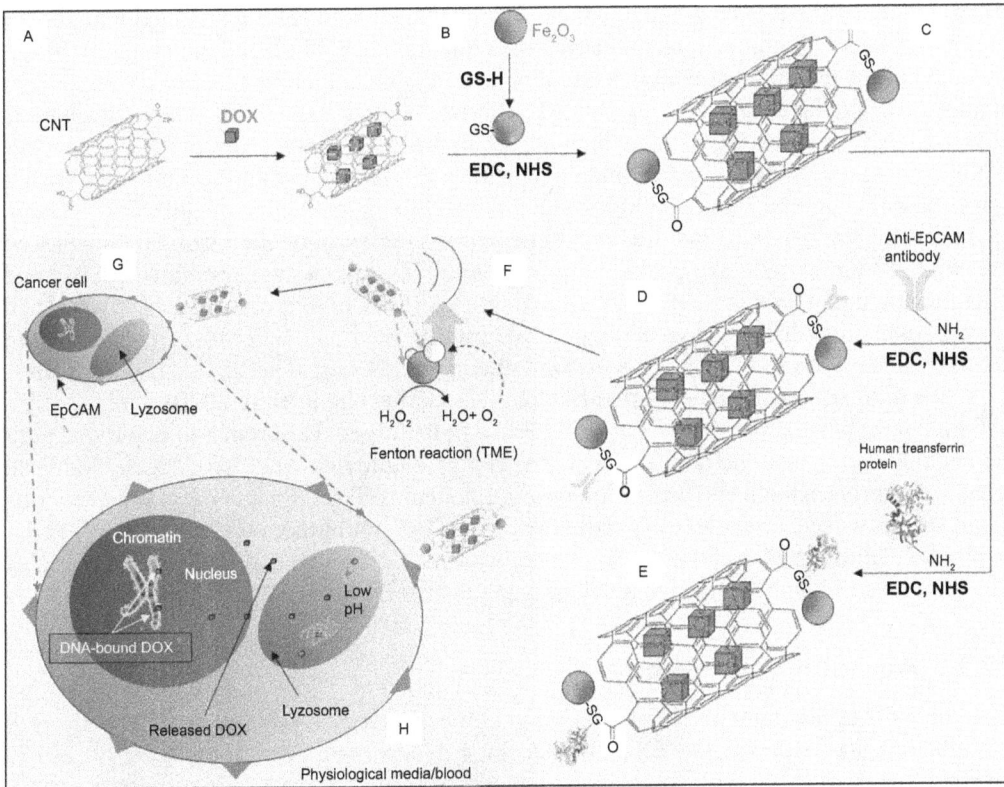

Figure 12.1 Iron-based nano-motor MNC for delivery of anticancer drug DOX to cancerous cells. A–E: Synthesis of CNT-based nanocarrier with conjugation of iron oxide gatekeeper to retain anticancer drug payload. Further conjugation to targeting moieties, anti-EpCAM, and transferrin ensures selectivity. F–H: Administration of CNT–iron oxide MNCs to CRC model cells. MNCs are propelled due to Fenton reaction-mediated buoyancy due to endogenously present hydrogen peroxide producing oxygen bubbles. Anti-EpCAM antibodies on MNCs bond with EpCAM proteins on the cancer cell surface, leading to MNC internalization into the cell. The MNC is processed in the lysosomes in acidic media in the presence of proteolytic enzymes to cleave the glutathione linkage and detach the iron oxide nanoparticle uncapping the CNT. DOX is subsequently released which enters the nucleus and binds to the DNA leading to cytotoxic results.

Source: (Andhari et al. 2020)

12.3 METAL/CARBON NANOPARTICLES IN MICROBOT/NANOBOT DESIGN

MNCs have been formulated with several metallic entities, the most commonly used metals being iron, gold, and silver. This chapter discusses iron oxide-based MNCs in various aspects; however, the most notable example of the use of iron oxide NPs involves its polyfunctional role in targeting, mobility, and ability to be guided to active site (Andhari et al. 2020; Wavhale et al. 2021). Gold-based MNCs have been used frequently for photo or magneto-thermal effects to manage tumors/cancers in a preclinical setting (White et al. 2017; Parchur et al. 2018; Emami et al. 2019). Functionalized gold MNCs have been used to some effect to deliver anticancer drugs to cancer cells. Chitosan and folic acid functionalized Au NPs are effective against colorectal cells (Lee, Kim, and Park 2022; Emami et al. 2019).

An excellent example of complex architectural MNCs targeting colorectal cancer cells employs a triple-layered gold nanorod (GNR) core coated with DOX. The GNR base is covered with mesoporous silica, which serves to hold the anticancer drug. Further, the GNR–mSiO2–DOX MNC was conjugated with pH-sensitive polyhistidine to aid with a cytoplasmic accumulation of the drug-loaded MNC. A high molecular weight PEG (Tocopherol polyethylene glycol 1000) was subsequently assembled onto the secondary surface of the MNC to aid in cellular uptake and retention. DOX release was instigated by pH and NIR-mediated photothermal (PT) activity resulting in very potent anticancer activity. Dual trigger responsive drug release made possible by this gold MNC could be considered an advanced candidate for precise personalized anticancer therapy. The MNC however suffers from a lack of selectivity, accentuated by the PEG coating (Jiang et al. 2020).

Gravitating more toward gold-based MNCs, a more targeted approach to colorectal anticancer therapy is demonstrated by a recent report employing gold nanocages (GNC) to trigger photothermal effects in addition to immunogenic cell targeting. GNCs encapsulating a specific anticancer drug, namely galunisertib (TGF-β inhibitor), labeled with anti-PDL1 monoclonal antibodies and macrophage membranes for camouflage was shown to be highly specific as well as highly effective in triggering a potent PT effect (Wang et al. 2021).

12.3.1 Autonomous motion

The value of contemporary MNCs is reflected in the transition of conventional nanomaterials toward micro-nano-robots or nanobots (Li, Esteban-Fernández de Ávila, et al. 2017). Currently, autonomous mobility remains an active area of research. Micro/nanobots are devices on the micro and nanoscale which are endowed with chemical or physical attributes that allow them to self-propel and self-administer. The limitation here is physics itself; at a very low Reynolds number, motion in aqueous media is free from inertial forces. Propulsion becomes a challenge at this scale (Wang 2021; Wang and Pumera 2015). Varied avenues have been proposed for the propulsion of micro/nanobots or motors. Synthesized micro/nanomotors can be self-propelled by chemical means or propelled by external energy, including magnetic/electrical fields, acoustics, and light. The propulsion of bio-hybrid micro/nanobots based on biological materials from nature has also been documented (Wang 2021; Wang and Pumera 2015; Zhang et al. 2013; Wang et al. 2006; Paxton, Sen, and Mallouk 2005).

Besides chemical fuels for locomotion, micro-nanobots (MNBs) are designed to exploit surface tension, self-electrophoresis, and bubble propellant mechanisms (Paxton, Sen, and Mallouk 2005). Endogenous fuels such as hydrogen peroxide have been well studied for effectiveness in propulsion as well as tumor penetration (Andhari et al. 2020; Wavhale et al. 2022, 2021). The Fenton reaction exploited here relies on disturbed redox balance in cancerous tissues.

12.3.2 Biocompatibility of nanomaterials

The choice of materials to be deployed for NPs synthesis can be driven by multiple factors including economics; however, the predominant determinant is usually biocompatibility, as the success of the nanomaterial is directly tied to the bio-acceptability of the material in all its presentations—shape, charge state, architecture, and mass. Substantial work has been carried out to show, however, that mass is not the primary aspect driving the interaction between the NP and the biological entity (Naahidi et al. 2013). Surface characteristics and chemistry dominate the nature of interactions here while also implying that the material contact and its influence can be controlled and customized.

As the focus of this chapter remains on MNCs, pertinent metal-centric NP/MNB insights will be presented here. Nearly all MNCs are capable of producing an inflammatory response, as described earlier, due to the intervention of the reticuloendothelial system (RES/MPS). The response is governed by MNC size, surface properties, composition, and even shape. Literature consensus reveals that the size of the MNC is indirectly proportional to its toxicity (Naahidi et al. 2013). For the reasons discussed earlier, MNCs are designed with gold and silver NPs, and occasionally with platinum, yielding relatively nontoxic MNCs. Reactive metals are frequently used, depending on their application such as copper NPs in thermally induced anticancer activity. Specifically, copper selenide nanocrystals are highly effective in treating colorectal cancer cells (Hessel et al. 2011). Metal particles while excellent contrast agents require immune-evasive strategies to be clinically useful, hence the low approval rate by the USFDA in the domain of MNCs. Currently, iron oxide remains one of the few NPs approved as a contrast agent and in iron deficiency treatment.

In addition to polymetal nanocomposites discussed in this chapter, combinations of polymetal MNCs and biomaterials like reduced graphene oxide be highly effective in dual strategy anticancer effect (Ahamed et al. 2022). The combination of molybdenum and zinc oxide NPs is shown to be superior to zinc oxide alone not only in anticancer activity but also in the protection of healthy cells to minimize the side effects of anticancer therapy. In addition, co-formulation routes to incorporate reduced graphene oxide resulted in enhanced anticancer effects while accentuating the protection of health cells. The incorporation of reduced graphene oxide is also efficacious with zinc oxide alone (Ahamed et al. 2020).

Along the lines of benign MNC behavior, more simplistic applications of magnetic MNCs include examples where PEGylated iron oxide NPs are loaded with anticancer agents for relatively selective drug delivery. One such example is a PEGylated magnetic NP loaded with a naturally sourced anticancer agent, gallic acid (Rosman et al. 2018). The resulting Fe_3O_4–PEG–GA MNC is highly effective in cancer cell lines including CRC-sourced HT-29 cell line while having significantly lower toxicity demonstrated in "normal" 3T3 cell line. This differential cytotoxic effect, if consistent in universal applications, may be considered to be a promising lead in the anticancer therapy of colorectal cancer to reduce the incidence of side effects.

12.3.3 Payload considerations

NPs are considered as having a limitation on their carrying/payload capacity. As described earlier, a reduction in size, especially in the nanometer scale brings a severe restriction in movement as well as drug-loading capacity (Wang 2021). This brick-wall issue, however, has a minor workaround, namely, in the architecture of the NPs/MNCs. As described elsewhere, various designs and architectures of NPs have varying effects on their biocompatibility, drug loading, surface modifications, and bioavailability (Mitchell et al. 2021). Detailed work has been reported on lipidic NPs and associated lipid-based MNCs showing relatively greater payload-carrying capacity than typical NPs (Sercombe et al. 2015; Fonseca-Santos,

Gremião, and Chorilli 2015; Fenton et al. 2018; Li et al. 2022). While lipidic MNCs are superior in this respect, liposome-like lipid NPs are known to have poor drug-loading issues, likely due to their ionizable nature.

Polymeric NPs suffer from aggregation issues and potential toxicity, rendering only a few NPs approval-worthy by the FDA. However, these are versatile and are available in diverse forms with highly tunable attributes. The modifications to surface properties can be used to improve the drug loading of polymeric MNCs (Patra et al. 2018). Various structural variants are observed in polymeric MNCs, such as nano-capsules, micelles, polymersomes, and so on, all of which are instrumental in protecting the cargo and self-stability. However, research is ongoing to improve drug-loading efficiencies.

Unlike lipidic and polymeric NPs, inorganic NPs/MNCs have unique properties that allow them to be used for stand-alone attributes, such as photothermal activity. Despite structural designs including nanocages, nanoshells, and nanorods, inorganic NPs are not known for their exceptional carrying capacity (Wang et al. 2019).

12.4 SYNTHETIC STRATEGIES FOR NANOCOMPOSITES

MNCs in a contemporary setting are enabled with mobility leading to their recognition as micro/nano-motors. As discussed earlier, MNBs are heavily studied for a range of applications including drug delivery, cell capture, smart functions, tunable mobility, large molecular cargo delivery, enhanced tissue or tumor penetration. Table 12.1 presented here depicts a few examples of metal nanocomposites with motor capabilities.

12.4.1 Synthetic methods

An interesting instance of metallo-MNBs is the tubular micro-rocket developed using chemical vapor deposition of various metals (Balasubramanian et al. 2011). A sacrificial photo-resist layer on a silicon wafer is coated sequentially with titanium, iron, silver, and platinum. Removal of the photo-resist layer rolls the multi-layer sheet into tubes that serve as micro-rockets. The micro-rockets are decorated with antibodies using N-hydroxysuccinimide (NHS) and N-(3-dimethylaminopropyl)-N'-ethylcarbodiimide hydrochloride (EDC) to form an amide bond.

Another example of MNCs is the fashioning of micro-rockets by entrapment of iron oxide nanoparticles inside carbon nanotubes (CNT) (Banerjee et al. 2015). Fe_3O_4 NPs were entrapped inside the CNT cavities by co-precipitating ferrous and ferric salts in the presence of ethylene glycol and ammonia. Thereafter, transferrin (Tf) was conjugated using the EDC coupling reaction as shown earlier, resulting in Tf–CNT–Fe_3O_4 micro-rockets. Here, Fe_3O_4 NPs act as nanocatalysts driving the forward-directed motion.

Table 1.1 describes a multi-metal, multi-phoretic propulsive, pH-responsive nano-motor for biomedical applications (Xing et al. 2020). The nanomotor was constructed by loading Au NPs-cys on aminated MSNs through a Cu^{2+} coordination bridge. Then the monolayer of MSNs-NH_2-Cu_2^+@AuNPs-cys was modified by Pt NPs via physical vapor deposition to obtain Janus (MSNs)-Pt@Au (JMPA) nanomotor. The JMPA nanomotors undergo three types of phoretic propulsion: self-diffusiophoresis, self-electrophoresis, and self-thermophoresis.

Simple synthetic methods are popular for green routes and economically viable approaches for anticancer MNCs. An interesting lead is the incorporation of phytic acid,

Table 12.1 MNCs with nanomotor features for advanced functionality

Carrier segment/ advanced materials	Motor segment/ substrate	Stimuli	Maximum speed/ MSD/De (medium of measurement)	Applications	References
MSNs	Pt, Au NPs	H_2O_2 and irradiation at 670 nm	D_e = 2.97 $\mu m^2/s$ (3% H_2O_2) D_e with and without irradiation = 1.26 $\mu m^2/s$ and 0.56 $\mu m^2/s$ (DI water)	Smart nanomachines responsive to biological stimuli with multiple and tunable propulsion modes	(Xing et al. 2020)
PSBMA	L-cysteine	H_2S (CBS-based cysteine desulfuration)	Speed = 5–6 $\mu m/s$ (cell suspension)	Inhibition of tumor growth	(Wan et al. 2021)
β-CD, CdS quantum dots	Ni and Pt tubular micro rockets	H_2O_2 and magnetic field	Speed = 165 $\mu m/s$ (aqueous solution of H_2O_2)	Supramolecular cargo platforms in the field of analytical chemistry, sensors, electronic devices, and biomedicine	(Muñoz, Urso, and Pumera 2022)
MWCNTs	Fe_3O_4 NPs (D = 16 nm)	H_2O_2 and magnetic field	Speed = 8.0 mm/s (serum), 2.3 mm/s (DMEM), 0.9 mm/s (PBS)	Targeted cellular delivery of drugs and tissue penetration	(Andhari et al. 2020)
Liposomes	MTB	Magnetic field	*Externally Modulated*	Delivery of a wide variety of therapeutic cargoes; to overcome persistent transport barriers	(Gwisai et al. 2022)
HPAM	L-arginine	H_2O_2 and NO (ROS)	$Speed_{avg}$ = 8 $\mu m/s$	Treatment of various diseases in different tissues including blood vessels and tumors	(Wan et al. 2019)
Janus hollow MSN	Urease and Fe	Urea and magnetic field	Speed = 5 body lengths/s (~10 $\mu m/s$) (aqueous solution of urea)	Transportation of cargoes of different sizes to target locations	(Ma et al. 2016)

Note: MSD: Mean squared displacement; D_e: diffusion co-efficient; PSBMA: Poly(sulfobetaine methacrylate); H2S: hydrogen sulfide; CBS: cystathionine β-synthase, CdS: Cadmium sulfide; β-CD: β-Cyclodextrin; HPAM: Hyperbranched polyamide; NO: nitric oxide; ROS: Reactive oxygen species, MTB: magneto-tactic bacteria; Fe: iron, Au NPs: Gold nanoparticles; Ni; Nickel; Pt: Platinum.

another naturally sourced anticancer agent in magnetic NPs to yield a combination of cytotoxic and cytostatic effects (Tan et al. 2018). The use of chitosan–iron oxide NPs enhances cellular uptake and magnetic targeting of tumor sites. Phytic acid has been demonstrated to induce a complex intrinsic pathway for apoptosis as well as a cell cycle arrest to induce a potent interplay of cytotoxic and cytostatic effects.

12.4.2 Drug choice and drug loading

MNCs and associated micro-nanomotors are unique vehicles for drug delivery, especially for anticancer therapeutics. While practical advantages for drug delivery using NPs or MNCs are numerous such as reduction of drug dose leading to lower side effects, site-specific delivery, site-triggered drug delivery, and so on, a very practical advantage of NPs is the delivery of poorly soluble drugs. A typical example is the anticancer drug paclitaxel (PTX), a natural plant alkaloid that is a highly effective cytostatic agent acting by interfering with microtubule assembly. While the drug is the first line of treatment in cancers such as breast cancer and advanced cancer of the ovary, it is not included in the anticancer agents commonly prescribed for colorectal cancer. The primary reason behind this poor adoption phenomenon is its significantly low aqueous solubility. Clinical data shows a 10^6-fold lower serum levels compared to the drug infusion rate, suggesting poor bioavailability and pharmacokinetics. PTX is subsequently observed to be bound to serum proteins to the extent of nearly 98% (Wishart et al. 2006; Weaver 2014). The drug efficacy is consequently challenged by physicochemical factors, despite numerous reports demonstrating the anticancer efficacy of PTX (Pham, Saelim, and Tiyaboonchai 2020; Ito et al. 2023; Phung et al. 2020; Yakati et al. 2022; Lv et al. 2017; Kennedy et al. 2000). Combination with other agents has shown additive effectiveness as well (Kennedy et al. 2000; Hu et al. 2021; Rao et al. 2020; Fujie et al. 2005; Wang et al. 2017). Consequently, the proper deployment of PTX has been hindered by inefficient drug delivery. Work by several groups has demonstrated the successful incorporation of PTX in multiple nano-formulations such as lipid particles, MNCs, PEGylated NPs, nanoemulsions, nanorods, and so on. The anticancer activity of PTX-loaded NPs in pre-clinical tests is efficacious and tunable in terms of drug loading.

Recent work by Lv et al. (2017) has shown the effectiveness of lower doses of PTX in demonstrating cytostatic activity in colorectal cancer cells. Nano-carriers are particularly effective in controlling drug release at the cellular or tumor level. The effectiveness of metallic nanocomposites in selective targeting of colorectal cancer cells and the use of metallic properties in self-propulsion in addition to drug-release control measures using amide-linkage is exemplified in a recent report showcasing the utility of iron oxide MNCs (Andhari et al. 2020). The relatively low drug-loading capacity is compensated by the additional traits of MNCs that enhance their cytotoxic activity, namely the ability to penetrate tumor cells and the ability to release the drug in a manner consistent with endocytosis and lysosomal processing.

The drug-loading efficiencies of NPs prepared by various methods vary and are also determined by the chemistry and geometry of the NPs. In the case of NPs that are designed to deliver PTX, for instance, the loading efficiencies vary in the range of 24–87%, covering a diverse portfolio of polymeric micelles, liposomes, nanocrystals, solid lipid particles, nanoemulsions, and beta-cyclodextrin NPs (Le et al. 2021; Li, Huang, et al. 2017; Biswas et al. 2013; Bilensoy et al. 2008; Pant et al. 2021; Dhaundiyal et al. 2016). Due to the hydrophobic nature of PTX, these NP formulations were preferred over MNCs. PTX loading in iron oxide based-MNCs has been reported to be between 30% (Hua et al. 2010) and 85% (Ahmed et al. 2017). Ironically, higher PTX loading in the iron oxide particles led to poor activity, likely due to steric hindrance. A lowered drug loading (18–25%) led to a virtual doubling of activity as assessed by tubulin-interaction assays (Hua et al. 2010). Drug loading of 5-fluorouracil on iron oxide MNCs is lower in comparison (~13%) (Dabaghi, Quaas, and Hilger 2020; Dabaghi et al. 2021). Despite the lower drug payload, the combination therapy of drug exposure and magnetic hyperthermia resulted in decreased tumor growth. Other instances of optimal drug loading include electrostatically loaded oxaliplatin on biomimetic iron oxide MNCs. Oxaliplatin payload in this report reached a maximum of 41% offering safe and reliable anticancer therapy for CRC (Jabalera et al. 2019).

Simplistic payload improvements are attained by the usage of polymeric conjugates on MNCs. In combination with magnetic properties, for instance, the incorporation of natural biomaterials with drug-loading capacity can be used effectively in this respect. Magnetic traits exploited by iron oxide MNCs as discussed extensively in this chapter are a popular focal point for many anticancer NPs. Iron oxide NPs can be combined with multiple substrates to enable their magnetization for a synergistic effect. Natural product-derived cellular nano-whiskers or nanocrystalline needles have been used as substrates to conjugate iron oxide nanoparticles to generate magnetic nano-whiskers. These magnetic nano-whiskers are further coated with chitosan to enable greater cellular interaction as well as to serve as drug carriers for 5FU (Yusefi et al. 2023). The resulting MNC can be magnetically guided to tumor sites and help in reducing the nonspecific drug activity of 5FU, a potent and commonly used antimetabolite anticancer drug molecule. The nano-form cellulose further provided properties of pH-dependent swelling and related drug release.

Interestingly, drug-free MNCs are also being explored, for instance, with the use of polymetal combinations. More rare metals like ytterbium (Yb) have been exploited in CRC anticancer therapy development (Kumar et al. 2019). Core-shell NPs made from sodium-Yb composites with the ability to utilize fluorescence resonance energy transfer phenomenon to generate PT effect as well as spontaneous oxidative stress have been demonstrated in combination with enzymatically controlled activation. The MNC, conjugated with a natural polymer, guar gum, is protected by the same until intestinal microbiota with specific enzymes to digest guar gum is encountered, resulting in activation of the MNCs, subject to PT effect and oxidative stress for synergistic anticancer effect. The long-term safety and efficacy of such a unique dosing strategy remain to be evaluated.

Aside from the common usage of gold nanocomposites in devising colorectal anticancer therapy, other metals have proven their utility, such as tin oxide (Wei et al. 2021) along with alginate salt and PEG to deliver isothiocyanate to modulate oxidative stress and inflammation. The tin core was coated with alginate followed by PEG to be finally encapsulated by the natural product which was shown to significantly reduce inflammatory biomarkers.

Similarly, copper NPs have been used to have remarkable effects in targeting cancer cells in TMEs. A smart self-assembling copper-based nanocrystal system has been demonstrated to be triggered within certain TME zones that produce H_2S (Chang et al. 2020). This intelligently designed system relies on copper oxide nanoparticles in conjunction with calcium carbonate to exploit the pH gradient to generate a "time-bomb" that is formed in situ as the calcium carbonate is depleted in the acid media to allow the copper-based system to be sulfurated by TME-endogenous H2S to generate Cu31S16 nanocrystals that are active in the PT mode under near-infrared radiation. The potent nanocrystals are doubly efficacious due to their effect in inducing a vaccine-like immune response.

12.4.3 Toxicological studies

Nanomaterial studies are typically accentuated with particle size analysis, morphological studies, and the determination of zeta potential. Toxicological studies are necessary to evaluate biocompatibility and in the case of anticancer drugs, cytotoxicity of MNCs and drug

Table 12.2 List of common colorectal carcinoma cell lines (non-exhaustive) segregated by morphological distinction and origin

Undifferentiated cells	Primary tumor site	Reference	Colon-like cells	Primary tumor site	Reference
CaCo2	Colon	Yakati et al. 2022	CL-34	Ascending colon	Berg et al. 2017
CL-11	Descending colon	Berg et al. 2017	CL-40	Ascending colon	Berg et al. 2017
Col15	Ascending colon	Berg et al. 2017	Colo205	Colon	Berg et al. 2017
Colo320	Sigmoid colon	Berg et al. 2017	EB	Colonic carcinoma	Berg et al. 2017
Colo678	Colon	Berg et al. 2017	FRI	Colonic carcinoma	Berg et al. 2017
DLD-1	Colon	Fujie et al. 2005	HCC2998	Colon	Berg et al. 2017
HCT116	Ascending colon	Andhari et al. 2020,	HT29	Colon	Jabalera et al. 2019
HCT15	Colon	Jabalera et al. 2019	IS3	Ascending colon	Berg et al. 2017
IS1	Ascending colon	Berg et al. 2017	KM12	Colon	Berg et al. 2017
LoVo	Colon	Wang et al. 2017	LS1034	Caecum	Berg et al. 2017
RKO	Colon	Yakati et al. 2022	LS174T	Colon	Berg et al. 2017
SW48	Colon	Berg et al. 2017	NCI-H508	Caecum	Berg et al. 2017
SW480	Descending colon	Jabalera et al. 2019	SW1116	Colon	Berg et al. 2017
SW620	Descending colon	Berg et al. 2017	SW1463	Rectum	Berg et al. 2017
SW837	Rectum	Berg et al. 2017	SW403	Colon	Berg et al. 2017
TC71	Sigmoid colon	Berg et al. 2017	V9P	Rectum	Berg et al. 2017
			WiDr	Colon	Berg et al. 2017

Note: Highlighted cell lineages indicate more commonly used CRC cell models.

conjugates. Most toxicology assays are performed in cultures of relevant cell lines, such as HCT-116, Caco-2, and HT-29 for colorectal cancer (Table 12.2).

Colon-like cells were noted to bear higher expression levels of carcinoembryonic antigen (CEA) while undifferentiated cells bore hallmarks of epithelial–mesenchymal (EMT) transition and upregulation of TGFB-1 and 2 (Berg et al. 2017). These differences in cell lines need to be considered when choosing the appropriate cell line for drug toxicological studies. Cell viability assays are routinely carried out with MTT or CCK8 which rely on the formation of formazan dye in metabolically active cells (Yang et al. 2020). CCK8 is reportedly a less-toxic substrate compared to MTT or XTT and may thus be preferred to avoid false additive results. Simpler viability tests are often performed with Trypan or Alamar blue stains. Cell proliferation assays are performed to assess the cytostatic/cytotoxic effect of experimental agents in cell cultures for 72 hours (Garcia-Pinel et al. 2020).

As a general outcome of well-designed MNCs, toxicological and proliferation assays depict a synergistic effect of nanomaterials with anticancer drugs. For instance, in the study utilizing biomimetic liposomes loaded with oxaliplatin described earlier (Jabalera et al. 2019), cytotoxicity is noted in all drug+vehicle trials in multiple CRC cell lines while drug-loaded MNC yields greater cytotoxicity that allows for a reduction in drug loading to reach IC_{50}-equivalent to oxaliplatin alone. To magnetic MNCs loaded with anticancer agents, the magnetic fluid hyperthermia (MFH) phenomenon is superior in cytotoxic effect, when compared to drug treatment alone (Hardiansyah et al. 2014; Palzer et al. 2021; Rodríguez-Luccioni et al. 2011). Interestingly, toxicological assays for biomimetic magnetic liposomes yield negative results implying a high degree of biocompatibility (Hardiansyah et al. 2014).

12.5 COLORECTAL CANCER MODELS

The biological evaluation of MNCs and micro-nanomotors is dependent on multiple factors such as the nature of cancer tissue, drug release mechanism, physicochemical traits of drugs, test media, and so on. By far, preclinical testing is dependent mostly on the appropriate selection of test models. Depending on the precise requirement, choices can be made among simple cell cultures, 3D cell cultures, tumor spheroids/organoids, or animal *in vivo* studies.

12.5.1 2d cell culture/monolayer models

The principal biological activity in preclinical screening is reliant on conventional (2D cell culture) cell line testing. As described earlier, anticancer activity in relevant tissue-origin cell culture is assayed to demonstrate cytotoxic/cytostatic influence. The nature of the cell line may frequently determine the outcome of the study. As mentioned earlier, care needs to be taken to select appropriate cell lines for demonstrative studies as illustrated in the report exploring the effect of phytic acid MNC in CRC cell lines HCT-116 and HT29 (Tan, Norhaizan, and Chan 2018). The IC_{50} observed in HCT116 cells, in the temporal study, is 30–50% higher as compared to the IC_{50} calculated for the HT29 cells. A practical utility of the 2D cell platform is to determine IC_{50} values to enable dose adjustment during preclinical trials.

12.5.2 Tumor spheroid models/3d cell culture

Tumor spheroids can be raised by multiple methods such as culturing on agarose substrate, cell-repellent u-bottomed well plates (Dey et al. 2021), or culturing in the hanging drop (Andhari et al. 2020). Spheroids offer a tumor surrogate spanning up to 1 mm or more in diameter which can help to model the behavior of MNCs in an *in vitro* manner. Prepared from clonal cells, the spheroid represents a cultured tumor as an aggregated mass of cells but devoid of tissue micro-vasculature or stromal mass. Recent advances in tumor spheroid research have enabled the vascularization of spheroids which renders *in vitro* modeling of tumors feasible (Dey et al. 2021). Spheroids enable a more sophisticated analysis of MNCs and nanomaterials in general, for instance, by allowing penetration studies of self-propelled or exogenously powered micro-nanomotors. Analogous to proliferation analysis, tumor integrity studies allow the exploration of anticancer agents by looking at tumor growth/collapse, shedding of cells, expansion, etc. 3D analysis techniques such as confocal microscopy offer a compelling avenue to study tumor-targeted pharmacodynamics.

3D cell cultures are enabled by substrate architecture. Tissue engineering advances have provided access to unique 3D scaffolds such as proteins, gels, fibers, 3D surfaces, and so on (Song and Munn 2011; Bobade et al. 2020). Unlike spheroids, 3D cell cultures are not aggregate cultures but a clone of cells functionally similar to 2D cultures in a manner of cellular attachment and dependence on substrate properties such as stiffness, porosity, hydrophobic nature, and so on. Each culture type offers unique advantages that can be exploited depending on the nature of the study.

12.5.3 Animal models/xenografts

Xenograft cultures of CRC in rodent models are frequently employed to assess anticancer activity in preclinical studies. Typically, xenografting involves the introduction of cancerous cells into a rodent host by employing additional stabilizing additives. For instance, the use of matrigel™ to house HT29 cells is being injected in an immune-deficient mouse (athymic nude mouse) (Dabaghi, Quaas, and Hilger 2020). The methodology provides an opportunity to study human cancerous cells in a rodent model, typically in a tumor form. Tumor volume is measured as an index of drug activity. The xenograft model allows the study of drug-MNCs in an *in vivo* setting that involves systemic circulation, bioavailability issues, drug metabolism, and excretion dimensions that mimic real-world challenges that are vital to assess drug performance. CRC xenografts have even been reported in zebrafish models, where screening of anticancer drug candidates was shown to affect tumor volume (Maradonna et al. 2022).

12.5.4 Lab-on-a-chip

Perhaps, the greatest innovation in recent times is the development of the lab-on-a-chip platform for *in vitro* evaluation of disease models. Cancer models raised in the L-o-C platform have been instrumental in recapitulating *in vivo* tumor microenvironments (Liu et al. 2021). The unique microfluidic platform allows for maintaining circulation-like media turnover and further helps in the assessment of downstream fluidic outflow to study physiological changes, biomarkers, drug changes, etc. Specific L-o-C models mimicking CRC have shown exceptional promise in demonstrating the similarity of genetic, physiological, and heterogeneity patterns with native *in vivo* CRC. Given the poor drug benefit analysis when studied in simple 2D cell cultures, the L-o-C platform offers significant similitude to provide near-*in vivo* responses to drug treatments (Komen et al. 2021; Strelez et al. 2021; Carvalho et al. 2019; Hachey et al. 2021).

12.6 STRATEGIES OF COLORECTAL CANCER CELL/TUMOR TARGETING

Based on targeted therapeutic solutions such as clinical antibodies (Xie, Chen, and Fang 2020) to CRC-selective cell biomarkers, strategies for CRC cell and tumor targeting by MNCs have evolved to exploit unique physiological or cellular features of CRC cells and tumor clusters. CRC-selective cell surface proteins have been explored to serve as targets for MNCs and NPs in general, for instance, epithelial cell adhesion molecules (EpCAM). Expressed in profusion by most cancerous cell types, anti-EpCAM antibodies are highly successful in selectively attaching to cancerous cells. This trait has been exploited by researchers to design CRC-targeting magnetic MNCs (Andhari et al. 2020; Wavhale et al. 2021) as well as nano-devices to capture circulating tumor cells (Khandare et al. 2020; Khandare et al. 2022).

Transferrin receptors (TfR) are routinely overexpressed in CRC cell lines, as well as in some other cancer cell types. Due to its high affinity for transferrin (Tf), TfR is ideally suited for targeting by nanocarriers (Wavhale et al. 2022; Wavhale et al. 2021). Additional research also demonstrated the ability of Tf-mediated micro-nano-devices to capture circulating cancer cells (Banerjee et al. 2015; Banerjee et al. 2013; Hazra et al. 2020).

Other markers such as FAT1 also serve as targeting moieties for MNCs. Antibody-coated gold MNCs against FAT1 are highly effective for tumor targeting (Grifantini et al. 2018; Fan et al. 2015). Carcinoembryonic antigen (CEA) is currently under active exploration for targetability by nanomaterials (Sousa, Oliveira, and Sarmento 2019; Pereira et al. 2018) Various other cell surface markers such as folate receptors described earlier, CD44, and so on are under investigation for MNC-mediated tumor targeting (Patel et al. 2021; Kesharwani et al. 2022).

Relatively untargeted approaches abound in MNC literature as well. Creative approaches toward CRC anticancer therapeutics involve the use of unusual metal-mineral bases in MNC form, such as montmorillonite with mucoadhesive but nontoxic traits in delivering therapeutic agents to cancer cells. The combination of montmorillonite with serum albumin significantly enhances mucosal barrier penetration which is exploited to deliver a detergent, saponin, to cancer cells (Akbal et al. 2018). Saponin is demonstrated to invade cancer sites and assert cytotoxicity in a dose-dependent manner, possibly by disruption of membrane integrity and enzyme functions.

12.7 CONCLUSIONS AND PATH FORWARD

The greater body of literature on metal nanocomposites in colorectal cancer therapeutics shows significant diversity in the composition and design of MNCs. Furthermore, MNCs benefit from combinatorial advantages of their metal nucleus such as magneto- or photothermal properties as well as radiocontrast to inherent cytotoxic properties. The cumulative activity of MNCs allows greater therapeutic benefit that any single MNC component alone, as described earlier. Synthetic methods for MNC production allow for the coupling of targeting agents and additional nanomaterials for enhanced cargo-carrying or propulsion support. The diversity of targeting moieties currently employed and the flexibility of decorating MNCs with any targeting molecule renders MNCs highly customizable. In the context of colorectal cancer, recognized as one of the deadlier cancers, MNCs that allow for highly selective drug delivery would be crucial to enhance the antitumor effect and reduce side effects that impact patient compliance directly. The ability to adjust payload capacity by alteration of surface chemistry or inclusion of nanocarriers helps to fine-tune the drug loading of the target tissue, such as the intra-tumoral matrix. MNCs can be designed to selectively target not just tissues or cell types but also subcellular compartments. MNCs can be designed to take advantage of compartmental internalization and utilize the lysosomal or endosomal acidic environment to process internalized MNCs to either release their payload for cytotoxic/cytostatic effect or exact inherent cytotoxic activity such as magnetically induced thermal effect.

The limited efficacy of conventional drug treatments such as 5FU in combination with irinotecan, with or without anti-growth factor (EGFR/VEGF) antibodies or immune-checkpoint inhibitors (PD-1/PD-L1) and their adverse effects necessitate an alternative approach to colorectal cancer care. Metastasized CRC is significantly more likely to reduce survival outcomes, indicating an urgent need for targeted, reliable, and efficacious therapeutic options. MNCs as described in this chapter present an attractive and economically feasible avenue to address this need and slow the progress of disease in a manner consistent with the maintenance of quality of life. The uptake of MNCs by macrophages and cells within the reticuloendothelial system allows clearance of MNCs from circulation and their subsequent removal from tissues, thereby circumventing the potential accumulation of metallic nanoparticles in the body. Currently, lipid nanoparticles and liposome formulations are approved by FDA for clinical use. While iron oxide nano-formulations are considered a safer option by FDA, there is a definitive lack of MNC-centric clinical trials to explore their utility in CRC. A revised interpretation of the effectiveness of MNCs in CRC is warranted, and while xenograft studies abound in literature, exploration of MNCs in clinical trials may reveal their practical utility in CRC.

REFERENCES

Abdelghani, Meher Ben, Christophe Borg, Louis-Marie Dourthe, Gael Deplanque, Julien Taïeb, Jean-Philippe Metges, Philippe Laplaige, Veronique Lotz, Amele Amrate, and Gérard Lledo. 2015. "Aflibercept in combination with FOLFIRI for the second-line treatment of patients with metastatic colorectal cancer: First interim safety data from AFEQT trial." *Journal of Clinical Oncology* no. 33 (3_suppl):661. doi: 10.1200/jco.2015.33.3_suppl.661.

Ahamed Maqusood, Akhtar Mohd Javed, Khan M.A. Majeed, Alhadlaq Hisham A. 2020. "Enhanced anticancer performance of eco-friendly-prepared Mo-ZnO/RGO nanocomposites: Role of oxidative stress and apoptosis." *ACS Omega* no. 7:7103–7115.

Ahamed Maqusood, Akhtar Mohd Javed, Khan M.A. Majeed, Alhadlaq Hisham A. 2022. "Facile green synthesis of ZnO-RGO nanocomposites with enhanced anticancer efficacy." *Methods* no. 199:28–36.

Ahmed, M. S. U., A. B. Salam, C. Yates, K. Willian, J. Jaynes, T. Turner, and M. O. Abdalla. 2017. "Double-receptor-targeting multifunctional iron oxide nanoparticles drug delivery system for the treatment and imaging of prostate cancer." *Int J Nanomedicine* no. 12:6973–6984.

Akbal Oznur, Vural Tayfun, Malekghasemi Soheil, Bozdoğan Betül and Denkbaş Emir Baki. 2018. "Saponin loaded montmorillonite-human serum albumin nanocomposites as drug delivery system in colorectal cancer therapy." *Applied Clay Science* no. 166: 214–222.

Andhari, Saloni S., Ravindra D. Wavhale, Kshama D. Dhobale, Bhausaheb V. Tawade, Govind P. Chate, Yuvraj N. Patil, Jayant J. Khandare, and Shashwat S. Banerjee. 2020. "Self-propelling targeted magneto-nanobots for deep tumor penetration and pH-responsive intracellular drug delivery." *Scientific Reports* no. 10 (1):4703. doi: 10.1038/s41598-020-61586-y.

Angelopoulou, Athina, Argiris Kolokithas-Ntoukas, Christos Fytas, and Konstantinos Avgoustakis. 2019. "Folic acid-functionalized, condensed magnetic nanoparticles for targeted delivery of doxorubicin to tumor cancer cells overexpressing the folate receptor." *ACS Omega* no. 4 (26):22214–22227. doi: 10.1021/acsomega.9b03594.

Antoniotti, Carlotta, Daniele Rossini, Filippo Pietrantonio, Aurélie Catteau, Lisa Salvatore, Sara Lonardi, Isabelle Boquet, Stefano Tamberi, Federica Marmorino, Roberto Moretto, Margherita Ambrosini, Emiliano Tamburini, Giampaolo Tortora, Alessandro Passardi, Francesca Bergamo, Alboukadel Kassambara, Thomas Sbarrato, Federica Morano, Giuliana Ritorto, Beatrice Borelli, Alessandra Boccaccino, Veronica Conca, Mirella Giordano, Clara Ugolini, Jacques Fieschi, Alexia Papadopulos, Clémentine Massoué, Giuseppe Aprile, Lorenzo Antonuzzo, Fabio Gelsomino, Erika Martinelli, Nicoletta Pella, Gianluca Masi, Gabriella Fontanini, Luca Boni, Jérôme Galon, and Chiara Cremolini. 2022. "Upfront FOLFOXIRI plus bevacizumab with or without atezolizumab in the treatment of patients with metastatic colorectal cancer (Atezo-TRIBE): A multicentre, open-label, randomised, controlled, phase 2 trial." *The Lancet Oncology* no. 23 (7):876–887. doi: 10.1016/s1470-2045(22)00274-1.

Apmann, K., R. Fulmer, A. Soto, and S. Vafaei. 2021. "Thermal conductivity and viscosity: Review and optimization of effects of nanoparticles." *Materials* no. 14 (5).

Balasubramanian, S., D. Kagan, C. M. Hu, S. Campuzano, M. J. Lobo-Castañon, N. Lim, D. Y. Kang, M. Zimmerman, L. Zhang, and J. Wang. 2011. "Micromachine-enabled capture and isolation of cancer cells in complex media." *Angew Chem Int Ed Engl* no. 50 (18):4161–4164.

Banerjee, Shashwat S., Archana Jalota-Badhwar, Sneha D. Satavalekar, Sujit G. Bhansali, Naval D. Aher, Russel R. Mascarenhas, Debjani Paul, Somesh Sharma, and Jayant J. Khandare. 2013. "Transferrin-mediated rapid targeting, isolation, and detection of circulating tumor cells by multifunctional magneto-dendritic nanosystem." *Advanced Healthcare Materials* no. 2 (6):800–805. https://doi.org/10.1002/adhm.201200164.

Banerjee, Shashwat S., Archana Jalota-Badhwar, Khushbu R. Zope, Kiran J. Todkar, Russel R. Mascarenhas, Govind P. Chate, Ganesh V. Khutale, Atul Bharde, Marcelo Calderon, and Jayant J. Khandare. 2015. "Self-propelled carbon nanotube based microrockets for rapid capture and isolation of circulating tumor cells." *Nanoscale* no. 7 (19):8684–8688. doi: 10.1039/c5nr01797a.

Berg, Kaja C. G., Peter W. Eide, Ina A. Eilertsen, Bjarne Johannessen, Jarle Bruun, Stine A. Danielsen, Merete Bjørnslett, Leonardo A. Meza-Zepeda, Mette Eknæs, Guro E. Lind, Ola Myklebost, Rolf I. Skotheim, Anita Sveen, and Ragnhild A. Lothe. 2017. "Multi-omics of 34 colorectal cancer cell lines—a resource for biomedical studies." *Molecular Cancer* no. 16 (1):116. doi: 10.1186/s12943-017-0691-y.

Bilensoy, Erem, Oya Gürkaynak, Mevlut Ertan, Murat Şen, and A. Atilla Hıncal. 2008. "Development of nonsurfactant cyclodextrin nanoparticles loaded with anticancer drug paclitaxel." *Journal of Pharmaceutical Sciences* no. 97 (4):1519–1529. https://doi.org/10.1002/jps.21111.

Biswas, Swati, Onkar S. Vaze, Sara Movassaghian, and Vladimir P. Torchilin. 2013. "Polymeric Micelles for the delivery of poorly soluble drugs." In Dennis Douroumis and Alfred Fahr (eds.), *Drug Delivery Strategies for Poorly Water-Soluble Drugs*, 411–476. West Sussex, United Kingdom: John Wiley & Sons.

Bobade, C. D., S. Nandi, N. R. Kale, S. S. Banerjee, Y. N. Patil, and J. J. Khandare. 2020. "Cellular regeneration and proliferation on polymeric 3D inverse-space substrates and the effect of doxorubicin." *Nanoscale Adv* no. 2 (6):2315–2325.

Bose, Sminu, Cissy Zhang, and Anne Le. 2021. "Glucose metabolism in cancer: The Warburg effect and beyond." In *The Heterogeneity of Cancer Metabolism*, edited by Anne Le, 3–15. Cham: Springer International Publishing.

Carvalho, M. R., D. Barata, L. M. Teixeira, S. Giselbrecht, R. L. Reis, J. M. Oliveira, R. Truckenmüller, and P. Habibovic. 2019. "Colorectal tumor-on-a-chip system: A 3D tool for precision onco-nanomedicine." *Sci Adv* no. 5 (5).

Cassim, Shamir, Milica Vučetić, Maša Ždralević, and Jacques Pouysségur. 2020. "Warburg and beyond: The power of mitochondrial metabolism to collaborate or replace fermentative glycolysis in cancer." *Cancers* no. 12 (5):1119.

Castillo, J., V. Bernard, F. A. San Lucas, K. Allenson, M. Capello, D. U. Kim, P. Gascoyne, F. C. Mulu, B. M. Stephens, J. Huang, H. Wang, A. A. Momin, R. O. Jacamo, M. Katz, R. Wolff, M. Javle, G. Varadhachary, I. I. Wistuba, S. Hanash, A. Maitra, and H. Alvarez. 2018. "Surfaceome profiling enables isolation of cancer-specific exosomal cargo in liquid biopsies from pancreatic cancer patients." *Annals of Oncology* no. 29 (1):223–229. doi: 10.1093/annonc/mdx542.

Chang Mengyu, Hou Zhiyao, Jin Dayong, Zhou Jiajia, Wang Man, Wang Meifang, Shu Mengmeng, Ding Binbin, Li Chunxia and Lin Jun. 2020. "Colorectal tumor microenvironment-activated bio-decomposable and metabolizable $Cu_2O@CaCO_3$ nanocomposites for synergistic oncotherapy". *Advanced Materials*: 2004647.

Dabaghi, M., R. Quaas, and I. Hilger. 2020. "The treatment of heterotopic human colon xenograft tumors in mice with 5-fluorouracil attached to magnetic nanoparticles in combination with magnetic hyperthermia is more efficient than either therapy alone." *Cancers* no. 12 (9).

Dabaghi, M., S. M. M. Rasa, E. Cirri, A. Ori, F. Neri, R. Quaas, and I. Hilger. 2021. "Iron oxide nanoparticles carrying 5-fluorouracil in combination with magnetic hyperthermia induce thrombogenic collagen fibers, cellular stress, and immune responses in heterotopic human colon cancer in mice." *Pharmaceutics* no. 13 (10).

Daniels, Tracy R., Tracie Delgado, Gustavo Helguera, and Manuel L. Penichet. 2006. "The transferrin receptor part II: Targeted delivery of therapeutic agents into cancer cells." *Clinical Immunology* no. 121 (2):159–176. https://doi.org/10.1016/j.clim.2006.06.006.

Delaunoit, Thierry, Richard M. Goldberg, Daniel J. Sargent, Roscoe F. Morton, Charles S. Fuchs, Brian P. Findlay, Sachdev P. Thomas, Muhammad Salim, Paul L. Schaefer, Philip J. Stella, Erin Green, and James A. Mailliard. 2004. "Mortality associated with daily bolus 5-fluorouracil/leucovorin administered in combination with either irinotecan or oxaliplatin." *Cancer* no. 101 (10):2170–2176. https://doi.org/10.1002/cncr.20594.

Dey, M., B. Ayan, M. Yurieva, D. Unutmaz, and I. T. Ozbolat. 2021. "Studying tumor angiogenesis and cancer invasion in a three-dimensional vascularized breast cancer micro-environment." *Adv Biol* no. 5 (7):15.

Dhaundiyal, Ankit, Sunil K. Jena, Sanjaya K. Samal, Bhavin Sonvane, Mahesh Chand, and Abhay T. Sangamwar. 2016. "Alpha-lipoic acid—stearylamine conjugate-based solid lipid nanoparticles for tamoxifen delivery: Formulation, optimization, in-vivo pharmacokinetic and hepatotoxicity study." *Journal of Pharmacy and Pharmacology* no. 68 (12):1535–1550. https://doi.org/10.1111/jphp.12644.

Diaz, Luis A., Jr., Kai-Keen Shiu, Tae-Won Kim, Benny Vittrup Jensen, Lars Henrik Jensen, Cornelis Punt, Denis Smith, Rocio Garcia-Carbonero, Manuel Benavides, Peter Gibbs, Christelle de la Fourchardiere, Fernando Rivera, Elena Elez, Dung T. Le, Takayuki Yoshino, Wen Yan Zhong, David Fogelman, Patricia Marinello, and Thierry Andre. 2022. "Pembrolizumab versus chemotherapy for microsatellite instability-high or mismatch repair-deficient metastatic colorectal cancer (KEYNOTE-177): Final analysis of a randomised, open-label, phase 3 study." *The Lancet Oncology* no. 23 (5):659–670. doi: 10.1016/s1470-2045(22)00197-8.

Division of Cancer Prevention and Control, CDC. 2023. *U.S. Cancer Statistics Colorectal Cancer Stat Bite.* US Department of Health and Human Services, June 6, 2022 [cited 21 March 2023]. Available from www.cdc.gov/cancer/uscs/about/stat-bites/stat-bite-colorectal.htm.

Elez, Elena, and Iosune Baraibar. 2022. "Immunotherapy in colorectal cancer: An unmet need deserving of change." *The Lancet Oncology* no. 23 (7):830–831. doi: 10.1016/s1470-2045(22)00324-2.

Emami, F., A. Banstola, A. Vatanara, S. Lee, J. O. Kim, J. H. Jeong, and S. Yook. 2019. "Doxorubicin and anti-PD-L1 antibody conjugated gold nanoparticles for colorectal cancer photochemotherapy." *Mol Pharm* no. 16 (3):1184–1199.

Fan, L., S. Campagnoli, H. Wu, A. Grandi, M. Parri, E. De Camilli, G. Grandi, G. Viale, P. Pileri, R. Grifantini, C. Song, and B. Jin. 2015. "Negatively charged AuNP modified with monoclonal antibody against novel tumor antigen FAT1 for tumor targeting." *J Exp Clin Cancer Res* no. 34 (1):15–214.

Fenton, Owen S., Katy N. Olafson, Padmini S. Pillai, Michael J. Mitchell, and Robert Langer. 2018. "Advances in biomaterials for drug delivery." *Advanced Materials* no. 30 (29):1705328.

Ferrara, Napoleone, Kenneth J. Hillan, Hans-Peter Gerber, and William Novotny. 2004. "Discovery and development of bevacizumab, an anti-VEGF antibody for treating cancer." *Nature Reviews Drug Discovery* no. 3 (5):391–400. doi: 10.1038/nrd1381.

Fonseca-Santos, Bruno, Maria Palmira Daflon Gremião, and Marlus Chorilli. 2015. "Nanotechnology-based drug delivery systems for the treatment of Alzheimer's disease." *International Journal of nanomedicine* no. 10:4981.

Fujie, Yujiro, Hirofumi Yamamoto, Chew Yee Ngan, Akimitsu Takagi, Taro Hayashi, Rei Suzuki, Koji Ezumi, Ichiro Takemasa, Masataka Ikeda, Mitsugu Sekimoto, Nariaki Matsuura, and Morito Monden. 2005. "Oxaliplatin, a potent inhibitor of survivin, enhances paclitaxel-induced apoptosis and mitotic catastrophe in colon cancer cells." *Japanese Journal of Clinical Oncology* no. 35 (8):453–463. doi: 10.1093/jjco/hyi130.

Garcia-Pinel, B., Y. Jabalera, R. Ortiz, L. Cabeza, C. Jimenez-Lopez, C. Melguizo, and J. Prados. 2020. "Biomimetic magnetoliposomes as oxaliplatin nanocarriers: In vitro study for potential application in colon cancer." *Pharmaceutics* no. 12 (6).

Grifantini, R., M. Taranta, L. Gherardini, I. Naldi, M. Parri, A. Grandi, A. Giannetti, S. Tombelli, G. Lucarini, L. Ricotti, S. Campagnoli, E. De Camilli, G. Pelosi, F. Baldini, A. Menciassi, G. Viale, P. Pileri, and C. Cinti. 2018. "Magnetically driven drug delivery systems improving targeted immunotherapy for colon-rectal cancer." *J Control Release* no. 280:76–86.

Gwisai, T., N. Mirkhani, M. G. Christiansen, T. T. Nguyen, V. Ling, and S. Schuerle. 2022. "Magnetic torque-driven living microrobots for increased tumor infiltration." *Sci Robot* no. 7 (71):26.

Hachey, Stephanie J., Silva Movsesyan, Quy H. Nguyen, Giselle Burton-Sojo, Ani Tankazyan, Jie Wu, Tuyen Hoang, Da Zhao, Shuxiong Wang, Michaela M. Hatch, Elizabeth Celaya, Samantha Gomez, George T. Chen, Ryan T. Davis, Kevin Nee, Nicholas Pervolarakis, Devon A. Lawson, Kai Kessenbrock, Abraham P. Lee, John Lowengrub, Marian L. Waterman, and Christopher C. W. Hughes. 2021. "An in vitro vascularized micro-tumor model of human colorectal cancer recapitulates in vivo responses to standard-of-care therapy." *Lab on a Chip* no. 21 (7):1333–1351. doi: 10.1039/d0lc01216e.

Hardiansyah, A., L. Y. Huang, M. C. Yang, T. Y. Liu, S. C. Tsai, C. Y. Yang, C. Y. Kuo, T. Y. Chan, H. M. Zou, W. N. Lian, and C. H. Lin. 2014. "Magnetic liposomes for colorectal cancer cells therapy by high-frequency magnetic field treatment." *Nanoscale Res Lett* no. 9 (1):9–497.

Hazra, Raj Shankar, Narendra Kale, Gourishankar Aland, Burhanuddin Qayyumi, Dipankar Mitra, Long Jiang, Dilpreet Bajwa, Jayant Khandare, Pankaj Chaturvedi, and Mohiuddin Quadir. 2020. "Cellulose mediated transferrin nanocages for enumeration of circulating tumor cells for head and neck cancer." *Scientific Reports* no. 10 (1):10010. doi: 10.1038/s41598-020-66625-2.

Hessel, C. M., V. P. Pattani, M. Rasch, M. G. Panthani, B. Koo, J. W. Tunnell, and B. A. Korgel. 2011. "Copper selenide nanocrystals for photothermal therapy." *Nano Lett* no. 11 (6):2560–2566.

Hind, D., P. Tappenden, I. Tumur, S. Eggington, and P. Sutcliffe. 2008. "The use of irinotecan, oxaliplatin and raltitrexed for the treatment of advanced colorectal cancer: Systematic review and economic evaluation (review of NICE Guidance No. 33)." In MA Hopkins, MA Goeree (eds.), *Health Technology Assessment*. Winchester, England: National Institute of Health and Care Research.

Hu, Y., K. Zhang, X. Zhu, X. Zheng, C. Wang, X. Niu, T. Jiang, X. Ji, W. Zhao, L. Pang, Y. Qi, F. Li, L. Li, Z. Xu, W. Gu, and H. Zou. 2021. "Synergistic inhibition of drug-resistant colon cancer growth with PI3K/mTOR dual inhibitor BEZ235 and nano-emulsioned paclitaxel via reducing multidrug resistance and promoting apoptosis." *Int J Nanomedicine* no. 16:2173–2186.

Hua, M. Y., H. W. Yang, C. K. Chuang, R. Y. Tsai, W. J. Chen, K. L. Chuang, Y. H. Chang, H. C. Chuang, and S. T. Pang. 2010. "Magnetic-nanoparticle-modified paclitaxel for targeted therapy for prostate cancer." *Biomaterials* no. 31 (28):7355–7363.

Ito, Ichiaki, Abdelrahman M. G. Yousef, Princess A. Dickson, Keith F. Fournier, Natalie Wall Fowlkes, and John Paul Y. C. Shen. 2023. "Antitumor activity of intraperitoneal (IP) paclitaxel to mucinous appendiceal adenocarcinoma in orthotopic patient-derived xenograft model." *Journal of Clinical Oncology* no. 41 (4_suppl):151. doi: 10.1200/JCO.2023.41.4_suppl.151.

Jabalera, Y., B. Garcia-Pinel, R. Ortiz, G. Iglesias, L. Cabeza, J. Prados, C. Jimenez-Lopez, and C. Melguizo. 2019. "Oxaliplatin-biomimetic magnetic nanoparticle assemblies for colon cancer-targeted chemotherapy: An in vitro study." *Pharmaceutics* no. 11 (8).

Jiang Yajun, Guo Zhaoyang, Fang Jing, Wang Beibei, Lin Zhiqiang, Chen Zhe-Sheng, Chen Yan, Zhang Ning, Yang Xiaoying, and Gao Wei. 2020. "A multi-functionalized nanocomposite constructed by gold nanorod core with triple-layer coating to combat multidrug resistant colorectal cancer." *Materials Science & Engineering C* no. 107: 110224.

Jurczyk, M., K. Jelonek, M. Musiał-Kulik, A. Beberok, D. Wrześniok, and J. Kasperczyk. 2021. "Single- versus dual-targeted nanoparticles with folic acid and biotin for anticancer drug delivery." *Pharmaceutics* no. 13 (3).

Kennedy, Andrew S., George H. Harrison, Carl M. Mansfield, Xiao Juan Zhou, Jing Fan Xu, and Elizabeth K. Balcer-Kubiczek. 2000. "Survival of colorectal cancer cell lines treated with paclitaxel, radiation, and 5-FU: Effect of TP53 or hMLH1 deficiency." *International Journal of Cancer* no. 90 (4):175–185. https://doi.org/10.1002/1097-0215(20000820)90:4<175::AID-IJC1>3.0.CO;2-W.

Kesharwani, Prashant, Rahul Chadar, Afsana Sheikh, Waleed Y. Rizg, and Awaji Y. Safhi. 2022. "CD44-targeted nanocarrier for cancer therapy." *Frontiers in pharmacology* no. 12. doi: 10.3389/fphar.2021.800481.

Khandare, Jayant, Smriti Arora, Balram Singh, Alain D'Souza, Nitin Singh, Narendra Kale, Shubham Bhide, Amrut Ashturkar, Aravindan Vasudevan, and Gourishankar Aland. 2020. "Device for the enumeration and continuous removal of circulating tumor cells in improving overall survival of epithelial cancer patients." *Journal of Clinical Oncology* no. 38 (15_suppl):e15043–e15043. doi: 10.1200/JCO.2020.38.15_suppl.e15043.

Khandare, Jayant, Alain D'Souza, Smriti Arora, Balram Singh, Gourishankar Aland, Narendra Kale, Isha Gore, Arnav Deshmukh, Rick Kamble, Vikas Jadhav, Pankaj Chaturvedi, Actorius Innovations, and Research. 2022. "Extracorporeal microchannel device to capture and eliminate circulating tumor cells from cancer patient's blood." *Journal of Clinical Oncology* no. 40 (16_suppl):e14522–e14522. doi: 10.1200/JCO.2022.40.16_suppl.e14522.

Komen, Job, Sanne M. van Neerven, Albert van den Berg, Louis Vermeulen, and Andries D. van der Meer. 2021. "Mimicking and surpassing the xenograft model with cancer-on-chip technology." *eBioMedicine* no. 66. doi: 10.1016/j.ebiom.2021.103303.

Kumar Balmiki, Murali Aparna, A. B. Bharath and Giri Supratim. 2019. "Guar gum modified upconversion nanocomposites for colorectal cancer treatment through enzyme-responsive drug release and NIR-triggered photodynamic therapy." *Nanotechnology* no. 30: 315102.

Lachowicz, J. I., L. I. Lecca, F. Meloni, and M. Campagna. 2021. "Metals and metal-nanoparticles in human pathologies: From exposure to therapy." *Molecules* no. 26 (21).

Le, Ngoc Thuy Trang, Dinh Tien Dung Nguyen, Ngoc Hoi Nguyen, Cuu Khoa Nguyen, and Dai Hai Nguyen. 2021. "Methoxy polyethylene glycol—cholesterol modified soy lecithin liposomes for poorly water-soluble anticancer drug delivery." *Journal of Applied Polymer Science* no. 138 (7):49858. https://doi.org/10.1002/app.49858.

Ledermann, J. A., S. Canevari, and T. Thigpen. 2015. "Targeting the folate receptor: Diagnostic and therapeutic approaches to personalize cancer treatments." *Annals of Oncology* no. 26 (10):2034–2043. doi: 10.1093/annonc/mdv250.

Lee, John K., Nathanael J. Bangayan, Timothy Chai, Bryan A. Smith, Tiffany E. Pariva, Sangwon Yun, Ajay Vashisht, Qingfu Zhang, Jung Wook Park, Eva Corey, Jiaoti Huang, Thomas G. Graeber, James Wohlschlegel, and Owen N. Witte. 2018. "Systemic surfaceome profiling identifies target antigens for immune-based therapy in subtypes of advanced prostate cancer." *Proceedings of the National Academy of Sciences* no. 115 (19):E4473–E4482. doi: 10.1073/pnas.1802354115.

Lee, Y. J., Y. J. Kim, and Y. Park. 2022. "Folic acid and chitosan-functionalized gold nanorods and triangular silver nanoplates for the delivery of anticancer agents." *Int J Nanomedicine* no. 17:1881–1902.

Li, Jinxing, Berta Esteban-Fernández de Ávila, Wei Gao, Liangfang Zhang, and Joseph Wang. 2017. "Micro/nanorobots for biomedicine: Delivery, surgery, sensing, and detoxification." *Science Robotics* no. 2 (4):eaam6431. doi: 10.1126/scirobotics.aam6431.

Li, Sixuan, Yizong Hu, Andrew Li, Jinghan Lin, Kuangwen Hsieh, Zachary Schneiderman, Pengfei Zhang, Yining Zhu, Chenhu Qiu, Efrosini Kokkoli, Tza-Huei Wang, and Hai-Quan Mao. 2022. "Payload distribution and capacity of mRNA lipid nanoparticles." *Nature Communications* no. 13 (1):5561. doi: 10.1038/s41467-022-33157-4.

Li, Zi Ling, You Sheng Huang, Xiang Yuan Xiong, Xiang Qin, and Yue Yuan Luo. 2017. "Synthesis, characterisation and in vitro release of paclitaxel-loaded polymeric micelles." *Micro & Nano Letters* no. 12 (3):191–194. https://doi.org/10.1049/mnl.2016.0690.

Liberti, M. V., and J. W. Locasale. 2016. "The Warburg effect: How does it benefit cancer cells?" *Trends Biochem Sci* no. 41 (3):211–218.

Liu, Xingxing, Jiaru Fang, Shuang Huang, Xiaoxue Wu, Xi Xie, Ji Wang, Fanmao Liu, Meng Zhang, Zhenwei Peng, and Ning Hu. 2021. "Tumor-on-a-chip: From bioinspired design to biomedical application." *Microsystems & Nanoengineering* no. 7 (1):50. doi: 10.1038/s41378-021-00277-8.

Locci, Emanuela, Ilaria Pilia, Roberto Piras, Sergio Pili, Gabriele Marcias, Pierluigi Cocco, Fabio De Giorgio, Manuele Bernabei, Valentina Brusadin, Laura Allegrucci, Alessandra Bandiera, Ernesto d'Aloja, Enrico Sabbioni, and Marcello Campagna. 2019. "Particle background levels in human tissues—PABALIHT project. Part I: A nanometallomic study of metal-based micro- and nanoparticles in liver and kidney in an Italian population group." *Journal of Nanoparticle Research* no. 21 (3):45. doi: 10.1007/s11051-019-4480-y.

Lv, Chaoxiang, Hao Qu, Wanyun Zhu, Kaixiang Xu, Anyong Xu, Baoyu Jia, Yubo Qing, Honghui Li, Hong-Jiang Wei, and Hong-Ye Zhao. 2017. "Low-dose paclitaxel inhibits tumor cell growth by regulating glutaminolysis in colorectal carcinoma cells." *Frontiers in pharmacology* no. 8. doi: 10.3389/fphar.2017.00244.

Ma, Xing, Xu Wang, Kersten Hahn, and Samuel Sánchez. 2016. "Motion control of urea-powered biocompatible hollow microcapsules." *ACS Nano* no. 10 (3):3597–3605. doi: 10.1021/acsnano.5b08067.

Maradonna, Francesca, Camilla M. Fontana, Fiorenza Sella, Christian Giommi, Nicola Facchinello, Chiara Rampazzo, Micol Caichiolo, Seyed Hossein Hoseinifar, Luisa Dalla Valle, Hien Van Doan, and Oliana Carnevali. 2022. "A zebrafish HCT116 xenograft model to predict anandamide outcomes on colorectal cancer." *Cell Death & Disease* no. 13 (12):1069. doi: 10.1038/s41419-022-05523-z.

Mitchell, Michael J., Margaret M. Billingsley, Rebecca M. Haley, Marissa E. Wechsler, Nicholas A. Peppas, and Robert Langer. 2021. "Engineering precision nanoparticles for drug delivery." *Nature Reviews Drug Discovery* no. 20 (2):101–124. doi: 10.1038/s41573-020-0090-8.

Muñoz, Jose, Mario Urso, and Martin Pumera. 2022. "Self-propelled multifunctional microrobots harboring chiral supramolecular selectors for 'enantiorecognition-on-the-fly'." *Angewandte Chemie International Edition* no. 61 (14):e202116090. https://doi.org/10.1002/anie.202116090.

Naahidi, Sheva, Mousa Jafari, Faramarz Edalat, Kevin Raymond, Ali Khademhosseini, and P. Chen. 2013. "Biocompatibility of engineered nanoparticles for drug delivery." *Journal of Controlled Release* no. 166 (2):182–194. https://doi.org/10.1016/j.jconrel.2012.12.013.

Nandi, Semonti, Narendra Kale, Ashwini Patil, Shashwat Banerjee, Yuvraj Patil, and Jayant Khandare. 2020. "A graphene-sandwiched DNA nano-system: Regulation of intercalated doxorubicin for cellular localization." *Nanoscale Advances* no. 2 (12):5746–5759. doi: 10.1039/d0na00575d.

Nawaz, Fathima Zahra, and Edward T. Kipreos. 2022. "Emerging roles for folate receptor FOLR1 in signaling and cancer." *Trends in Endocrinology & Metabolism* no. 33 (3):159–174. doi: 10.1016/j.tem.2021.12.003.

Orentas, Rimas, James Yang, Xinyu Wen, Jun Wei, Crystal Mackall, and Javed Khan. 2012. "Identification of cell surface proteins as potential immunotherapy targets in 12 pediatric cancers." *Frontiers in Oncology* no. 2. doi: 10.3389/fonc.2012.00194.

Palzer, J., L. Eckstein, I. Slabu, O. Reisen, U. P. Neumann, and A. A. Roeth. 2021. "Iron oxide nanoparticle-based hyperthermia as a treatment option in various gastrointestinal malignancies." *Nanomaterials* no. 11 (11).

Pant, Tejal, Ganesh Gaikwad, Dhiraj Jain, Prajakta Dandekar, and Ratnesh Jain. 2021. "Establishment and characterization of lung co-culture spheroids for paclitaxel loaded Eudragit® RL 100 nanoparticle evaluation." *Biotechnology Progress* no. 37 (6):e3203. https://doi.org/10.1002/btpr.3203.

Parchur, Abdul Kareem, Gayatri Sharma, Jaidip M. Jagtap, Venkateswara Rao Gogineni, Peter S. LaViolette, Michael J. Flister, Sarah Beth White, and Amit Joshi. 2018. "Vascular interventional radiology-guided photothermal therapy of colorectal cancer liver metastasis with theranostic gold nanorods." *ACS Nano* no. 12 (7):6597–6611. doi: 10.1021/acsnano.8b01424.

Parkin, D. M., F. Bray, J. Ferlay, and P. Pisani. 2001. "Estimating the world cancer burden: Globocan 2000." *Int J Cancer* no. 94 (2):153–156.

Patel, N., L. Ghali, I. Roitt, L. P. Munoz, and R. Bayford. 2021. "Exploiting the efficacy of Tyro3 and folate receptors to enhance the delivery of gold nanoparticles into colorectal cancer cells in vitro." *Nanoscale Adv* no. 3 (18):5373–5386.

Patra, Jayanta Kumar, Gitishree Das, Leonardo Fernandes Fraceto, Estefania Vangelie Ramos Campos, Maria del Pilar Rodriguez-Torres, Laura Susana Acosta-Torres, Luis Armando Diaz-Torres, Renato Grillo, Mallappa Kumara Swamy, and Shivesh Sharma. 2018. "Nano based drug delivery systems: Recent developments and future prospects." *Journal of Nanobiotechnology* no. 16 (1):1–33.

Paxton, Walter F., Ayusman Sen, and Thomas E. Mallouk. 2005. "Motility of catalytic nanoparticles through self-generated forces." *Chemistry—A European Journal* no. 11 (22):6462–6470. https://doi.org/10.1002/chem.200500167.

Pereira, I., F. Sousa, P. Kennedy, and B. Sarmento. 2018. "Carcinoembryonic antigen-targeted nanoparticles potentiate the delivery of anticancer drugs to colorectal cancer cells." *Int J Pharm* no. 549 (1–2):397–403.

Pham, Duy Toan, Nuttawut Saelim, and Waree Tiyaboonchai. 2020. "Paclitaxel loaded EDC-cross-linked fibroin nanoparticles: A potential approach for colon cancer treatment." *Drug delivery and Translational Research* no. 10 (2):413–424. doi: 10.1007/s13346-019-00682-7.

Phung, Cao Dai, Thien Giap Le, Van Hai Nguyen, Thi Trang Vu, Huong Quynh Nguyen, Jong Oh Kim, Chul Soon Yong, and Chien Ngoc Nguyen. 2020. "PEGylated-paclitaxel and dihydroartemisinin nanoparticles for simultaneously delivering paclitaxel and dihydroartemisinin to colorectal cancer." *Pharmaceutical Research* no. 37 (7):129. doi: 10.1007/s11095-020-02819-7.

Rao, S., F. Sclafani, C. Eng, R. A. Adams, M. G. Guren, D. Sebag-Montefiore, A. Benson, A. Bryant, C. Peckitt, E. Segelov, A. Roy, M. T. Seymour, J. Welch, M. P. Saunders, R. Muirhead, P. O'Dwyer, J. Bridgewater, S. Bhide, R. Glynne-Jones, D. Arnold, and D. Cunningham. 2020. "International rare cancers initiative multicenter randomized phase II trial of cisplatin and fluorouracil versus carboplatin and paclitaxel in advanced anal cancer: InterAAct." *J Clin Oncol* no. 38 (22):2510–2518.

Rodríguez-Luccioni, H. L., M. Latorre-Esteves, J. Méndez-Vega, O. Soto, A. R. Rodríguez, C. Rinaldi, and M. Torres-Lugo. 2011. "Enhanced reduction in cell viability by hyperthermia induced by magnetic nanoparticles." *Int J Nanomedicine* no. 6:373–380.

Rosman, Raihana, Saifullah Bullo, Maniam Sandra, Dorniani Dena, Hussein Mohd Zobir, and Fakurazi Sharida. 2018. "Improved anticancer effect of magnetite nanocomposite formulation of GALLIC acid (Fe_3O_4-PEG-GA) against lung, breast and colon cancer cells." *Nanomaterials* no. 8: 83.

Rustum, Y. M. 1990. "Biochemical rationale for the 5-fluorouracil leucovorin combination and update of clinical experience." *J Chemother* no. 1:5–11.

Saltz, Leonard B. 2001. "Irinotecan: A new agent comes of age." *The Oncologist* no. 6 (1):65. doi: 10.1634/theoncologist.6-1-65.

Scaranti, Mariana, Elena Cojocaru, Susana Banerjee, and Udai Banerji. 2020. "Exploiting the folate receptor α in oncology." *Nature Reviews Clinical Oncology* no. 17 (6):349–359. doi: 10.1038/s41571-020-0339-5.

Sercombe, Lisa, Tejaswi Veerati, Fatemeh Moheimani, Sherry Y. Wu, Anil K. Sood, and Susan Hua. 2015. "Advances and challenges of liposome assisted drug delivery." *Frontiers in pharmacology* no. 6:286.

Shahnazari, Mina, Saeid Afshar, Mohammad Hassan Emami, Razieh Amini, and Akram Jalali. 2023. "Novel biomarkers for neoplastic progression from ulcerative colitis to colorectal cancer: A systems biology approach." *Scientific Reports* no. 13 (1):3413. doi: 10.1038/s41598-023-29344-y.

Shin, Seoyoun, Hyeonseok Yoon, and Jyongsik Jang. 2008. "Polymer-encapsulated iron oxide nanoparticles as highly efficient Fenton catalysts." *Catalysis Communications* no. 10 (2):178–182. https://doi.org/10.1016/j.catcom.2008.08.027.

Song, J. W., and L. L. Munn. 2011. "Fluid forces control endothelial sprouting." *Proc Natl Acad Sci U S A* no. 108 (37):15342–15347.

Sousa, A. R., M. J. Oliveira, and B. Sarmento. 2019. "Impact of CEA-targeting nanoparticles for drug delivery in colorectal cancer." *J Pharmacol Exp Ther* no. 370 (3):657–670.

Strelez, C., S. Chilakala, K. Ghaffarian, R. Lau, E. Spiller, N. Ung, D. Hixon, A. Y. Yoon, R. X. Sun, H. J. Lenz, J. E. Katz, and S. M. Mumenthaler. 2021. "Human colorectal cancer-on-chip model to study the microenvironmental influence on early metastatic spread." *iScience* no. 24 (5):21.

Sung, H., J. Ferlay, R. L. Siegel, M. Laversanne, I. Soerjomataram, A. Jemal, and F. Bray. 2021. "Global cancer statistics 2020: GLOBOCAN estimates of incidence and mortality worldwide for 36 cancers in 185 countries." *CA Cancer J Clin* no. 71 (3):209–249.

Tan, B. L., M. E. Norhaizan, and L. C. Chan. 2018. "An intrinsic mitochondrial pathway is required for Phytic Acid-Chitosan-Iron Oxide Nanocomposite (Phy-CS-MNP) to induce G_0/G_1 cell cycle arrest and apoptosis in the human colorectal cancer (HT-29) cell line." *Pharmaceutics* no. 10 (4).

Tian, Jun, Jonathan H. Chen, Sherry X. Chao, Karin Pelka, Marios Giannakis, Julian Hess, Kelly Burke, Vjola Jorgji, Princy Sindurakar, Jonathan Braverman, Arnav Mehta, Tomonori Oka, Mei Huang, David Lieb, Maxwell Spurrell, Jill N. Allen, Thomas A. Abrams, Jeffrey W. Clark, Andrea C. Enzinger, Peter C. Enzinger, Samuel J. Klempner, Nadine J. McCleary, Jeffrey A. Meyerhardt, David P. Ryan, Matthew B. Yurgelun, Katie Kanter, Emily E. Van Seventer, Islam Baiev, Gary Chi, Joy Jarnagin, William B. Bradford, Edmond Wong, Alexa G. Michel, Isobel J. Fetter, Giulia Siravegna, Angelo J. Gemma, Arlene Sharpe, Shadmehr Demehri, Rebecca Leary, Catarina D. Campbell, Omer Yilmaz, Gad A. Getz, Aparna R. Parikh, Nir Hacohen, and Ryan B. Corcoran. 2023. "Combined PD-1, BRAF and MEK inhibition in BRAFV600E colorectal cancer: A phase 2 trial." *Nature Medicine* no. 29 (2):458–466. doi: 10.1038/s41591-022-02181-8.

Vanhoefer, U., A. Harstrick, W. Achterrath, S. Cao, S. Seeber, and Y. M. Rustum. 2001. "Irinotecan in the treatment of colorectal cancer: Clinical overview." *J Clin Oncol* no. 19 (5):1501–1518.

Wan, M., Z. Liu, T. Li, H. Chen, Q. Wang, T. Chen, Y. Tao, and C. Mao. 2021. "Zwitterion-based hydrogen sulfide nanomotors induce multiple acidosis in tumor cells by destroying tumor metabolic symbiosis." *Angew Chem Int Ed Engl* no. 60 (29):16139–16148.

Wan, Mimi, Huan Chen, Qi Wang, Qian Niu, Ping Xu, Yueqi Yu, Tianyu Zhu, Chun Mao, and Jian Shen. 2019. "Bio-inspired nitric-oxide-driven nanomotor." *Nature Communications* no. 10 (1):966. doi: 10.1038/s41467-019-08670-8.

Wang, Hong, and Martin Pumera. 2015. "Fabrication of micro/nanoscale motors." *Chemical Reviews* no. 115 (16):8704–8735. doi: 10.1021/acs.chemrev.5b00047.

Wang, Jianxin, Andrea M. Potocny, Joel Rosenthal, and Emily S. Day. 2019. "Gold nanoshell-linear tetrapyrrole conjugates for near infrared-activated dual photodynamic and photothermal therapies." *ACS Omega* no. 5 (1):926–940.

Wang, Joseph. 2021. "Will future microbots be task-specific customized machines or multi-purpose 'all in one' vehicles?" *Nature Communications* no. 12 (1):7125. doi: 10.1038/s41467-021-26675-0.

Wang, Siyu, Song Yue, Cao Kunxia, Zhang Lingxiao, Fang Xuedong, Chen Fangfang, Feng Shouhua, Yan Fei. 2021. "Photothermal therapy mediated by gold nanocages composed of anti-PDL1 and galunisertib for improved synergistic immunotherapy in colorectal cancer." *Acta Biomaterialia* no. 134: 621–632.

Wang, Yang, Rose M. Hernandez, David J. Bartlett, Julia M. Bingham, Timothy R. Kline, Ayusman Sen, and Thomas E. Mallouk. 2006. "Bipolar electrochemical mechanism for the propulsion of catalytic nanomotors in hydrogen peroxide solutions." *Langmuir* no. 22 (25):10451–10456. doi: 10.1021/la0615950.

Wang, Yijia, Chunze Zhang, Shiwu Zhang, Zhenying Zhao, Jiawen Wang, Jiali Song, Yue Wang, Jun Liu, and Shaobin Hou. 2017. "Kanglaite sensitizes colorectal cancer cells to Taxol via NF-κB inhibition and connexin 43 upregulation." *Scientific Reports* no. 7 (1):1280. doi: 10.1038/s41598-017-01480-2.

Wang, Zi-Hao, Wen-Bei Peng, Pei Zhang, Xiang-Ping Yang, and Qiong Zhou. 2021. "Lactate in the tumour microenvironment: From immune modulation to therapy." *eBioMedicine* no. 73. doi: 10.1016/j.ebiom.2021.103627.

Warburg, Otto. 1956. "On the origin of cancer cells." *Science* no. 123 (3191):309–314. doi: 10.1126/science.123.3191.309.

Wavhale, Ravindra D., Kshama D. Dhobale, Shraddha Patil, Govind P. Chate, Bhausaheb V. Tawade, Yuvraj N. Patil, and Shashwat S. Banerjee. 2022. "Self-propelled catalytically powered dual-engine magnetic nanobots for rapid and highly efficient capture of circulating fetal trophoblasts." *Advanced Materials Interfaces* no. 9 (22):2200522. https://doi.org/10.1002/admi.202200522.

Wavhale, Ravindra D., Kshama D. Dhobale, Chinmay S. Rahane, Govind P. Chate, Bhausaheb V. Tawade, Yuvraj N. Patil, Sandesh S. Gawade, and Shashwat S. Banerjee. 2021. "Water-powered self-propelled magnetic nanobot for rapid and highly efficient capture of circulating tumor cells." *Communications Chemistry* no. 4 (1):159. doi: 10.1038/s42004-021-00598-9.

Weaver, Beth A. 2014. "How Taxol/paclitaxel kills cancer cells." *Molecular Biology of the Cell* no. 25 (18):2677–2681. doi: 10.1091/mbc.e14-04-0916.

Wei Wei, Li Rongxian, Liu Qinghang, Devanathadesikan Seshadri Vidya, Veeraraghavan Vishnu Priya, Surapaneni Krishna Mohan, Rengarajan Thamaraiselvan. 2021. Amelioration of oxidative stress, inflammation and tumor promotion by Tin oxide-Sodium alginate-Polyethylene glycol-Allyl isothiocyanate nanocomposites on the 1,2-Dimethylhydrazine induced colon carcinogenesis in rats. *Arabian Journal of Chemistry* no. 14: 103238.

White, S. B., D. H. Kim, Y. Guo, W. Li, Y. Yang, J. Chen, V. R. Gogineni, and A. C. Larson. 2017. "Biofunctionalized hybrid magnetic gold nanoparticles as catalysts for photothermal ablation of colorectal liver metastases." *Radiology* no. 285 (3):809–819.

Wishart, D. S., C. Knox, A. C. Guo, S. Shrivastava, M. Hassanali, P. Stothard, Z. Chang, and J. Woolsey. 2006. "DrugBank: A comprehensive resource for in silico drug discovery and exploration." *Nucleic Acids Res* no. 34 (Database issue).

Wu, Haijun, Qiongyan Zou, Hong He, Yu Liang, Mingjun Lei, Qin Zhou, Dan Fan, and Liangfang Shen. 2019. "Long non-coding RNA PCAT6 targets miR-204 to modulate the chemoresistance of colorectal cancer cells to 5-fluorouracil-based treatment through HMGA2 signaling." *Cancer Medicine* no. 8 (5):2484–2495. https://doi.org/10.1002/cam4.1809.

Xie, Yuan-Hong, Ying-Xuan Chen, and Jing-Yuan Fang. 2020. "Comprehensive review of targeted therapy for colorectal cancer." *Signal Transduction and Targeted Therapy* no. 5 (1):22. doi: 10.1038/s41392-020-0116-z.

Xing, Yi, Mengyun Zhou, Tailin Xu, Songsong Tang, Yang Fu, Xin Du, Lei Su, Yongqiang Wen, Xueji Zhang, and Tianyi Ma. 2020. "Core@Satellite Janus nanomotors with pH-responsive multiphoretic propulsion." *Angewandte Chemie International Edition* no. 59 (34):14368–14372. https://doi.org/10.1002/anie.202006421.

Yakati, Venu, Swathi Vangala, Vijay Sagar Madamsetty, Rajkumar Banerjee, and Gopikrishna Moku. 2022. "Enhancing the anticancer effect of paclitaxel by using polymeric nanoparticles decorated with colorectal cancer targeting CPKSNNGVC-peptide." *Journal of Drug Delivery Science and Technology* no. 68:103125. https://doi.org/10.1016/j.jddst.2022.103125.

Yamazaki, Tomoko, Andrew J. Gunderson, Miranda Gilchrist, Mark Whiteford, Maria X. Kiely, Amanda Hayman, David O'Brien, Rehan Ahmad, Jeffrey V. Manchio, Nathaniel Fox, Kayla McCarty, Michaela Phillips, Evelyn Brosnan, Gina Vaccaro, Rui Li, Miklos Simon, Eric Bernstein, Mary McCormick, Lena Yamasaki, Yaping Wu, Ashley Drokin, Trevor Carnahan, Yy To, William L. Redmond, Brian Lee, Jeannie Louie, Eric Hansen, Matthew C. Solhjem, Julie Cramer, Walter J. Urba, Michael J. Gough, Marka R. Crittenden, and Kristina H. Young. 2022. "Galunisertib plus neoadjuvant chemoradiotherapy in patients with locally advanced rectal cancer: A single-arm, phase 2 trial." *The Lancet Oncology* no. 23 (9):1189–1200. doi: 10.1016/s1470-2045(22)00446-6.

Yang, S. J., S. Y. Tseng, C. H. Wang, T. H. Young, K. C. Chen, and M. J. Shieh. 2020. "Magnetic nanomedicine for CD133-expressing cancer therapy using locoregional hyperthermia combined with chemotherapy." *Nanomedicine* no. 15 (26):2543–2561.

Yusefi Mostafa, Shameli Kamyar, Soon Lee-Kiun Michiele, Teow Sin-Yeang, Moeini Hassan, Ali Roshafima Rasit, Kia Pooneh, Jie Chia Jing, Abdullah Nurul Hidayah. 2023. "Chitosan coated magnetic cellulose nanowhisker as a drug delivery system for potential colorectal cancer treatment." *International Journal of Biological Macromolecules* no. 233: 123388.

Zhang, Hua, Wentao Duan, Lei Liu, and Ayusman Sen. 2013. "Depolymerization-powered autonomous motors using biocompatible fuel." *Journal of the American Chemical Society* no. 135 (42):15734–15737. doi: 10.1021/ja4089549.

Zwicke, G. L., G. A. Mansoori, and C. J. Jeffery. 2012. "Utilizing the folate receptor for active targeting of cancer nanotherapeutics." *Nano Rev* no. 3 (10):7.

Chapter 13

Nanotherapeutics for ovarian cancer using metal nanocomposites

Sandesh Lodha, Shrikant Joshi, Bhavin Vyas, Gajanan Kalyankar, Hetal Patel, and Ditixa Desai

LIST OF ABBREVIATIONS

BHT	Butylated hydroxyl toluene
BSA	Bovine serum albumin
CT	Computed tomography
CVP	Cell viability percentage
DOX	Doxorubicin
FDA	Food and Drug Administration
FSH	Follicle-stimulating hormone
FTE	Fallopian tube epithelium
GCL	Glutamate-cysteine ligase
GSH	Glutathione
HGSOC	High-grade serous ovarian cancer
HIF-1	Hypoxia-inducible factor 1
ICIs	Immune checkpoint inhibitors
MAPKs	Mitogen-activated protein kinases
MDA	Malonaldehyde
MKP3	Mitogen-activated protein kinase phosphatase 3
MNPs	Magnetic nanoparticles
MOFs	Metal–organic frameworks
GNRs	Gold nanorods
MRI	Magnetic resonance imaging
NIR	Near-infrared
OC	Ovarian cancer
OSE	Ovarian surface epithelium
PAA	Polyacrylic acid
PAM	Polyacrylamide
PEG	Poly ethylene glycol
PEI	Polyethyleneimine
PEITC	Phenethyl isothiocyanate
PTT	Photothermal therapy
PTX	Paclitaxel
PVA	Polyvinyl alcohol
PVAc	Polyvinyl acetate
PVP	Polyvinylpyrrolidone

DOI: 10.1201/9781032621135-13

rGo–Ag   Reduced graphene oxide–silver
ROS   Reactive oxygen species

13.1 INTRODUCTION

In 2020, ovarian cancer (OC) ranked third among gynecological cancers worldwide. Almost 90% of all occurrences of OC are ovarian carcinoma, making it the most prevalent kind. OC has five primary histological subtypes, the most prevalent of which is high-grade serous carcinoma. OC is the most fatal gynecological cancer because it is frequently discovered at an advanced stage. OC often has a bad prognosis, with a patient at an advanced stage having a 17% 5-year survival rate (Huang et al. 2022). Surgery is currently the gold standard of care for those with OC, followed by chemotherapy and radiation. Unfortunately, long-term remission or increased survival rates are not usually the results of this treatment (Padhi and Behera 2021).

Nanoparticles have been shown to have an impact on several aspects of cell life, including communication between cells, stress response, and apoptosis (programmed cell death). Metal nanocomposites have been heralded as a revolutionary development in the field of cancer treatment. By combining metal particles (Ranch et al. 2021) with other materials, scientists have been able to create nanocomposites that are smaller than a cell and can be used to target and destroy cancer cells without affecting healthy cells. In addition to their increased precision, metal nanocomposites also offer a range of advantages over traditional treatments, such as increased potency, reduced side effects, and increased safety. This makes them an attractive option for cancer treatment, and research has already shown them to be effective against a range of cancers, including prostate, breast, and OC (Behera and Padhi 2022). In terms of how they work, metal nanocomposites use their small size to penetrate cancer cells and deliver a powerful dose of drugs and radiation. The metal particles are attracted to the cancer cells and bind to them, delivering the drugs and radiation directly to the cancer cells. This allows for a more targeted approach to cancer treatment, which can reduce the impact on healthy cells while still being effective against the targeted cancer cells.

The use of metal nanocomposites in cancer treatment is still in its early stages, but the potential for this technology is great. With further research, metal nanocomposites could revolutionize the way cancer is treated, offering a more precise and effective approach that reduces the side effects and risk of other cancer treatments.

13.2 PATHOGENESIS OF OVARIAN CANCER

OC is a complicated illness with a multifactorial etiology that includes hereditary as well as environmental variables. The pathogenesis of OC is unknown, but recent data indicates that it develops from epithelial cells lining the ovarian surface or fallopian tube fimbriae. Several theories for OC pathogenesis are backed by experimental data and clinical reports. The following are some of the main theories for the pathogenesis of OC.

1. *Ovarian surface epithelium (OSE) hypothesis*: According to the OSE theory, OC develops from the OSE, a layer of epithelial cells that surrounds the ovary. Chronic ovulation and inflammation, according to this theory, can cause repeated harm and

repair of the OSE, leading to mutations in genes and epigenetic modifications that result in the development of OC (Fathalla 1971; Auersperg et al. 2001).

2. *Fallopian tube epithelium (FTE) hypothesis*: The FTE theory suggests that the FTE, which is continuous with the OSE, causes high-grade serous OC (HGSOC), the most prevalent and life-threatening subtype of OC. According to this theory, HGSOC develops from dysplastic tumors in the FTE and spreads to the ovary and peritoneum (Crum et al. 2007; Karst and Drapkin 2010).

3. *Inflammation hypothesis*: Chronic inflammation, induced by factors such as endometriosis or pelvic inflammatory disease, is thought to add to the development of OC by promoting DNA damage, oxidative stress, and abnormal signaling cascades (Wei, William, and Bulun 2011; Macciò and Madeddu 2012).

4. *Gonadotropin hypothesis*: According to the gonadotropin theory, elevated amounts of gonadotropins like follicle-stimulating hormone (FSH) and luteinizing hormone (LH) can encourage OC by promoting the development and survival of ovarian surface epithelial cells and causing DNA damage (Choi et al. 2007; Parchwani et al. 2022).

5. *Stem cell hypothesis*: According to the stem cell theory, OC develops from stem cells in the ovary that can regenerate themselves and transform into many different cell forms, including cancer cells (Ponnusamy and Batra 2008; Shah and Landen 2014).

Several genetic abnormalities in OC have also been found, including mutations in the BRCA1, BRCA2, TP53, PTEN, and PIK3CA genes (Kurman and Shih 2010). Mutations in genes such as BRCA1, BRCA2, and TP53, both inherited and somatic, have been linked to an increased risk of developing OC (Bolton et al. 2012; Kinde et al. 2013). DNA methylation, histone modifications, and non-coding RNA are examples of epigenetic alterations that can control gene expression and contribute to the development of OC (Esteller 2008). Further, several signaling pathways, including the PI3K/AKT/mTOR, RAS/RAF/MEK/ERK, and WNT/β-catenin pathways, have been implicated in OC pathogenesis (Steelman et al. 2011). These pathways' dysregulation can result in altered gene expression and enhance carcinogenic signaling. MicroRNAs (miRNAs), short non-coding RNAs that influence gene expression, are overexpressed in OC and may play a role in carcinogenesis (Rattanapan et al. 2020). These genetic changes can disrupt a variety of biological mechanisms, including DNA repair, cell cycle control, apoptosis, and signal transduction, leading to the development and spread of OC.

Smoking, obesity, and hormonal variables, in addition to genetic factors, play a part in the pathogenesis of OC (Faber et al. 2013). Obesity and smoking have been linked to a higher chance of OC in postmenopausal women (Beehler et al. 2006). Environmental variables such as asbestos and talc have been linked to a higher occurrence of OC (Camargo et al. 2011; Cramer et al. 2016). Hormonal variables, such as the use of oral contraceptives and hormone replacement treatment, can also influence OC risk (Riman, Persson, and Nilsson 1998).

There is also evidence that various subtypes of OC may have varying pathogenetic pathways. For example, high-grade serous OC, the most prevalent and aggressive subtype, is frequently linked with TP53 mutations and changes in DNA repair pathways, whereas low-grade serous OC may be caused by KRAS or BRAF mutations (Vang, Shih, and Kurman 2009).

Furthermore, a new study has emphasized the involvement of the tumor microenvironment in the pathogenesis of OC. Tumor development, infiltration, and spread can be influenced by the tumor microenvironment, which includes stromal cells, immune cells, and extracellular matrix components. Immune dysregulation, including impaired immune surveillance and immunosuppression, has been linked to the development of OC (Hanahan and Weinberg 2011). There is also evidence that various subtypes of OC may have unique pathogenetic pathways. For example, high-grade serous OC, the most prevalent and aggressive subtype, is frequently linked with TP53 mutations and changes in DNA repair pathways, whereas low-grade serous OC may be caused by KRAS or BRAF mutations (Vang, Shih, and Kurman 2009).

The etiology of OC has been linked to oxidative stress, which is defined as an imbalance between reactive oxygen species (ROS) and antioxidant defense systems. Multiple studies have shown that OC cells have increased amounts of ROS, which can cause DNA damage and contribute to tumor start, progression, and chemoresistance (Dunyaporn, Jerome, and Peng 2009). High amounts of ROS in OC cells were linked to higher expression of hypoxia-inducible factor 1 (HIF-1), a transcription factor that regulates genes implicated in angiogenesis and tumor development, according to one research (Hielscher and Gerecht 2015; Semenza 2010). Another research discovered that oxidative stress-mediated loss of mitogen-activated protein kinase phosphatase 3 (MKP3) increases tumorigenicity and chemoresistance in OC cells (Chan et al. 2008). MKP3 is a phosphatase that regulates the action of mitogen-activated protein kinases (MAPKs), which are signaling proteins involved in a variety of cellular processes such as cell growth, differentiation, and apoptosis. Previous research has indicated that MKP3 functions as a tumor inhibitor in OC.

The researchers discovered that oxidative stress caused by hydrogen peroxide (H_2O_2) decreased MKP3 mRNA in OC cells. MKP3 deficiency was linked to higher MAPK pathway activation and expression of genes implicated in cancer and chemoresistance (Chan et al. 2008). ROS accumulation caused by phenethyl isothiocyanate (PEITC), a natural substance found in cruciferous veggies, preferentially kills OC cells via unfolded protein response (UPR)-mediated apoptosis (Hong et al. 2015). PEITC-induced ROS buildup has been shown to produce ER stress, which activates the UPR pathway in OC cells. The UPR pathway is a cellular reaction to ER stress that is essential for protein homeostasis and cell viability. Prolonged or extreme ER stress, on the other hand, can cause UPR-mediated death. According to the findings, PEITC-induced ROS buildup resulted in protracted and severe ER stress, which activated the UPR-mediated apoptosis pathway and resulted in OC cell death. Importantly, normal ovarian cells were found to be less prone to PEITC-induced ROS accumulation and UPR-mediated apoptosis, implying that this substance may be harmful to only cancer cells (Hong et al. 2015). Hanna et al. (2021) also showed that folic acid-coated tin oxide nanoparticles (FAS-NPs) cause mitochondria-mediated apoptosis in human OC cells (Hanna and Saad 2021). FAS-NPs were found to be preferentially taken up by OC cells that overexpressed the folate receptor alpha (FR-), which is frequently overexpressed in OC. FAS-NPs were then shown to concentrate in the mitochondria of OC cells, resulting in mitochondrial dysfunction and cytochrome c leakage into the cytosol. Cytochrome c release activates caspases, which are proteases that play an important part in the implementation of apoptosis. Caspase activation and DNA fragmentation were detected in OC cells treated with FAS-NPs, suggesting that the nanoparticles caused apoptosis in these cells. FAS-NPs were found to have little toxicity against normal ovary cells, implying that they may have selective toxicity against cancer cells (Hanna and Saad 2021). In summary, these results indicate that FAS-NPs may have therapeutic promise for OC by causing mitochondria-mediated apoptosis in cancer cells via FR-targeting. Furthermore, several studies have shown that antioxidants like N-acetylcysteine and vitamin C can prevent OC cell development and make cells more susceptible to chemotherapy by lowering oxidative stress levels (Kwon 2021). To summarize, oxidative stress promotes tumor start, development, and chemoresistance during the pathogenesis of OC. Developing novel therapies for this disease by targeting oxidative stress pathways could be a viable strategy.

In the end, an improved understanding of the complex pathogenesis of OC is required to create more effective prevention and therapy methods. Advances in genomics, epigenomics, and immunology are shedding light on the molecular processes underlying OC, potentially leading to the creation of tailored treatments.

13.3 CHALLENGES WITH CURRENT THERAPY FOR OVARIAN CANCER TREATMENT

OC is a complicated and challenging disease to treat. OC treatment currently consists of an amalgam of surgery and chemotherapy. Despite improvements in therapy, OC remains a difficult disease to treat because of its high incidence of recurrence and often late presentation.

The aim of surgery as the first-line therapy for OC is to remove as much of the tumor as feasible. Even with optimum debulking surgery, the microscopic disease often persists, necessitating adjuvant treatment to target any residual cancer cells (Lheureux, Braunstein, and Oza 2019).

Chemotherapy is typically administered following surgery to reduce the risk of recurrence. Platinum-based drugs such as carboplatin and cisplatin are widely used in OC chemotherapy regimens, often in conjunction with a taxane such as paclitaxel. While chemotherapy can be successful, it also has serious adverse effects such as nausea, vomiting, fatigue, and hair loss (Markman 2019).

- One of the difficulties with present-day OC treatment is the emergence of drug resistance. OC cells can develop resistance to chemotherapy, making long-term recovery challenging to achieve. Changes in drug absorption and efflux, changes in DNA damage repair mechanisms, and changes in apoptotic signaling cascades can all lead to resistance (Lheureux, Braunstein, and Oza 2019).
- Another concern with current treatment is the harm of chemotherapy. Chemotherapy medications can cause tiredness, nausea, vomiting, and hair loss, which negatively impact the patient's quality of life. Furthermore, chemotherapy can cause long-term side effects such as neuropathy, renal damage, and cardiac failure (Markman 2019).
- Another problem is a shortage of successful OC-targeted treatments. While targeted treatments have proven to be effective in treating other kinds of cancer, such as breast and lung cancer, they have not been as effective in treating OC. This is due in part to the heterogeneity of OC, which makes identifying targets that are unique to all kinds of OC challenging (Willmott and Fruehauf 2010).
- Furthermore, the use of neoadjuvant chemotherapy (NACT) for metastatic OC has been controversial. While NACT can decrease tumor burden and possibly allow for full surgical resection, it can also raise the chance of residual illness and lower chemotherapy efficacy due to drug resistance (Wright et al. 2016).
- Another barrier to OC therapy is a dearth of efficient screening instruments for early diagnosis. There is currently no regular screening test for OC, and signs like bloating, pelvic or abdominal discomfort, and trouble eating can be ambiguous and nonspecific, resulting in a delayed diagnosis. This delay in diagnosis can have an influence

on therapy efficacy and, eventually, patient outcomes (Lheureux, Braunstein, and Oza 2019).

- Finally, more personalized and precision medicine methods for OC therapy are required. This would entail tailoring therapy to a patient's tumor's unique traits, such as its molecular makeup, to maximize efficacy while minimizing toxicity (Lheureux, Braunstein, and Oza 2019).

In conclusion, while surgery and chemotherapy are currently the standards of care for OC, there are several challenges associated with this approach, including drug resistance, toxicity, a lack of effective screening tools, and the need for more personalized and precision medicine approaches. To address these obstacles, more studies and the creation of novel treatments and diagnostic instruments will be needed to obtain better results for women with OC.

13.4 CLASSIFICATION AND APPLICATIONS OF METAL NANOCOMPOSITES

Metal nanocomposites are materials composed of a metal matrix and nanoscale reinforcements, which can be other metals, ceramics, or polymers. There are various types of metal nanocomposites, and here are some examples.

13.4.1 Metal–carbon nanocomposites

Because of their substantial specific surface area, good porosity, strong electron conductivity, and comparatively high chemical inertness, carbon materials like activated carbon, carbon nanotubes, carbon nanofibers, fullerenes, and graphene are employed. Because of these materials' superior electrical, thermal, optical, and mechanical characteristics as well as their high surface-to-volume ratios, their composition with metal nanoparticles has received a great deal of attention (Sharma et al. 2020). Examples include gold–carbon nanotube, platinum–graphene, silver–graphene nanocomposites, and so on.

Choi and coworkers (2018) assessed the apoptotic efficacy of graphene oxide–silver nanocomposite (rGO–Ag) in OC cells and their subgroups. The first objective was to employ R-phycoerythrin to make graphene oxide, reduced graphene oxide, silver nanoparticles, and nanocomposite (rGO–Ag). The second goal was to assess the cytotoxic capability of synthesized metal nanoparticles/nanocomposites in tumor cells (OC cells and its stem cells) along with the investigation of possible mechanisms of cytotoxicity using various cellular assays. The ultimate goal was to evaluate the impact of rGO–Ag, salinomycin, and their combined impact on the cytotoxicity of tumor cells. Salinomycin, a novel agent, has the potential to have cutting-edge and effective chemotherapeutic activity since it specifically targets OC stem cells (OvCSCs) as well as therapy-resistant cancer cells. The IC_{50} value of rGO–Ag for dose-dependent inhibition of cell viability of A2780 OC cells was found severalfolds lower and showed a more noticeable inhibitory effect in comparison to other tested nanomaterials. Such positive outcomes were made possible by the anchoring of 10 nm-sized nanoscale silver particles on the face of two-dimensional graphene sheets. Further, the cell viability study was demonstrated on diverse subgroups of OvCSCs, (ALDH$^+$CD133$^+$, ALDH$^+$CD133$^-$, ALDH$^-$CD133$^+$, and ALDH$^-$CD133$^-$) isolated from OC cells to corroborate the efficacy of rGO–Ag nanocomposite. Surprisingly, relative to other subgroups, ALDH+/CD133+ cells showed a more profound effect toward rGO–Ag. Similar results were obtained from clonogenic assay which assesses the colony formation of cells during the proliferation state. Although the same amounts of rGO–Ag were administered to each subgroup, varying responses were observed, which showed loss of viability in increasing order for ALDH⁻CD133⁺, ALDH⁺CD133⁻, ALDH⁻CD133⁻, and ALDH⁺CD133⁺. Many cellular enzyme assays, such as adenylate kinase, glucose-6-phosphate dehydrogenase, and lactate dehydrogenase (LDH), are used to ascertain the mechanism of toxicity. By measuring the amount of LDH leakage in the cell culture media at 24 hours in four separate subpopulations of cells treated with rGO–Ag, cell toxicity was determined. Every cell subpopulation released LDH into the medium with ALDH⁺CD133⁺ showing the greatest sensitivity. The mechanism of reduced cell viability by rGO–Ag is depicted in Figure 13.1. The significant cytotoxic capability was demonstrated by rGO–Ag against ALDH⁺CD133⁺ cells that are highly tumorigenic. Moreover, the rGO–Ag plus salinomycin combination caused potential cytotoxicity at low concentrations and can be selectively targeted to tumorigenic ALDH⁺CD133⁺ OvCSCs.

A similar study was steered by Gurunathan et al. (2015), who synthesized reduced graphene oxide–silver (rGO–Ag) nanocomposites using *Tilia amurensis* plant extracts as natural reducing and stabilizing agents. According to the author, A2780 cancer cells had their cell viability significantly inhibited, and their production of ROS and LDH had increased. Additionally, ROS markers such as intracellular malonaldehyde (MDA) and glutathione (GSH) were also evaluated. When the cells were incubated with the nanocomposite, the levels of MDA showed that lipid peroxidation considerably increased. It's interesting to

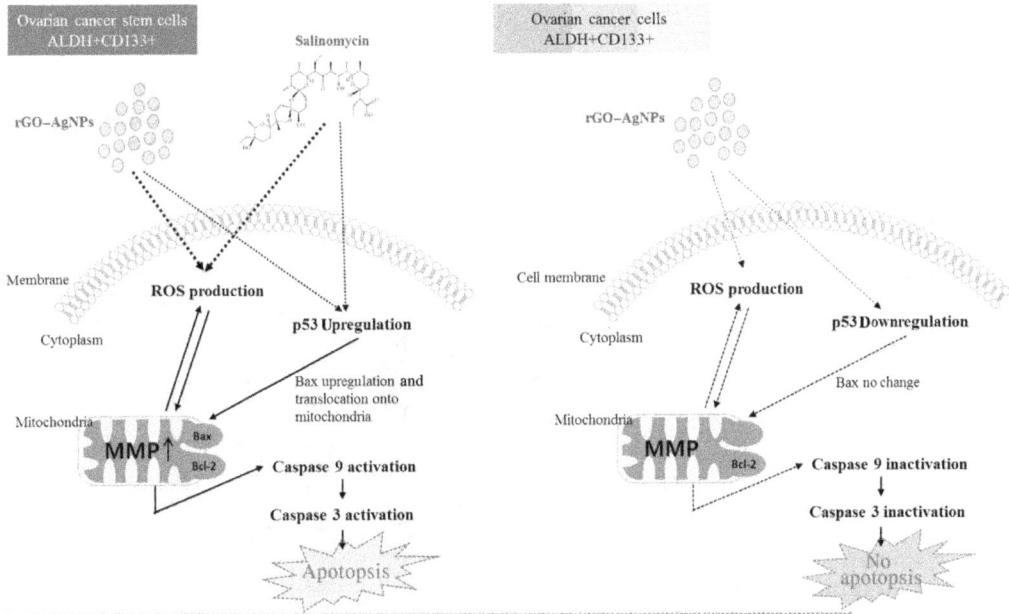

Figure 13.1 Hypothetical model explaining the working mechanism of rGO–Ag, and salinomycin to induce toxicity and apoptosis in ovarian cancer cells and OvCSCs.

note that although the GSH level within the cells treated with the rGO–Ag nanocomposite was fourfold lower than that of the control cells, the quantities of intracellular MDA in the treated cells were double that of the control cells. This supports the hypothesis put forth in earlier publications that oxidative stress may result from the disparity between the molecules MDA and GSH. These experiments showed a synergistic effect from the combined nanomaterials, indicating that the efficiency of the combined process exceeded the sum of the efficiency of the individual nanomaterials (Gurunathan et al. 2015).

Naskar et al. (2018) described the one-pot, low temperature (95°C), surfactant-free synthesis (Figure 13.2) of nano gold-integrated zinc oxide-chemically converted graphene nanocomposite (AZG) which was non-covalently immobilized with bovine serum albumin (BSA). The nanocomposite Au–ZnO–graphene immobilized with BSA was assessed for cell viability by MTT assay using a human ovarian teratocarcinoma cell line, PA1. It is understood that a significant phosphate concentration in the continuous double layer of lipid molecules of biological membranes can cause the cell to become negatively charged. Because of this fact, there exists an attractive force (electrostatic) between cells and positively charged Au nanoparticles ultimately resulting in the accumulation of Au particles onto the surface of the cell creating nanopores. The aforementioned mechanism could significantly lower cell viability, and the phenomenon connected to surface charges may be the cause of the improved cytotoxicity of nanocomposite with higher gold content. To prevent this interaction, the AZG nanocomposite was immobilized with BSA protein. Immobilization of AZG nanocomposite reduced the ZnO nanoparticle dissolution and weakened the electrostatic contact between the cell membrane and the gold nanoparticles at the same time. This combined impact may result in less cell membrane deterioration and higher cell viability.

Compared to individual metal oxide (Fe_3O_4, TiO_2, SnO_2, and ZnO) nanocomposites, graphene–metal oxide nanocomposites have proven to be more successful in treating cancer.

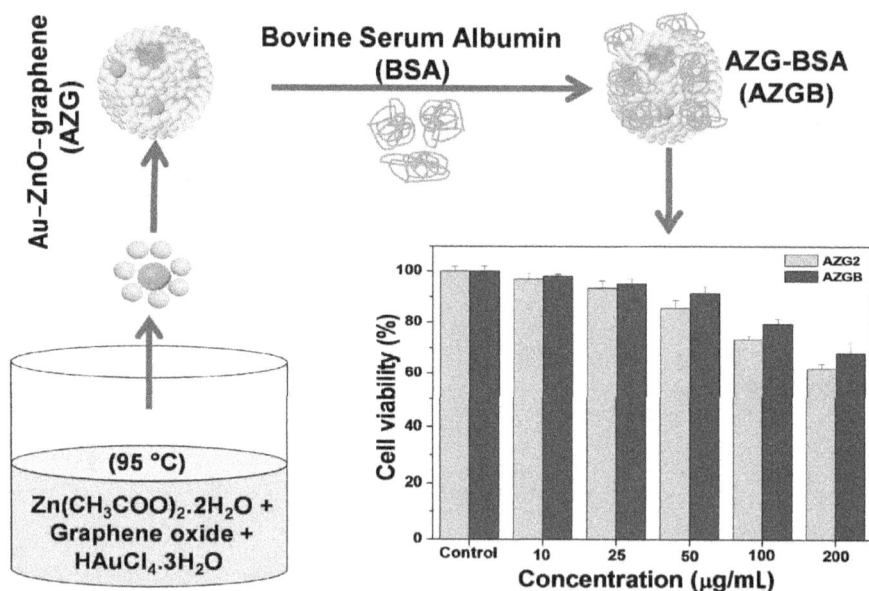

Figure 13.2 Synthesis of Au–ZnO–graphene nanocomposite and its biological characteristics.

Naskar et al. developed graphene-linked zinc oxide nanocomposites which exhibited a problem of agglomeration of the particles. To address this issue, polyethylene glycol (PEG) was coupled with a graphene–metal oxide nanocomposite. ZnO–CCG (chemically converted graphene) was coupled with PEG by solvothermal synthesis at 95°C to produce a ZGP (ZnO–CCG–PEG) nanocomposite having the size of 56 nm. The authors have also prepared nano-ZnO (ZO), ZnO–CCG (ZG), and ZnO–PEG(ZP) for comparison. The coupling of PEG onto ZnO–CCG was confirmed by various analytical techniques. XRD, Raman, and FTIR studies proved the conversion of graphene oxide into CCG which lowered the oxygen functional group. The chemical bonding of PEG with CCG was confirmed by Raman and FTIR studies. As per the results of field emission scanning and transmission electroscope, less agglomeration of particles was observed in the case of ZGP compared to other systems (ZP, ZG, and ZO) which again proved the interaction of PEG with ZG nanocomposite. The elemental analysis result showed that C, O, and Zn were present in the nanocomposite due to the presence of CCG, PEG, and ZnO. The authors performed *in vitro* cytotoxicity study of ZO, ZP, ZG, and ZGP using a human OC cell line (OAW42) in which 80% viable cells were observed using a 50 µg/mL dose of the composite. The order of the reduction in cell viability percentage (CVP) was ZGP > ZP > ZG > ZO. The highest cytotoxicity of ZO was justified by the cellular uptake of Zn^{+2}. Excess oxygen functional groups of CCG are responsible for the cytotoxic effect of ZG. In the case of ZP, PEG prevented the dissolution of the metal oxides and prevented cytotoxicity. The interaction of PEG with oxygen functional groups present on CCG and free hydroxyl groups on ZnO particles was responsible for providing the highest cell viability to the composite. Microstructural investigations had shown that the agglomeration of ZnO nanoparticles had decreased, and this had a positive effect on cell survival.

13.4.2 Metal–polymer/enzyme nanocomposites

Nanoscale inorganic particles that are usually 10–100 nm in at least one dimension is dispersed in an organic polymer matrix to greatly improve the polymer's performance properties. An innovative replacement for traditionally filled polymers is polymer nanocomposites. Filler dispersion nanocomposites exhibit noticeably superior characteristics when compared to pure polymers or their conventional composites because of their nanoscale diameters. They consist of improved strength and modulus, superior barrier properties, increased solvent and heat protection, and decreased flammability (Reddy 2011). Examples of polymers that have been used for functionalizing metal nanocomposites include polyvinyl alcohol (PVA), polyethyleneimine (PEI), polyethylene glycol (PEG), polyacrylamide (PAM), polyacrylic acid (PAA), polyvinylpyrrolidone (PVP), polyvinyl acetate (PVAc), and so on. Examples of metal–polymer nanocomposites are silver–polyethylene, gold–polystyrene, copper–polyvinyl chloride nanocomposites, and so on (Padhi et al. 2022).

Baskar et al. (2017) developed a nanocomposite (SNA) consisting of asparaginase-bound silver nanoparticles to confirm anticancer activity. Enzyme action was prominent in the case of asparaginase-linked silver nanoparticles compared to crude asparaginase. The mean size of the SNA was found to be 60–80 nm. The presence of primary and secondary amide/amine functional groups in the system was confirmed by FTIR study, and these sites were acting as a binding site for asparaginase to silver nanoparticles. As per EDS analysis, C, O, and silver were present in a weight percentage of 60.46%, 19.56%, and 13.38%, respectively. The existence of asparaginase in the SNA was confirmed by such elemental analysis. Asparagine synthase helps normal cells make asparagine for regular growth activities. Because they lack asparagine synthase, cancerous cells develop by utilizing the freely circulating asparagine. Asparaginase produces aspartic acid and ammonia from free asparagine and so malignant cells don't get asparagine which results in the death of the cell. Asparaginase and silver nanoparticles worked together to create metalloproteins that increased anticancer efficacy by inhibiting the growth of tumor cells over time and inducing apoptosis and necrosis in the tumor cells. As per the results of the MTT assay, the SNP showed better cytotoxicity than free asparaginase on cell line, A2780. The cytotoxicity values were 11.64%, 12.98%, 13.37%, and 15.21% for SNA concentrations 25, 50, 75, and 100 µg/mL, respectively.

Metal chlorides and polyvinylpyrrolidone (PVP) were used as precursors to prepare $MnZnFe_2O_4$ ferrites by co-precipitate technique. PVP was used to prevent the agglomeration of particles. Fourier transform infrared spectroscopy confirmed the chemical interaction between PVP and metallic hydroxide of ferrite nanoparticles. As per the MTT assay, PVP-coupled $MnZnFe_2O_4$ in the presence of a magnetic field proved a significant reduction in the viability of treated SKOV-3 (ovarian cancer) cells. The developed nanocomposites also exhibited a growth inhibitory effect on cancer cells and triggered death via mitochondrial damage through activated P53 (Kareem et al. 2020).

Han and his team produced microbubbles (MBs) coated Cu-Se nanoparticles (MBs@Cu-Se NPs), coupled with ultrasound to potentially treat ovarian carcinoma cells. Several rare earth metals have been reported to possess antitumor activity, one of which is selenium (Se). Se causes considerable toxicity to cancer cells while exhibiting modest toxicity to healthy cells. It causes the cancer cells to undergo ROS-mediated apoptosis pathways, which has an anticancer impact, and according to studies, this effect is exaggerated when exposed to ultrasound. A crucial microelement for humans, copper (Cu) is important for many physiological processes, boosting immunity and reducing the incidence of illness. Nanosized Cu has a better anticancer impact while being less hazardous than other Cu compounds. At low concentrations, Cu can reduce the quantity of ROS to protect healthy cells. In this work, synthesized copper–selenium nanoparticles (Cu–Se NPs) were coated with lipid microbubbles (MBs). Such nanocomposites provide stability to colloidal, hydrophobic Cu–Se NPs, prolong their circulation in the blood, and impart biocompatibility (Han et al. 2020).

The Cu–Se, MBs@Cu–Se NPs, and MBs@Cu–Se NPs with ultrasound have *in vitro* cancer potential that effectively kills the OC cells (A2780 and CisRA2780) and lowers cell viability depending on the dose (Figure 13.3). The ovarian tumors in the MBs@Cu–Se NPs with ultrasound showed noticeably increased levels of apoptosis, which was validated by the various biochemical assays and nuclear staining with Hoechst-33,258. The improved cell-killing properties are primarily the result of the synergistic interactions between MBs@Cu–Se NPs and ultrasound (Han et al. 2020).

13.4.3 Metal–ceramic nanocomposites

By adding nanocrystalline oxide/non-oxide ceramic particles or their hybrid combination, the toughness, creep strength, and thermoresistance of ceramic–matrix composites have dramatically improved (Banerjee and Manna 2013). Due to improved processing techniques that enable these materials to transition from the research laboratory scale to the commercial level, interest in ceramic nanocomposites is expanding (Palmero 2015). Examples include aluminum oxide–nickel, titanium carbide–tungsten, silicon carbide–copper nanocomposites, among others.

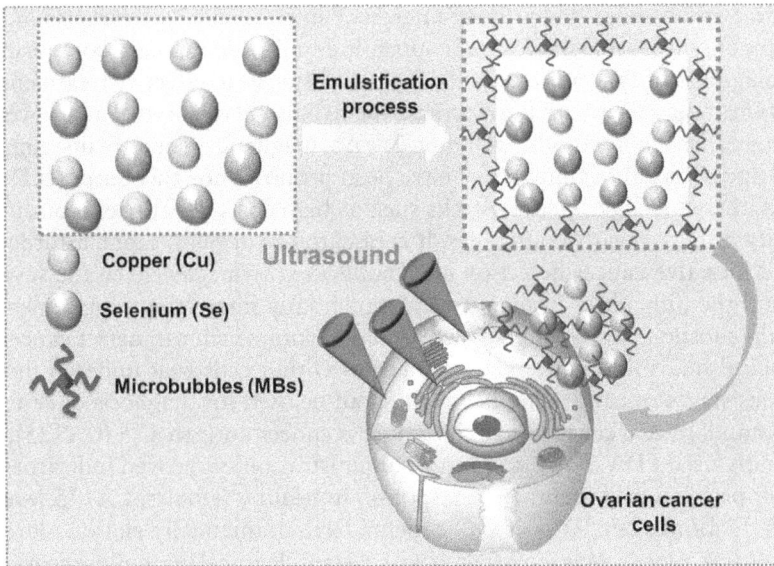

Figure 13.3 A graphical sketch of the synthesized of MBs@Cu–Se NP for ovarian cancer radiation therapy.

A biosensor made of gold–silica (Au–SiO$_2$) nanocomposites was developed by Mishra Y.K. et al. and can be used to identify OC cells. The atom beam co-sputtering method was used to prepare Au–SiO$_2$ nanocomposite in which gold nanoparticles were incorporated into the silica matrix. As per the results of transmission electron microscopy, the size of Au–SiO$_2$ nanocomposite particles was found to be 1.8–5.4 nm. The nanocomposite having 9.7% gold was used for the identification of human OC (HOC) cells. The blue shift of 6 nm in the surface plasmon resonance peak position was observed in human OC (HOC) cells treated with Au–SiO$_2$ nanocomposites (in MCDB + M199 media at a concentration of 1.2X106 cells/mL) (Mishra et al. 2007).

To effectively deliver doxorubicin to OC cells which have drug-resistant properties, Arora et al. (2012) developed nanocomposites of iron oxide–titanium dioxide. The mechanism of doxorubicin's mechanical attachment to the titanium dioxide surface involved a weak bond that was broken by the acidification of cell endosomes. By avoiding Pgp-mediated drug export, doxorubicin was released, transported through the intracellular milieu, and then entered the cell nucleus. Using confocal fluorescent imaging, fluorescence, and flow cytometry, it was shown that DNCs can regulate the uptake and distribution of transferrin in cells. The fact that clathrin-mediated endocytosis increased transferrin uptake suggests that both DNCs and nanocomposites might obstruct the clearance of transferrin from cells.

13.4.4 Metal–semiconductor nanocomposites

Due to their high electron conduction capabilities and broad-spectrum light response from the UV to visible range, semiconductor materials have been widely used for electrochemistry, energy catalysis, and sensor management applications. Semiconductor doping and surface modification are promising methods for enhancing the performance of semiconductor nanostructures and have gained a lot of interest (Fan et al. 2017). Examples include gold–titanium dioxide, silver–zinc oxide, and copper–indium–selenium nanocomposites.

The late diagnosis of cancer is one of the biggest obstacles to effective treatment. So timely detection of the disease is essential to increase the likelihood of survival and lower the death rate from cancer. Several methods, including X-rays, magnetic resonance imaging, histology, and clinical and physical examinations, have been proposed for this purpose. Despite their effectiveness, these tactics have drawbacks such as high costs, lengthy execution times, and the possibility of false-positive outcomes. It is essential to develop rapid, affordable, subtle, focused, and lucrative cancer detection and monitoring strategies given the severity of the condition and the difficulties in diagnosis. A potentially interesting replacement for more traditional diagnostic methods is provided by biosensors which can detect cancer biomarkers in biological fluids. Biomarkers are molecules within cells that undergo modifications during various stages of cancer and hence they can be used for diagnosis of cancer.

The gold standard marker for ovarian tumors is cancer antigen 125 (CA125), one of the most frequently used FDA (Food and Drug Administration)-approved indicators for monitoring cancer progress and therapeutic response. In healthy females, CA125 levels are normally below 35 U/mL, but 90% of OC patients have dramatically elevated levels. Among various biosensor types, nanomaterials-based electrochemical immunosensors with their high sensitivity, specificity, and simplicity represent a promising technology that enables low-cost and quick examination.

Foroozandeh et al. (2023) synthesized the graphitic carbon nitride–magnetite–polyaniline (g-C_3N_4/Fe_3O_4/PANI) composite to increase the sensor's sensitivity for the detection of CA125 using chemical polymerization and coprecipitation techniques. Due to the specific surface area, bio-acceptability, low cell toxicity, and chemical inertness of g-C_3N_4, a semiconductor, this two-dimensional nanomaterial have drawn a lot of interest as a nanomaterial in the biological and biosensing domains. To overcome the low conductivity of g-C_3N_4, polyaniline (PANI), a conductive polymer, has been employed as a converter. It possesses unique opto-photoelectric properties with high conductivity, structural flexibility, ease of manufacturing, low cost, favorable electrochemical activity, and compatible carrier for many biological molecules. Along with g-C_3N_4 and PANI, magnetite (Fe_3O_4) owing to its distinctive qualities similar to g-C_3N_4, magnetic nanoparticles (MNPs) were also incorporated as a carrier in the composite for capturing and detecting biomolecules in biosensors.

The key characteristic that sets a biosensor apart from other devices is its bioreceptors, or biological recognition components, which are enzyme, antibody/antigen, nucleic acid sequence, cell structure, and biologically similar molecules (Sun and Zu 2015). In this study, the author has employed the CA125 aptamer (Kd = 12 nM) using $3'SH_2$ modification (Figure 13.4).

Biosensors that use aptamers, <100 nucleotides, produced *in vitro* using the SELEX process, are called aptasensors (Karunakaran, Rajkumar, and Bhargava 2015). This study used an electrochemical aptasensor based on voltammetry wherein a gold electrode's surface was treated with polyaniline (g-C_3N_4/Fe_3O_4/PANI), graphitic carbon nitrides, and magnetic nanoparticles to keep aptamer chains in place (GE). The nanoprobe was prepared as follows (Figure 13.5): The surface of the gold electrode was polished by using alumina powder. Thereafter, the prepared solutions (g-C_3N_4, g-C_3N_4/Fe_3O_4, and g-C_3N_4/Fe_3O_4/PANI) with a concentration of 5 mg/mL were drop-cast over the electrode's surface and allowed to dry at room temperature while illuminated by a standard lamp. The dried material on the electrode surface was then given a glutaraldehyde treatment. Once more, aptamer solution

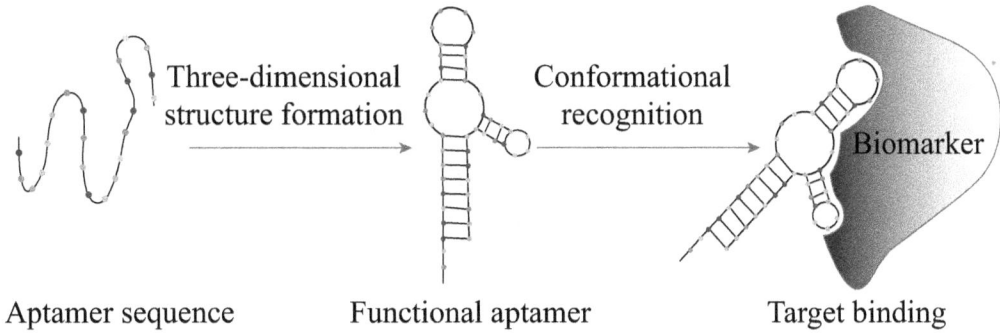

Figure 13.4 Schematic diagram of aptamer conformational recognition of targets to form an aptamer-target complex.

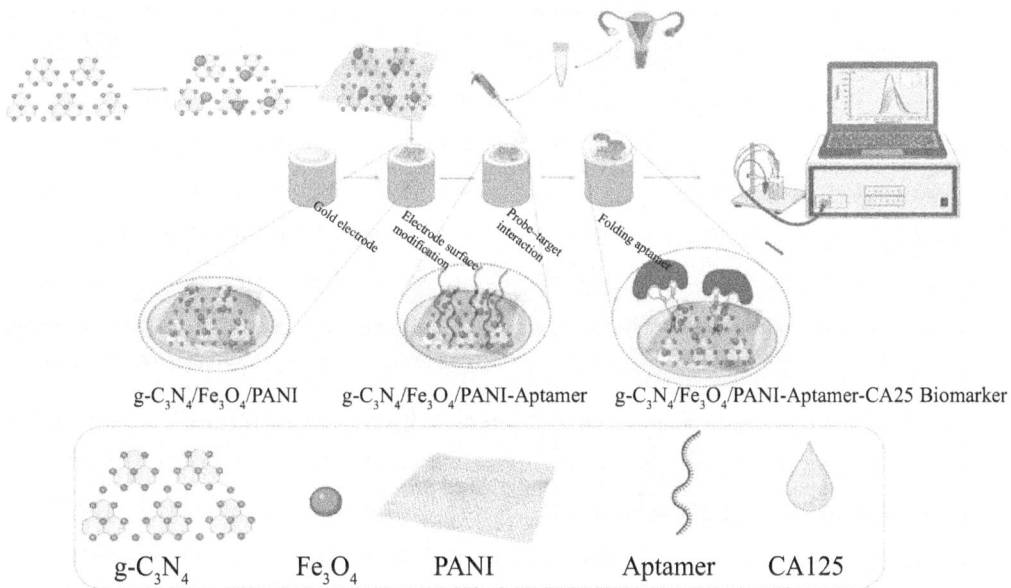

Figure 13.5 The g-C3N4/Fe3O4/PANI nanocomposite modified with aptamer for CA125 detection.

(lyophilized powder reconstituted in phosphate buffer saline) was drop-cast onto it after ten minutes; the pH was restored to 7.4 and kept at 20°C. The unbound aptamer was washed with PBS. The analyte was then poured onto the surface of the electrode at a concentration of 25 U/mL. The sample was incubated for 30 minutes after covering it with a cap. The excess unbound analyte was washed with PBS.

To demonstrate the application of the developed aptasensor, human serum samples (tested positive and negative with CA125) were also measured. The developed sensor was exposed to other proteins, such as glutamate-cysteine ligase (GCL), PSA, urea, and CA 15–3, to test the suggested method's selectivity which is depicted in Figure 13.6. This suggested that the

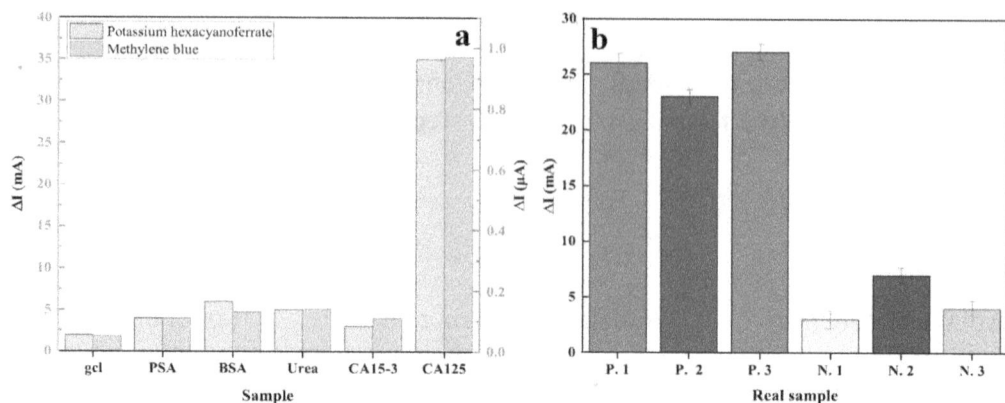

Figure 13.6 (a) Selectivity of designed nanoprobe for GCL (90 mg/dL), PSA (2.5 ng/mL), BSA (7 μg/mL), Urea (7 μg/mL), CA 15–3 (10 U/mL), and CA125 (4 U/mL), (b) the current difference of the modified electrode with six human blood serums, including three patients (sample P. 1, P. 2, and P. 3) and three normal samples (sample N. 1, N. 2, and N. 3) by SWV method in $[Fe(CN)_6]^{-3/-4}$ (0.2 mM) and methylene blue 2.5 mM (PBS pH 7.4).

aptamer and CA125 biomarkers had complementary morphologies, which led to their effective interaction and thereby a stable tertiary structure. The electroactive probe employed was the hexacyanoferrate redox system, which typically functioned at 0.3 V (vs. Ag/AgCl) and had a detection limit of 0.418 U/mL in the hexacyanoferrate medium. Methylene blue was used as a redox-active medium to reduce the electrochemical potential for the detection of CA125. The electrochemical aptasensor had a broad dynamic linear range (DLR) extending between 5 U/mL and 60 U/mL, with a regression coefficient of 0.993 and a limit of detection (LOD) of <0.3 U/mL after the detection potential was decreased to 0.2 V. Figure 13.6b reflects the difference in peak current between the serum sample and the analyte-free modified electrode as determined by SWV voltammograms. However, multi-aptamer biosensors and noble metal nanoparticles are suggested as ways to change the electrode surface to increase the effectiveness of biosensors.

13.4.5 Metal-oxide nanocomposites

Due to their structure, nanocomposites have particular, sometimes unique, physical and chemical properties that can be applied in a wide range of fields, including engineering, medicine, and ecology as well as the creation of new materials in the construction industry. Metal-oxide nanocomposites are an extremely promising type of nanomaterial that is being studied more and more in contemporary literature (Dontsova, Nahirniak, and Astrelin 2019). Examples: Copper oxide-graphene, iron oxide-aluminum, and titanium oxide-titanium nanocomposites etc.

Xu and his colleagues decorated gold nanoparticles (Au NPs) over L-arginine-modified magnetic nanoparticles and evaluated their properties over ovarian tumor cells. To determine the cytotoxic effects of prepared Au NPs/L-arginine/Fe_3O_4 nanocomposite, an MTT assay was conducted. In this study, the cells were cured with various doses of the Au NPs/L-arginine/Fe_3O_4 nanocomposite (Figure 13.7) on normal human umbilical vein endothelial

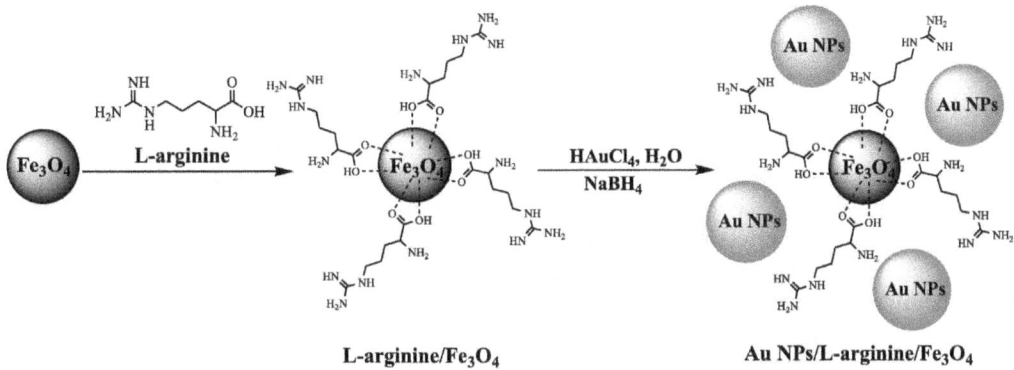

Figure 13.7 Scheme for preparation of Au NPs/L-arginine/Fe_3O_4 nanocomposite.

cells (HUVEC) and malignant cell lines (SK-OV-3, SW-626, and PA-1) for 48 hours, which was found to be 149, 162, and 213 mg/mL, respectively. These effects are due to the antioxidant property of nanocomposite which was evident by the (2,2-diphenyl-1-picrylhydrazyl) DPPH test. The IC_{50} of prepared nanocomposite (Au NPs/L-arginine/Fe_3O_4) and control, butylated hydroxytoluene (BHT), against DPPH free radicals was found to be 180 and 125 µg/mL, respectively. Because inflammation and oxidative stress are so closely related to tumor formation, a substance with anti-inflammatory or antioxidant capabilities can act as an antitumor agent (Xu et al. 2021). Ji et al. (2022) conducted similar studies on novel dual core-shell type nanoparticles (Fe_3O_4@CS-Starch/Cu). Here the copper nanoparticles (Cu NPs) are supported by a chitosan–starch composite and functionalized with Fe_3O_4. The IC_{50} values for the bio-nanocomposite and control (BHT) were 198 and 155 µg/mL, respectively, against DPPH free radicals while the IC_{50} values of Fe_3O_4@CS-Starch/Cu for OC (PA-1, SW-626, and SK-OV-3) cell lines were found to be 248, 189, and 281 µg/mL, respectively.

The work conducted by Javid aimed to create bio-acceptable, governable, free-flowing, and long-circulating (3-aminopropyl) triethoxysilane (APTES)-PEG (AP) modified superparamagnetic iron oxide nanoparticles (SPIONPs), for stimuli (magnetic)-sensitive targeted drug delivery to tumor cells, to achieve the pharmacokinetic parameters, i.e., improved drug retention time and bioavailability at tumor site. The bare and AP-modified SPIONPs were produced by the coprecipitation method, thereby loading drugs, doxorubicin (DOX), and paclitaxel (PTX) into the nanocomposites by physical diffusion. When compared to the same concentrations of free drugs, 10 mg/mL of DOX and PTX-loaded AP-SPIONPs significantly increased cytotoxicity and decreased the expression of antiapoptotic proteins. In immunocompromised female Balb/c mice, the *in vivo* antiproliferative efficacy of the current formulation resulted in ovarian tumor reduction from 2,920 to 143 mm^3 after 40 days (Javid et al. 2014).

Metal–organic frameworks (MOFs), which are porous crystalline solids comprising metal ions and organic linkers, are promising carriers for use in drug delivery systems (DDS) because of their high surface area, high porosity, and exceptional biocompatibility. Such an approach to conglomerate porosity of MOFs and magnetism within nanoparticles generates a novel carrier for targeted drug delivery to cancer cells (Bazzazzadeh et al. 2020).

13.5 BENEFITS OF METAL NANOCOMPOSITES IN TREATING OVARIAN CANCER

Metal nanocomposites have several potential benefits for treating OC. First, they can target specific cells or tissues more effectively than conventional treatments, which can help reduce the side effects associated with chemotherapy and other treatments. Second, they are highly biocompatible and can be tailored to the individual patient's needs. Third, they can be used to deliver drugs more effectively to cancer cells, which can help improve the efficacy of treatments. Finally, they can be used to detect cancer cells more accurately, which can help physicians diagnose and monitor the progression of the disease more effectively (Barani et al. 2021).

13.6 CHALLENGES IN USING METAL NANOCOMPOSITES FOR OVARIAN CANCER

The use of metal nanocomposites for OC presents several challenges, namely

    i. Biocompatibility
    ii. Targeting specificity
    iii. *In vitro* and *in vivo* instability
    iv. Optimization issues regarding imaging
    v. Drug delivery issues
    vi. Aggregation and agglomeration
    vii. Immune responses

13.6.1 Biocompatibility

Metal nanocomposites must be biocompatible, i.e., they should not induce harmful reactions in the body. This requires a thorough evaluation of the toxicity of the materials and the nanocomposites. There are several examples of biocompatibility issues with the use of metal nanocomposites for OC, including:

1. *Toxicity of metal ions*: Some metal nanocomposites release toxic metal ions into the body, which can cause damage to cells and tissues. For example, the release of silver ions from silver nanocomposites has been shown to cause toxicity to normal cells and tissues, leading to a decline in their efficacy for OC applications (Attarilar et al. 2020; Manuja et al. 2021).

2. *Formation of reactive oxygen species*: Some metal nanocomposites can induce the formation of ROS, which can cause oxidative stress and damage to cells and tissues. This has been reported for metal nanocomposites made of gold, silver, and iron (Manuja et al. 2021).

3. *Accumulation in organs*: Some metal nanocomposites can accumulate in specific organs, such as the liver, spleen, and kidneys, and cause toxicity. This can be caused by the physical and chemical properties of the nanocomposites or their interactions with the immune system (Chandrakala, Aruna, and Angajala 2022; Yao et al. 2019).

4. *Interference with normal physiological functions*: Some metal nanocomposites can interfere with normal physiological functions, such as the normal functioning of

enzymes, hormones, and neurotransmitters. This can lead to a range of adverse effects, including metabolic disorders, cardiovascular disease, and neurological disorders (Manuja et al. 2021).

13.6.2 Targeting specificity

One of the main challenges of using metal nanocomposites for OC is to ensure that they specifically target cancer cells and not healthy cells. This requires the development of nanocomposites with specific targeting moieties that can recognize and bind to cancer cells. The challenges in achieving targeting specificity when using metal nanocomposites for OC include the following:

1. *Inefficient delivery to cancer cells*: Metal nanocomposites may not effectively reach the cancer cells due to the lack of specificity in their interactions with the cells. This can result in low delivery efficiency and reduced efficacy of the nanocomposites for OC treatment (Nie et al. 2009).
2. *Off-target toxicity*: Metal nanocomposites may also cause toxicity to normal cells and tissues, leading to side effects and reducing their therapeutic efficacy. This can occur if the nanocomposites are not specifically targeted to the cancer cells and instead interact with normal cells (Damasco et al. 2020).
3. *Lack of cancer-specific receptors*: OC cells may not have the specific receptors necessary for the metal nanocomposites to bind to and enter the cells. This can result in low delivery efficiency and reduced efficacy of the nanocomposites for OC treatment (Liu et al. 2020; Martinelli, Pucci, and Ciofani 2019; Pantshwa et al. 2020).
4. *Competition with other cellular components*: Metal nanocomposites may compete with other cellular components for binding sites on the cancer cells, reducing their specificity and efficacy (Liu et al. 2020).
5. *Interference with normal physiological functions*: Some metal nanocomposites can interfere with normal physiological functions, such as the normal functioning of enzymes, hormones, and neurotransmitters. This can lead to a range of adverse effects, including metabolic disorders, cardiovascular disease, and neurological disorders (Liu et al. 2020).

13.6.3 *In vitro* and *in vivo* instability

Metal nanocomposites must be stable *in vitro* and *in vivo* to ensure their efficacy. This requires the development of nanocomposites with enhanced stability against degradation, aggregation, and clearance from the body. There are several *in vitro* and *in vivo* stability issues associated with metal nanocomposites used for OC, including:

1. *Aggregation*: Metal nanocomposites may aggregate *in vitro* and *in vivo*. Aggregation can be caused by physical and chemical interactions between the nanocomposites and the surrounding environment (Ashraf et al. 2018).
2. *Degradation*: Metal nanocomposites may degrade *in vivo*, leading to a decline in their efficacy and increased toxicity. Degradation can be caused by the interactions of the nanocomposites with biological fluids, such as blood and tissues, or by the action of enzymes or immune cells (Dzulkharnien and Rohani 2022).

3. *Photodegradation*: Some metal nanocomposites may photodegrade *in vitro* and *in vivo*. Photodegradation can be caused by the interaction of the nanocomposites with light (Karlsson et al. 2008; Nel et al. 2006).
4. *Leaching*: Metal nanocomposites may leach metal ions into the surrounding environment. Leaching can be caused by physical or chemical interactions between the nanocomposites and the environment (Biswas 2021).

13.6.4 Imaging

Metal nanocomposites are often used for imaging purposes, but the imaging properties of the nanocomposites must be optimized for OC applications. This requires the development of nanocomposites with improved imaging contrast, specificity, and sensitivity. There are several optimization issues for the imaging properties of metal nanocomposites used for the treatment of OC, including:

1. *Signal-to-noise ratio*: The signal-to-noise ratio of the metal nanocomposites may be low, leading to poor imaging quality and reduced accuracy in detecting OC (Arunadevi 2022; Shang, Xu, and Nienhaus 2019).
2. *Tissue penetration*: Some metal nanocomposites may not penetrate deep into the tissue, leading to limited imaging depth and reduced accuracy in detecting OC (Arunadevi 2022; Shang, Xu, and Nienhaus 2019).
3. *Background signal*: The background signal from tissues and organs may interfere with the imaging signal from the metal nanocomposites, leading to poor image quality and reduced accuracy (Arunadevi 2022; Shang, Xu, and Nienhaus 2019).
4. *Size*: The size of the metal nanocomposites may impact their imaging properties, with larger nanocomposites having better imaging properties but lower specificity for cancer cells (Arunadevi 2022; Shang, Xu, and Nienhaus 2019).
5. *Chemical composition*: The chemical composition of the metal nanocomposites may impact their imaging properties, with some compositions leading to better imaging properties but increased toxicity (Arunadevi 2022; Shang, Xu, and Nienhaus 2019).

13.6.5 Drug delivery

Metal nanocomposites are also used for drug delivery purposes, but the drug delivery efficiency of the nanocomposites must be optimized for OC applications. This requires the development of nanocomposites with enhanced drug-loading capacity, release rate, and specificity. The following factors affect drug delivery efficiency and efficacy in treating OC:

1. *Release rate*: The release rate of the drug from the metal nanocomposites may not be optimal (Chandrakala, Aruna, and Angajala 2022; Dzulkharnien and Rohani 2022).
2. *Target specificity*: The metal nanocomposites may not be specific enough to target only cancer cells (Dzulkharnien and Rohani 2022).
3. *Endocytosis*: The metal nanocomposites may not effectively enter cancer cells through endocytosis (Dzulkharnien and Rohani 2022).
4. *Drug stability*: The drug may not be stable in the metal nanocomposite (Dzulkharnien and Rohani 2022).

5. *Drug loading (encapsulation efficiency)*: Drug loading may be poor and need various approaches to improve it (He et al. 2021).

13.6.6 Aggregation and agglomeration during synthesis

Aggregation and agglomeration are two common phenomena that can occur during the production process of metal nanocomposites. Aggregation refers to the clumping together of particles, whereas agglomeration is the process by which these particles stick together to form larger clusters or aggregates. These two processes can have a significant impact on the properties and performance of metal nanocomposites, and therefore it is important to understand and control them during the production process (Ashraf et al. 2018).

13.6.7 Immune response

Some studies have shown that metal nanocomposites can induce both innate and adaptive immune responses. Innate immune responses involve the activation of immune cells such as macrophages and dendritic cells, which recognize and engulf foreign particles. Metal nanocomposites have been shown to activate these immune cells and stimulate the production of cytokines and chemokines, which are involved in inflammation and immune cell recruitment (Luo, Chang, and Lin 2019).

Adaptive immune responses involve the activation of T and B cells, which can recognize and respond to specific antigens. Metal nanocomposites have been shown to induce antigen-specific immune responses, which could potentially be harnessed for vaccine development. However, the mechanisms underlying these immune responses are not well understood, and further research is needed to fully characterize them (Roach, Stefaniak, and Roberts 2019).

It is important to note that the immune responses produced by metal nanocomposites can vary depending on factors such as their size, shape, composition, and surface properties. For example, gold nanocomposites have been shown to induce less inflammation than silver nanocomposites, possibly due to differences in their surface chemistry (Talarska, Boruczkowski, and Żurawski 2021).

These challenges highlight the need for continued research and development to optimize the use of metal nanocomposites for OC. Efforts should be made to develop nanocomposites that are biocompatible, stable, and specifically target cancer cells while delivering drugs and imaging agents.

13.7 ADVERSE EFFECTS/TOXICITIES ASSOCIATED WITH THE USE OF METAL NANOCOMPOSITES

The use of metal nanocomposites may induce toxicity in the body, resulting in harmful effects on cells and tissues. This can be caused by the release of toxic metal ions from the nanocomposites or the formation of ROS. The toxicity due to metal nanocomposite can be classified into two categories: local toxicity and systemic toxicity. The local toxicity can be caused by physical or chemical interactions between the nanocomposites and the tissues, such as inflammation, tissue damage, and oxidative stress. While in systemic toxicity the metal ions are released into the bloodstream affecting liver, kidneys, and other organs. In addition, metal nanocomposites may also cause toxicity by interfering with normal biological processes, such as cellular metabolism and immune function (Manuja et al. 2021; Padhi and Behera 2022).

Metal nanocomposites can elicit an immune response, resulting in the formation of antibodies against the nanocomposites. This can lead to a decline in the efficacy of the nanocomposites over time. Metal nanocomposites may be recognized as foreign bodies by the immune system, leading to the formation of antibodies against the nanocomposites. Eventually they may induce an inflammatory response, leading to the activation of immune cells such as macrophages and T cells. This may result in increased tissue damage and reduced drug delivery efficiency. Metal nanocomposites may suppress the immune system, reducing the ability of the immune system to target and eliminate cancer cells (Roach, Stefaniak, and Roberts 2019).

Metal nanocomposites can induce allergic reactions, such as skin rashes and itching. This can be caused by the release of allergens from the nanocomposites or their interactions with the immune system. In severe cases, an allergic reaction can progress to anaphylaxis, a life-threatening condition that can cause difficulty breathing, low blood pressure, and shock (Roach, Stefaniak, and Roberts 2019).

It is important to note that not all metal nanocomposites have the same potential for side effects, and the specific side effects depend on the properties and composition of the nanocomposites. Moreover, further research is needed to better understand the side effects of metal nanocomposites and to develop strategies to minimize or eliminate these side effects.

13.8 CONCLUSION AND FUTURE PERSPECTIVE

OC is one of the most lethal gynecologic malignancies, and current treatments encounter fundamental difficulties including chemotherapy medicines that exhibit toxicity and resistance. The absence of efficient targeted therapy for OC is still another problem. Researchers in this area are already working on promising techniques like individualized medicine, tailored drug delivery systems, and combination therapies. Yet to improve pharmaceutical administration, reduce toxicity, and deal with the issue of resistance, more sophisticated techniques must be developed. To maximize their safety and effectiveness, nanocomposites with the following strategies for treating OC may be used in the future.

Sánchez-Ramírez et al. (2020) reported the potential of biodegradable photo-responsive nanoparticles (NPs) as a platform for the simultaneous delivery of chemotherapy agents, photothermal agents, and photosensitizers for the treatment of OC. These NPs can be designed to be responsive to light, enabling targeted drug release and therapy. They can encapsulate chemotherapy agents such as doxorubicin, photothermal agents such as gold nanorods, and photosensitizers such as porphyrins. The biodegradability of these NPs reduces the risk of toxicity, and their photo-responsiveness allows for targeted drug release and therapy, reducing side effects.

The use of gold nanorods (GNRs) for photothermal therapy (PTT) to treat OC is discussed in this chapter. Because of their special optical characteristics, GNRs can absorb near-infrared (NIR) light and turn it into heat, which can kill cancer cells. To increase their effectiveness and lessen the possibility of spreading to healthy tissues, GNRs can be functionalized with targeting ligands, loaded with chemotherapeutic drugs, and included in scaffold materials. A prospective platform for the creation of cutting-edge and efficient OC therapeutics is represented by GNRs. Nevertheless, GNRs represent a promising platform for the development of novel and effective therapies for OC (Chuanzhi et al. 2020).

Near-infrared II light has a longer wavelength than NIR I and can penetrate deeper into tissues; it can target cancer cells with more accuracy and achieve better heat conversion efficiency. Photothermal therapy (PTT) caused by NIR-II can heat tumor tissues, which kills cells by denaturing proteins, rupturing cell membranes, and damaging DNA. PTT-induced hyperthermia can also result in ER stress, which can improve chemotherapy effectiveness by lowering the expression of efflux pumps and boosting drug accumulation. PTT plus chemotherapy may work synergistically to increase the death of tumor cells and enhance therapeutic effects (Kong et al. 2021).

Immune checkpoint inhibitors (ICIs) have demonstrated efficacious outcomes in the treatment of OC. Immunotherapy does not always work for all patients, and resistance can grow over time. To increase the effectiveness of immunotherapy, combination therapies are being investigated, such as combining ICIs with chemotherapy, targeted therapy, or radiation therapy. Moreover, the use of biomarkers can assist in identifying individuals who are more likely to respond to ICIs, such as tumor mutation burden and programmed death-ligand 1 (PD-L1) expression (Maiorano et al. 2021). Beyond systemic distribution, nanoparticle-based immunotherapies can enhance medication delivery to immune cells in the tumor microenvironment. These treatments can also administer immunostimulatory substances, such as agonists of the Toll-like receptor, to boost the immune response against cancer cells (Li et al. 2020).

Cancer medicines with low solubility or high toxicity have been demonstrated to have improved therapeutic efficacy when administered with pH-responsive nanocarriers. They can also lessen adverse effects by limiting drug exposure to healthy tissue and have the potential to overcome treatment resistance by delivering medications directly to cancer cells.

As pH-responsive nanocarriers, polymeric nanoparticles like poly(lactic-co-glycolic acid) (PLGA) and polyethyleneimine (PEI) are also frequently utilized (Padhi et al. 2022). It is possible to program the release of medications from these nanoparticles in response to pH, temperature, or other environmental cues. To deliver medications to only cancer cells, they can also be functionalized by targeting ligands like antibodies or peptides (Alsawaftah et al. 2022).

Image-guided therapy involves using imaging methods, such as computed tomography (CT) or magnetic resonance imaging (MRI), to direct the delivery of therapies to particular parts of the body. Both cancer treatment and diagnosis are possible using HER2-targeted theranostic nanoparticles. The HER2 receptor, which is overexpressed in many cancer types, including OC, is the target of these nanoparticles. These nanoparticles can deliver medications specifically to cancer cells while preserving healthy cells by focusing on HER2. Targeted drug delivery and image-guided therapy using HER2-targeted theranostic nanoparticles can improve treatment outcomes in the case of OC, which is frequently heterogeneous, meaning that it consists of various types of cancer cells with varying degrees of aggressiveness and response to therapy. Overall, HER2-targeted theranostic nanoparticles are a promising new tool for image-guided therapy and targeted drug administration in the treatment of heterogeneous OC. To ascertain the effectiveness and safety of this strategy in clinical settings, more research is necessary (Satpathy et al. 2019).

Despite substantial advancements in the use of nanomaterials for cancer therapy, more study is required to improve the design of nanoparticle and nanocomposite-based treatments, assess their efficacy and safety, and create suitable clinical protocols. Nonetheless, employing nanocomposite is a viable method for creating novel and efficient cancer treatments. Women with OC may experience significantly better outcomes if it is possible to properly target cancer cells and deliver medications or other treatments right to the cancer cells. The recent trends are encouraging and suggest that nanotechnological techniques will play an increasingly important role in the battle against OC, even though much more study is required to fully develop these approaches.

REFERENCES

Alsawaftah, Nour M, Nahid S Awad, William G Pitt, and Ghaleb A Husseini. 2022. "Ph-Responsive Nanocarriers In Cancer Therapy." *Polymers* No. 14 (5):936. Doi: Https://Doi.Org/10.3390/Polym14050936.

Arora, Hans C, Mark P Jensen, Ye Yuan, Aiguo Wu, Stefan Vogt, Tatjana Paunesku, and Gayle E Woloschak. 2012. "Nanocarriers Enhance Doxorubicin Uptake In Drug-Resistant Ovarian Cancer Cellsnanocarriers Enhance Clathrin-Mediated Endocytosis." *Cancer Research* No. 72 (3):769–778. Doi: Https://Doi.Org/10.1158/0008-5472.CAN-11-2890.

Arunadevi, N. 2022. "Metal Nanocomposites for Advanced Futuristic Biosensing Applications." *Materials Letters* No. 309:131320. Doi: Https://Doi.Org/10.1016/J.Matlet.2021.131320.

Ashraf, Muhammad Aqeel, Wanxi Peng, Yasser Zare, and Kyong Yop Rhee. 2018. "Effects of Size and Aggregation/Agglomeration of Nanoparticles on the Interfacial/Interphase Properties and Tensile Strength of Polymer Nanocomposites." *Nanoscale Research Letters* No. 13 (1):1–7. Doi: Https://Doi.Org/10.1186/S11671-018-2624-0.

Attarilar, Shokouh, Jinfan Yang, Mahmoud Ebrahimi, Qingge Wang, Jia Liu, Yujin Tang, and Junlin Yang. 2020. "The Toxicity Phenomenon and the Related Occurrence in Metal and Metal Oxide Nanoparticles: A Brief Review from the Biomedical Perspective." *Frontiers In Bioengineering and Biotechnology* No. 8:822. Doi: Https://Doi.Org/10.3389/Fbioe.2020.00822.

Auersperg, Nelly, Alice ST Wong, Kyung-Chul Choi, Sung Keun Kang, and Peter CK Leung. 2001. "Ovarian Surface Epithelium: Biology, Endocrinology, and Pathology." *Endocrine Reviews* No. 22 (2):255–288. Doi: Https://Doi.Org/10.1210/Edrv.22.2.0422.

Banerjee, Rajat, And Indranil Manna. 2013. *Ceramic Nanocomposites*: Elsevier.

Barani, Mahmood, Muhammad Bilal, Fakhara Sabir, Abbas Rahdar, and George ZJ Life Sciences Kyzas. 2021. "Nanotechnology in Ovarian Cancer: Diagnosis and Treatment." No. 266:118914. Doi: Https://Doi.Org/10.1016/J.Lfs.2020.118914.

Baskar, G, Garrick Bikku George, and M Chamundeeswari. 2017. "Synthesis and Characterization of Asparaginase Bound Silver Nanocomposite against Ovarian Cancer Cell Line A2780 and Lung Cancer Cell Line A549." *Journal of Inorganic and Organometallic Polymers And Materials* No. 27:87–94. Doi: Https://Doi.Org/10.1007/S10904-016-0448-X.

Bazzazzadeh, Amin, Babak Faraji Dizaji, Nazanin Kianinejad, Arezo Nouri, and Mohammad Irani. 2020. "Fabrication of Poly (Acrylic Acid) Grafted-Chitosan/Polyurethane/Magnetic MIL-53 Metal Organic Framework Composite Core-Shell Nanofibers for Co-Delivery of Temozolomide and Paclitaxel against Glioblastoma Cancer Cells." *International Journal Of Pharmaceutics* No. 587:119674. Doi: Https://Doi.Org/10.1016/J.Ijpharm.2020.119674.

Beehler, Gregory P, Manveen Sekhon, Julie A Baker, Barbara E Teter, Susan E Mccann, Kerry J Rodabaugh, and Kirsten B Moysich. 2006. "Risk of Ovarian Cancer Associated with BMI Varies by Menopausal Status." *The Journal Of Nutrition* No. 136 (11):2881–2886. Doi: Https://Doi.Org/10.1093/Jn/136.11.2881.

Behera, Anindita, and Santwana Padhi. 2022. "Ph-Sensitive Polymeric Nanoparticles for Cancer Treatment." *Polymeric Nanoparticles for the Treatment of Solid Tumors*, 401–425. Doi: 10.1007/978-3-031-14848-4_15

Biswas, Anjana. 2021. "Nanocomposite of Ceria and Trititanate Nanotubes as an Efficient Defluoridating Material for Real-Time Groundwater: Synthesis, Regeneration, and Leached Metal Risk Assessment." *ACS Omega* No. 6 (47):31751–31764. Doi: Https://Doi.Org/10.1021/Acsomega.1c04424.

Bolton, Kelly L, Georgia Chenevix-Trench, Cindy Goh, Siegal Sadetzki, Susan J Ramus, Beth Y Karlan, Diether Lambrechts, Evelyn Despierre, Daniel Barrowdale, and Lesley Mcguffog. 2012. "Association Between BRCA1 And BRCA2 Mutations and Survival in Women with Invasive Epithelial Ovarian Cancer." *Jama* No. 307 (4):382–389. Doi: Https://Doi.Org/10.1001/Jama.2012.20.

Camargo, M Constanza, Leslie T Stayner, Kurt Straif, Margarita Reina, Umaima Al-Alem, Paul A Demers, And Philip J Landrigan. 2011. "Occupational Exposure to Asbestos and Ovarian Cancer: A Meta-Analysis." *Environmental Health Perspectives* No. 119 (9):1211–1217. Doi: Https://Doi.Org/10.1289/Ehp.1003283.

Chan, David W, Vincent WS Liu, George SW Tsao, Kwok-Ming Yao, Toru Furukawa, Karen KL Chan, and Hextan YS Ngan. 2008. "Loss Of MKP3 Mediated By Oxidative Stress Enhances Tumorigenicity and Chemoresistance of Ovarian Cancer Cells." *Carcinogenesis* No. 29 (9):1742–1750. Doi: Https://Doi.Org/10.1093/Carcin/Bgn167.

Chandrakala, V, Valmiki Aruna, and Gangadhara Angajala. 2022. "Review on Metal Nanoparticles as Nanocarriers: Current Challenges and Perspectives in Drug Delivery Systems." *Emergent Materials*:1–23. Doi: Https://Doi.Org/10.1007/S42247-021-00335-X.

Choi, Jung-Hye, Alice ST Wong, He-Feng Huang, and Peter CK Leung. 2007. "Gonadotropins And Ovarian Cancer." *Endocrine Reviews* No. 28 (4):440–461. Doi: Https://Doi.Org/10.1210/Er.2006-0036.

Choi, Yun-Jung, Sangiliyandi Gurunathan, and Jin-Hoi Kim. 2018. "Graphene Oxide—Silver Nanocomposite Enhances Cytotoxic And Apoptotic Potential Of Salinomycin in Human Ovarian Cancer Stem Cells (Ovcscs): A Novel Approach for Cancer Therapy." *International Journal Of Molecular Sciences* No. 19 (3):710. Doi: Https://Doi.Org/10.3390/Ijms19030710.

Chuanzhi, LIU, GONG Ping, Yuping Liang, WANG Zuobin, And WANG Li. 2020. "Application of Gold Nanorods for Photothermal Therapy." *Materials Science* No. 26 (3):243–248. Doi: Https://Doi.Org/10.5755/J01.Ms.26.3.21577.

Cramer, Daniel W, Allison F Vitonis, Kathryn L Terry, William R Welch, and Linda J Titus. 2016. "The Association Between Talc Use and Ovarian Cancer: A Retrospective Case—Control Study in Two US States." *Epidemiology (Cambridge, Mass.)* No. 27 (3):334. Doi: Https://Doi.Org/10.1097/EDE.0000000000000434.

Crum, Christopher P, Ronny Drapkin, Alexander Miron, Tan A Ince, Michael Muto, David W Kindelberger, and Yonghee Lee. 2007. "The Distal Fallopian Tube: A New Model for Pelvic Serous Carcinogenesis." *Current Opinion In Obstetrics And Gynecology* No. 19 (1):3–9. Doi: Https://Doi.Org/10.1097/GCO.0b013e328011a21f.

Damasco, Jossana A, Saisree Ravi, Joy D Perez, Daniel E Hagaman, and Marites P Melancon. 2020. "Understanding Nanoparticle Toxicity to Direct a Safe-By-Design Approach in Cancer Nanomedicine." *Nanomaterials* No. 10 (11):2186. Doi: Https://Doi.Org/10.3390/Nano10112186.

Dontsova, Tetiana A, Svitlana V Nahirniak, and Ihor MJ Astrelin. 2019. "Metaloxide Nanomaterials and Nanocomposites of Ecological Purpose." *Journal of Nanomaterials* No. 2019:1–31. Doi: Https://Doi.Org/10.1155/2019/5942194.

Dunyaporn, Trachootham, Alexandre Jerome, and Huang Peng. 2009. "Targeting Cancer Cells By ROS-Mediated Mechanisms: A Radical Therapeutic Approach." *Nat. Rev. Drug Discov* No. 8:579–591. Doi: Https://Doi.Org/10.1038/Nrd2803.

Dzulkharnien, Nur Syafiqah Farhanah, And Rosiah Rohani. 2022. "A Review on Current Designation of Metallic Nanocomposite Hydrogel in Biomedical Applications." *Nanomaterials* No. 12 (10):1629. Doi: Https://Doi.Org/10.3390/Nano12101629.

Esteller, Manel. 2008. "Epigenetics in Cancer." *New England Journal of Medicine* No. 358 (11):1148–1159. Doi: Https://Doi.Org/10.1056/Nejmra072067.

Faber, Mette T, Susanne K Kjær, Christian Dehlendorff, Jenny Chang-Claude, Klaus K Andersen, Estrid Høgdall, Penelope M Webb, Susan J Jordan, Australian Cancer Study, and Australian Ovarian Cancer Study Group. 2013. "Cigarette Smoking and Risk of Ovarian Cancer: A Pooled Analysis Of 21 Case—Control Studies." *Cancer Causes & Control* No. 24:989–1004. Doi: Https://Doi.Org/10.1007/S10552-013-0174-4.

Fan, Jin-Xuan, Miao-Deng Liu, Chu-Xin Li, Sheng Hong, Di-Wei Zheng, Xin-Hua Liu, Si Chen, Hong Cheng, and Xian-Zheng Zhang. 2017. "A Metal—Semiconductor Nanocomposite As An Efficient Oxygen-Independent Photosensitizer For Photodynamic Tumor Therapy." *J Nanoscale Horizons* No. 2 (6):349–355. Doi: Https://Doi.Org/10.1039/C7NH00087A.

Fathalla, Mahmoud Fahmy. 1971. "Incessant Ovulation—A Factor in Ovarian Neoplasia?" *The Lancet* No. 298 (7716):163. Doi: Https://Doi.Org/10.1016/S0140-6736(71)92335-X.

Foroozandeh, Amin, Majid Abdouss, Hossein Salaramoli, Mehrab Pourmadadi, and Fatemeh Yazdian. 2023. "An Electrochemical Aptasensor Based on G-C3N4/FE3O4/Pani Nanocomposite Applying Cancer Antigen_125 Biomarkers Detection." *Process Biochemistry* 127: 82–91. Doi:10.1016/J.Procbio.2023.02.004.

Gurunathan, Sangiliyandi, Jae Woong Han, Jung Hyun Park, Eunsu Kim, Yun-Jung Choi, Deug-Nam Kwon, and Jin-Hoi Kim. 2015. "Reduced Graphene Oxide—Silver Nanoparticle Nanocomposite: A Potential Anticancer Nanotherapy." *International Journal of Nanomedicine* No. 10:6257. Doi: Https://Doi.Org/10.2147/IJN.S92449.

Han, Wei, Xia Liu, Lingyan Wang, And Xuemei Zhou. 2020. "Engineering of Lipid Microbubbles-Coated Copper and Selenium Nanoparticles: Ultrasound-Stimulated Radiation of Anticancer Activity Ian Human Ovarian Cancer Cells." *Process Biochemistry* No. 98:113–121. Doi: Https://Doi.Org/10.1016/J.Procbio.2020.07.013.

Hanahan, Douglas, and Robert A Weinberg. 2011. "Hallmarks of Cancer: The Next Generation." *Cell* No. 144 (5):646–674. Doi: Https://Doi.Org/10.1016/J.Cell.2011.02.013.

Hanna, Demiana H, And Gamal R. Saad. 2021. "Induction of Mitochondria Mediated Apoptosis in Human Ovarian Cancer Cells By Folic Acid Coated Tin Oxide Nanoparticles." *Plos One* No. 16 (10):E0258115. Doi: Https://Doi.Org/10.1371/Journal.Pone.0258115.

He, Siyu, Li Wu, Xue Li, Hongyu Sun, Ting Xiong, Jie Liu, Chengxi Huang, Huipeng Xu, Huimin Sun, and Weidong Chen. 2021. "Metal-Organic Frameworks for Advanced Drug Delivery." *Acta Pharmaceutica Sinica B* No. 11 (8):2362–2395. Doi: Https://Doi.Org/10.1016/J.Apsb.2021.03.019.

Hielscher, Abigail, and Sharon Gerecht. 2015. "Hypoxia And Free Radicals: Role in Tumor Progression and the Use of Engineering-Based Platforms to Address These Relationships." *Free Radical Biology and Medicine* No. 79:281–291. Doi: Https://Doi.Org/10.1016/J.Freeradbiomed.2014.09.015.

Hong, Yoon-Hee, Md Hafiz Uddin, Untek Jo, Boyun Kim, Jiyoung Song, Dong Hoon Suh, Hee Seung Kim, And Yong Sang Song. 2015. "ROS Accumulation By PEITC Selectively Kills Ovarian Cancer Cells Via UPR-Mediated Apoptosis." *Frontiers In Oncology* No. 5:167. Doi: Https://Doi.Org/10.3389/Fonc.2015.00167.

Huang, Junjie, Wing Chung Chan, Chun Ho Ngai, Veeleah Lok, Lin Zhang, Don Eliseo Lucero-Prisno, Wanghong Xu, Zhi-Jie Zheng, Edmar Elcarte, Mellissa Withers, Martin C. S. Wong, and On Behalf of NCD Global Health Research Group of Association of Pacific Rim Universities. 2022. "Worldwide Burden, Risk Factors, and Temporal Trends of Ovarian Cancer: A Global Study." No. 14 (9):2230. Doi: Https://Doi.Org/10.3390/Cancers14092230.

Javid, Amaneh, Shahin Ahmadian, Ali Akbar Saboury, Seyed Mehdi Kalantar, Saeed Rezaei-Zarchi, and Sughra Shahzad. 2014. "Biocompatible APTES—PEG Modified Magnetite Nanoparticles: Effective Carriers of Antineoplastic Agents To Ovarian Cancer." *Applied Biochemistry And Biotechnology* No. 173:36–54. Doi: Https://Doi.Org/10.1007/S12010-014-0740-6.

Ji, Na, Chunyan Dong, and Jingjing Jiang. 2022. "Evaluation of Antioxidant, Cytotoxicity, and Anti-Ovarian Cancer Properties of The Fe3O4@ CS-Starch/Cu Bio-Nanocomposite." *Inorganic Chemistry Communications* No. 140:109452. Doi: Https://Doi.Org/10.1016/J.Inoche.2022.109452.

Kareem, Sahira Hassan, Amel Muhson Naji, Zainab J Taqi, and Majid S Jabir. 2020. "Polyvinylpyrrolidone Loaded-Mnznfe2o4 Magnetic Nanocomposites Induce Apoptosis in Cancer Cells Through Mitochondrial Damage And P53 Pathway." *Journal of Inorganic and Organometallic Polymers and Materials* No. 30 (12):5009–5023. Doi: Https://Doi.Org/10.1007/S10904-020-01651-1.

Karlsson, Hanna L, Pontus Cronholm, Johanna Gustafsson, and Lennart Moller. 2008. "Copper Oxide Nanoparticles are Highly Toxic: A Comparison Between Metal Oxide Nanoparticles and Carbon Nanotubes." *Chemical Research in Toxicology* No. 21 (9):1726–1732. Doi: Https://Doi.Org/10.1021/Tx800064j.

Karst, Alison M, and Ronny Drapkin. 2010. "Ovarian Cancer Pathogenesis: A Model in Evolution." *Journal Of Oncology* No. 2010. Doi: Https://Doi.Org/10.1155/2010/932371.

Karunakaran, Chandran, Raju Rajkumar, and Kalpana Bhargava. 2015. "Introduction to Biosensors." In *Biosensors And Bioelectronics*, 1–68. Elsevier.

Kinde, Isaac, Chetan Bettegowda, Yuxuan Wang, Jian Wu, Nishant Agrawal, Ie-Ming Shih, Robert Kurman, Fanny Dao, Douglas A Levine, and Robert Giuntoli. 2013. "Evaluation of DNA from the Papanicolaou Test to Detect Ovarian and Endometrial Cancers." *Science Translational Medicine* No. 5 (167):167ra4–167ra4. Doi: Https://Doi.Org/10.1126/Scitranslmed.3004952.

Kong, Qingduo, Dengshuai Wei, Peng Xie, Bin Wang, Kunyi Yu, Xiang Kang, and Yongjun Wang. 2021. "Photothermal Therapy Via NIR II Light Irradiation Enhances DNA Damage and Endoplasmic Reticulum Stress for Efficient Chemotherapy." *Frontiers In Pharmacology* No. 12:670207. Doi: Https://Doi.Org/10.3389/Fphar.2021.670207.

Kurman, Robert J, and Ie-Ming Shih. 2010. "The Origin and Pathogenesis of Epithelial Ovarian Cancer-A Proposed Unifying Theory." *The American Journal Of Surgical Pathology* No. 34 (3):433. Doi: Https://Doi.Org/10.1097/PAS.0b013e3181cf3d79.

Kwon, Youngjoo. 2021. "Possible Beneficial Effects of N-Acetylcysteine for Treatment of Triple-Negative Breast Cancer." *Antioxidants* No. 10 (2):169. Doi: Https://Doi.Org/10.3390/Antiox10020169.

Lheureux, Stephanie, Marsela Braunstein, and Amit M Oza. 2019. "Epithelial Ovarian Cancer: Evolution Of Management in the Era of Precision Medicine." *CA: A Cancer Journal for Clinicians* No. 69 (4):280–304. Doi: Https://Doi.Org/10.3322/Caac.21559.

Li, Yuan, Yan Gao, Xi Zhang, Hongyan Guo, and Huile Gao. 2020. "Nanoparticles in Precision Medicine for Ovarian Cancer: From Chemotherapy to Immunotherapy." *International Journal of Pharmaceutics* No. 591:119986. Doi: Https://Doi.Org/10.1016/J.Ijpharm.2020.119986.

Liu, Chen-Guang, Ya-Hui Han, Ranjith Kumar Kankala, Shi-Bin Wang, and Ai-Zheng Chen. 2020. "Subcellular Performance of Nanoparticles in Cancer Therapy." *International Journal of Nanomedicine*:675–704. Doi: Https://Doi.Org/10.2147/IJN.S226186.

Luo, Yueh-Hsia, Louis W Chang, and Pinpin Lin. 2019. "Metal-Based Nanoparticles and the Immune System: Activation, Inflammation, and Potential Applications." In *Immune Aspects of Biopharmaceuticals And Nanomedicines*, 699–730. Jenny Stanford Publishing.

Macciò, Antonio, and Clelia Madeddu. 2012. "Inflammation and Ovarian Cancer." *Cytokine* No. 58 (2):133–147. Doi: Https://Doi.Org/10.1016/J.Cyto.2012.01.015.

Maiorano, Brigida Anna, Mauro Francesco Pio Maiorano, Domenica Lorusso, and Evaristo Maiello. 2021. "Ovarian Cancer in the Era of Immune Checkpoint Inhibitors: State of the Art and Future Perspectives." *Cancers* No. 13 (17):4438. Doi: Https://Doi.Org/10.3390/Cancers13174438.

Manuja, Anju, Balvinder Kumar, Rajesh Kumar, Dharvi Chhabra, Mayukh Ghosh, Mayank Manuja, Basanti Brar, Yash Pal, BN Tripathi, and Minakshi Prasad. 2021. "Metal/Metal Oxide Nanoparticles: Toxity Concerns Associated With Their Physical State and Remediation for Biomedical Applications." *Toxicology Reports* No. 8:1970–1978. Doi: Https://Doi.Org/10.1016/J.Toxrep.2021.11.020.

Markman, Maurie. 2019. "Pharmaceutical Management of Ovarian Cancer: Current Status." *Drugs* No. 79 (11):1231–1239. Doi: Https://Doi.Org/10.1007/S40265-019-01158-1.

Martinelli, Chiara, Carlotta Pucci, and Gianni Ciofani. 2019. "Nanostructured Carriers as Innovative Tools for Cancer Diagnosis and Therapy." *APL Bioengineering* No. 3 (1):011502. Doi: Https:// Doi.Org/10.1063/1.5079943.

Mishra, YK, S Mohapatra, DK Avasthi, D Kabiraj, NP Lalla, JC Pivin, Himani Sharma, Rajarshi Kar, and Neeta Singh. 2007. "Gold—Silica Nanocomposites for the Detection of Human Ovarian Cancer Cells: A Preliminary Study." *Nanotechnology* No. 18 (34):345606. Doi: Https://Doi. Org/10.1088/0957-4484/18/34/345606.

Naskar, Atanu, Susanta Bera, Rahul Bhattacharya, Sib Sankar Roy, and Sunirmal Jana. 2018. "Effect of Bovine Serum Albumin Immobilized Au-Zno—Graphene Nanocomposite On Human Ovarian Cancer Cell." *Journal Of Alloys And Compounds* No. 734:66–74. Doi: Https://Doi. Org/10.1016/J.Jallcom.2017.11.029.

Nel, Andre, Tian Xia, Lutz Madler, and Ning Li. 2006. "Toxic Potential of Materials at the Nano-level." *Science* No. 311 (5761):622–627. Doi: Https://Doi.Org/10.1126/Science.1114397.

Nie, Hemin, Shih Tak Khew, Lai Yeng Lee, Kai Ling Poh, Yen Wah Tong, and Chi-Hwa Wang. 2009. "Lysine-Based Peptide-Functionalized PLGA Foams for Controlled DNA Delivery." *Journal Of Controlled Release* No. 138 (1):64–70. Doi: Https://Doi.Org/10.1016/J.Jconrel.2009.04.027.

Padhi, Santwana, and Anindita Behera. 2021. "Advanced Drug Delivery Systems in the Treatment of Ovarian Cancer." In *Advanced Drug Delivery Systems in the Management of Cancer*, pp. 127–139. Academic Press. Doi: 10.1016/B978-0-323-85503-7.00020-1.

Padhi, Santwana, and Anindita Behera. 2022. "Cellular Internalization and Toxicity of Polymeric Nanoparticles." In *Polymeric Nanoparticles for the Treatment of Solid Tumors*, pp. 473–488. Cham: Springer International Publishing. doi: 10.1007/978-3-031-14848-4_17.

Padhi, Santwana, Sweta Priyadarshini Pradhan, and Anindita Behera. 2022. "Methods to Formulate Polymeric Nanoparticles." In *Polymeric Nanoparticles for the Treatment of Solid Tumors*, pp. 51–74. Cham: Springer International Publishing. doi: 10.1007/978-3-031-14848-4_2

Palmero, Paola. 2015. "Structural Ceramic Nanocomposites: A Review of Properties and Powders' Synthesis Methods." *J Nanomaterials* No. 5 (2):656–696. https://doi.org/10.3390/Nano5020656.

Pantshwa, Jonathan M, Pierre PD Kondiah, Yahya E Choonara, Thashree Marimuthu, and Viness Pillay. 2020. "Nanodrug Delivery Systems for the Treatment of Ovarian Cancer." *Cancers* No. 12 (1):213. Doi: Https://Doi.Org/10.3390/Cancers12010213.

Parchwani, Deepak, Sagar Jayantilal Dholariya, Sohil Takodara, Ragini Singh, Vivek Kumar Sharma, Alpana Saxena, Digishaben D Patel, and Madhuri Radadiya. 2022. "Analysis of Prediagnostic Circulating Levels of Gonadotropins and Androgens with Risk of Epithelial Ovarian Cancer." *Journal Of Laboratory Physicians* No. 14 (1):47–56. Doi: Https://Doi. Org/10.1055/S-0041-1741443.

Ponnusamy, Moorthy P, and Surinder K Batra. 2008. "Ovarian Cancer: Emerging Concept on Cancer Stem Cells." *Journal of Ovarian Research* No. 1:1–9. Doi: Https://Doi.Org/10.1186/1757-2215-1-4.

Ranch, Ketan M, Manish R Shukla, Furqan A Maulvi, and Ditixa T Desai. 2021. "Carbon-Based Nanoparticles and Dendrimers for Delivery of Combination Drugs." In *Nanocarriers for the Delivery of Combination Drugs*, pp. 227–257. Elsevier.

Rattanapan, Yanisa, Veerawat Korkiatsakul, Adcharee Kongruang, Teerapong Siriboonpiputtana, Budsaba Rerkamnuaychoke, and Takol Chareonsirisuthigul. 2020. "Microrna Expression Profiling of Epithelial Ovarian Cancer Identifies New Markers of Tumor Subtype." *Microrna* No. 9 (4):289–294. Doi: Https://Doi.Org/10.2174/2211536609666200722125737.

Reddy, Boreddy. 2011. *Advances in Nanocomposites: Synthesis, Characterization and Industrial Applications*: Bod—Books On Demand.

Riman, Tomas, Ingemar Persson, and Staffan Nilsson. 1998. "Hormonal Aspects of Epithelial Ovarian Cancer: Review of Epidemiological Evidence." *Clinical Endocrinology* No. 49 (6):695–707. Doi: Https://Doi.Org/10.1046/J.1365-2265.1998.00577.X.

Roach, Katherine A, Aleksandr B Stefaniak, and Jenny R Roberts. 2019. "Metal Nanomaterials: Immune Effects and Implications of Physicochemical Properties on Sensitization, Elicitation, and Exacerbation of Allergic Disease." *Journal Of Immunotoxicology* No. 16 (1):87–124. Doi: Https://Doi.Org/10.1080/1547691X.2019.1605553.

Sánchez-Ramírez, Dante R, Rossina Domínguez-Ríos, Josué Juárez, Miguel Valdés, Natalia Hassan, Antonio Quintero-Ramos, Alicia Del Toro-Arreola, Silvia Barbosa, Pablo Taboada, and Antonio Topete. 2020. "Biodegradable Photoresponsive Nanoparticles for Chemo-, Photothermal- and Photodynamic Therapy of Ovarian Cancer." *Materials Science And Engineering: C* No. 116:111196. Doi: Https://Doi.Org/10.1016/J.Msec.2020.111196.

Satpathy, Minati, Liya Wang, Rafal J Zielinski, Weiping Qian, Y Andrew Wang, Aaron M Mohs, Brad A Kairdolf, Xin Ji, Jacek Capala, and Malgorzata Lipowska. 2019. "Targeted Drug Delivery and Image-Guided Therapy of Heterogeneous Ovarian Cancer Using HER2-Targeted Theranostic Nanoparticles." *Theranostics* No. 9 (3):778. Doi: Https://Doi.Org/10.7150/Thno.29964.

Semenza, Gregg L. 2010. "HIF-1: Upstream and Downstream of Cancer Metabolism." *Current Opinion In Genetics & Development* No. 20 (1):51–56. Doi: Https://Doi.Org/10.1016/J.Gde.2009.10.009.

Shah, Monjri M, and Charles N Landen. 2014. "Ovarian Cancer Stem Cells: Are They Real and Why Are They Important?" *Gynecologic Oncology* No. 132 (2):483–489. Doi: Https://Doi.Org/10.1016/J.Ygyno.2013.12.001.

Shang, Li, Jie Xu, and Gerd Ulrich Nienhaus. 2019. "Recent Advances in Synthesizing Metal Nanocluster-Based Nanocomposites for Application in Sensing, Imaging and Catalysis." *Nano Today* No. 28:100767. Doi: Https://Doi.Org/10.1016/J.Nantod.2019.100767.

Sharma, Poonam, R Krishnapriya, Pragati R Sharma, and Rakesh K Sharma. 2020. "Recent Advances in Synthesis of Metal—Carbon Nanocomposites and Their Application in Catalytic Hydrogenation Reactions." *J Advanced Heterogeneous Catalysts Volume 1: Applications At The Nano-Scale* 403–458. Doi: Https://Doi.Org/10.1021/Bk-2020–1359.Ch014.

Steelman, Linda S, William H Chappell, Stephen L Abrams, C Ruth Kempf, Jacquelyn Long, Piotr Laidler, Sanja Mijatovic, Danijela Maksimovic-Ivanic, Franca Stivala, and Maria C Mazzarino. 2011. "Roles of the Raf/MEK/ERK and PI3K/PTEN/Akt/Mtor Pathways in Controlling Growth and Sensitivity to Therapy-Implications for Cancer and Aging." *Aging (Albany NY)* No. 3 (3):192. Doi: Https://Doi.Org/10.18632/Aging.100296.

Sun, Hongguang, and Youli Zu. 2015. "A Highlight of Recent Advances in Aptamer Technology and Its Application." *Molecules* No. 20 (7):11959–11980. Doi: Https://Doi.Org/10.3390/Molecules200711959.

Talarska, Patrycja, Maciej Boruczkowski, and Jakub Żurawski. 2021. "Current Knowledge of Silver and Gold Nanoparticles in Laboratory Research—Application, Toxicity, Cellular Uptake." *Nanomaterials* No. 11 (9):2454. Doi: Https://Doi.Org/10.3390/Nano11092454.

Vang, Russell, Ie-Ming Shih, and Robert J Kurman. 2009. "Ovarian Low-Grade and High-Grade Serous Carcinoma: Pathogenesis, Clinicopathologic and Molecular Biologic Features, and Diagnostic Problems." *Advances In Anatomic Pathology* No. 16 (5):267. Doi: Https://Doi.Org/10.1097/PAP.0b013e3181b4fffa.

Wei, Jian-Jun, Josette William, and Serdar Bulun. 2011. "Endometriosis and Ovarian Cancer: A Review of Clinical, Pathologic, and Molecular Aspects." *International Journal of Gynecological Pathology: Official Journal of the International Society of Gynecological Pathologists* No. 30 (6):553. Doi: Https://Doi.Org/10.1097/PGP.0b013e31821f4b85.

Willmott, Lyndsay J, and John P Fruehauf. 2010. "Targeted Therapy in Ovarian Cancer." *Journal of Oncology* No. 2010. Doi: Https://Doi.Org/10.1155/2010/740472.

Wright, Alexi A, Kari Bohlke, Deborah K Armstrong, Michael A Bookman, William A Cliby, Robert L Coleman, Don S Dizon, Joseph J Kash, Larissa A Meyer, and Kathleen N Moore. 2016. "Neoadjuvant Chemotherapy for Newly Diagnosed, Advanced Ovarian Cancer: Society of Gynecologic

Oncology and American Society of Clinical Oncology Clinical Practice Guideline." *Gynecologic Oncology* No. 143 (1):3–15. Doi: Https://Doi.Org/10.1016/J.Ygyno.2016.05.022.

Xu, Jing, Li Li, Jun Zhang, and Yifei Min. 2021. "Decorated of Au Nps Over L-Arginine-Modified Fe3O4 Nanoparticles As a Novel Nanomagnetic Composite for the Treatment of Human Ovarian Cancer." *Arabian Journal of Chemistry* No. 14 (8):103283. Doi: Https://Doi.Org/10.1016/J.Arabjc.2021.103283.

Yao, Ying, Yiteng Zang, Jing Qu, Meng Tang, and Ting Zhang. 2019. "The Toxicity of Metallic Nanoparticles on Liver: The Subcellular Damages, Mechanisms, and Outcomes." *International Journal of Nanomedicine*:8787–8804.

Chapter 14

Nanotherapeutics for breast cancer using metal nanocomposites

Devesh U. Kapoor, Deepak Sharma, Pratishtha Sharma, Rupesh K. Gautam, and Naitik D. Trivedi

LIST OF ABBREVIATIONS

3D-MCTSs	3-D Multicellular Tumor Spheroids
Ag–DOX NC	Gold–Doxorubicin Nanocomposite
Ag NCs	Silver Nanocomposites
Ag NPs	Silver NPs
AMJ13	Arabian and Middle Eastern breast cancer cell line
ALB	Albumin
Arg-Gly-Asp, RGD	Arginine-Glycine-Aspartic
Au/GO	Gold/Graphene Oxide
Au NCs	Gold Nanocomposites
Au NS	Gold Nanostar
BC	Breast Cancer
BSA	Bovine Serum Albumin
Ce NCs	Cerium-Based Nanocomposites
CM	*Curcuma mangga*
CNTs	Carbon Nanotubes
Cu NCs	Copper-Based Nanocomposites
CU NPs	Copper NPs
DM1	Maitansine
DOX	Doxorubicin
DTX	Docetaxel
FeO NCs	Iron Oxide Nanocomposites
FESEM	Field-Emission Scanning Electron Microscopy
GO	Graphene Oxide
HDF1	Human Fibroblast Cell Line
HER2	Epidermal Growth Factor Receptor 2 Antibodies
HRGO	Reduced Graphene Oxide
IC50	Inhibitory Concentration
ICG	Indocyanine Green
LDH	Lactate Dehydrogenase
LPSR	Localized Plasmon Surface Resonance
l-PSR	Plasmon Surface Resonance
MBI	Molecular Breast Imaging
MNCs	Metal-Based Nanocomposites

DOI: 10.1201/9781032621135-14

MNPs	Metallic NPs
MRI	Magnetic Resonance Imaging
MWCNTs	Multiwalled CNTs
NF	Nanofibers
NIR	Near-Infrared
NPs	Plasmonic NPs
NRs	Nanorods
PD-1	Programmed Cell Death-1
PN	*Prunus Nepalensis*
PRs	Progesterone and Estrogen Receptors
PTT	Photothermal Therapy
PTX	Paclitaxel
QDs	Quantum Dots
RGD	Arginylglycylaspartic acid
ROS	Reactive Oxygen Species
SAR	Specific Absorption Rate
TiO_2 NCs	Titanium Dioxide Nanocomposite
TNBC	Triple Negative Breast Cancer
ZnO NCs	Zinc Oxide Nanocomposites

14.1 INTRODUCTION

14.1.1 Breast cancer

The most prevalent form of cancer in women is breast cancer. The breast of women has different cells and based on the cell types and functioning, breast cells can be of three types: Lobules or the milk-secreting glands; ducts—the transporter of milk from lobules to nipple, and the connective tissue (fibrous and fatty tissue) responsible for the size and shape of the breast. Breast cancer growth is comparative in people (male/female); be that as it may, cancer in men's breasts is all the more as often as possible, positive to chemical receptors, and might be extra delicate to hormonal-based therapy. The gamble gives off an impression of being higher with acquired BRCA2 as opposed to mutations of the BRCA1 gene. Men will generally be analyzed at a later sickness stage and a more established age than females. Growths in men's bosom are bound to communicate the progesterone and estrogen receptors (PRs) and lesser inclined to overexpose Her-2/neu compared to female partners. However, in males, it is frequently analyzed at a later stage usually stage III or IV of cancer, and the show is normally a bump or areola reversal. Public drives are progressively expected to give data and backing to male bosom disease patients (Nounou et al., 2015). Depending upon the type of breast cells invaded, breast cancer can be classified as (Figure 14.1) (Khan et al., 2022):

1. *Intrusive Ductal Carcinoma*: It begins in ducts and spreads to other parts of the breast and then metastasizes to the body.
2. *Intrusive Lobular Carcinoma*: It begins in lobules and spreads to other parts of the breast and then to the body.

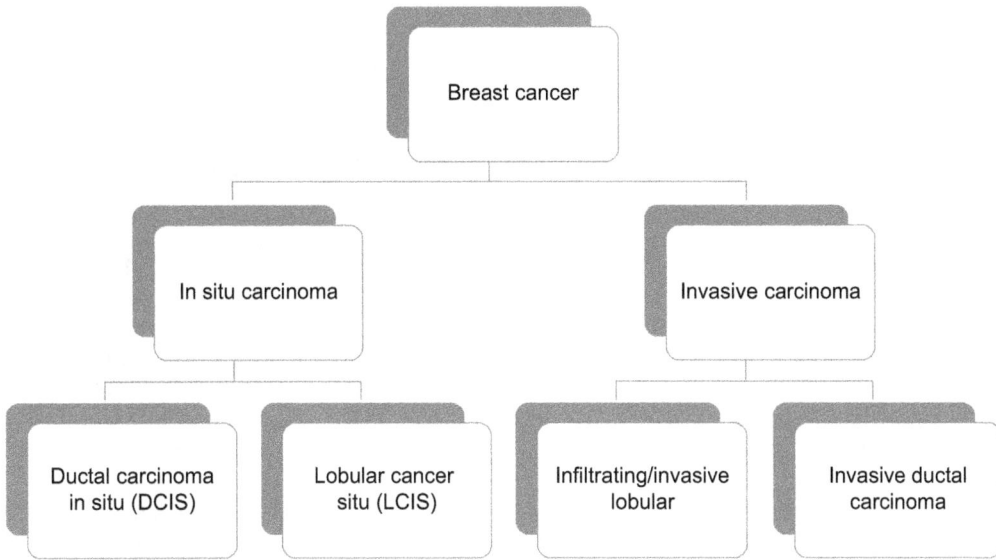

Figure 14.1 Classification of breast cancer.

Some other prevalent types of breast cancer include but are not limited to (Karakas, 2011) the following.

Inflammatory Breast Cancer: It is one of the rarest and most offensive cancer-causing blockages of lymph vessels of breast skin, leading to swelling and inflammation.

Paget Disease of the Breast: It is the cancer of the areola (dark color skin surrounding the nipple) and nipple skin. It is one of the rarest cancer types, which can be seen in men also but is rarest. Initial symptoms seem to be similar to that of eczema or psoriasis.

Mucinous Carcinoma: It is a special type of cancer subtype as per WHO, which starts in the mucous cells and is one of the most invasive cancers, it may invade any cell type responsible for the production of mucus but majorly invades the breast cells.

As per a report published by WHO, by the end of 2020, the number of alive women sufferings was more than 7.8 million (Hawrot et al., 2021), although, being the most ubiquitous cancer type among women due to its heterogeneity and different development stages, necessitates enhanced consideration of the disease. The treatment success rate is high nowadays due to modern methodologies for early and better diagnosis, empowering treatment using niftier technology and therapeutics, which are target-specific and even cell-specific.

In 2020, approximately 2.3 million new cases of breast cancer surpassed the incidences of lung cancer, becoming the leader of all cancers, with a share of about 12% among all cancer cases (Figure 14.2). From 1965 to 1985, the rise in incidences of breast cancer in India is almost 50%; in 2016; the reported breast cancer cases in females in India was more than 5,00,000, which has increased in the last 26 years (1990–2016) by 39%. Figure 14.3 depicts the rate of death due to different types of cancer.

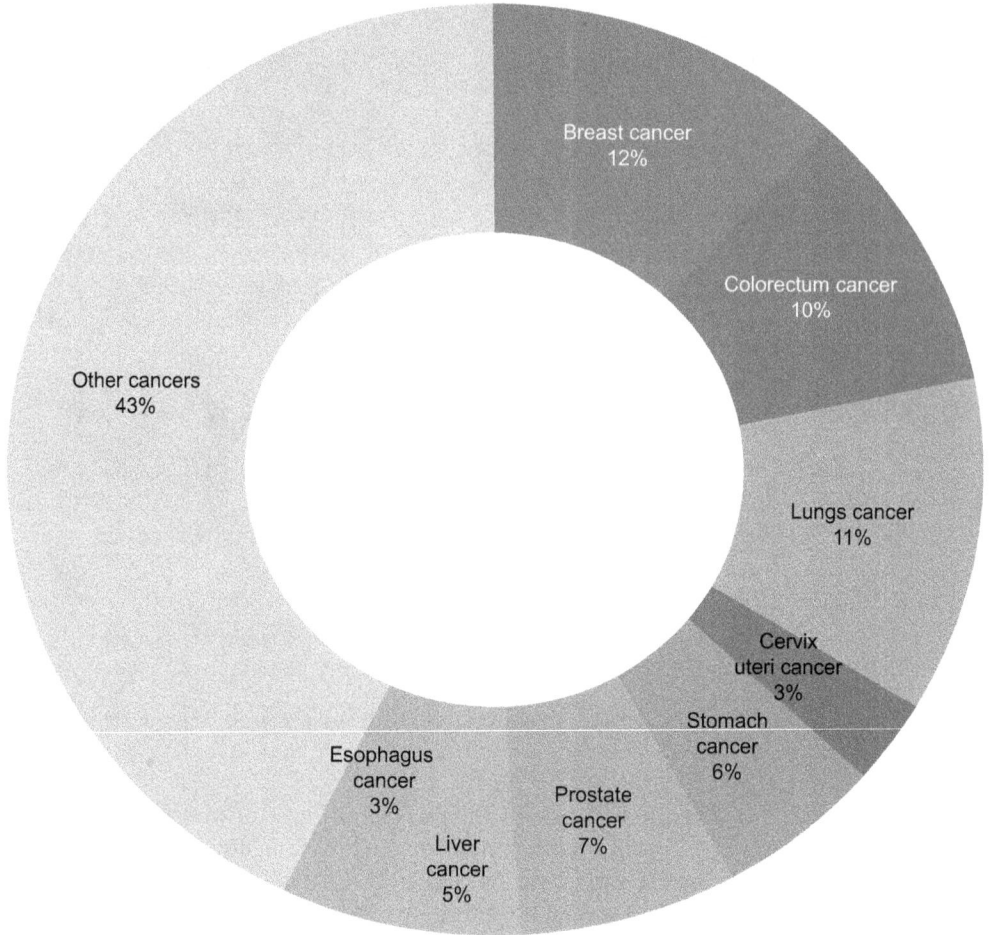

Figure 14.2 Representation of the percentage of new incidences of cancer (irrespective of age and sex) (in 2020 as per WHO).

With the impact of urbanization on the lifestyle of women, the number of breast cancer cases is increasing day by day. The major factors responsible for increasing the chances of pre- and post-menopausal breast cancer include, but are not limited to, alcohol consumption, high-fat-diet, greater weight, higher consumption of contraceptive pills, hormonal-based replacement therapy, and stagnant and stressful lifestyle of urban women while lactation and heavy physical exercise may decrease the chances. In addition to the factors related to urbanization, some of the factors which cannot be avoided but may lead to a higher risk of breast cancer include an age of more than 50 years, menopause after the age of 55 years, bigger size of breast having more connective tissues than fatty tissues which also interfere with early diagnosis, family history, and exposure to radiations for longer term (Bowen et al., 2021).

Malignant growth of breast tumor have undistinguishable gamble for the bosom disease especially in ladies old enough over 40 years, heftiness, with high liquor use, and family ancestry. One of the vital stages in the excursion for individuals with side effects who

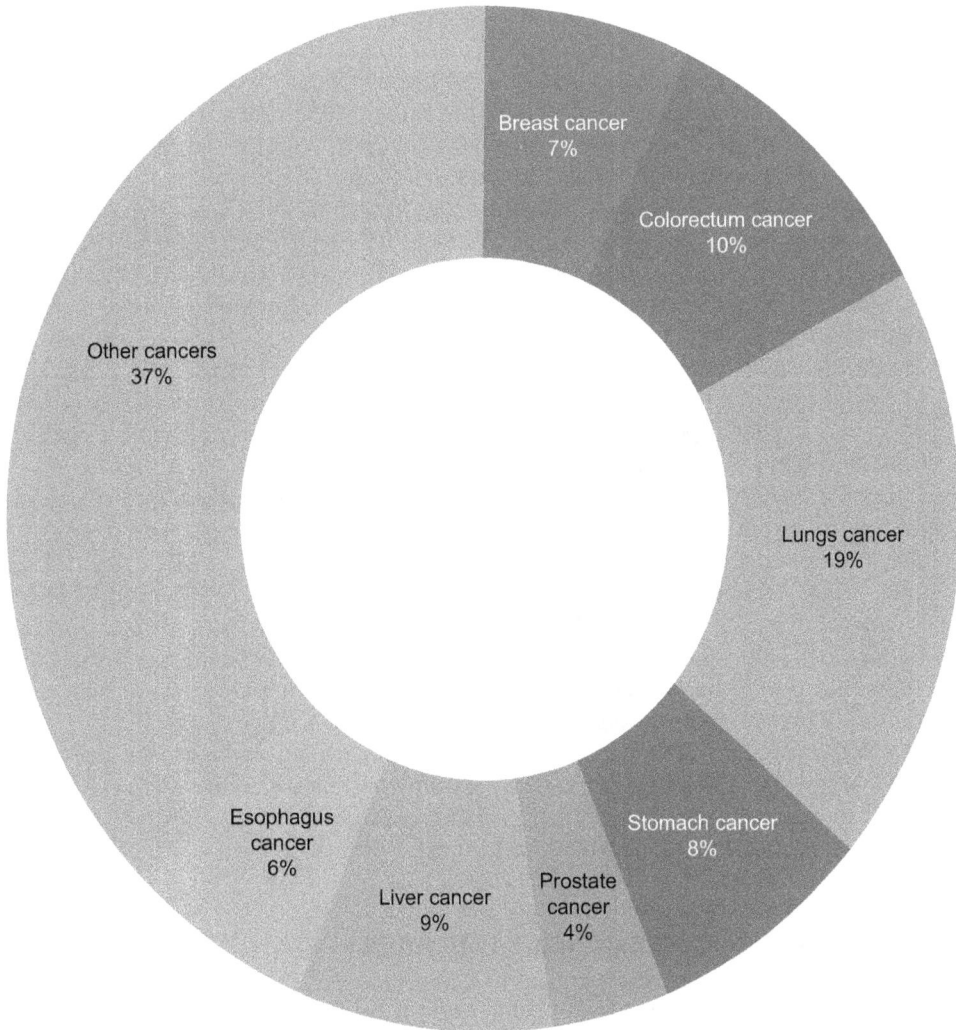

Figure 14.3 Representation of the percentage of deaths (irrespective of age and sex) due to cancer (in 2020, as per WHO).

proceed to foster malignant growth is the "analytical span." The success of the journey of cancer treatment is majorly dependent on the diagnostic interval which indicates the first-ever occurrence of tumor signs and symptoms till it proper diagnosis (Hamilton, 2010). The overall time for diagnosis of cancer gets reduced, and the diagnosis interval also reduces, which leads to early care and treatment of the cancer patient resulting in better outcomes of cancer treatment (Tørring et al., 2011).

14.1.2 Diagnosis of breast cancer

- *Mammography*: This technique involves X-ray imaging of the female breast. Digitalization of X-ray diagnosis has improved the screening of breast cancer aided with computers or algorithm-based detection, artificial intelligence, and machine learning, helping the radiologist to identify even the minor conceivable irregularities marked on the mammograms.

- *Magnetic Resonance Imaging (MRI)* is used to produce powerful images over-riding the complications associated with radiation. Breast MRI is not a routine procedure for the detection of BC but can be used as an adjunct to mammography in case of false positive results, which otherwise may lead to a biopsy.
- *Contrast Media* including nucleolin-targeted liposomal NPs were patented by US-owned Imperial Innovations Limited, Medical Research Council, in 2010, for MRI of cancer tumors. In this patent they described the use of gadolinium III, as a differentiation mediator, stacked on the innovative liposomal definition involving folates (Kamaly, Kalber, Thanou, Bell, & Miller, 2009).
- Turetschek et al., in their study, used ultra-small super-paramagnetic iron oxide NPs that can be utilized for the quantitative portrayal of growth micro vessels. Evaluations of trans endothelial porousness have corresponded with histologic cancer grade and thus, can survey disease miniature vessel attributes (Turetschek et al., 2001).
- *Molecular Breast Imaging (MBI)*: This involves the use of radioactive tracers with nuclear medicine scanners. This test is also known as specific gamma imaging, scintimammography, sestamibi test, or miraluma test. Compared to MRI, MBI is highly specific and capable of detecting minor breast abnormalities (O'Connor, Rhodes, & Hruska, 2009).
- *Breast Biopsy:* It is the only foolproof technique for the identification of breast cancer. To enhance the result success rate, breast biopsy is accompanied by breast imaging and other breast examinations (Triple test) (Palmer & Tsangaris, 1993).
- *Needle Biopsy*: This includes two types for breast cancer diagnosis—fine needle aspiration cytology (a less invasive technique involving the withdrawal of cells from the lesion of the breast using a fine needle) and core needle biopsy, on the other hand, withdraws a small mass of tissue (rice grain size) (Van Goethem et al., 2006).

14.2 CONVENTIONAL TREATMENT FOR BREAST CANCER

In addition to radiation therapy and chemotherapy, targeted therapy and breast cancer surgery are the major conventional treatment for breast cancer (Dhankhar et al., 2010). The first line of treatment for breast cancer is surgery, which is then followed by chemotherapy in addition to radiation therapy. Selected circumstances may be treated with hormonal (targeted) therapy after surgery to minimize the reappearance of the tumor cells. Depending from case to case, chemotherapy may vary; all women suffering from breast cancer do not require chemotherapy. Chemotherapy is preferred usually in the early diagnosis stage, as an adjuvant to surgery or radiation therapy to minimize the cancer cells left after the surgery or radiation therapy (adjuvant therapy); in some cases, chemotherapy is preferred before surgery to shrink the size of the tumor and bargain the chances of widespread surgery (non-adjuvant therapy). Radiation therapy is an effective course of treatment, and efficacy increases in association with chemotherapy but is associated with side effects like soreness, redness of the skin, itching, skin problems, and so on (Stephenson et al., 2019).

Chemotherapy is usually effective, but the use of more than one drug increases the effectiveness of the treatment. The choice of drugs for combination therapy varies from patient to patient and the choice of the prescriber, still some of the drugs used for adjuvant and non-adjuvant treatment include taxanes, 5-fluorouracil, carboplatin, cyclophosphamide, and anthracyclines (doxorubicin); but in the case of metastatic breast cancer, in addition to taxanes and anthracyclines, ixabepilone, vinorelbine, platinum agents, eribulin, gemcitabine, and capecitabine can be used (Le et al., 2019).

Radiation therapy can be done in several cases of breast cancer, it can be an added treatment to surgery. Radiation therapy is used in cases where breast surgery is not possible or in cases with inflammatory breast cancer. Sometimes, mastectomy is trailed by radiation therapy to minimize the chances of reoccurrence. Breast conservation therapy or lumpectomy is the surgical procedure where only the infected part of the breast is removed and the remaining breast is left intact. This surgery is followed by radiation therapy to minimize the pain as well as to reduce the chances of reoccurrence of cancer. Radiation therapy may be of the entire breast, using external beam radiation of the whole breast or part of the breast; includes the direction of internal or external radiations to a part of the breast only (Braunstein et al., 2020).

Notwithstanding chemotherapy, certain drugs are utilized to focus on the proteins on the bosom disease cells to ease back their development or to kill them perpetually; they might be helpful in situations where chemotherapy doesn't work by any means or its viability is problematic. They might be monoclonal immunizer-based immunotherapy, which might upgrade the invulnerable framework as well. Designated treatment might incorporate targeted treatment for HER2-positive cancer of the breast, targeted treatment for chemical receptor-positive malignant growth of the breast, targeted treatment for ladies with BRCA quality transformations, and designated treatment for triple-negative malignant growth in the breast.

In addition to chemotherapy, certain medicines are used to target the proteins on the breast cancer cell to slow their growth or to kill them forever, they may be useful in cases where chemotherapy does not work at all or its efficacy is questionable. They may be monoclonal antibody-based immunotherapy, which may enhance the immune system too. The treatment might incorporate designated treatment for HER2-positive bosom disease, designated treatment for chemical receptor-positive malignant growth of breast, designated treatment for women with BRCA quality changes, and designated treatment for triple-negative malignant growth of breast (Costa et al., 2020).

Breast cancer is immunogenic; immunotherapy can play a vital role in breast cancer treatment and which can be proved by the improved survival rate in patients of HER2+ breast cancer, with the advent of certain monoclonal antibodies including pertuzumab and trastuzumab, and the recent one, Margetuximab. The efficacy of the treatment can be increased with antibodies–drug conjugates, like T-DM1 (ado-trastuzumab-emtansine) where the antitumor advantages of trastuzumab, a humanized anti-human epidermal growth factor receptor 2 antibodies (HER2) is tagged with Maitansine (DM1), an anti-tubercular agent, acting as a potent microtubule-disturbing agent. In addition to this, immune checkpoint inhibitors (ICI), transfigured the treatment strategies, for malignant cancers, especially in the cases of triple-negative breast cancer (TNBC). Studies suggest an increase in response toward immune therapy was observed in TNBC patients under ICI-based therapy. Clinical trials show surprising results in patients with high programmed cell death-1 (PD-1) and programmed cell death ligand 1 (PD-L1), undergoing ICI therapy monotherapy or in mix with chemotherapy, to improve the adequacy of the treatment (Debien et al., 2023; Henriques, Mendes, & Martins, 2021).

14.3 NANOTECHNOLOGY-BASED APPROACHES TO TREAT BREAST CANCER

With the rapidly developing field of science and technology, protein engineering, and material science, nanotechnology can be a promising diagnosis and treatment methodology to cure breast cancer due to their ability to be target specific, with prolonged circulation and better accumulation, better bioavailability, increased biocompatibility, reduction in degradation of active constituent during circulation and fewer side effects (Jain et al., 2020).

NPs bearing large surface area, being small in size, can entrap various drugs and other small molecules, like DNA strands, RNA strands, proteins, antibodies, hormones, and so on for enhanced therapeutic or diagnostic effects. Malignant growth treatments are currently restricted to a surgical procedure, radiation therapies, and/or chemotherapy but with a large number of side effects bearing risk involving harm to normal tissues or inadequate annihilation of the malignant growth. Nanotechnologies improve the chemotherapy capabilities directly and make them more selective toward the cancerous cell; surgeries can be more precise; and the efficacy of radiation therapy can be enhanced. All these can amplify the survival probability while reducing the risk to the cancer patient (Padhi and Behera, 2020).

14.3.1 Nanotechnology in chemotherapy

For decades, nanotechnology aims to work on the adequacy, diminish the aftereffects, and improve the selectivity of the target cells to deliver the drugs to breast cancer cells during chemotherapy, to make the treatment more effective, mapped to the nano-size which proportionally increases the surface area, enhancing the overall therapeutic index of the encapsulated or surface conjugated drugs (Behera and Padhi, 2020). Selective delivery is tagged with the boosted penetrability and withholding effect. Various nanoparticle-based drugs used for the treatment of malignant growth of breast, including Genexol-PM®, Lipoplatin®, Abraxane®, Doxil®, and Onivyde® are proven to be effective as per the clinical data (Ganesan et al., 2021).

O'Brien et al. (2004) studied the adequacy of PEGylated liposomal doxorubicin HCl against doxorubicin, bearing less cardiotoxicity during first-line therapy of metastatic instances of malignant growth of breast disease. Montero et al. (2011) discussed the boosted activity of albumin-bound [nab] paclitaxel NPs (Abraxane®) over traditional paclitaxel (Taxol®).

14.3.2 Nanotechnology in radiation therapy

Radiation treatment incorporates the utilization of radiation for killing the malignant growth cells and contracting cancer, by modifying the DNA, which doesn't recognize tumor cells between the normal cells. Nanotechnology decreases the harmful impacts of radiation treatment, making them more particular. This can be accomplished by combining radiations with high nuclear weight NPs, which prompts improved photoelectric and Crompton impacts. On infusing, the core of NPs (generally made of lanthanide or hafnium-doped high-Z center) gets gathered inside the abnormally growing cells and when illuminated with X-rays, radiate noticeable light photons locally, which thus sets off the external layer of NPs to create singlet oxygen, which is cytotoxic for cancer cells (Tamanoi, Matsumoto, Doan, Shiro, & Saitoh, 2020).

14.3.3 NPs in gene therapy

The extension can be seen in the half-life of highly unstable and prone to degradation nucleic acid, during circulation either by encapsulating in a nanoparticle or by surface modification of a nanoparticle. Genexol-PM from Samyang Biopharm (2007, South Korea) is a novel nanoparticle system (PEG-PLA polymeric micelle) containing the drug paclitaxel for the management of malignant growth of breast cells, the same drug in the form of nanoparticle-bound albumin as Abraxane from Abraxis/Celgene in 2005 was official for the management of malignant growth in breast as a novel nanoparticle technique. Similarly, doxorubicin liposomes, in the name of Myocet from Cephalon, Lipo-Dox from Taiwan Liposome, and Doxil from Johnson and Johnson, were approved nanotherapeutics in the year 2000 (EU), 1998 (Taiwan) and 1995, 1999, 2003, 2007 (EU and Canada) respectively (Lee et al., 2008; Shi et al., 2017).

The improvement in the field of NPs incorporated metal-based NPs, which were utilized as theranostic items for the therapy of diseases. For the radiation, doses are upgraded and for imaging process, gold nanospheres were prepared by reducing tetrachloroauric (III) acid trihydrate (HAuCl4.3H2O) with trisodium citrate. Additionally, gold nanorods and nanoshells were reported for the therapy of different types of diseases, including malignant growth of the breast (Shang, Zhou, Zhang, Shi, & Zhong, 2021).

14.4 DIFFERENT NANO-FORMULATIONS FOR THE THERAPY OF BREAST CANCER

14.4.1 Liposomes

Liposomes are spherical nano-formulations, composed of cell membrane-compatible products, containing both hydrophilic and hydrophobic drugs. However, little information is available on hybrid liposomes and metallic nanostructures (Musielak, Potoczny, Boś-Liedke, & Kozak, 2021). Bromma et al. (2019) designed gold-trapped NPs linked with liposomes for the management of malignant growth of the breast, to enhance the local radiation dose, and treatment efficacy, with enhanced survival rate. Liu et al. (2021) designed, bovine serum albumin (BSA), indocyanine green (ICG), and doxorubicin (DOX)-stacked natural liposomes with an inorganic center for effective photothermal treatment of malignant growth of breast.

Shang et al. (2021) developed consolidated thermosensitive phospholipids onto the outer layer of anisotropic gold NPs to further improve drug conveyance, with conceivable extra applications for *in vivo* imaging and photothermal treatment of malignant growth. Lipid-covered nanohybrids stacked with docetaxel (DTX) were prepared using a film development, hydration, and sonication technique. Chithrani *et al.* (2010) have shown that there are thousand-fold increases in the uptake of gold NPs when they are combined with liposomes. Chowdhury and his coworkers (2020) formulated liposomes (aptamer-labeled) containing doxorubicin for selectively treating Her-2+ cancer of breast cells (Chowdhury et al., 2020). Boulikas discussed the success story of cisplatin liposomes for the treatment of malignant growth of breast cells (Boulikas, 2009). Rau et al. combined 500 mg of cyclophosphamide and 500 mg of 5-fluorouracil with PEGylated liposomal doxorubicin for effective and safe salvage therapy, especially in breast cancer patients, where progression is seen after treatment with a taxane (Rau et al., 2015).

14.4.2 Dendrimers

Dendrimers are nano-sized, tree-like, branched, 3D structures with various functional groups at the periphery and a core in the center. Dendrimer metal nanocomposites are structures containing the properties of both organic and inorganic molecules, where the organic portion acts as a base for the organization of inorganic ions. Dendrimers, in addition to acting as a template for the metal composite, further stabilize the same by chemical reactions (Pantapasis et al., 2017).

Wei *et al.* (2016) synthesized a gold nanostar (Au NS) stabilized with cyclic arginine-glycine-aspartic (Arg-Gly-Asp, RGD) peptide-modified amine-terminated generation 3 poly(amidoamine) dendrimers for gene delivery. These metal nanocomposite dendrimers exhibited theranostic (combination of investigation and therapeutic) properties. Lu *et al.* (2018) developed iron-oxide tagged, stabilized gold nano-flowers dendrimers as multimodal theranostic agents for selective diagnosis and imaging of tumor cells by MRI, CT, or through photo-acoustic imaging process combined with photothermal therapy or radiotherapy for tumor cells.

14.4.3 Carbon nanotubes

Carbon nanotubes can act suitably to transport biomolecules and small molecules for anticancer therapy, effectively and precisely at the target site. Plasmonic photothermal treatment is a method for cancer treatment where photon energy is quickly changed over into heat by a couple of radiative and non-radiative events. Gold NPs and carbon nanotubes are plasmonic NPs. Moreover, carbon nanotubes could infiltrate cells. Gold NPs were utilized to manufacture multiwalled carbon nanotubes, which could support its adequacy in disease treatment as per plasmonic photothermal treatment (Mohammed, Mohammed, & Al-Rawi, 2022).

14.5 METAL NANOCOMPOSITES FOR BREAST CANCER THERAPY

Various metal nanocomposites are designed successfully for the specific, effective, and targeted therapy of breast cancer, as they serve as a theranostic agent and can be used multidirectionally. Commonly used metals having the potential to be tagged with NPs for breast cancer treatment are gold, silver, nickel, copper, iron, iron oxide, zinc, cerium, titanium, barium, magnesium, bismuth, platinum, and palladium (Mohammed et al., 2022).

14.5.1 Gold NPs

Gold NPs, having high tissue penetrability and maintenance, capacity to retain light, surface plasmon resonance, and capacity to bind with drugs (Pradhan et al., 2022a), make them appropriate for malignant growth treatment. Their disease-focusing capacity can be upgraded by surface functionalization or covering. These metal NPs can be utilized for demonstrative, bio-imaging, and prognostic reasons notwithstanding their enemy of growth impact. Gold NPs can be combined utilizing different customary and current procedures including green synthesis (Lee, Chatterjee, Lee, & Krishnan, 2014; Pradhan et al., 2022b, 2023; Sa et al., 2023).

Granja et al. (2021) discussed gold NPs as the best photothermal therapy (PTT) agents, having high photothermal translation proficiency, biocompatibility, and adaptability with the easily modified surface. Mkandawire and his coworkers (2015) reported the antiapoptotic activity of gold NPs by targeting mitochondria in human breast cancer cell lines. Super green fluorescent protein variants conjugated with gold NPs were targeted to the amino terminus of the mitochondrial restriction signal. Gold NPs formed were further complexed with cationic maltotriose-conjugated poly (propylene imine) third-generation dendrimers. Elwakkad et al. (2023) reported that gold NPs prepared from *Saccharomyces cerevisiae* in conjugation with plasmonic, photothermal treatment was found to be effective in breast cancer treatment in comparison to conventional treatment.

14.5.2 Silver NPs

Silver NPs have a high potential against breast tumors, specifically triple-negative breast cancer (TNBC). Swanner et al. (2019) performed careful structural and functional, biocompatibility, and viability studies to evaluate the potential for silver NPs for the treatment of TNBC. Silver NPs are selectively cytotoxic to the TNBC owing to their shape and size and are not cytotoxic to non-malignant epithelial cells of the breast (Swanner et al., 2019).

Tao et al. (2021) synthesized silver NPs bearing potent cytotoxic ability toward cancer cells irrespective of their sensitivity toward tamoxifen, with an average cell size of about 63 nm and a circular shape.

14.5.3 Platinum NPs

Platinum NPs have a high potential for cancer cells, similar to silver NPs, but exhibit anticancer activity only at 50 nm size and have no anticancer activity at a size of 5 nm and 20 nm. Ruiz et al. (2022) developed platinum-encapsulated poly(lactic-co-glycolic acid) NPs, with surface modification to enhance the circulation time and specificity, for triple-negative breast cancer cells.

14.5.4 Palladium NPs

Redox reactions can be catalyzed using palladium as a catalyst. In nanoparticle form, the efficacy of palladium increases due to increased surface area. Palladium NPs possess cytotoxic activity, in addition to antibacterial activity (Yaqoob, Adnan, Rameez Khan, & Rashid, 2020).

Ramalingam et al. (2020) prepared profoundly powerful PVP-functionalized palladium NPs for the therapy of human breast cancer cells (MCF7) by in situ strategy. The PVP-functionalized palladium NPs produced an excess of ROS leading to mitochondrial dysfunction and DNA damage with induced apoptosis through activation of caspase3/7 enzymatic action (Ramalingam, Raja, & Harshavardhan, 2020).

14.6 ROLE OF METAL NANOCOMPOSITES IN BREAST CANCER TREATMENT

14.6.1 Silver nanocomposites (Ag NCs)

Because of their characteristics and therapeutic value in many disease treatments (e.g. retinal neovascularization), Ag NCs have gained much attention amid the nano products in the sector of nanotherapeutics. Ag NPs have loading ability as a carrier in the healthcare sector. In industries, for developing photocatalytic sensors functional groups can be add-on in the silver NPs, or phytochemicals can be used for purification purposes (Wei et al., 2015). Moreover, due to their numerous uses in various sectors, silver NPs have become most wanted in industries. Akter et al. (2018) focused on the increasing uses of Ag NPs and their adverse effects on the environment and living organisms.

Many researchers have focused on cytotoxicity produced by Ag NPs and their mechanism by using different cell lines. Ag NPs due to their physicochemical properties, such as size, shape, composition, and agglomeration with a biological process, can alter mitochondrial function by penetrating and accumulating in the mitochondrial membrane. Hence, Ag NPs have characteristics that are essential to their usage as biocides as well as their suitability for cleansing the environment (Akter et al., 2018).

Gurunathan et al. (2013) studied the cell death mechanism and cytotoxic effects of Ag NPs on MDA-MB-231 breast cancer cells (BCC). They developed a green approach for manufacturing Ag NPs by using *Bacillus funiculus* culture supernatant. The cytotoxicity was determined by cell growth, enzymatic rate, and oxidative stress. Ag NPs were applied to MDA-MB-231 BCC in a dose range (5–25 g/mL) for 24 hours. Using the MTT assay, they found that Ag NPs in a dose-dependent way reduced tumor cell growth. Ag NPs induced cytotoxicity against MDA-MB-231 cells in a dose-dependent manner by activating reactive oxygen species (ROS), lactate dehydrogenase (LDH), caspase-3, and cell death, which was subsequently validated by nuclear fragment. The findings showed that Ag NPs might be used as a replacement agent in the treatment of BCC.

By specifically targeting Ag NPs as a medication for tumor cells and creating an advanced anticancer agent, Azizi et al. (2017) proved the effectiveness of their unique nanocomposite. The anticancer characteristics of the newly produced albumin-coated silver NPs (ASNPs) were tested against the human breast cancer cell line MDA-MB 231. The findings indicated that apoptosis was the mechanism used for cell death. The structural alterations of the cells were observed by inverted fluorescent microscopy and by DNA ladder pattern on gel electrophoresis. ASNP with a diameter of 90 nm and zeta potential of approximately –20 mV has been preferentially absorbed by tumor cells. The lethal dose (LD_{50}) of ASNPs toward MDA-MB 231 (5 M) was found to be 30 times larger than that of regular WBCs (152 M). According to the findings, ASNPs might be an excellent chemotherapeutic treatment.

By using the biological synthesis method, *Padina tetrastromatica* seaweed extract was used by Selvi et al. (2016) to synthesize and study the cytotoxicity of the Ag NPs against MCF-7 cell lines. The biosynthesized Ag NPs were spherical with a size range of 40–50 nm. Also, the biologically produced Ag NPs showed dose-dependent cytotoxicity against the MCF-7 human BCC, and the inhibitory concentration (IC_{50}) for Ag NCP against MCF-7 was found after a 24-hour incubation period.

14.6.2 Gold nanocomposites (Au NCs)

Gold is a diverse material that has been employed in medicinal applications for ages. It has been known for its anti-corrosive, antioxidant, and bacteriostatic properties. Its properties at the nanoscale helped a lot to develop more advanced applications, and it allows various functional groups to conjugate like targeted antibodies or drug products (Vines, Yoon, Ryu, Lim, & Park, 2019). Au NPs have an effective surface resonance and hence are excellent photothermal agents used in photothermal therapy (Kalyane, Polaka, Vasdev, & Tekade, 2022).

Gold–graphene oxide nanocomposites have been employed in therapy but also have unique applications such as bio-imaging and cancer detection (Adil et al., 2021). Au NPs have several benefits including plasmon resonance and the ability to bind with a vast category of biomolecules to Au NPs has demonstrated localized surface plasmon resonance (LSPR). The photoacoustic and photothermal activity of Au NPs is because of their absorbance at specific wavelengths. Therefore, they find applications in medical imaging and hyperthermic cancer therapy. Modification in the physical characteristics of Au NPs can effectively change their LSPR properties, hence modifying their optical properties. Au NPs enhance the distribution of drugs in the body through the leaky blood vessels of tumors and can be excreted through urine (Vines et al., 2019).

Nanocomposites made of functionalized, highly reduced graphene oxide (HRGO) and Au NPs were studied by Adil et al. (2021) for their capacity to induce apoptosis and for their anticancer effects (AP-HRGO-G). Doxorubicin was used as a positive control in studies involving numerous human cancer cell lines, including the lungs (A549), liver (HepG2), and BCC (MCF-7); all the samples were examined for anticancer activity. To improve the sample's solubility and bioavailability, HRGO was fabricated with 1-aminopyrine (1-AP) as a stabilizing agent. By providing chemically distinct binding sites, the ligand also made it easier for Au NPs to develop uniformly on the surface of HRGO. The functionalization increased the physical stability and dispersibility of the AP-HRGO-G nanocomposite. Functionalized nanocomposites demonstrated stronger apoptotic ability than non-functionalized samples, according to a comparison of anticancer investigations of pure HRGO, non-functionalized HRGO-G, and 1-AP-functionalized AP-HRGO-G samples, whereas pristine HRGO was ineffective against all cell types examined for cancer. After 48 hours of treatment, both HRGO-G and AP-HRGO-G elicited a decrease in cell viability that is concentration-dependent across all tested cell lines, with MCF-7 cells responding much more than the other cells. As a result, MCF-7 cells were chosen for thorough studies that included a cell cycle analysis, an apoptosis assay, and measurements of ROS. According to their findings, AP-HRGO-G induced apoptosis in human BCC (Fathy et al., 2018).

A pulsatile chemo-photothermal treatment for triple-negative breast cancer (TNBC) using a gold-doxorubicin nanocomposite (Au-DOXNC) that the CD44 receptor was reported by Kalyane et al. (2022). Electrostatic contacts were used to load the doxorubicin, which resulted in great entrapment and loading efficiency (>75%). After NIR laser irradiation (808 nm), Au-DOXNC demonstrated significant photothermal reaction and reversible photothermal stability. Au-DOXNC also showed a pH-dependent drug release and responsiveness to lasers, indicating its suitability for chemo-photothermal therapy, particularly in the tumor microenvironment. Cell viability, cellular uptake, ROS production, and apoptosis experiments revealed that Au-DOXNC was selectively localized in cancer cells and had a substantial lethal impact against MDA-MB-231 BCC. Moreover, the Au-DOXNC induced ferroptosis in MDA-MB-231 cells. The presence of Au-DOXNC-mediated thermal ablation was confirmed by flow cytometry, which was characterized by a significant increase in the formation of ROS and apoptosis. It has been demonstrated that the photothermal ablation of cancer cells that are responsive to the NIR-808 laser was more effective than nonresponsive to the NIR-808 laser, indicating a critical function for photothermal ablation. Findings showed that the newly developed Au-DOXNC was a novel laser-guided chemo-photothermal ablation treatment of cancer cells.

Graphene oxide (GO) and gold/graphene oxide (Au/GO) nanocomposites were synthesized using the simple chemical method by Ramazani et al. (2018), and their effectiveness as excellent nanocarriers for curcumin delivery was proven. Curcumin was used as a model medication for loading through p-p stacking and hydrophobic contact on the GO. The MCF7 BCC line and HEK293 (human embryonic kidney 293) cells were used to test the cytotoxicity of GO, Au/GO, and curcumin-loaded Au/GO nanocomposite (Au/GO/Cur). The nanocomposites were further investigated for toxicity against brine shrimp (*Artemia salina*) larvae and membrane rupture in human RBC. In comparison to GO, the Au/GO/Cur system demonstrated cancer cell-specific characteristics with no discernible toxicity on normal healthy cells following 48 and 72 hours of incubation at varied doses. Moreover, the Au/GO/Cur nanocomposite did not have a substantial fatal impact on the brine shrimp larva (LC_{50} = 657.35 µg/mL). The nanocomposite did not influence the integrity of RBC membranes, confirming its hemocompatibility and biocompatibility. As a result, the produced nanocomposite system provided a new formulation that combined the distinctive features of a biodegradable material for biomedical applications.

Ranjan et al. (2022) create a sensitive electrochemical immunosensor for detecting CD44 antigen, a BCC biomarker. The glassy carbon electrode was immobilized with graphene oxide, ionic liquid, and Au NPs (GO-IL-AuNPs). Because of the availability of oxygen, GO facilitated antibody immobilization. Nevertheless, 1-butyl-3-methylimidazolium tetrafluoroborate ($BMIM.BF_4$) and Au NPs promoted electron transport and improved an effective surface area, improving immunosensor efficacy. EIS detection and differential pulse voltammetry methods were used to detect CD44 antigens quantitatively. The suggested immunosensor has been demonstrated to have high detection performance in both phosphate-buffered saline (PBS) and serum samples. The immunosensor's linear detection range for CD44 antigen was 5.0 fg/mL to 50.0 fg/mL under ideal circumstances, and the limit of detection was 2.0 and 1.90 fg/mL in PBS, as measured by DPV and EIS, respectively. Furthermore, the immunosensor had good sensitivity and specificity and was used to detect CD44 antigens in clinical samples.

Yee et al. (2019) demonstrated the use of Au NPs made from *Curcuma mangga* (CM) extract in the photothermal killing of BCC (MCF-7). CM-AuNPs outperformed citrate-Au NPs in terms of photothermal heating efficiency when subjected to a 532 nm laser. In addition, it was found that treatment of MCF-7 cells with CM-AuNPs and laser ablation for 120 seconds significantly reduced cell viability (72%) compared to 13% with citrate-AuNPs. The CM-AuNP-dependent photothermal-induced MCF-7 cell death was largely brought on by an apoptotic mechanism, according to flow cytometry results. These results demonstrated that CM-AuNPs may be employed as therapeutic agents in the treatment of cancer.

14.6.3 Zinc oxide nanocomposites (ZnO NCs)

Zinc serves four important biological functions: structural, signaling, catalytic, and regulatory. Zinc is a divalent cation, and its valency impacts its stability and reactivity, as well as its binding in biological systems. Zinc metal is present in all six classes of IUPAC and is involved in the operation of about 300 enzymes. It is essential for cell development, division, and homeostasis. Nearly all RNA polymerases (I, II, and III) are metalloenzymes, and zinc is required for DNA and RNA stability. As a result, zinc deficiency may result in cellular malfunction or cancer/progression of cancer. Cells get into S-phase in the absence of zinc, and growth factor effects on cell proliferation increase as zinc labile concentration increased (Grattan & Freake, 2012).

Evidence by Arab-Bafrani et al. (2021) suggested that three-dimensional multicellular tumor spheroids (3D-MCTSs) precisely imitated *in vivo* tumor cells' responses to therapies. As a result, the group investigated the efficacy of pure zinc oxide NPs (ZnO NPs) and chitosan-ZnO bio-nanocomposites (CS-ZnO BNCs) to enhanced radio sensitization of MDA-MB-231 BCC in the 3D-MCTSs model. A simple coprecipitation approach was used to make ZnO NPs and CS-ZnO BNCs. FE-SEM scans demonstrated that homogeneous spherical ZnO NPs with an approximate diameter of 35 nm were effectively disseminated over chitosan. MDA-MB-231 MCTSs grown in a nonadherent culture plate exhibited functional aspects of an *in-vivo* tumor. The emphasis of such a culture approach over traditionally utilized 2D monolayer (or parental) cell culture was the mimicking of the tumor microenvironment. Using an MTT-colorimetric test, the toxicity of CS-ZnO BNCs and ZnO NPs against MDA-M231 BCCs was assessed, which revealed that CS-ZnO BNCs were more biocompatible than pure ZnO NPs (even at high concentrations of 100 µg/mL). Survival fraction examination of cells exposed to clinical X-ray irradiation (6 MV) revealed that 3D MCTSs were more radioresistant than parental cells. Moreover, the addition of CS-ZnO BNCs significantly decreased the irradiated MCTSs' ability to form clones, much like monolayer cells would. For MCTSs and monolayer cells, the sensitivity enhancement ratios (SER) were found to be 1.5 and 1.63, respectively. It was discovered that CS-ZnO BNCs not only enhanced the amount of radiation-induced complex DNA breaks and apoptotic death in MCTSs but also hampered DNA repair mechanisms. It was found that CS-ZnO BNCs in a nontoxic concentration could significantly increase the radiosensitivity of resistant MCTSs, making them an excellent *in vitro* tumor model. CS-ZnO BNCs were found to be a promising option for reducing BCC radiation resistance.

ZnO/GO nanocomposite (NC) was created by Nagaraj et al. (2020) by combining green-synthesized zinc oxide NPs (ZnO NPs) with graphene oxide (GO) nanosheets (Hummers' process). Evaluation of the antibacterial effects of the manufactured nanomaterials ZnO NPs, GO, and ZnO/GO NC on infections in humans revealed significant bacterial resistance against *Staphylococcus aureus*, with ZnO/GO NC acting as a particularly potent antibacterial agent. Using lung cancer cells (A549) and BCC (MCF-7), *in vitro*, cytotoxicity tests were performed with ZnO NPs, GO and ZnO/GO NC. An MTT assay revealed that ZnO/GO NC (IC_{50} = 15 µg/mL) had much higher anticancer activity against the MCF-7 cell line than ZnO NPs (IC_{50} = 24 µg/mL) and GO (IC_{50} = 23 µg/mL). Hence, the investigation demonstrated the high potency of ZnO/GO NC as an effective cancer treatment agent.

Hira et al. (2018) used a precipitation process to create a pectin–guar gum–zinc oxide (PEC–GG–ZnO) nanocomposite. PEC–GG–ZnO was utilized for the first time as an immunomodulator to enhance the capacity of human peripheral blood lymphocytes to eradicate cancer cells (PBL). The experiment on lymphocyte proliferation showed an increase in immunomodulatory properties of PEC–GG–ZnO with concentration from 25 µg/mL to 200 µg/mL. ELISA detection verified a considerable increase in IFNs, IL-2, and TNF-α cytokine production, and flow cytometry analysis demonstrated higher expression of CD3, CD8, and CD56 following PEC–GG–ZnO therapy compared to PEC and GG treatment. Furthermore, we discovered that nanocomposite pretreatment human PBL had higher cytotoxicity against lung (A549) and breast cancer (MCF-7) cells than untreated PBL. Increasing the effector: target ratio from 2.5:1 to 20:1 led to an increase in cancer cell mortality, according to the micro cytotoxicity experiment. Therefore, the available data supported the immunostimulatory properties of the novel nanocomposite PEC–GG–ZnO and implied potential anticancer activity.

14.6.4 Iron oxide nanocomposites (FeO NCs)

For cancer applications, iron oxide NPs (FeO NPs) have received increased attention since they are magnetic and can target specific tumor cells and can be used to enhance MRI. FeO NP delivery systems have been used to deliver peptides, DNA molecules, chemotherapeutics, radioactive drugs as well as chemicals that generate hyperthermia. Recent iron oxide-based nanoparticle delivery systems have focused on the treatment of infections, arthritis, inflammation, and photodynamic therapy. Moreover, they are relatively safe for healthy cells and can be excreted readily from the body after use. They can be activated by infrared and selectively kill cancer cells over healthy cells since cancer cells have increased sensitivity to heat. However, it has been suggested that the aforementioned selenium is much more toxic to cancer cells than FeO NP (Hauksdóttir & Webster, 2018).

Salimi et al. (2018) created FeO NPs by coprecipitation and coated them with a polyamidoamine dendrimer of the fourth generation (G4). G4@FeONPs at various concentrations were tested for cytotoxicity in a BCC (MCF7) and a human fibroblast cell line (HDF1). Prussian blue staining was used to investigate the hemolysis, stability, and interactions of G4@FeONPs with MCF7 cells. Using data from measurements and modeling, the heat generation and specific absorption rate (SAR) at 200 kHz and 300 kHz were calculated. MCF7 and HDF1 cells were treated with G4@FeONPs for 2 hours before being placed in the magnetic coil for 120 minutes. T1- and T2-weighted magnetic resonance images were used to perform relaxometry studies with varying doses of G4@FeONPs. The *in vitro* toxicity tests revealed that the produced G4@FeONPs were not harmful. During magnetic hyperthermia, the viability of MCF7 cells treated with G4@FeONPs reduced considerably. The same group reported the anticancer activity of G4@FeONPs in breast cancer-bearing Bagg albino strain C (BALB/c) mice. Upon incubation with G4@FeONPs and exposure to an alternating magnetic field (AMF), the viability of breast cancer cells was significantly decreased as a result of apoptosis and an increase in the Bax (Bcl-2 associated X)/Bcl-2(B-cell lymphoma 2) ratio. The liver, lungs, and tumor tissues were examined histopathologically and immunohistochemically in both treated and control animals. The tissues from the control animals had metastatic breast cancer cells, but the tissues from the treatment group's liver and lungs were free of these cells. Additionally, the current study found that magnetic hyperthermia therapy suppressed tumor development by enhancing cancer cell death and decreasing tumor angiogenesis (Salimi, Sarkar, Hashemi, & Saber, 2020).

Samantha et al. developed hydrogel nanocomposites that can control the amount of heat and chemotherapeutic medication administered (paclitaxel). Tissue heating (hyperthermia) between 41 °C and 45 °C has been shown to increase the efficacy of cancer treatment when combined with radiation and/or chemotherapy. The examined nanocomposites are stealth poly (ethylene glycol) (PEG)-based iron oxide systems imprisoned within the hydrogel matrix. It was discovered that the hydrogel nanocomposites could be heated in an intermittent magnetic field. It has been demonstrated that the cross-linking of the hydrogel network regulated the heating of the hydrogel systems. The hydrogels with lower swelling ratios heated more than those with higher ratios. The amount of drug released from hydrogel systems followed a non-Fickian release, with the amount depending on the topology of the hydrogel network. To determine if paclitaxel and hyperthermia together would have a synergistic cytotoxic effect, three cell lines—M059K (glioblastoma), MDA MB 231 (breast cancer), and A549 (lung adenocarcinoma)—were subjected to one of three treatments: hyperthermia alone, paclitaxel alone, or both. The efficacy of paclitaxel effectiveness was increased by heat in A549 cells; however, M059K and MDA MB 231 cells did not exhibit the same reaction (Meenach, Shapiro, Hilt, & Anderson, 2013).

14.6.5 Copper-based nanocomposites (Cu NCs)

A lot of research has been done on copper oxide NPs (CuO NPs) because of their fascinating physical, biological, and pharmacological characteristics. Due to their numerous applications, metal oxide NPs have lately emerged as a potential study field. CuO NPs have a strong anticancer potential due to apoptosis induction, enhanced ROS production, and MMP loss may be potential mechanisms of action (Zughaibi et al., 2022).

Azizi et al. (2017) fabricated copper nanocomposites that would have excellent cytotoxicity and negligible side effects against invasive breast cancer cells. The researchers developed an albumin (ALB) nanocarrier for the efficient delivery of copper NPs (Cu NPs). A standard MTT assay was employed to compare the anticancer efficacy of the ALB-CuNPs for normal cells (MCF-10A) and breast cancer cells (MDA-MB 231). Flow cytometry, inverted and fluorescent microscopy, and gel electrophoresis were employed to investigate the mechanism of cell death caused by ALB-CuNPs. The ALB-CuNPs affected ROS generation in MDA-MB 231 cell lines were also investigated. The ALB-CuNPs were found to be suitable for extravasation into tumor cells. When compared to CuNPs, ALB-CuNPs considerably reduced cancer cell viability while being less toxic to normal cells. The ROS levels in the cancer cell line (MDA-MB 231) were more after being treated with ALB-CuNPs when compared to untreated cells. After 24 hours, the surge in ROS production showed that ALB-CuNPs induced apoptosis. The characteristics of ALB-CuNPs, such as intact albumin structure, superior toxicity for cancer cells compared to normal cells, and apoptosis induction as a cell death mechanism, demonstrated that this nanocomposite is an excellent candidate for use as a chemotherapeutic agent against invasive MDA-MB-231.

Zughaibi et al. (2022) assessed the antitumor potential of CuO NPs prepared from the extract of pumpkin seed against MDA-MB-231. The researchers evaluated the formulation by employing MTT assay, morphological alterations, and alteration in mitochondrial membrane potential (MMP). The cell viability was dose-dependently reduced by biogenic CuO NPs, with an IC_{50} value of 20 µg/mL. In MDA-MB-231 cells, an IC_{50} dose of CuO NPs caused noteworthy morphological changes such as shrinkage, blebbing of the membrane, and deformation in shape. CuO NPs treatment also resulted in a significant dose-dependent increase in ROS production and MMP modulation. It was observed that CuO NPs demonstrated significant anticancer activity in MDA-MB-231 cell lines. Though, additional validation of research data in *in vivo* and *ex vivo* models is required before this nano-formulation can be employed to treat human breast cancer.

Mahmood et al. (2022) fabricated CuO NPs by employing plant extract of *Annona muricata L.* (*A. muricata*) and evaluated the antitumor potential for breast cancer cell lines (MCF-7 and AMJ-13). The human breast epithelial cell line (HBL-100) and the AMJ-13 were used to assess the antiproliferative characteristics of the produced NPs. The CuO NPs decreased AMJ-13 and MCF-7 cell proliferation. For various concentrations or testing intervals, HBL-100 cells were not appreciably suppressed. The findings implied that the synthesized CuO NPs inhibited the growth of particular cell lines associated with breast cancer. After 24 hours of incubation with CuO NPs, cancer cells showed an IC_{50} of 18.08 µg/mL for AMJ-13. For MCF-7 cancer cell lines, however, it was (19.20 µg/mL). It shows that CuO NPs resulted in apoptosis of cancer cells. Furthermore, CuO NPs treatment increased the formation of LDH, which was likely brought on by cell membrane disruption that led to leaks containing cellular components including LDH. So, according to the research findings, the CuO NPs reduced the anti-proliferative effects by inducing apoptosis, which results in cell death.

Due to improved medication effectiveness and decreased toxicity in the nanosize-mediated drug delivery model, the green synthesis of NPs from bioactive compounds has garnered an extensive variety of applications. Biresaw & Taneja (2022) fabricated Cu NPs using an extract of *Prunus nepalensis* (PN) fruits. The objective of the study was to ascertain the anticancer activity of PN-CuNPs in both healthy (MCFA10) and malignant (MCF-7) human breast cell lines. The expression of apoptotic marker genes (p53, P21, P14/P19, and caspase-3) was examined in MCF-7 cells that were treated with 100–200 µg of PN-CuNPs for 72 hours. The researchers demonstrated that PN-CuNPs promoted apoptotic gene expression in a dose-dependent fashion. The MCF-7 cells exposed to PN-CuNPs revealed a considerable elevation of p53, Bax, caspase-3, and caspase-9 and downregulation of the mRNA expression of Ras and Myc genes in the real-time PCR information. According to the study, PN-CuNPs caused apoptosis in MCF-7 cells by upregulating tumor suppressor genes and downregulating oncogenes.

Sharma et al. (2020) investigated the cytotoxicity of Cu NPs on MCF-7 human breast cancer cells as a result of ROS (reactive oxygen species). By using the MTT assay, the dose-dependent toxicity of Cu NPs was assessed for concentrations (0.001–100 µg/mL). The Cu NPs showed strong dose-dependent toxicity, which was caused by the mitochondrial damage in MCF-7. The MCF-7 cells treated with Cu NPs showed cytotoxicity due to cell membrane breakage, shrinkage, and oxidative stress brought on by ROS.

14.6.6 Cerium-based nanocomposites (Ce NCs)

The insulating and ionic oxide cerium oxide (CeO_2) is commonly known as ceria. Ceria has no specific biological implication, but its NPs with ionic form (Ce^{3+}) exhibit pertinent biological effects as bactericidal, bacteriostatic, antiemetic, and antitumor agents. The anticancer activity of ceria is due to the generation of ROS in the tumor cells (Abbas et al., 2015).

Atif et al. (2021) fabricated CeNCs doped with manganese using a hydrothermal route. The researchers evaluated the anticancer potential of MN-CeNCs for breast cancer cell lines (MCF-7). The MN-CeNCs exhibited considerable anticancer activity against MCF-7 cell lines, making them an appropriate candidate for targeted cancer therapy. When MN was added to the Ce NCs, which is related to the production of highly reactive oxygen species, the anticancer efficacy of the compound was greatly improved.

Sulak et al. (2022) prepared Ce NPs by employing a fresh extract of green walnut shell in a microwave setting. The RT-PCR, p53, and Annexin V-FITC detection, crystal purple staining, and NF-Kb, luciferase reporter assays were accomplished to assess the mechanism of action of Ce NPs in MCF-7. In MCF-7 cells, the prepared Ce NPs caused apoptosis and had cytotoxic effects. Moreover, it was demonstrated that Ce NPs increased the expression of p53 and decreased the gene expression of NF-κB. The current findings led to the conclusion that CeNPs caused cell death by decreasing NF-κB-mediated transcription and induced apoptosis via acting on p53 at the transcriptional level.

Thakur et al. (2022) formulated Ce NPs functionalized with folic acid to deliver a natural flavonoid known as morin against the MCF-7 in a controlled fashion. By creating coordination bonds between metal ions and ligands, morin was effectively equipped onto the surface of

FA-CeNPs to create Morin–FA–CeNPs nanohybrids. The nanohybrid, as it was created, had a roughly spherical form with an average diameter of about 4–5 nm. Moreover, the nanohybrid demonstrated pH-responsive drug-release behavior along with a drug-loading content of less than 10%. The prepared nanohybrids significantly reduced the viability of MCF-7 cell lines by inducing intracellular ROS, which in turn led to mitochondria-dependent apoptosis. The nano-hybrids also appeared to have an anti-migratory impact. The Morin–FA–CeNPs nanohybrids had the strongest anticancer impact in tumor-bearing mice, according to the *in vivo* results, when compared to free morin and FA-CeNPs. Additionally, serum biochemical markers did not significantly change as a result of the Morin–FA–CeNPs nanohybrid. The researchers per-formed molecular docking research of the morin protein with the BCL-2 protein to establish a potential mechanistic pathway. The results showed that morin can bind to Bcl-2 with an affin-ity of 8.1 kcal/mol. The combined cytotoxic action of both morin and nanoceria is primarily responsible for the strong anticancer activity of the nanohybrid on MCF-7 cell lines.

14.6.7 Titanium dioxide nanocomposite (TiO$_2$ NCs)

The current upsurge in scientific attention to titanium dioxide (also known as titania, TiO$_2$), a non-organic chemical, can be attributed to its photoactivity. TiO$_2$ generates different types of ROS after UV light is shone on it in aqueous solutions (ROS). The PDT employs the capacity to generate ROS and afterward causes cell death to treat different kinds of illnesses including psoriasis and cancer (Ziental et al., 2020).

Zandvakili et al. (2022) formulated TiO$_2$ nanotubes (NTs) coated, doped, and deposited with Ag. The titanium foil used to make the TiO$_2$ NTs was anodized. The catalytic behavior and performance of Ag-TiO$_2$NTs were examined by photocatalytic degradation employing methyl orange solution. The cancer cells that were capable of proliferating were placed on Ag-TiO$_2$NTs of various shapes and sizes. The toxicity differences of the prepared Ag-TiO$_2$NTs with UV radiation and non-UV circumstances were evaluated. The investigation showed that under UV radiation, Ag-TiO$_2$NTs had an impact on the cell populations.

Mund et al. (2014) formulated TiO$_2$ NPs by using modified propanol drying step. To fabricate paclitaxel (PTX) with TiO$_2$ NCs, these TiO$_2$ NPs were conjugated with the cyto-toxic drug PTX. The loading efficiency of the drug of the propanol-induced drying step of TiO$_2$ NPs was found to be higher than that of the air-dried TiO$_2$ NPs, showing a difference of 71.70%. The loaded PTX was released *in vitro* in a pH-dependent manner, in an acidic pH. The release of the drug was greater in acidic pH than in physiological pH. The NCs' time-dependent internalization exhibited considerable enhancement in uptake by surging incubation from 3 hours to 24 hours, confirmed by cell uptake studies and flow cytometry. When cells were exposed to PTX alone, their viability was only 41.5%; however, after 24 hours, cells exposed to PTX–TiO$_2$NCs had a viability of 22.6%, indicating greater cytotoxic efficacy. Studies on apoptosis showed that cells exposed to PTX–TiO$_2$NCs had more signifi-cant apoptotic bodies than cells treated with PTX alone.

14.7 CONCLUSION

One of the potential next-generation anticancer therapies is MNCs because of their extraor-dinarily flexible physical and chemical properties. Noble MNCs have distinctive plasmonic capabilities that make it possible to track therapeutic nano-complex carriers inside the body with greater accuracy and fewer adverse effects than conventional therapies. The possibility of using abrasive and environmentally harmful substances and solvents has also been avoided via green synthesis. New criteria for the clinical use of MNCs to treat breast cancer and drug delivery, as well as novel methods to assess the effectiveness and safety measures of such MNCs, must be created by the regulatory bodies.

REFERENCES

Abbas, F., Jan, T., Iqbal, J., Ahmad, I., Naqvi, M. S., & Malik, M. (2015). Facile synthesis of ferromagnetic ni doped CEO_2 nanoparticles with enhanced anticancer activity. Applied Surface Science *357*, 931–936.

Adil, S. F., Shaik, M. R., Nasr, F. A., Alqahtani, A. S., Ahmed, M. Z., Qamar, W., . . . Siddiqui, M. R. H. (2021). Enhanced apoptosis by functionalized highly reduced graphene oxide and gold nanocomposites in MCF-7 breast cancer cells. Acs Omega *6*(23), 15147–15155.

Akter, M., Sikder, M. T., Rahman, M. M., Ullah, A. A., Hossain, K. F. B., Banik, S., . . . Kurasaki, M. (2018). A systematic review on silver NPs-induced cytotoxicity: Physicochemical properties and perspectives. Journal of Advance Research *9*, 1–16.

Arab-Bafrani, Z., Zabihi, E., Jafari, S. M., Khoshbin-Khoshnazar, A., Mousavi, E., Khalili, M., & Babaei, A. (2021). Enhanced radiotherapy efficacy of breast cancer multicellular tumor spheroids through in-situ fabricated chitosan-zinc oxide bio-nanocomposites as radio-sensitizing agents. International Journal of Pharmaceutics *605*, 120828.

Atif, M., Iqbal, S., Fakhar-e-Alam, M., Mansoor, Q., Alimgeer, K., Fatehmulla, A., . . . Ahmad, S. (2021). Manganese-doped cerium oxide nanocomposite as a therapeutic agent for MCF-7 adenocarcinoma cell line. Saudi Journal of Biological Sciences *28*(2), 1233–1238.

Azizi, M., Ghourchian, H., Yazdian, F., Dashtestani, F., & Zeinabad, H. A. (2017). Cytotoxic effect of albumin coated copper nanoparticle on human breast cancer cells of MDA-MB 231. Plos One *12*(11), e0188639.

Behera, A., & Padhi, S. (2020). Passive and active targeting strategies for the delivery of the camptothecin anticancer drug: A review. Environmental Chemistry Letters *18*(5), 1557–1567.

Biresaw, S. S., & Taneja, P. (2022). Copper NPs green synthesis and characterization as anticancer potential in breast cancer cells (MCF7) derived from *Prunus nepalensis* phytochemicals. Material Today: Proceedings *49*, 3501–3509.

Boulikas, T. (2009). Clinical overview on Lipoplatin™: A successful liposomal formulation of cisplatin. Expert Opinion on Investigational Drugs *18*(8), 1197–1218.

Bowen, D. J., Fernandez Poole, S., White, M., Lyn, R., Flores, D. A., & Haile, H. G (2021). The role of stress in breast cancer incidence: Risk factors, interventions, and directions for the future. International Journal of Environmental Research and Public Health *18*(04), 1871.

Braunstein, L. Z., Gillespie, E. F., Hong, L., Xu, A., Bakhoum, S. F., Cuaron, J., . . . Powell, S. (2020). Breast radiation therapy under COVID-19 pandemic resource constraints: Approaches to defer or shorten treatment from a comprehensive cancer center in the United States. Advances in Radiation Oncology *5*(4), 582–588.

Bromma, K., Rieck, K., Kulkarni, J., O'Sullivan, C., Sung, W., Cullis, P., . . . Chithrani, D. B. (2019). Use of a lipid nanoparticle system as a trojan horse in delivery of gold NPs to human breast cancer cells for improved outcomes in radiation therapy. Cancer Nanotechnology *10*, 1–17.

Chithrani, D. B., Dunne, M., Stewart, J., Allen, C., & Jaffray, D. A. (2010). Cellular uptake and transport of gold NPs incorporated in a liposomal carrier. Nanomedicine: Nanotechnology, Biology and Medicine *6*(1), 161–169.

Chowdhury, N., Chaudhry, S., Hall, N., Olverson, G., Zhang, Q. J., Mandal, T., . . . Kundu, A. (2020). Targeted delivery of doxorubicin liposomes for Her-2+ breast cancer treatment. AAPS PharmaSciTech *21*, 1–12.

Costa, B., Amorim, I., Gärtner, F., & Vale, N. (2020). Understanding breast cancer: From conventional therapies to repurposed drugs. European Journal of Pharmaceutical Sciences *151*, 105401.

Debien, V., De Caluwé, A., Wang, X., Piccart-Gebhart, M., Tuohy, V. K., Romano, E., & Buisseret, L. (2023). Immunotherapy in breast cancer: An overview of current strategies and perspectives. NPJ Breast Cancer *9*(1), 7.

Dhankhar, R., Vyas, S. P., Jain, A. K., Arora, S., Rath, G., Goyal, A. K. (2010). Advances in novel drug delivery strategies for breast cancer therapy. Artificial Cells, Blood Substitutes, and Biotechnology *38*(5), 230–249.

Elwakkad, A., Gamal el Din, A. A., Saleh, H. A., Ibrahim, N. E., Hebishy, M. A., & Mourad, H. H. (2023). Gold NPs combined baker's yeast as a successful approach for breast cancer treatment. Journal of Genetic Engineering and Biotechnology *21*(1), 1–18.

Fathy, M. M., Mohamed, F. S., Elbialy, N., & Elshemey, W. M. (2018). Multifunctional chitosan-capped gold NPs for enhanced cancer chemo-radiotherapy: An in vitro study. Physica Medica *48*, 76–83.

Ganesan, K., Wang, Y., Gao, F., Liu, Q., Zhang, C., Li, P., . . . Chen, J. (2021). Targeting engineered NPs for breast cancer therapy. Pharmaceutics *13*(11), 1829.

Granja, A., Pinheiro, M., Sousa, C. T., & Reis, S. (2021). Gold nanostructures as mediators of hyperthermia therapies in breast cancer. Biochemical Pharmacology *190*, 114639.

Grattan, B. J., & Freake, H. C. (2012). Zinc and cancer: Implications for LIV-1 in breast cancer. Nutrients *4*(7), 648–675.

Gurunathan, S., Han, J. W., Eppakayala, V., Jeyaraj, M., & Kim, J.-H. (2013). Cytotoxicity of biologically synthesized silver nanoparticles in MDA-MB-231 human breast cancer cells. BioMed Research International, 1–10.

Hamilton, W. (2010). Cancer diagnosis in primary care. Journal of General Practice *60*(571), 121–128.

Hauksdóttir, H. L., & Webster, T. J. (2018). Selenium and iron oxide nanocomposites for magnetically-targeted anti-cancer applications. Journal of Biomedical Nanotechnology *14*(3), 510–525.

Hawrot, K., Shulman, L. N., Bleiweiss, I. J., Wilkie, E. J., Frosch, Z. A., Jankowitz, R. C., . . . Laughlin, A. I. (2021). Time to treatment initiation for breast cancer during the 2020 COVID-19 pandemic. JCO Oncology Practice *17*(9), 534–540.

Henriques, B., Mendes, F., & Martins, D. (2021). Immunotherapy in breast cancer: When, how, and what challenges? Biomedicines *9*(11), 1687.

Hira, I., Kumar, A., Kumari, R., Saini, A. K., & Saini, R. V. (2018). Pectin-guar gum-zinc oxide nanocomposite enhances human lymphocytes cytotoxicity towards lung and breast carcinomas. Materials Science and Engineering *90*, 494–503.

Jain, V., Kumar, H., Anod, H. V., Chand, P., Gupta, N. V., Dey, S., . . . Kesharwani, S. S. (2020). A review of nanotechnology-based approaches for breast cancer and triple-negative breast cancer. Journal of Controlled Release *326*, 628–647.

Kalyane, D., Polaka, S., Vasdev, N., & Tekade, R. K. (2022). CD44-receptor targeted gold-doxorubicin nanocomposite for pulsatile chemo-photothermal therapy of triple-negative breast cancer cells. Pharmaceutics *14*(12), 2734.

Kamaly, N., Kalber, T., Thanou, M., Bell, J. D., & Miller, A. D. (2009). Folate receptor targeted bimodal liposomes for tumor magnetic resonance imaging. Bioconjugate Chemistry *20*(4), 648–655.

Karakas, C. (2011). Paget's disease of the breast. Journal of Carcinogenesis *10*.

Khan, M. M., Tazin, T., Zunaid Hussain, M., Mostakim, M., Rehman, T., & Singh, S. (2022). Breast tumor detection using robust and efficient machine learning and convolutional neural network approaches. Computational Intelligence and Neuroscience.

Le, B. T., Raguraman, P., Kosbar, T. R., Fletcher, S., Wilton, S. D., & Veedu, R. N. (2019). Antisense oligonucleotides targeting angiogenic factors as potential cancer therapeutics. Molecular Therapy-Nucleic Acids *14*, 142–157.

Lee, J., Chatterjee, D. K., Lee, M. H., & Krishnan, S. (2014). Gold NPs in breast cancer treatment: Promise and potential pitfalls. Cancer Letters *347*(1), 46–53.

Lee, K. S., Chung, H. C., Im, S. A., Park, Y. H., Kim, C. S., & Kim, S. B. (2008). Multicenter phase II trial of genexol-PM, a cremophor-free, polymeric micelle formulation of paclitaxel, in patients with metastatic breast cancer. Breast Cancer Research and Treatment 108, 241–250.

Liu, H., Zhuang, F., Zhang, C., Ai, W., Liu, W., & Zhou, X. (2021). Functionalized organic: Inorganic liposome nanocomposites for the effective photo-thermal therapy of breast cancer. Frontiers in Materials 8, 710187.

Lu, S., Li, X., Zhang, J., Peng, C., Shen, M., & Shi, X. (2018). Dendrimer-stabilized gold nanoflowers embedded with ultrasmall iron oxide NPs for multimode imaging: Guided combination therapy of tumors. Advanced Science 5(12), 1801612.

Mahmood, R. I., Kadhim, A. A., Ibraheem, S., Albukhaty, S., Mohammed-Salih, H. S., Abbas, R. H., . . . AlMalki, F. A. (2022). Biosynthesis of copper oxide NPs mediated annona muricata as cytotoxic and apoptosis inducer factor in breast cancer cell lines. Scientific Reports 12(1), 16165.

Meenach, S. A., Shapiro, J. M., Hilt, J. Z., & Anderson, K. W. Polymer Edition. (2013). Characterization of PEG: Iron oxide hydrogel nanocomposites for dual hyperthermia and paclitaxel delivery. Journal of Biomaterials Science 24(9), 1112–1126.

Mkandawire, M., Lakatos, M., Springer, A., Clemens, A., Appelhans, D., Krause-Buchholz, U., . . . Mkandawire, M. (2015). Induction of apoptosis in human cancer cells by targeting mitochondria with gold NPs. Nanoscale 7(24), 10634–10640.

Mohammed, S. N., Mohammed, A. M., & Al-Rawi, K. F. (2022). Novel combination of multi-walled carbon nanotubes and gold nanocomposite for photothermal therapy in human breast cancer model. Steroids 186, 109091.

Montero, A. J., Adams, B., Diaz-Montero, C. M., & Glück, S. (2011). Nab-paclitaxel in the treatment of metastatic breast cancer: A comprehensive review. Expert Review of Clinical Pharmacology 4(3), 329–334.

Mund, R., Panda, N., Nimesh, S., & Biswas, A. (2014). Novel titanium oxide NPs for effective delivery of paclitaxel to human breast cancer cells. Journal of Nanoparticle Research 16, 1–12.

Musielak, M., Potoczny, J., Boś-Liedke, A., & Kozak, M. (2021). The combination of liposomes and metallic NPs as multifunctional nanostructures in the therapy and medical imaging: A review. International Journal of Molecular Sciences 22(12), 6229.

Nagaraj, E., Shanmugam, P., Karuppannan, K., Chinnasamy, T., & Venugopal, S. (2020). The biosynthesis of a graphene oxide-based zinc oxide nanocomposite using Dalbergia latifolia leaf extract and its biological applications. New Journal of Chemistry 44(5), 2166–2179.

Nounou, M. I., ElAmrawy, F., Ahmed, N., Abdelraouf, K., Goda, S., & Syed-Sha-Qhattal, H. (2015). Breast cancer: Conventional diagnosis and treatment modalities and recent patents and technologies. Breast Cancer: Basic and Clinical Research 9, BCBCR. S29420.

O'Brien, M. E., Wigler, N., Inbar, M., Rosso, R., Grischke, E., Santoro, A., . . . Ackland, S. (2004). Reduced cardiotoxicity and comparable efficacy in a phase III trial of pegylated liposomal doxorubicin HCl (CAELYX™/Doxil®) versus conventional doxorubicin for first-line treatment of metastatic breast cancer. Annals of Oncology 15(3), 440–449.

O'Connor, M., Rhodes, D., & Hruska, C. (2009). Molecular breast imaging. Expert Review of Anticancer Therapy 9(8), 1073–1080.

Padhi, S., & Behera, A. (2020). Nanotechnology based targeting strategies for the delivery of Camptothecin. In Sustainable Agriculture Reviews 44: Pharmaceutical Technology for Natural Products Delivery Vol. 2 Impact of Nanotechnology, (pp. 243–272): Springer.

Palmer, M. L., & Tsangaris, T. N. (1993). Breast biopsy in women 30 years old or less. The American Journal of Surgery 165(6), 708–712.

Pantapasis, K., Anton, G. C., Bontas, D. A., Sarghiuta, D., Grumezescu, A. M., & Holban, A. M. (2017). Bioengineered nanomaterials for chemotherapy In Nanostructures for Cancer Therapy, (pp. 23–49): Elsevier.

Pradhan, S. P., Sahoo, S., Behera, A., Sahoo, R., & Sahu. P. K. (2022b). Memory amelioration by hesperidin conjugated gold nanoparticles in diabetes induced cognitive impaired rats. Journal of Drug Delivery Science and Technology 69, 103145.

Pradhan, S. P., Swain, S., Sa, N., Pilla, S. N., Behera, A., Sahu. P. K., . . . Si, S. C. (2022a). Photocatalysis of environmental organic pollutants and antioxidant activity of flavonoid conjugated gold nanoparticles. Spectrochimica Acta Part A: Molecular and Biomolecular Spectroscopy 282, 121699.

Pradhan, S. P., Tejaswani, P., Sa, N., Behera, A., Sahoo, R., & Sahu. P. K. (2023). Mechanistic study of gold nanoparticles of vildagliptin and vitamin E in diabetic cognitive impairment. Journal of Drug Delivery Science and Technology, 104508.

Ramalingam, V., Raja, S., & Harshavardhan, M. (2020). In situ one-step synthesis of polymer-functionalized palladium NPs: An efficient anticancer agent against breast cancer. Dalton Trans 49(11), 3510–3518. doi:10.1039/c9dt04576g

Ramazani, A., Abrvash, M., Sadighian, S., Rostamizadeh, K., & Fathi, M. (2018). Preparation and characterization of curcumin loaded gold/graphene oxide nanocomposite for potential breast cancer therapy. Research on Chemical Intermediates 44, 7891–7904.

Ranjan, P., Abubakar Sadique, M., Yadav, S., Khan, R. (2022). An electrochemical immunosensor based on gold-graphene oxide nanocomposites with ionic liquid for detecting the breast cancer CD44 biomarker. ACS Applied Materials & Interfaces 14(18), 20802–20812.

Rau, K. M., Lin, Y. C., Chen, Y. Y., Chen, J. S., Lee, K. D., Wang, C. H., & Chang, H. K. (2015). Pegylated liposomal doxorubicin (Lipo-Dox®) combined with cyclophosphamide and 5-fluorouracil is effective and safe as salvage chemotherapy in taxane-treated metastatic breast cancer: An open-label, multi-center, non-comparative phase II study. BMC Cancer 15(1), 1–8.

Ruiz, A. L., Arribas, E. V., & McEnnis, K. (2022). Poly (lactic-co-glycolic acid) encapsulated platinum NPs for cancer treatment. Material Advances 3(6), 2858–2870.

Sa, N., Tejaswani, P., Pradhan, S. P., Alkhayer, K. A., Behera, A., & Sahu. P. K. (2023). Antidiabetic and antioxidant effect of magnetic and noble metal nanoparticles of clitoria ternatea. Journal of Drug Delivery Science and Technology, 104521.

Salimi, M., Sarkar, S., Hashemi, M., & Saber, R. (2020). Treatment of breast cancer-bearing BALB/c mice with magnetic hyperthermia using dendrimer functionalized iron-oxide NPs. Nanomaterials 10(11), 2310.

Salimi, M., Sarkar, S., Saber, R., Delavari, H., Alizadeh, A. M., & Mulder, H. T. (2018). Magnetic hyperthermia of breast cancer cells and MRI relaxometry with dendrimer-coated iron-oxide NPs. Cancer Nanotechnology 9, 1–19.

Selvi, B. C. G., Madhavan, J., & Santhanam, A. (2016). Cytotoxic effect of silver NPs synthesized from Padina tetrastromatica on breast cancer cell line. Advances in Natural Sciences: Nanoscience and Nanotechnology 7(3), 035015.

Shang, L., Zhou, X., Zhang, J., Shi, Y., & Zhong, L. (2021). Metal NPs for photodynamic therapy: A potential treatment for breast cancer. Molecules 26(21), 6532.

Sharma, P., Goyal, D., Baranwal, M., & Chudasama, B. (2020). ROS-induced cytotoxicity of colloidal copper NPs in MCF-7 human breast cancer cell line: An in vitro study. Journal of Nanoparticle Research 22, 1–11.

Shi, Y., Zhu, H., Ren, Y., Li, K., Tian, B., & Han, J. (2017). Preparation of protein-loaded PEG-PLA micelles and the effects of ultrasonication on particle size. Colloid and Polymer Science 295, 259–266.

Stephenson, A., Eggener, S. E., Bass, E. B., Chelnick, D. M., Daneshmand, S., Feldman, D., . . . Liauw, S. L. (2019). Diagnosis and treatment of early stage testicular cancer: AUA guideline. Journal of Urology 202(2), 272–281.

Sulak, M., Turgut, G. C., & Sen, A. (2022). Cerium oxide NPs biosynthesized using fresh green walnut shell in microwave environment and their anticancer effect on breast cancer cells. Chemistry & Biodiversity 19(8), e202200131.

Swanner, J., Fahrenholtz, C. D., Tenvooren, I., Bernish, B. W., Sears, J. J., Hooker, A., . . . Donati, G. L. (2019). Silver NPs selectively treat triple-negative breast cancer cells without affecting non-malignant breast epithelial cells in vitro and in vivo. FASEB BioAdvances 1(10), 639.

Tamanoi, F., Matsumoto, K., Doan, T. L. H., Shiro, A., & Saitoh, H. (2020). Studies on the exposure of gadolinium containing NPs with monochromatic X-rays drive advances in radiation therapy. Nanomaterials 10(7), 1341.

Tao, L., Chen, X., Sun, J., & Wu, C. (2021). Silver NPs achieve cytotoxicity against breast cancer by regulating long-chain noncoding RNA XLOC_006390-mediated pathway. Toxicology Research 10(1), 123–133.

Thakur, N., Sadhukhan, P., Kundu, M., Singh, T. A., Hatimuria, M., Pabbathi, A., . . . Sil, P. C. (2022). Folic acid-functionalized cerium oxide NPs as smart nanocarrier for pH-responsive and targeted delivery of Morin in breast cancer therapy. Inorganic Chemistry Communications 145, 109976.

Tørring, M., Frydenberg, M., Hansen, R., Olesen, F., Hamilton, W., & Vedsted, P. (2011). Time to diagnosis and mortality in colorectal cancer: A cohort study in primary care. British Journal of Cancer 104(6), 934–940.

Turetschek, K., Roberts, T. P., Floyd, E., Preda, A., Novikov, V., Shames, D. M., . . . Brasch, R. C. (2001). Tumor microvascular characterization using ultrasmall superparamagnetic iron oxide particles (USPIO) in an experimental breast cancer model. Journal of Magnetic Resonance Imaging: An Official Journal of Magnetic Resonance in Medicine 13(6), 882–888.

Van Goethem, M., Tjalma, W., Schelfout, K., Verslegers, I., Biltjes, I., & Parizel, P. (2006). Magnetic resonance imaging in breast cancer. European Journal of Surgical Oncology 32(9), 901–910.

Vines, J. B., Yoon, J. H., Ryu, N. E., Lim, D. J., & Park, H. (2019). Gold NPs for photothermal cancer therapy. Frontier in Chemistry 7, 167.

Wei, L., Lu, J., Xu, H., Patel, A., Chen, Z. S., & Chen, G. (2015). Silver NPs: Synthesis, properties, and therapeutic applications. Drug Discovery Today 20(5), 595–601.

Wei, P., Chen, J., Hu, Y., Li, X., Wang, H., Shen, M., . . . Shi, X. (2016). Dendrimer-stabilized gold nanostars as a multifunctional theranostic nanoplatform for CT imaging, photothermal therapy, and gene silencing of tumors. Advanced Healthcare Materials 5(24), 3203–3213.

Yaqoob, S. B., Adnan, R., Rameez Khan, R. M., & Rashid, M. (2020). Gold, silver, and palladium NPs: A chemical tool for biomedical applications. Frontiers In Chemistry 8, 376.

Yee Foo, Y., Saw, W. S., Periasamy, V., Chong, W. Y., Abd Malek, S. N., Tayyab, S. J. M., . . . Letters, N. (2019). Green synthesised-gold NPs in photothermal therapy of breast cancer. Micro & Nano Letters 14(5), 470–474.

Zandvakili, A., Moradi, M., Ashoo, P., Pournejati, R., Yosefi, R., Karbalaei-Heidari, H. R., . . . Behaein, S. (2022). Investigating cytotoxicity effect of ag-deposited, doped and coated titanium dioxide nanotubes on breast cancer cells. Materials Today Communications 32, 103915.

Ziental, D., Czarczynska-Goslinska, B., Mlynarczyk, D. T., Glowacka-Sobotta, A., Stanisz, B., Goslinski, T., . . . Sobotta, L. (2020). Titanium dioxide NPs: Prospects and applications in medicine. Nanomaterials 10(2), 387.

Zughaibi, T. A., Mirza, A. A., Suhail, M., Jabir, N. R., Zaidi, S. K., Wasi, S., . . . Tabrez, S. (2022). Evaluation of anticancer potential of biogenic copper oxide NPs (CuO NPs) against breast cancer. Journal of Nanomaterials, 1–7.

Chapter 15

Nanotherapeutics for liver cancer using metal nanocomposites

Hitesh Malhotra, Anjoo Kamboj, and Rupesh K. Gautam

LIST OF ABBREVIATIONS

ASGPR   Asialoglycoprotein receptor antibody
CCl$_4$ Carbon tetrachloride
CS   Chitosan
DEN   N-nitroso diethylamine
DOX   Doxorubicin
DSPE   Distearoyl phosphoethanolamine
ERP   Enhanced permeability and retention
FR   Folate receptor
GA   Gum arabic
GPC3   Glypican-3
HBV   Hepatitis B virus
HCC   Hepatocellular carcinoma
MAPKs   Mitogen-activated protein kinases
MNP   Magnetic nanoparticle
MRI   Magnetic resonance imaging
NPs   Nanoparticles
PLGA   Poly (lactic-co-glycolic) acid
RGD   Arginylglycylaspartic acid
ROS   Reactive oxygen species
SiNP   Silica nanoparticles
SR-B1   Scavenger receptor class B type I
TfR   Transferrin receptor
VEGF   Vascular endothelial growth factor

15.1 INTRODUCTION

The most frequent type of primary liver cancer, i.e., hepatocellular carcinoma (HCC) is the most widespread cancer worldwide (Wong et al., 2018). It has a high death rate and kills more than 600,000 people per year. HCC patients are mainly detected at advanced illness stages when there are few and poor therapy choices available because of the disease's stealthy development pattern (Dhanasekaran et al., 2012). Alcoholic liver disease, Hepatitis C, steatohepatitis, and hemochromatosis (iron overload in the liver) are a few examples

DOI: 10.1201/9781032621135-15

of chronic liver illnesses that can lead to cirrhosis. Given that the hepatitis B virus (HBV) can promote oncogenesis by integrating into the DNA of liver cells, HCC can potentially develop in the presence of HBV infection even without cirrhosis (Nault et al., 2015). Afla-toxin B1 released from *Aspergillus* fungus is also a well-known fungal toxin that can be consumed through contaminated staple foods and is a major contributor to carcinoma in Africa and Asia (Kew, 2013). In addition, hypoglycemic medications (Zhou et al., 2016; Kawaguchi et al., 2010) are linked to a higher risk of developing HCC except for biguanides (Singh et al., 2013) and thiazolidinediones (Chang et al., 2012) which have a lower risk.

Early cancer detection can considerably increase the likelihood of survival (Kakushadze et al., 2017). This is not the case, though, particularly in the case of HCC, which has a wide range of symptoms and frequently co-occurs with other illnesses, making it tough to distinguish HCC from its linked conditions. According to statistics, only 25% of all HCC patients had an early stage diagnosis (Tsuchiya et al., 2015; Farinati et al., 2009). HCC is one of the worst types of cancer since patients with advanced, incurable stages of the disease are given palliative treatment. HCC symptoms include stomach ache, jaundice, hepatomeg-aly, anorexia, and diarrhea. Additionally, some patients regularly suffer from HCC-related side effects, such as bleeding in the abdomen and peritoneum, hepato-portal vein invasion, thrombosis, and hepatic vein occlusion (Dimitroulis et al., 2017). Additionally, when HCC metastasizes to the bone patients may endure terrible agony (Christian-Miller and Frenette, 2018). HCC has a wide range of treatment choices, but which one is selected will depend on the stage of the disease and the characteristics of each patient. The most popular forms of treatment for HCC are ablation, surgery, liver transplantation, transarterial chemoembo-lization, radiation, and chemotherapy. Different restrictions are placed on these traditional therapies. The current conventional methods for treating HCC are being hampered by sev-eral factors, including increasing patient resistance to chemotherapy, unintended effects of radiotherapy, unintentional metastasis due to surgery, and uncomfortable chronic immu-nosuppressants due to grafting (Luqmani, 2005). However, as the patient is diagnosed as having the initial stage of HCC, surgery and transplantation do stand improved odds of survival (Wall and Marotta, 2000).

The cirrhosis-related consequences, such as hemorrhage, ascites, encephalopathy, and hepatorenal syndrome, are frequently to blame for the mortality rather than the tumor itself. Therapeutic approaches should now focus on boosting the host immune system and more specifically targeting tumor cells. Additionally, these medications have serious side effects, especially in patients who also have concurrent cirrhosis and liver disease. Chemo-therapeutic medications target particular carcinogenic pathways to reduce systemic side effects. Unfortunately, HCC cannot be developed through a single dominant mechanism.

Nanotechnology has emerged as a rapidly developing topic and a cutting-edge way to address the difficulties currently facing HCC therapy. This is mainly due to its drug delivery capabilities, such as precise targeting, improved pharmaceutical properties, simultaneous administration of multiple drugs, imaging of drug delivery sites, and the therapeutic activ-ity of nanomaterials, such as gold nanoshells (Shi et al., 2017; Min et al., 2017). The fast complexation of particle carriers by Kupffer cells, however, significantly restricts the forma-tion of tumoral hepatocytes and gravely impairs treatment efficacy. Consequently, a viable approach for treating HCC is targeted delivery to the hepatic tumor through overexpressed receptors (Pranatharthiharan et al., 2017).

Due to their enormous relative surface area and capacity for electromagnetic wave absorp-tion, nanomaterials have recently played a crucial role in the diagnosis and treatment of liver neoplasms. Nanomaterials have benefits over other biological materials in the diagnosis and therapy of liver cancer, including good tissue compatibility, a high safety factor, high controllability, and ease of use and modification. The hepatic sinusoid microenvironment is particularly suited for the entry and retention of nanomaterials, providing inherent benefits for the use of nanomaterials. This is in contrast to other solid tumors such as lung cancer, stomach cancer, and colon cancer. A lot of focus has been placed on the fact that liver cancer therapy delivered with precision medicine using nanotechnology could effectively prevent recurrence and reduce the metastasis (Sang et al., 2019). Furthermore, several new applications for nanomaterials have been widely reported in recent years, including as carriers for oligonucleotide therapy, thermal energy conversion medium, and hazardous nanoparticle therapy for liver cancer cells.

Small, solid particles called nanoparticles (NPs) range in size from 1 nm to 100 nm, and they contain an anticancer drug that is either trapped in the center or binds to the surface (Mohanraj and Chen, 2006; Padhi and Behera, 2020). The use of NPs for anticancer medication delivery holds enormous potential and has potential applications across the spectrum of medicine (Behera and Padhi, 2020). Small particles have a wide surface area, which increases the ability of antineoplastic drugs with low solubility to dissolve. By adjusting NPs' size, surface properties, and particle charge, it is possible to direct them to a specific tumor site (Kayser et al., 2005; Gu et al., 2018). Nanoparticulate drug delivery methods also stop drug molecule degradation and minimize unintended toxicities related to anticancer drugs. By avoiding all the biological and physical barriers that often prevent conventional medications from working, NPs can prove successful in treating cancer.

Many different NPs have been altered to deliver drugs specifically to HCC. According to their chemical makeup, NPs used to treat HCC are often categorized as macromolecular organic structures or inorganic, such as ceramic and metal NPs. The surface characteristics of NPs and the active chemical groups of targeting agents (ligands) can modify the NPs surface, primarily by covalent bonding, to produce the required therapeutic efficacy (Ghosh et al., 2009). In this context, the combination of various materials in a nanoparticle drug delivery system, such as polymer-coated metallic or ceramic NPs, has also been studied. Sorafenib was the first systemic medicine to show efficacy in advanced HCC and has been the standard of care as first-line therapy for over ten years (Marisi et al., 2018; Raoul et al., 2019). Atezolizumab and bevacizumab recently got FDA approval, and they significantly outperformed sorafenib in terms of overall and progression-free survival results in metastatic HCC (Lee et al., 2019). Bevacizumab is a monoclonal antibody that targets vascular endothelial growth factor (VEGF), preventing tumor angiogenesis and reversing T-cell suppression, whereas atezolizumab exclusively targets death ligand 1 and reverses T-cell inhibition (Herbst et al., 2014).

Progress in HCC nanomedicine

Since it was initially introduced in 1974, nanotechnology has steadily advanced, particularly in the field of cancer research (Taniguchi, 1974). Numerous research has been undertaken and is still being conducted to address the difficulties in precisely targeted drug delivery to HCC, given the growing interest in HCC nanomedicine. Although the majority of NPs accumulate in the hepatocytes, making it a target in HCC more profoundly. The pharmacokinetics in a cirrhotic liver greatly alters and thus is a hurdle because the liver serves as the major site of drug processing and so most medicines must pass through it. A further challenge is focusing on tumoral hepatocytes specifically. Therefore, a targeted therapy that targets only the afflicted cells is preferred to reduce the toxicity of the treatment (Blanco et al., 2015). The most current method is gene engineering (Azangou-Khyavy et al., 2020). Afterward, there has been advancement in the medicine of interest to be delivered by the aforementioned nanocarriers. Genome engineering offers a novel strategy for treating HCC via nanotechnology by introducing nucleic acids (Montaño-Samaniego et al., 2020).

15.2 BIOCOMPATIBILITY AND TOXICITY OF NANOCOMPOSITES

It is essential to assess the biocompatibility and toxicity of NP before using them as drug delivery agents. This will help to ensure that the medications are released to the desired areas without any problems or toxicity (Padhi and Behera, 2022). The form, size, and surface features of NPs, as well as the surroundings, have a significant impact on their compatibility. The immune system can be stimulated and suppressed by NPs. NPs have demonstrated adjuvant characteristics in the context of immunostimulation, preventing the development of any secondary tumors (Naahidi et al., 2013). According to Gref et al. (1994), macrophages quickly eliminate NPs that lack surface changes to prevent opsonin adsorption (proteins that aid in the body's phagocytosis process, which helps remove foreign objects from the body). Because of its metabolic products, glycolic acid and lactic acid, PLGA has proven to be a compatible polysaccharide for the available nano-formulations (Bahadar et al., 2016).

Oxidative stress, alterations in cellular shape, platelet accumulation, and toxicity from metal residues are a few of the problems connected to the use of carbon nanotubes. Like gold, cationic NPs can traverse the blood–brain barrier and exhibit blood coagulation and hemolysis properties (De Jong and Borm, 2008). When compared to other nanoparticles (NPs), metal NPs like silver are more hazardous because they cause lactate dehydrogenase leakage and reactive oxygen species (ROS) to form. Numerous investigations using silica NPs have found that silica NPs generate inflammatory biomarkers, oxidative stress, and other hepatotoxic consequences.

15.3 NANO-ANTIOXIDANTS

Radicals with a single unpaired electron in their outermost shell, ROS are extremely reactive. Because of an increase in metabolic activity, oncogene activity, or peroxisome activity, high amounts of ROS are seen in malignant cells (De Jong and Borm, 2008). Biologically, ROS are utilized to defend us against viral and bacterial infections while antioxidants keep ROS levels in check. But altered signaling pathways and neuro carcinogenesis can result from an imbalance between ROS and antioxidants (Liou and Storz, 2010; Yaswen et al, 2015). Antioxidants can be employed as a cancer therapy by removing ROS from the tumor microenvironment. However, barriers to employing antioxidants as an anticancer therapeutic include poor membrane permeability, poor aqueous solubility, decomposition rate, and storage issues.

Nanotechnology has been used to find solutions to these problems. The so-called nano-antioxidants are NPs that can be employed as nano-carriers for antioxidants (Pradhan et al., 2022a, 2023; Sa et al., 2023). Carbon nanotubes and a few other nanoparticles based on carbon are examples of nano-antioxidants (Usmani et al., 2018). Nano curcumin was delivered as an anticancer drug in DEN/CCL$_4$-induced carcinoma in rodents, and both biochemical and histological investigations revealed significant improvements in HCC-bearing rats following nano-curcumin treatment (Eftekhari et al., 2018).

Rutin, a naturally occurring flavonoid with antioxidant and anticancer effects, was utilized to load poly (lactic-co-glycolic acid) (PLGA) NPs in *in vivo* and *in vitro* investigations by Pandey et al. (2018). Liver nodules and decolorized tissue, indicative of liver cancer, were present in the hepatic tissue of DEN-induced rats. Rutin-treated DEN-induced rats showed improvement in their physiology and color, whereas rutin-PLGA NPs showed modest decolorization and the absence of all nodules. According to these findings, DEN-induced HCC in rats was successfully treated with rutin-PLGA NPs.

15.4 MACROMOLECULAR NANOCOMPOSITES

These NPs are composed of big molecules and can be either natural or artificial. They have a variety of functional groups available due to their macrostructure, which can be used to load the anticancer medication as well as other targeted groups (Figure 15.1). Since they exhibit a notable therapeutic role in drug delivery, macromolecular NPs are one of the most researched nanomaterial systems for a variety of biomedical applications (Masood, 2016).

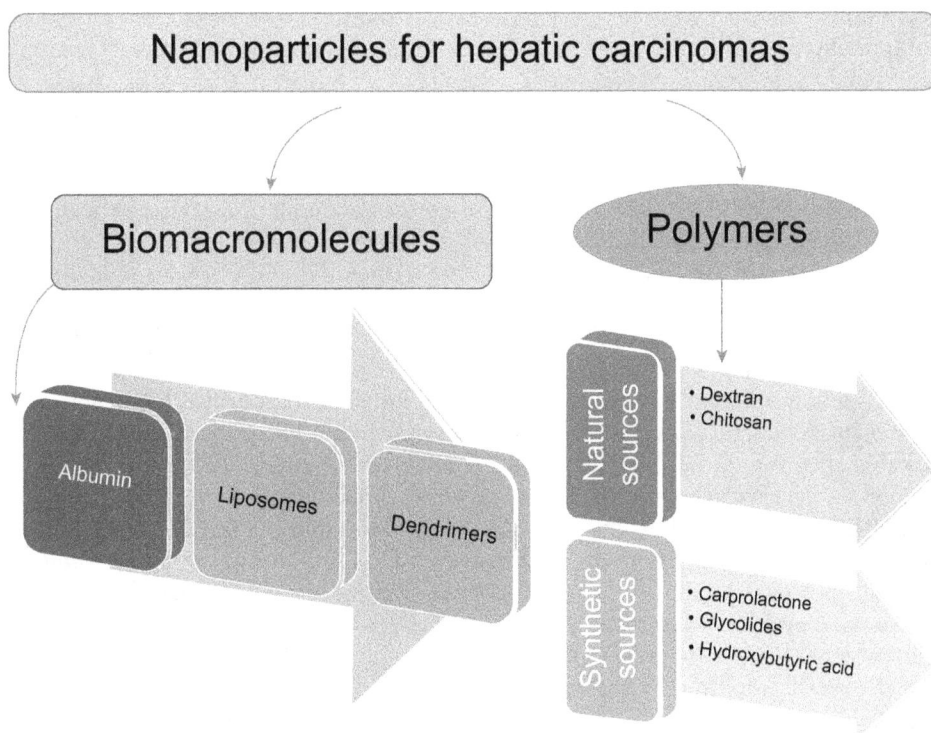

Figure 15.1 Different types of NPs for HCC.

The NPs' alluring qualities, like biocompatibility and biodegradability, enable them to effectively serve as drug delivery systems and carriers. The medication's solubility, bioavailability, and retention time are all increased by their capacity to complete the drug, which also shields the drug from early deterioration. Additionally, it improves their intracellular penetration and tissue selectivity, boosting therapeutic efficacy (Kumari et al., 2010). Active targeting is a type of targeted medication delivery in which the complexed anticancer agent is more selective to the tissues due to surface modification with targeting ligands. Passive targeting depends on the enhanced permeability and retention (EPR) effect, which causes NPs to concentrate on the tumor area because the tumor vasculature is porous and there is no lymphatic outflow (Khalid and El-Sawy, 2017).

15.4.1 Albumin

An essential part of the human body, albumin is a natural protein. The albumin molecule is a perfect choice for drug delivery applications due to its versatility and natural origin. Albumin is a flexible carrier in cancer therapies due to its capacity to stay in inflamed tumor cells and a variety of groups for drug binding. To track their mobility and location inside malignant cells, these NPs can also be combined with different colors (Kratz, 2008). When albumin-bound paclitaxel was used instead of taxane medicines, Zhou et al. (2011) noticed a more prominent obliteration of the hepatic cancerous mass, which was demonstrated by higher efficacy in HCC xenograft models. By using a passive targeting strategy, the controlled release of significant amounts of the medicines cinobufagin and sodium ferulate from albumin NPs was assessed. Their release was ascribed to the covalent cross-linking of glutaraldehyde. Additionally, it was discovered that albumin combined with zinc sulfide NPs inhibited the growth of hepatoma at a dosage of 0.36 mg/mL (Cao et al., 2011).

15.4.2 Liposomes

When one or more layers of lipids arrange into a sphere and contain water, they produce liposomes, also known as vesicles or just lipids. Liposomes are vesicles made of cholesterol and nontoxic, amphiphilic phospholipids that are frequently used to encapsulate lipophilic and hydrophilic medicinal compounds. Water-soluble medications are contained within the inner core of liposomes while lipid-miscible medications are contained within the bilayer (Jiang et al., 2007). The fact that liposomes' composition is dependent on elements of the human physiological membrane made it easier for the FDA to approve them, which is one of their main advantages. Additionally, liposomes are very biocompatible, nontoxic, and remarkably effective at internalizing into cells (Hann and Prentice, 2001).

To improve the effectiveness of chemotherapeutics, liposomes (NPs) have been frequently employed as medicinal carriers (Souto, 2011). One example is the PEGylated liposomes that contain doxorubicin (DOX) for the treatment of HCC. Clinical trials have shown that PEGylated liposomal DOX (PLD) is an effective treatment for HCC cases by increasing hepatic uptake and localization to the liver and spleen. When compared to pure medication, liposomal DOX had similar anticancer activity and less toxicity, according to Rahman et al. (1980). Additionally, the liposomal DOX formulation led to a higher therapeutic index, increased effectiveness, and effective administration of higher drug dosages. Lyso-thermosensitive DOX liposomes have also been found to be effective in treating medium-to-large-sized hepatic tumors by Poon and Borys (2009). This formulation caused tissue necrosis and irreversible tissue damage because it accumulated in the tumor locations and the release depended upon temperature. It also investigated how to target ligands onto liposomes using surface engineering. To target ligand attachment, a variety of functional groups can be added to the water-soluble moiety of lipids. Distearoyl phosphoethanol-amine (DSPE), one of the commonly utilized lipids, was used as an example. It has an active amine group that permits the covalent attachment of targeted ligands to the lipid at the hydrophilic head's end. Folate (Amin, 2008), transferrin (Doi et al., 2008), the RGD peptide (Jiang et al., 2010), and antibodies (Sapra and Allen, 2004) are examples of target-ing ligands that have had their surfaces changed for liposome-based targeted drug delivery. Using SP94-conjugated, DOX-encapsulated liposomes, the SP94 peptide's potential for drug delivery was subsequently assessed. It has been demonstrated that SP94-LD is superior in the treatment of mice receiving human hepatocellular cancer (Lo et al., 2008). It is thought that SP94-targeted PEG liposomal DOX binds to the surface of cancer cells and binds to them via the EPR effect before being internalized by ligand-mediated endocytosis (by the active targeting effect) (Zhang et al., 2016b).

15.4.3 Dendrimers

Highly branched synthetic macromolecules known as dendrimers are created progressively utilizing branched monomer units, and they exhibit essentially monodisperse and homoge-nous size distribution properties. When compared to the corresponding polymers, which are dispersed and primarily branched, these characteristics are thought to be favorable (Nazemi and Gillies, 2013). The core, branched monomers, and active chemical groups on the sur-face make up the characteristic architectural arrangement of dendrimers. The branching is present around the core in layers, with each layer being referred to as a generation (Aulenta et al., 2003). The fundamental use of this dendrimers structure is that it acts as biological scaffolding for a range of ligands (Parat et al., 2015).

The dendrimer nucleus undergoes multiple chemical syntheses which results in the for-mation of dendrimers with a greater surface area. Additionally, it is possible to create a wide variety of molecules that are capable of transporting a variety of medications thanks to chemical changes in dendrimers. For instance, hydrophobic medicines may be trapped in the core of aqueous dendrimers with a carboxylic acid on the surface. By increasing the survival of mice grafting with Huh7 cells, a plasmid vector encapsulated in amido-amine dendrimer was shown to be effective *in vivo* (Maruyama-Tabata et al., 2000). Dendrimers were discovered to be inferior to other nanosystems, though, since it was difficult to regu-late their drug delivery, they had trouble telling cancer cells from normal cells, and they had purifying problems.

15.4.4 Micelles

Micelles range in size from 5 nm to 100 nm and are a member of the amphiphilic col-loidal family. The amphiphilic/lipid molecules that make up the micelles have two clearly defined structural components: hydrophilic groups and hydrophobic groups. Both groups exhibit completely different polarities, which results in very different affinities for a par-ticular solvent (Husseini and Pitt, 2008). Drug degradation is guarded by the lipophilic core of micelles, and opsonization is avoided by the hydrophilic surface. Micelle carriers improve the solubility of the lipophilic part and enhance the retention time in the body.

For instance, naturally occurring anticancer compound like curcumin (Cur) was combined with the highly water-soluble polysaccharide gum arabic (GA) to create nanomicelles. When compared to free curcumin, this assembly significantly improved curcumin solubility and stability under physiological pH. The cell uptake experiment revealed that the GA–Cur conjugate transfers the medication to the cytosol of HepG2 cells (Sarika et al., 2015). A further illustration of a micelle assembly is the utilization of pluronic micelles in medication delivery systems. The pluronic micelles were smaller in size (10–100 nm) boosting their concentration in the tumor location and making them appropriate for injection (Hanahan and Weinberg, 2011). In a study, DOX, hematoporphyrin (HP), and pluronic F68 were coupled to create a straightforward, compatible, and diverse functional nanosystem. The FDA's endorsement of such constituents is their primary benefit. A core-shell nanostructure was produced in this nanoparticle system, where DOX was associated with HP to form a lipophilic core by the involvement of intermolecular forces. Results demonstrated that these HPDF micellar nanoparticles boosted the tumor-targeted transportation and exerted their synergistic effects by combining SDT and chemotherapy. They also enhanced the biostability of HP and DOX.

15.4.5 Polymeric nanoparticles

The polymers belonging to the polysaccharides class are found in nature and have good biocompatibility and versatility, allowing for further structural modification to incorporate a variety of medications. Dextran and chitosan are two examples. Chitin undergoes alkaline deacetylation to produce chitosan (CS), a naturally occurring linear bio-polyaminosaccharide. After cellulose, it is the most prevalent natural biopolymer (Lee et al., 2009; Padhi et al., 2022). Contrarily, dextran is a sophisticated branched polysaccharide that is frequently utilized in drug delivery applications. However, the FDA has not yet given its approval to chitosan or any of its derivatives. The –OH and –NH$_2$ functional groups, which are located along the chitosan chain, significantly affect how functional and reactive chitosan is to different kinds of pharmacological molecules. Additionally, the natural hydrophilic surface of chitosan can delay the body's macrophages from phagocytosing it, extending blood circulation (Moghimi et al., 2001).

Chitosan has been extensively used for the fabrication of hepatoma-directed nanoparticles (Qi et al., 2007). The development of nanoparticles was found to directly correlate with tumor inhibition in cancerous mice. This finding could be explained by the fact that larger nanoparticles can hold more drugs and that cells prefer to take up 100 nm particles over smaller ones. Chitosan is difficult to synthesize in an easily useable formulation because it is weakly soluble in water at physiological pH. The resulting solutions are thus extremely viscous with prominent chitosan chain aggregation. To overcome these challenges, low molecular weight chitosan has been employed in drug delivery because of its high aqueous solubility at physiological pH, thus increasing its suitability for use in medicine delivery. The combination of DOX and chitosan NPs was shown to be sufficiently deadly to human liver cancer (Jain and Jain, 2010).

By reacting chitosan and sulfosuccinimidobiotin, Yao et al. (2007) created a novel molecule in the form of biotinylated chitosan. In HepG2 cells, nanoparticles made with this substance had more cellular absorption than ordinary chitosan nanocarriers. It is thought that vascular endothelial growth factor receptor II inhibition or p53 pathway stimulation is the mechanism through which chitosan nanoparticles prevent HCC (Zu et al., 2010).

Curcumin was incorporated into dextran as a nano-sized medication carrier to increase its absorption and therapeutic effectiveness. The cytotoxicity of the medicine increased with increasing drug concentration in the dextran NPs, making it safe and effective for the delivery of curcumin to the liver (Anirudhan, 2016). Dextran was applied to the surfaces of magnetite nanoparticles in investigations to provide a multifunctional bioactive coating (El-Kharrag and Amin, 2017). This uniform coating was further embellished with crocetin, a key bioactive component of saffron (stigmas of the *Crocus sativus* flower) and an herbal treatment for HCC. Without affecting the magnetism of the magnetite core NPs, dextran coating served as an intermediary molecule for the immobilization of crocin (El-Kharrag and Amin, 2017).

15.4.6 Poly (hydroxyl acids)

Poly (hydroxyl) acids as well as their copolymers are examples of synthetic biodegradable polymers that have been extensively exploited in drug delivery applications. They are categorized as polyesters because of only little variations in their structural makeup but significantly different rates of deterioration. One biodegradable, FDA-approved substance that is widely employed in the field of cancer nanotechnology is PLGA. Due to the hydrophobic core of this polymer, PLGA more effectively entraps lipophilic medicines. Additionally, PLGA modification enhances its capacity for drug loading and hence broadens its range of potential uses. Ghosh et al. (2009) encapsulated curcumin in PLGA polymer to create nanocarriers for HCC, and it was discovered that this had a protective effect because malignant cells were killing themselves. A unique cationic lipopolymer was created by Liang et al. (2011) for the integration of IL-12, and it showed increased transfection efficiency and mouse survival. Similar to this, it was demonstrated that a unique recombinant plasmid internalized into such nanocarriers increased cancer cell targeting in addition to guarding against their enzymatic breakdown (He, 2004).

A new nanocomposite (FA–PEG–PLGA NP) with sorafenib co-encapsulated folate-conjugated PEG–PLGA was created. Folate-conjugated PEG NPs demonstrated prolonged drug release in an acidic medium that was pH dependent, particularly in tumor cells. When compared to free drugs or non-targeted ones, FA–PEG–PLGA NP significantly increased anticancer efficacy and successfully reduced tumor cell proliferation (Li, 2015).

15.4.7 Metal nanoparticles

In the nanoscale, metal nanoparticles differ from their bulk counterparts in terms of their physical and chemical properties. Due to their biocompatibility, gold and silver nanoparticles have been the subject of the majority of metallic nanoparticle investigations (Comfort et al., 2011; Behera et al., 2020; Padhi and Behera, 2021). Despite the nonmetallic nature of the element carbon, carbon nanotubes are frequently categorized as metallic nanoparticles. It could be explained by the fact that it has a molecular structure that makes it comparable to gold and silver as single-element drug carriers.

Due to their straightforward production and functionalization, as well as their exceptional optical and photothermal properties, gold nanoparticles have been used for biomedical applications (Huang and El-Sayed, 2010; Pradhan et al., 2022b;, Sa et al., 2023). Due to their ability to absorb and scatter visible laser light, gold nanoparticles act as a photothermal agent, generating localized heat that kills malignant cells. Amazing antibacterial properties have been demonstrated by silver nanoparticles. Additionally, biocompatible, stable, increased solubility, and high yield without aggregation are characteristics of biologically synthesized silver NP (Gurunathan et al., 2015). Ahmadian et al. (2018) concluded that silver NP showed enhanced pro-apoptotic bodies, suggesting that it may be a promising subject for the therapy of HCC after conducting an *in vitro* investigation on HepG2 cells.

Colloidal gold particles that are submicron in size are known as gold nanoparticles, and they are frequently found scattered in aqueous solutions. Since bFGF and VEGF165 are blocked by gold nanocarriers, these growth factors have been demonstrated to exhibit antiangiogenic characteristics (Cherukuri et al., 2010). Tomuleasa et al. (2012) reported the presence of gold nanoparticles boosted the liver cancer cells' susceptibility to the drugs capecitabine, doxorubicin, and cisplatin. When evaluated on the HCC cell line Hep3B, Au NPs were discovered which lack cytotoxic or antiproliferative properties (Gannon et al, 2008).

15.4.8 Ceramic nanoparticles

Metal oxides which make up the majority of ceramics are frequently used as biomaterials. Alumina, titania, zinc oxide, silica, and magnetite are some examples. For a variety of biological applications, ceramic NPs with porous structures have appeared as important substitutes for organic systems. Magnetite and silica nanoparticles are frequently utilized as nanocarriers for medication delivery, particularly in the treatment of cancer. The high compatibility of silica with diverse functional groups makes it a suitable candidate for diverse anticancer drugs.

Magnetite

Magnetite possesses high biodegradability, biocompatibility, and paramagnetic characteristics, and due to this magnetite NPs (MNPs) are particularly acceptable drug carriers. Magnetic resonance imaging (MRI) makes extensive use of its superparamagnetic properties (Revia and Zhang, 2016). MNP surfaces can hold a significant number of anticancer drugs due to their high surface area-to-volume ratio. These therapeutic chemicals are delivered to patients and released at the desired site through passive or active transportation, potentially reducing adverse effects. Polypeptides, lipids, complex carbohydrates, and other biomacromolecules can be first coated on MNPs to increase their biocompatibility and drug-carrying capacity (Muzzarelli and Muzzarelli, 2002). The developed MNP drug delivery systems offer advantages over other drug delivery methods, which create the possibility of overcoming the negative aspect of conventional therapy (Wedmore et al., 2006).

Chitosan MNP complex is considered as a safe and nontoxic biomaterial. Chitosan-coated magnetic particles can be utilized to quickly change the surfaces of MNPs by forming magnetic microspheres that directly interact with particular ligands (Moghimi et al., 2001). Furthermore, the natural hydrophilic surface of chitosan can prevent macrophage phagocytosis in the body, extending blood circulation. Another illustration is the efficient drug delivery and stabilization of iron oxide nanoparticles using oleic acid and its derivatives (Soenen et al., 2011). The activity is depicted using several powerful folic acid conjugations with diverse linkers, mechanisms, and cleavable bonds. Furthermore, it was claimed that the MNPs–folic acid complex could encourage the endocytosis of cancer cells by the folate receptor.

Silica

Applications for silica nanoparticles (Si NPs) in biomedicine have been investigated in various ways. Since crystalline silica NPs were discovered to cause chronic obstructive pulmonary disease, they were determined to be poisonous (Meijer et al., 2011). As an alternative, noncrystalline Si NPs were frequently used in therapeutics, particularly for multimodal imaging and drug delivery. Silica NPs have special qualities that make them suitable for use *in vivo* (Barbe et al., 2004), including a hydrophilic surface that encourages prolonged circulation, a chemistry that is adaptable for surface activity, compatibility, simplicity of industrial production, and low-cost manufacturing. Mesoporous Si NPs are also renowned for stable shape, specific and regulated size, and porosity. These qualities supported their usage in applications involving medication delivery. In general, porosity assists in generating high drug loads by trapping the drug within the porin channels (Mehmood et al., 2017). Nonporous Si NPs, on the other hand, are also employed for drug administration and are crucial for targeted molecular imaging and early diagnosis systems that have advanced to human clinical trials. Human HCC is 10,000 times more avid to SP94 peptide multivalent binding than are endothelial cells, hepatocytes, mononuclear cells, and B-/T-lymphocytes. The SP94 peptide's exquisite targeting specificity and increased multicomponent cargo delivery enable sensitive differentiation between targeted and normal tissue. To better understand the mechanism of active tumor targeting, the SP94 peptide is an excellent model (Ashley et al, 2011).

15.5 TARGETING NANOCOMPOSITES TO HCC

15.5.1 Asialoglycoprotein receptor (ASGPR)

In 1974, Morell and Ashwell made the initial recognition of ASGPR as a hetero-oligomer made up of two polypeptides with subunits HL-1 and HL-2, respectively (Shi et al., 2013). In addition to being broadly distributed on hepatocytes, it is present on the basolateral and sinusoidal hepatic plasma membranes, except the biliary duct membrane (D'souza and Devarajan, 2015). Additionally, the hepatic ASGPR serves as a receptor for extracellular glycoproteins that have terminal lactose, galactose, or galactosamine residues exposed, making it the ideal candidate for drug delivery to the liver (Wall and Hubbard, 1981). Some of the naturally occurring ligands for the ASGPR include Asialoorosomucoid, Asialoceruloplasmin, Asialofectin, and Asialotransferrin (Ishibashi et al., 1994; Rozema et al., 2007). It's noteworthy that two clinical trials have looked into the use of ASPGR targeting as a means of using receptors as a target, to reduce the chance of side effects in other tissues (Julyan et al., 1999). On the surface of hepatocytes, ASGPR1 expression does, however, show polarity and zonality, with high levels of expression on the membrane of the hepatocyte and diminished levels on the apical membrane. For galactose-mediated delivery, this is crucial (polarity). Similar to this, its expression is more pronounced in the portalobular regions of the hepatocytes than in the centrilobular regions (zonality) (Ise et al., 2004). Numerous studies have demonstrated how nanotechnology is used in HCC ASGPR targeting.

15.5.2 Glypican-3 (GPC3)

Glypican-3, a proteoglycan from the glypican family, is also known as heparan sulfate proteoglycan or HSPG (Filmus and Capurro, 2013). The 70 kDa GPC3 core protein possesses a cleavage site for furin in the protein cor0065. The cleavage of furin leads to the formation of 30 kDa and 40 kDa N-terminal fragments. Among other growth factors, they regulate Wnt signaling activity. Active targeting can circumvent medication delivery problems by utilizing a moiety that specifically targets over-expressed receptors present in cancer cells. Various studies reveal that GPC3 targeting is useful for treating HCC because the increased expression of the protein in the tumor is different from that in healthy and cirrhotic livers, where it is undetectable. Previous investigations that successfully targeted HCC cells with a humanized anti-GPC3 monoclonal antibody have shown the potential function against the disease (Ishiguro et al., 2008).

15.5.3 Transferrin receptor (TfR)

TfR is essential for regulating iron levels and fostering cell growth. The contact of the ligand and receptor causes endocytosis, which enables the internalization of the TfR–Fe complex that results in the release of iron and enables the receptors to be recycled back in an acidic environment (Kolhatkar et al., 2011). This is made possible by the high affinity of iron-bound transferrin for TfR. The liver, which is the main organ for iron storage, is closely associated with both the expression of TfR1 and iron metabolism. Prior studies have shown that HCC has an aberrant iron metabolism. Since dramatically elevated mRNA levels of genes involved in iron absorption have been discovered, TfR1 is a promising target for active targeting approaches (Crielaard et al., 2017). Transferrin (Tf) has been used widely as a targeting ligand up to this point, but its use is constrained by the high levels of endogenous Tf (Tang et al., 2019).

15.5.4 Folate receptor (FR)

The FR, a membrane-anchored 38 kDa glycoprotein with glycosylphosphatidylinositol, is overexpressed in a variety of cancers, including HCC. Malignant liver tissues have substantially higher FR expression than normal liver tissues, which has a high affinity for folic acid and makes it a prime goal for drug delivery to the hepatocytes (Liu et al., 2016; Zwicke et al., 2012). Folate has been scientifically shown to be crucial for the replication of DNA, mitosis, growth, and survival in rapidly proliferating cells and a deficiency leads to methylation and breaking of chromosomal strands (Kelemen, 2006). Folic acid's significant role in HCC could therefore be used as a targeted moiety for NPs.

15.5.5 Scavenger receptor class B type I (SR-BI)

SR-B1 is a multiligand membrane receptor protein that binds to several other lipoprotein receptors and has binding properties for lipid transportation to hepatic cells. Because of this, SR-B1 is crucial for preserving the stability, fluidity, and organization of the cell plasma in the liver and is also crucial for maintaining cholesterol homeostasis (Shen et al., 2018). Since SR-B1 is largely overexpressed on hepatocytes and is responsible for HDL uptake, HDL mimics are of significant interest as a potential, flexible, and successful approach to HCC (Lacko et al., 2002). Regularly, HDLs have come under the spotlight in the battle against HCC because of the increased expression of HDL receptors such as SR-B1, which the tumors require to quench their insatiable thirst for cholesterol for cell growth. Studies that target HDL use a variety of aspects of the HDL lifecycle in the liver (Henrich and

Thaxton, 2019). HDL mimics have shown advantages for nano dispersion because their size and surface properties resemble genuine HDL and also protect them from clearance in contrast to other foreign compounds (Yang et al., 2011). It has had remarkable success in the treatment of many types of cancer because of its capacity to act as a gateway for the delivery of therapeutic materials through HDL mimics like apoA-1; hence, it has the potential to deliver nanoparticles as a selective and specific HCC therapy.

15.5.6 Role of peptides in HCC therapy

Studies have discovered proteins that are uniquely produced by tumoral cells as a result of the expanding research into therapeutic alternatives for HCC. The two main challenges of tumoral specificity and selectivity in HCC targeting could be overcome by using these proteins to produce specific peptides (Jiang et al., 2006). Due to their small size, peptides have several distinctive properties that make them intriguing therapeutic agents. These characteristics include efficient tissue penetration and disregard for the host immune system, which results in fewer or no adverse effects and off-target effects (Savier et al., 2021). Additionally, peptides that specifically target overexpressed targets like GPC3 are being researched for use in the creation of an HCC vaccine (Nobuoka et al, 2013). To compensate for the relatively poor affinity of peptides and enhance anticancer efficacy, the use of peptides as a targeting modality in drug-encapsulated nano-formulations is anticipated in the future.

15.6 NANOCOMPOSITES IN THE THERAPY OF HCC

Numerous nanocomposites are now undergoing clinical trials or have received FDA approval. These nanocomposites target a variety of other disorders in addition to cancer, and they can also be used as a tool for early diagnostics. However, in this study, we examined nanomaterials created especially to attack HCC. Styrene-maleic acid copolymer and zinostatin stimaneocarzinostatin were created. It targeted the HCC cells passively and exhibited the EPR effect. In 1994, the Japanese FDA gave it the go-ahead (Weissig et al., 2014).

For patients with advanced HCC, sorafenib (BRAF inhibitor) is a frequently prescribed medication. However, sorafenib has recently been linked to many significant adverse effects. For instance, it has been demonstrated that sorafenib causes paradoxical activation of the mitogen-activated protein kinases (MAPKs) pathway, which is involved in several cellular activations, including the advancement of cancer (Peluso et al., 2019). To overcome this obstacle, Chen and his colleague (2017) created lipid-coated PLGA nanoparticles that codeliver sorafenib and the MEK inhibitor AZD6244 with a high encapsulation effectiveness of over 70% (Chen et al., 2017). These new nanoparticles were used for targeted distribution since they were modified to bear CTCE-9908 on their surface, an antagonist peptide ligand for CXCR4, which is frequently overexpressed in HCC and thus serves the ideal purpose for targeted delivery. Due to ligand dependence, Chen's group has demonstrated increased cellular absorption of these nanoparticles and demonstrated both *in vitro* and *in vivo* synergistic cytotoxic effects against HCC. As a result, the sorafenib-induced MAPK activation that has been documented has been successfully stopped in our study.

The idea of codelivery was also used by Zhao and his colleague in 2015, and both DOX and curcumin were engineered into lipid nanoparticles (Zhao et al., 2015). The well-known anthracycline antibiotic DOX has been used to treat a variety of malignancies, but its therapeutic effectiveness is compromised by its side effects. Curcumin, a dietary component of yellow dietary polyphenols found in *Curcuma longa*, has strong anticancer properties. These two bioactive substances were investigated in a mouse model of DEN-induced HCC while enclosed in lipid nanoparticles. The generated nanoparticle was smooth-surfaced, spherical, and had a homogeneous particle size of about 90 nm. For DOX and Cur, it also demonstrated outstanding encapsulation efficacy, which was above 90%. Combining *in vitro* and *in vivo* testing revealed that both the bioactive substances had a synergistic effect on the apoptosis, proliferation, and angiogenesis of HCC. It's interesting to note that the idea of a nanoparticle-based delivery system extends beyond chemotherapy to include cancer gene therapy. For instance, gold nanoparticles were created by Xue and his colleague in 2016 to deliver miR-375 (Xue et al., 2016). One of the small non-coding RNAs known as micro-RNAs (miRNAs), miR-375 is crucial for controlling the expression of genes. However, in certain cancer types, having defective miRNAs is critical for the course of the disease. MiR-375 is considerably downregulated in HCC, and its expression can be forced up, which reduces cell invasion and proliferation (Table 15.1).

PEG was applied to the surface of gold nanoparticles to stabilize the particles and create a covalent bond for the double-stranded miR-375 mimic that is marked with cy3 fluorescent dye for fluorescence imaging. The average size of the created gold nanoparticle is 53 nm.

Table 15.1 Nanocomposites targeting HCC

Type of nano-formulation	Active drug	Cell line	Mechanism	References
Micelles	Sorafenib	HepG2-Luc	Suppression of tumor	Su et al., 2018
Gadolinium Liposomes	Sorafenib	HepG2-H22	Enhance the image resolution and suppress cell growth	Xiao et al., 2016
Lipid–polymer hybrid	Sorafenib and doxorubicin	HepG2-H22	Tumor suppression	Zhang et al., 2016a
Silica-coated gold NP	Doxorubicin	SMMC-7721	Reduce the toxicity profile and tumor size	Wang et al., 2017
Hydroxyapatite shell and magnetic core of iron NP	Doxorubicin and curcumin	HEpG2	Reduce tumor growth	Manatunga et al., 2018
Chitosan NP	Doxorubicin	HCC SMMC-7721, HepG2-H22	Depresses tumor growth	Qi et al., 2015
Porous iodinated magnetic NP	Doxorubicin	Liver tumor in rabbits	Increase liver cell apoptosis	Jeon et al., 2016
Silica NP	Cetuximab and doxorubicin	HEpG2	Anticancer activity	Wang et al., 2017
Alginate/doxorubicin NP	Doxorubicin	HepG2 cells	Suppress tumor metastasis	Guo et al., 2013
PEG-modified phospholipid micelle	Sorafenib	HepG2 cells	Antitumor activity	Azzariti et al., 2017
Lipoprotein-modified silica NP	Docetaxel and thalidomide	HepG2	Increase anticancer activity	Ao et al., 2018

Once delivered to the targeted HCC tissues, miR-375 anti-tumor activity was successfully therapeutically effective, according to both *in vitro* and *in vivo* assessments of their generated nanoparticle. In a different study, the therapeutic benefits of combining cancer gene therapy and chemotherapeutic medication into a single nanocarrier were evaluated in comparison to the usual contemporaneous co-treatment of each medication alone.

P-glycoprotein (Pgp) oncogenic silencing was used in this work by Zhang and his team in 2016 (Zhang et al., 2016a). One of the well-known efflux transporters known to contribute to the multidrug resistance (MDR) phenomenon is P-gp, which actively removes drugs from cells, lowering drug accumulation and creating a considerable barrier to efficient chemotherapy (Su et al., 2018). The crocin-loaded magnetite nanoparticles performed significantly better in terms of their anticancer efficacy than free crocin when tested *in vivo* and *in vitro*. Further studies into the molecular mechanisms behind the effects of the magnetic nanoparticle-released crocin have also been conducted.

15.7 LIMITATIONS OF NANOCOMPOSITES

Using nanocomposites has various disadvantages, they are not error-free. The small size and larger surface area of NPs could cause sticking and friction issues. As a result of NPs' chemical reactivity, ROS that cause neurological disorders, oxidative stress, inflammation, and damage to DNA, proteins, and membranes that result in toxicity may increase. Numerous physical and chemical interactions can result from modifications to the form and size of nanocomposites. The surrounding environment, chemical composition, surface structure, the presence of functional groups, etc., all play a role in NP toxicity. Unexpected reactions and interactions may occur once the substance has entered the body. These NPs can pass cell membranes to enter capillaries and cross the blood–brain barrier. Although biodegradable nanocomposites are eliminated from the body, slow-degrading or nonbiodegradable nanoparticles can build up in the body and trigger inflammation. Anionic NPs have no harmful effects on the cardiovascular system; however, cationic NPs can cause hemolysis and blood coagulation (Shubhika et al., 2012).

15.8 CONCLUSIONS

Over the past few years, the number of HCC cases has dramatically increased. The diagnosis is typically made when the disease is well along, and the available treatments are no longer highly effective. The increased interest in cancer nanomedicine gives a glimmer of hope for the development of efficient tools for improving HCC therapies and diagnostics. Some of the nanoparticles that are thought to be beneficial as drug delivery agents include albumin, dendrimers, liposomes, micelles, ceramic, and metal nanoparticles. To gain a better idea of their therapeutic potential in comparison to currently utilized treatments, further research is required to understand the precise mechanism of action of various nanocomposites, their long-term effects, and drug clearance mechanisms, especially of freshly created drug carrier systems. Due to their origin, structure, content, functional groups, etc., nanocomposites have several limits while being effective, as demonstrated by biocompatibility and toxicity studies. Despite these drawbacks, the field of nano-research continues to grow. These studies have also produced nano-antioxidants, which are currently being investigated for use in the treatment of various malignancies. The advantageous EPR effect, which makes it simpler to introduce these nanoparticles within the cells, is the source of this rise in the study. However, there is still a long way to go before these NPs can be effectively used as therapeutic agents because numerous clinical trials must be conducted to rule out any chances of immunotoxicity, adverse reactions, and side effects.

REFERENCES

Ahmadian, E., Dizaj, S.M., Rahimpour, E., et al., 2018. "Effect of silver nanoparticles in the induction of apoptosis on human hepatocellular carcinoma (HepG2) cell line." Material Science and Engineering: C 93, 465–471.

Amin, A., 2008. "Ketoconazole-induced testicular damage in rats reduced by gentiana extract." Experimental and Toxicologic Pathology 59, 377–384.

Anirudhan, T.S., 2016. "Dextran based nanosized carrier for the controlled and targeted delivery of curcumin to liver cancer cells." International Journal of Biological Macromolecules 1, 222–235.

Ao, M, Xiao, X, Ao, Y., 2018. "Low density lipoprotein modified silica nanoparticles loaded with docetaxel and thalidomide for effective chemotherapy of liver cancer." Brazilian Journal of Medical and Biological Research 51.

Ashley, C.E., Carnes, E.C., Phillips, G.K. et al., 2011. "The targeted delivery of multicomponent cargos to cancer cells by nanoporous particle-supported lipid bilayers." Nature Material 10, 389.

Aulenta, F., Hayes, W., Rannard, S., 2003. "Dendrimers: A new class of nanoscopic containers and delivery devices." European Polymer Journal 39, 1741–1771.

Azangou-Khyavy, M., Ghasemi, M., Khanali, J., Boroomand-Saboor, M., Jamalkhah, M., Soleimani, M., et al, 2020. "CRISPR/Cas: From tumor gene editing to T cell-based immunotherapy of cancer." Frontiers in Immunology 11, 2062.

Azzariti, A, Iacobazzi, R.M., Fanizza, E, et al., 2017. "Sorafenib delivery nanoplatform based on superparamagnetic iron oxide nanoparticles magnetically targets hepatocellular carcinoma." Nano Research 10 (7), 2431–2448.

Bahadar, H., Maqbool, F., Niaz, K., Abdollahi, M., 2016. "Toxicity of nanoparticles and an overview of current experimental models." Iranian Biomedical Journal 20, 1.

Barbe, C., Bartlett, J., Kong, L., Finnie, K., et al., 2004. "Silica particles: A novel drug-delivery system". Advanced Materials 16, 1959–1966.

Behera, A., Mittu, B., Padhi, S., Patra, N., Singh, J., 2020. "Bimetallic nanoparticles: Green synthesis, applications, and future perspectives." In Multifunctional Hybrid Nanomaterials for Sustainable Agri-Food and Ecosystems, pp. 639–682. Elsevier.

Behera, A., Padhi, S., 2020. "Passive and active targeting strategies for the delivery of the camptothecin anticancer drug: A review." Environmental Chemistry Letters 18 (5), 1557–1567.

Blanco, E., Shen, H., Ferrari, M., 2015. "Principles of nanoparticle design for overcoming biological barriers to drug delivery." Nature Biotechnology 33, 941–951.

Cao, Y., Wang, H.J., Cao, C., et al., 2011. "Inhibition effects of protein-conjugated amorphous zinc sulfide nanoparticles on tumor cells growth." Journal of Nanoparticle Research 13, 2759–2767.

Chang, C.H., Lin, J.W., Wu, L.C., Lai, M.S., Chuang, L.M., Chan, K.A., 2012. "Association of thiazolidinediones with liver cancer and colorectal cancer in type 2 diabetes mellitus." Hepatology 55, 1462–1472.

Chen, Y., Liu, Y.C., Sung, Y.C., Ramjiawan, R.R., et al., 2017. "Overcoming sorafenib evasion in hepatocellular carcinoma using CXCR4-targeted nanoparticles to co-deliver MEK-inhibitors." Scientific Reports 9, 44123.

Cherukuri, P., Glazer, E.S., Curley, S.A., 2010. "Targeted hyperthermia using metal nanoparticles." Advances in Drug Delivery Reviews 62, 339–345.

Christian-Miller, N., Frenette, C., 2018. "Hepatocellular cancer pain: Impact and management challenges." Journal of Hepatocellular Carcinoma 75.

Comfort, K.K., Maurer, E.I., Braydich-Stolle, L.K., Hussain, S.M., 2011. "Interference of silver, gold, and iron oxide nanoparticles on epidermal growth factor signal transduction in epithelial cells." American Chemical Society Nano 5, 10000–10008.

Crielaard, B.J., Lammers, T., Rivella, S., 2017. "Targeting iron metabolism in drug discovery and delivery." Nature Reviews Drug Discovery 16, 400.

D'souza, A.A., Devarajan, P.V., 2015. "Asialoglycoprotein receptor mediated hepatocyte targeting: Strategies and applications." Journal of Controlled Release 203, 126–139.

De Jong, W.H., Borm, P.J., 2008. "Drug delivery and nanoparticles: Applications and hazards." International Journal of Nanomedicine and Nanosurgery 3, 133–149.

Dhanasekaran, R., Limaye, A., Cabrera, R., 2012. "Hepatocellular carcinoma: Current trends in worldwide epidemiology, risk factors, diagnosis, and therapeutics". Hepatic Medicine: Evidence and Research 4, 19.

Dimitroulis, D., Damaskos, C., Valsami, S., Davakis, S., et al., 2017. "From diagnosis to treatment of hepatocellular carcinoma: An epidemic problem for both developed and developing world." World Journal of Gastroenterology, 5282.

Doi, A., Kawabata, S., Iida, K. Yokoyama, K., et al., 2008. "Tumor-specific targeting of sodium borocaptate (BSH) to malignant glioma by transferrin-PEG liposomes: A modality for boron neutron captures therapy." Journal of Neurooncology 87, 287–294.

Eftekhari, A., Dizaj, S.M., Chodari, L., Sunar, S., et al., 2018. "The promising future of nanoantioxidant therapy against environmental pollutants induced-toxicities." Biomedicine & Pharmacotherapy 31, 1018–1027.

El-Kharrag, R., Amin, A., Soleiman, H., et al., 2017. "Development of a therapeutic model of precancerous liver using crocin-coated magnetite nanoparticles." International Journal of Oncology 50, 212–222.

Farinati, F., Sergio, A., Baldan, A., et al., 2009. "Early and very early hepatocellular carcinoma: When and how much do staging and choice of treatment really matter? A multi-center study." BMC Cancer, 33.

Filmus, J., Capurro, M., 2013. "Glypican-3: A marker and a therapeutic target in hepatocellular carcinoma." FEBS Journal 280, 2471–2476.

Gannon, C.J., Patra, C.R., Bhattacharya, R., et al., 2008. "Intracellular gold nanoparticles enhance non-invasive radiofrequency thermal destruction of human gastrointestinal cancer cells." Journal of Nanobiotechnology 6, 2.

Ghosh, D., Choudhury, S.T., Ghosh, S., et al., 2009. "Nanocapsulated curcumin: Oral chemopreventive formulation against diethylnitrosamine induced hepatocellular carcinoma in rat." Chemico-Biological Interactions 195, 206–214.

Gref, R., Minamitake, Y., Peracchia, M.T., et al, 1994. "Biodegradable long-circulating polymeric nanospheres." Science 263, 1600–1603.

Gu, M., Wang, X., Boon, T., Hooi, T.L., Tenen, D.G., Chow, E.K.H., 2018. "Nanodiamond based platform for intracellularspecific delivery of therapeutic peptides against hepatocellular carcinoma." Advances in Therapy 1 (8), 1–10.

Guo, H., Lai, Q., Wang, W., et al., 2013. "Functional alginate nanoparticles for efficient intracellular release of doxorubicin and hepatoma carcinoma cell targeting therapy." International Journal of Pharmaceutics 451, 1–11.

Gurunathan, S., Park, J.H., Han, J.W., Kim, J.H., 2015. "Comparative assessment of the apoptotic potential of silver nanoparticles synthesized by *Bacillus tequilensis* and *Calocybe indica* in MDA-MB-231 human breast cancer cells: Targeting p53 for anticancer therapy." International Journal of Nanomedicine 10, 4203.

Hanahan, D., Weinberg, R.A., 2011. "Hallmarks of cancer: The next generation." Cell 144, 646–674.

Hann, I.M., Prentice, H.G., 2001. "Lipid-based amphotericin B: A review of the last 10 years of use." International Journal of Antimicrobial Agents 17, 161–169.

He, Q., Liu, J., Sun, X., Zhang, Z.R., 2004. "Preparation and characteristics of DNA-nanoparticles targeting to hepatocarcinoma cells." World Journal of Gastroenterology 10, 660.

Henrich, S.E., Thaxton, C.S., 2019. "An update on synthetic high-density lipoprotein-like nanoparticles for cancer therapy." Expert Review on Anticancer Therapy 19, 515–528.

Herbst, R.S., Soria, J.C., Kowanetz, M., Fine, G.D., Hamid, O., Gordon, M.S., et al, 2014. "Predictive correlates of response to the anti-PD-L1 antibody MPDL3280A in cancer patients." Nature 515, 563–567.

Huang, X., El-Sayed, M.A., 2010. "Gold nanoparticles: Optical properties and implementations in cancer diagnosis and photothermal therapy." Journal of Advanced Research 1, 13–28.

Husseini, G.A., Pitt, W.G., 2008. "Micelles and nanoparticles for ultrasonic drug and gene delivery." Advanced Drug Delivery Reviews 60, 1137–1152.

Ise, H., Nikaido, T., Negishi, N., Sugihara, N., Suzuki, F., Akaike, T., et al., 2004. "Effective hepatocyte transplantation using rat hepatocytes with low asialoglycoprotein receptor expression." American Journal of Pathology 165, 501–510.

Ishibashi, S., Hammer, R.E., Herz, J., 1994. "Asialoglycoprotein receptor deficiency in mice lacking the minor receptor subunit." The Journal of Biological Chemistry 269, 27803–27806.

Ishiguro, T., Sugimoto, M., Kinoshita, Y., Miyazaki, Y., Nakano, K., Tsunoda, H., et al., 2008. "Anti-glypican 3 antibody as a potential antitumor agent for human liver cancer." Cancer Research 68, 9832–9838.

Jain, N.K., Jain, S.K., 2010. "Development and in vitro characterization of galactosylated low molecular weight chitosan nanoparticles bearing doxorubicin." American Association of Pharmaceutical Scientists PharmSciTech 1, 686–697.

Jeon, M.J., Gordon, A.C., Larson, A.C., Chung, J.W., Il, K.Y., Kim, D.H., 2016. "Transcatheter intra-arterial infusion of doxorubicin loaded porous magnetic nano-clusters with iodinated oil for the treatment of liver cancer." Biomaterials 88, 25–33.

Jiang, J., Yang, S.J., Wang, J.C., Yang, L.J., et al., 2010. "Sequential treatment of drug resistant tumors with RGD-modified liposomes containing siRNA or doxorubicin." European Journal of Pharmaceutics and Biopharmaceutics 76, 170–18.

Jiang, W., Kim, B.Y., Rutka, J.T., Chan, W.C., 2007. "Advances and challenges of nanotechnology-based drug delivery systems." Expert Opinion in Drug Delivery 4, 621–633.

Jiang, Y.Q., Wang, H.R., Li, H.P., Hao, H.J., Zheng, Y.L., Gu, J., et al., 2006. "Targeting of hepatoma cell and suppression of tumor growth by a novel 12mer peptide fused to superantigen TSST-1." Molecular Medicine 12, 81–87.

Julyan, P.J., Seymour, L.W., Ferry, D.R., Daryani, S., Boivin, C.M., Doran, J., et al., 1999. "Preliminary clinical study of the distribution of HPMA copolymers bearing doxorubicin and galactosamine." Journal of Controlled Release 57, 281–290.

Kakushadze, Z., Raghubanshi, R., Yu, W., 2017. "Estimating cost savings from early cancer diagnosis." Data 2, 30.

Kawaguchi, T., Taniguchi, E., Morita, Y., Shirachi, M., Tateishi, I., Nagata, E., et al, 2010. "Association of exogenous insulin or sulphonylurea treatment with an increased incidence of hepatoma in patients with hepatitis C virus infection." Liver International 30, 479–486.

Kayser, O., Lemke, A., Hernandez-Trejo, N., 2005. "The impact of nanobiotechnology on the development of new drug delivery systems." Current Pharmaceutical Biotechnology 6, 3–5.

Kelemen, L.E. 2006. "The role of folate receptor in cancer development, progression and treatment: Cause, consequence or innocent bystander?" International Journal Cancer 119, 243–250.

Kew, M.C., 2013. "Aflatoxins as a cause of hepatocellular carcinoma." Journal of Gastrointestinal and Liver Diseases 22, 305–310.

Khalid, M., El-Sawy, H.S., 2017. "Polymeric nanoparticles: Promising platform for drug delivery." International Journal of Pharmaceutics 528, 675–691.

Kolhatkar, R., Lote, A., Khambhati, H., 2011. "Active tumor targeting of nanomaterials using folic acid, transferrin and integrin receptors." Current Drug Discovery Technologies 8, 197–206.

Kratz, F., 2008. "Albumin as a drug carrier: Design of prodrugs, drug conjugates and nanoparticles." Journal of Controlled Release 132, 171–183.

Kumari, A., Yadav, S.K., Yadav, S.C., 2010. "Biodegradable polymeric nanoparticles-based drug delivery systems." Colloids and Surfaces B: Biointerfaces 75, 1–8.

Lacko, A.G., Nair, M., Paranjape, S., Johnson, S., McConathy, W.J., 2002. "High density lipoprotein complexes as delivery vehicles for anticancer drugs." Anticancer Research 22, 2045–2050.

Lee, D.W., Lim, H., Chong, H.N., Shim, W.S., 2009. "Advances in chitosan material and its hybrid derivatives: A review." Open Biomaterial Journal 1.

Lee, M., Ryoo, B.Y., Hsu, C.H., Numata, K., Stein, S., Verret, W., et al, 2019. "Randomised efficacy and safety results for atezolizumab (Atezo)+ bevacizumab (Bev) in patients (pts) with previously untreated, unresectable hepatocellular carcinoma (HCC)." The Annals of Oncology 30, 875.

Li, Y.J., Dong, M., Kong, F.M., Zhou, J.P., 2015. "Folate-decorated anticancer drug and magnetic nanoparticles encapsulated polymeric carrier for liver cancer therapeutics." International Journal of Pharmaceutics 489, 83–90.

Liang, G.F., Zhu, Y.L., Sun, B., Hu, F.H., Tian, T., Li, S.C., et al., 2011. "PLGA-based gene delivering nanoparticle enhance suppression effect of miRNA in HePG2 cells." Nanoscale Research Letters 6, 447.

Liou, G.Y., Storz, P., 2010. "Reactive oxygen species in cancer." Free Radical Research 44, 479–496.

Liu, M.C., Liu, L., Wang, X.R., Shuai, W.P., Hu, Y., Han, M., et al., 2016. "Folate receptor-targeted liposomes loaded with a diacid metabolite of norcantharidin enhance antitumor potency for H22 hepatocellular carcinoma both in vitro and in vivo." International Journal of Nanomedicine 11, 1395.

Lo, A., Lin, C.T., Wu, H.C., 2008. "Hepatocellular carcinoma cell-specific peptide ligand for targeted drug delivery." Molecular Cancer Therapy 7, 579–589.

Luqmani, Y.A., 2005. "Mechanisms of drug resistance in cancer chemotherapy." Medical Principles and Practice 14, 35–48.

Manatunga, D.C., de Silva, R.M., de Silva, K.M.N., et al., 2018. "Effective delivery of hydrophobic drugs to breast and liver cancer cells using a hybrid inorganic nanocarrier: A detailed investigation using cytotoxicity assays, fluorescence imaging and flow cytometry." European Journal of Pharmaceutics and Biopharmaceutics 128, 18–26.

Marisi, G., Cucchetti, A., Ulivi, P., Canale, M., Cabibbo, G., Solaini, L., et al., 2018. "Ten years of sorafenib in hepatocellular carcinoma: Are there any predictive and/or prognostic markers?" World Journal of Gastroenterology 24, 4152.

Maruyama-Tabata, H., Harada, Y., et al., 2000. "Effective suicide gene therapy in vivo by EBV-based plasmid vector coupled with polyamidoamine dendrimers." Gene Therapy 7, 53.

Masood, F., 2016. "Polymeric nanoparticles for targeted drug delivery system for cancer therapy." Materials Science and Engineering C 1, 569–578.

Mehmood, A., Ghafar, H., Yaqoob, S., Gohar, U.F., Ahmad, B., 2017. "Mesoporous silica nanoparticles: A review." Journal of Developing Drugs 6, 1000174.

Meijer, E., Kromhout, H., Heederik, D., 2011. "Respiratory effects of exposure to low levels of concrete dust containing crystalline silica." American Journal of Industrial Medicine 40, 133–140.

Min, Y., Roche, K.C., Tian, S., Eblan, M.J., McKinnon, K.P., Caster, J.M., et al., 2017. "Antigencapturing nanoparticles improve the abscopal effect and cancer immunotherapy." Nature Nanotechnology 12, 877.

Moghimi, S.M., Hunter, A.C., Murray, J.C., 2001. "Long-circulating and target-specific nanoparticles: Theory to practice." Pharmacology Reviews 53, 283–318.

Mohanraj, V.J., Chen, Y., 2006. "Nanoparticles-a review." Tropical Journal of Pharmaceutical Research 5 (1), 561–573.

Montaño-Samaniego, M., Bravo-Estupiñan, D.M., Méndez-Guerrero, O., Alarcon-Hernández, E., Ibanez-Hernandez, M., 2020. "Strategies for targeting gene therapy in cancer cells with tumor-specific promoters." Frontier in Oncology 10, 2671.

Muzzarelli, C., Muzzarelli, R.A., 2002. "Natural and artificial chitosan: Inorganic composites." Journal of Inorganic Biochemistry 92, 89–94.

Naahidi, S., Jafari, M., Edalat, F., et al., 2013. "Biocompatibility of engineered nanoparticles for drug delivery." Journal of Controlled Release 166, 182–194.

Nault, J.C., Datta, S., Imbeaud, S., Franconi, A., Mallet, M., Couchy, G., et al., 2015. "Recurrent AAV2-related insertional mutagenesis in human hepatocellular carcinomas." Nature Genetics 47, 1187–1193.

Nazemi, A., Gillies, E.R., 2013. "Dendritic surface functionalization of nanomaterials: Controlling properties and functions for biomedical applications." Brazilian Journal of Pharmaceutical Sciences 49, 15–32.

Nobuoka, D., Yoshikawa, T., Sawada, Y., Fujiwara, T., Nakatsura, T., 2013. "Peptide vaccines for hepatocellular carcinoma." Human Vaccines Immunotherapy 9, 210–212.

Padhi, S., Behera, A., 2020. "Nanotechnology based targeting strategies for the delivery of Camptothecin." Sustainable Agriculture Reviews 44: Pharmaceutical Technology for Natural Products Delivery Vol. 2 Impact of Nanotechnology, pp. 243–272. Springer.

Padhi, S., Behera, A., 2021. "Silver-based nanostructures as antifungal agents: Mechanisms and applications." In Silver Nanomaterials for Agri-Food Applications, pp. 17–38. Elsevier.

Padhi, S., Behera, A., 2022. "Cellular internalization and toxicity of polymeric nanoparticles." In Polymeric Nanoparticles for the Treatment of Solid Tumors, pp. 473–488. Springer International Publishing. doi: 10.1007/978-3-031-14848-4_17

Padhi, S., Behera, A., Hasnain, M.S., Nayak, A.K., 2022. "Chitosan-based drug delivery systems in cancer therapeutics." In Chitosan in Drug Delivery, pp. 159–193. Academic Press.

Pandey, P., Rahman, M., Bhatt, P.C., Beg, S., et al., 2018. "Implication of nano-antioxidant therapy for treatment of hepatocellular carcinoma using PLGA nanoparticles of rutin." Nanomedicine 13, 849–870.

Parat, A., Bordeianu, C., Dib, H., et al., 2015. "Dendrimer: Nanoparticle conjugates in nanomedicine." Nanomedicine 10, 977–992.

Peluso, I., Yarla, N.S., Ambra, R., Pastore, G., Perry, G., 2019. "MAPK signalling pathway in cancers: Olive products as cancer preventive and therapeutic agents." Seminars in Cancer Biology 56, 185–195.

Poon, R., Borys, N., 2009. "Lyso-thermosensitive liposomal doxorubicin: A novel approach to enhance efficacy of thermal ablation of liver cancer." Expert Opinion on Pharmacology and Therapeutics 10 (2), 333–343.

Pradhan, S.P., Sahoo, S., Behera, A., Sahoo, R., Sahu. P.K., 2022a. "Memory amelioration by hesperidin conjugated gold nanoparticles in diabetes induced cognitive impaired rats." Journal of Drug Delivery Science and Technology 69, 103145.

Pradhan, S.P., Swain, S., Sa, N., Pilla, S.N., Behera, A., Sahu. P.K., Si, S.C., 2022b. "Photocatalysis of environmental organic pollutants and antioxidant activity of flavonoid conjugated gold nanoparticles." Spectrochimica Acta Part A: Molecular and Biomolecular Spectroscopy 282, 121699.

Pradhan, S.P., Tejaswani, P., Sa, N., Behera, A., Sahoo, R., Sahu. P.K. 2023. "Mechanistic study of gold nanoparticles of vildagliptin and vitamin E in diabetic cognitive impairment." Journal of Drug Delivery Science and Technology, 104508.

Pranatharthiharan, S., Patel, M.D., Malshe, V.C., Pujari, V., Gorakshakar, A., Madkaikar, M., et al., 2017. "Asialoglycoprotein receptor targeted delivery of doxorubicin nanoparticles for hepatocellular carcinoma." Drug Delivery, 24, 20–29.

Qi, L., Xu, Z., Chen, M., 2007. "In vitro and in vivo suppression of hepatocellular carcinoma growth by chitosan nanoparticles." European Journal of Cancer 43, 184–193.

Qi, X., Rui, Y., Fan, Y., Chen, H., Ma, N., Wu, Z., 2015. "Galactosylated chitosan-grafted multiwall carbon nanotubes for pH-dependent sustained release and hepatic tumor-targeted delivery of doxorubicin in vivo." Colloids Surface B 133, 314–322.

Rahman, A. Kessler, A., More, N., et al., 1980. "Liposomal protection of Adriamycin-induced cardiotoxicity in mice." Cancer Research. 40, 1532–1537.

Raoul, J.L., Frenel, J.S., Raimbourg, J., Gilabert, M. 2019. "Current options and future possibilities for the systemic treatment of hepatocellular carcinoma." Hepatic Oncology 6, 11.

Revia, R.A., Zhang, M., 2016. "Magnetite nanoparticles for cancer diagnosis, treatment, and treatment monitoring: Recent advances." Material Today 19, 157–168.

Rozema, D.B., Lewis, D.L., Wakefield, D.H., Wong, S.C., Klein, J.J., Roesch, P.L., et al., 2007. "Dynamic PolyConjugates for targeted in vivo delivery of siRNA to hepatocytes." Proceedings of the National Academy of Sciences USA 104, 12982–12987.

Sa, N., Tejaswani, P., Pradhan, S.P., Alkhayer, K.A., Behera, A., Sahu. P.K., 2023. "Antidiabetic and antioxidant effect of magnetic and noble metal nanoparticles of clitoria ternatea." Journal of Drug Delivery Science and Technology, 104521.

Sang, W., Zhang, Z., Dai, Y., Chen, X., 2019. "Recent advances in nano materials-based synergistic combination cancer immunotherapy." Chemical Society Reviews 48 (14), 3771–3810.

Sapra, P., Allen, T.M., 2004. "Improved outcome when B-cell lymphoma is treated with combinations of immunoliposomal anticancer drugs targeted to both the CD19 and CD20 epitopes." Clinical Cancer Research 10, 2530–2537.

Sarika, P.R., James, N.R., Kumar, P.A., Raj, D.K., Kumary, T.V., 2015. "Gum arabic-curcumin conjugate micelles with enhanced loading for curcumin delivery to hepatocarcinoma cells." Carbohydrate Polymers 10, 167–174.

Savier, E., Simon-Gracia, L., Charlotte, F., Tuffery, P., Teesalu, T., Scatton, O., et al., 2021. "Bi-functional peptides as a new therapeutic tool for hepatocellular carcinoma." Pharmaceutics 13, 1631.

Shen, W.J., Azhar, S., Kraemer, F.B., 2018. "SR-B1: A unique multifunctional receptor for cholesterol influx and efflux." Annual Review of Physiology 80, 95–116.

Shi, B., Abrams, M., Sepp-Lorenzino, L., 2013. "Expression of asialoglycoprotein receptor 1 in human hepatocellular carcinoma." Journal of Histochemistry and Cytochemistry 61, 901–909.

Shi, J., Kantoff, P.W., Wooster, R., Farokhzad, O.C., 2017. "Cancer nanomedicine: Progress, challenges and opportunities." Nature Reviews Cancer 17, 20.

Shubhika, K., 2012. "Nanotechnology and medicine-the upside and the downside." International Journal of Drug Development and Research 5, 1.

Singh, S., Singh, P.P., Singh, A.G., Murad, M.H., Sanchez, W., 2013. "Anti-diabetic medications and the risk of hepatocellular cancer: A systematic review and meta-analysis." Official Journal of the American College of Gastroenterology 108, 881–891.

Soenen, S.J., Brisson, A.R., et al., 2011. "The labeling of cationic iron oxide nanoparticle resistant hepatocellular carcinoma cells using targeted magnetoliposomes." Biomaterials 32, 1748–1758.

Souto, E., 2011. "Lipid Nanocarriers in Cancer Diagnosis and Therapy." Smithers Rapra, 11–28.

Su, Y., Wang, K., Li, Y., et al., 2018. "Sorafenib-loaded polymeric micelles as passive targeting therapeutic agents for hepatocellular carcinoma therapy." Nanomedicine 13, 1009–1023.

Tang, J., Wang, Q., Yu, Q., Qiu, Y., Mei, L., Wan, D., et al., 2019. "A stabilized retro-inverso peptide ligand of transferrin receptor for enhanced liposome-based hepatocellular carcinoma-targeted drug delivery." Acta Biomaterial 83, 379–389.

Taniguchi, N. 1974. "On the basic concept of nanotechnology." In Proceedings of the International Conference on Production Engineering, Tokyo, Japan, August.

Tomuleasa, C., Soritau, O., Orza, A., et al., 2012. "Gold nanoparticles conjugated with cisplatin/doxorubicin/capecitabine lower the chemoresistance of hepatocellular carcinoma-derived cancer cells." Journal of Gastrointestinal and Liver Disease 21.

Tsuchiya, N., Sawada, Y., Endo, I., Saito, K., Uemura, Y., Nakatsura, T., 2015. "Biomarkers for the early diagnosis of hepatocellular carcinoma." World Journal of Gastroenterology, 10573.

Usmani, A., Mishra, A., Ahmad, M., 2018. "Nanomedicines: A theranostic approach for hepatocellular carcinoma." Artificial Cells, Nanomedicine, and Biotechnology 46, 680–690.

Wall, D.A., Hubbard, A.L., 1981. "Galactose-specific recognition system of mammalian liver: Receptor distribution on the hepatocyte cell surface." Journal of Cell Biology 90, 687–696.

Wall, W.J., Marotta, P.J., 2000. "Surgery and transplantation for hepatocellular cancer". Liver Transplant 6, 16–22.

Wang, J., Zhou, Y., Guo, S., et al., 2017. "Cetuximab conjugated and doxorubicin loaded silica nanoparticles for tumor-targeting and tumor microenvironment responsive binary drug delivery of liver cancer therapy." Material Science and Engineering C 76, 944–950.

Wedmore, I., McManus, J.G., Pusateri, A.E., Holcomb, J.B., 2006. "A special report on the chitosan-based hemostatic dressing: Experience in current combat operations." Journal of Trauma and Acute Care Surgery 60, 655–658.

Weissig, V., Pettinger, T.K., Murdock, N., 2014. "Nanopharmaceuticals (part 1): Products on the market." International Journal of Nanomedicine 9, 4357.

Wong, C.M., Tsang, F.H.C., Ng, I.O.L., 2018. "Non-coding RNAs in hepatocellular carcinoma: Molecular functions and pathological implications." Nature Reviews Gastroenterology & Hepatology 15, 137.

Xiao, Y., Liu, Y., Yang, S., et al., 2016. "Sorafenib and gadolinium co-loaded liposomes for drug delivery and MRI-guided HCC treatment." Colloids Surface B 141, 83–92.

Xue, W.J., Feng, Y., Wang, F., et al., 2016. "Asialoglycoprotein receptor-magnetic dual targeting nanoparticles for delivery of RASSF1A to hepatocellular carcinoma." Scientific Reports 26, 22149.

Yang, M., Jin, H., Chen, J., Ding, L., Ng, K.K., Lin, Q., et al., 2011. "Efficient cytosolic delivery of siRNA using HDL-mimicking nanoparticles." Small 7, 568–573.

Yao, Q., Hou, S.X., Zhang, X., et al., 2007. "Preparation and characterization of biotinylated chitosan nanoparticles." Yao Xue Xue Bao 42 (5), 557–556.

Yaswen, P., MacKenzie, K.L., Keith, W.N., et al., 2015. "Therapeutic targeting of replicative immortality." Seminars in Cancer Biology 35, S104-S128.

Zhang, C.G., Zhu, W.J., Liu, Y., et al., 2016a. "Novel polymer micelle mediated co-delivery of doxorubicin and P-glycoprotein siRNA for reversal of multidrug resistance and synergistic tumor therapy." Scientific Reports 31, 23859.

Zhang, J., Hu, J., Chan, H.F., Skibba, M., Liang, G., Chen, M. 2016b. "IRGD decorated lipid-polymer hybrid nanoparticles for targeted co-delivery of doxorubicin and sorafenib to enhance anti-hepatocellular carcinoma efficacy." Nanomedicine 12, 1303–1311.

Zhao, X., Chen, Q., Li, Y., Tang, H., Liu, W., Yang, X., 2015. "Doxorubicin and curcumin codelivery by lipid nanoparticles for enhanced treatment of diethylnitrosamine-induced hepatocellular carcinoma in mice." European Journal of Pharmaceutics and Biopharmaceutics 1, 27–36.

Zhou, Q., Ching, A.K., Leung, W.K., et al., 2011. "Novel therapeutic potential in targeting microtubules by nanoparticle albumin-bound paclitaxel in hepatocellular carcinoma." International Journal of Oncology 38, 721–731.

Zhou, Y.Y., Zhu, G.Q., Liu, T., Zheng, J.N., Cheng, Z., Zou, T.T., et al, 2016 "Systematic review with network meta-analysis: Antidiabetic medication and risk of hepatocellular carcinoma." Scientific Reports 6, 33743.

Zu, Y., Wen, Z., Xu, Z., 2010. "Chitosan nanoparticles inhibit the growth of human hepatocellular carcinoma xenografts through an antiangiogenic mechanism." Anticancer Research 30, 5103–5110.

Zwicke, G.L., Ali Mansoori, G., Jeffery, C.J., 2012. "Utilizing the folate receptor for active targeting of cancer nanotherapeutics." Nano Reviews 3, 18496.

Chapter 16

Nanotherapeutics for reversal of multidrug resistance in chemotherapy with metal nanocomposites

Devesh U. Kapoor, Dimpy Rani, Madan Mohan Gupta, and Deepak Sharma

LIST OF ABBREVIATIONS

2D	Two-dimensional
Ag NCs	Silver nanocomposites
ATP	Adenosine triphosphate
Au NCs	Gold nanocomposites
BSA	Bovine serum albumin
CDT	Chemodynamic therapy
Copper Nanoparticles	Cu nanoparticles or Cu NPs
CuO NPs	Copper oxide-based nanocomposites
DLS	Dynamic light scattering
DNA	Deoxyribonucleic acid
DOX	Doxorubicin
GO	Hydrophilic graphene oxide
GOx	Glucose oxidase
H_2O_2	Hydrogen peroxide
HA	Hyaluronic acid
HR-TEM	High-resolution transmission electron microscopy
HUVEC	Human umbilical vein endothelial cells
IONCs	Iron oxide nanocomposites
IONPs	Iron oxide nanoparticles
KP1019	Indazole trans-[tetrachlorobis(1H-indazole)ruthenate(III)]
MDR	Multidrug resistance
MIC	Minimum inhibitory concentration
MRI	Magnetic resonance imaging
NAMI-A	New anti-tumor metastasis inhibitor
NC	Nanocomposite
NPs	Nanoparticles
PEG	Polyethylene glycol
PTT	Photothermal therapy
PVP	Polyvinylpyrrolidone
RuNPs	Ruthenium nanoparticles
SDEDTC	Sodium diethyl dithiocarbamate
SMILE	Small incision lenticule extraction
TDNs	Titanium dioxide nanoparticles

DOI: 10.1201/9781032621135-16

TEM	Transmission electron microscopy
TF/TFR	Transferrin/transferrin receptor
TiO$_2$ NCs	Titanium dioxide nanocomposites
TLD1433	Rutherrin component
USEPA	United States Environmental Protection Agency
UV	Ultraviolet
XRD	X-ray diffraction
Zn–CuO NPs	Zinc–copper oxide-based nanocomposites

16.1 INTRODUCTION

The inclination to make the universal underlying materials, both nuclear and sub-atomic attributes, is required by the improved technologies of the 20th century. Yet as technology has advanced, there is a far greater need for new useful materials. The need for materials with specific combinations of innovative and distinctive features, such as catalytic-magnetic, conductive-transparent, magnetic-transparent, and so forth, is still high. In this manner, even while conventional composite materials and filled frameworks truly do benefit from the mix of ideal lattice attributes and favorable filler characteristics, they experience the ill effects of a split difference between execution and processability since a lot of filler is used. For a material to develop technology and civilization, it often has to have four fundamental qualities (Alfarouk, Stock et al. 2015; Bukowski, Kciuk et al. 2020; Emran, Shahriar et al. 2022).

Glass is capable of meeting this need, but its brittleness creates obstacles. In line with the foregoing, a material must be both strong and translucent, which makes using metals alone impracticable. Softness and resilience make up the third essential quality while the capacity to be dragged into fiber and filament makes up the fourth. Metallic materials can create fibers, but they aren't soft or resilient. On the other hand, polymeric materials lack hardness while being translucent, pliable, and capable of being pulled into fibers. In this sense, no conventional material can satisfy the aforementioned four fundamental properties on its own. Although all other conventional materials fell short of the requirements set forth by material scientists, the combination of both can produce the desired material with nano-dimension (Alfarouk, Stock et al. 2015) (Xue, Bendayan et al. 2020). Because of their better characteristics than those of single metal nanoparticles, nanocomposite (NC) materials certainly stand out and draw the attention of researchers in current years (Izdebska, Zielińska et al. 2019; Halder, Pradhan et al. 2022).

16.1.1 Multidrug resistance (MDR) against chemotherapy

One of the main sources of mortality around the globe is malignant growth. Cancer therapy remains the essential methodology for treating malignant growth regardless of the significant progressions in disease treatment throughout recent years. Many gatherings of routinely utilized chemotherapeutic medications can be isolated because of how they work (antimetabolites, alkylating specialists, mitotic spindle inhibitors, topoisomerase inhibitors, and others). Around 90% of disease patients who get traditional chemotherapeutics or state-of-the-art designated medications lose their lives due to multidrug resistance (MDR) (Yadav, Ambudkar et al. 2022). MDR is something similar to antibiotic resistance, demonstrated as the ability of cancerous cells to overcome the chemotherapy leading to several deaths every year, worldwide. In addition to drug release exterior to the cancerous cells leading to reduced absorption of anticancerous drugs, some other factors may be responsible for MDR in cancer therapy, including changes in the microenvironment of the cancerous cells, enhanced metabolism of drugs used for cancer treatment, gene mutation like ATP Binding Cassette-subfamily B and member 1 (the protein found within the cell wall of cancerous cells, majorly regulating the absorption, accumulation, and excretion of drugs and chemical compounds used against cancer treatment), gene demethylation suppressing the deposition of anticancer drugs within the cancerous cells, mutation at target site including epigenetic variations causing depression of DNA repair system, conjugation with glutathione leading to drug inactivation, reallocation of a drug at a subcellular level, and drug–drug/environment interaction (Figure 16.1). (Emran, Shahriar et al. 2022; Bukowski, Kciuk et al. 2020; Wang, Seebacher et al. 2017; Dallavalle, Dobričić et al. 2020; Wu, Yang et al. 2014; Bukowski, Kciuk et al. 2020).

16.1.2 Scenario of MDR in chemotherapy

Increased xenobiotic metabolism, improved drug efflux, growth factors, greater DNA repair ability, and hereditary variables are some of the processes behind MDR (gene mutations, amplifications, and epigenetic alterations). The goal of an ever-growing number of biomedical investigations is to develop chemotherapeutics that can avoid or reverse MDR. The objective of this chapter is to highlight innovative possible antitumor medications that have been developed to circumvent these resistance mechanisms, as well as the most recent

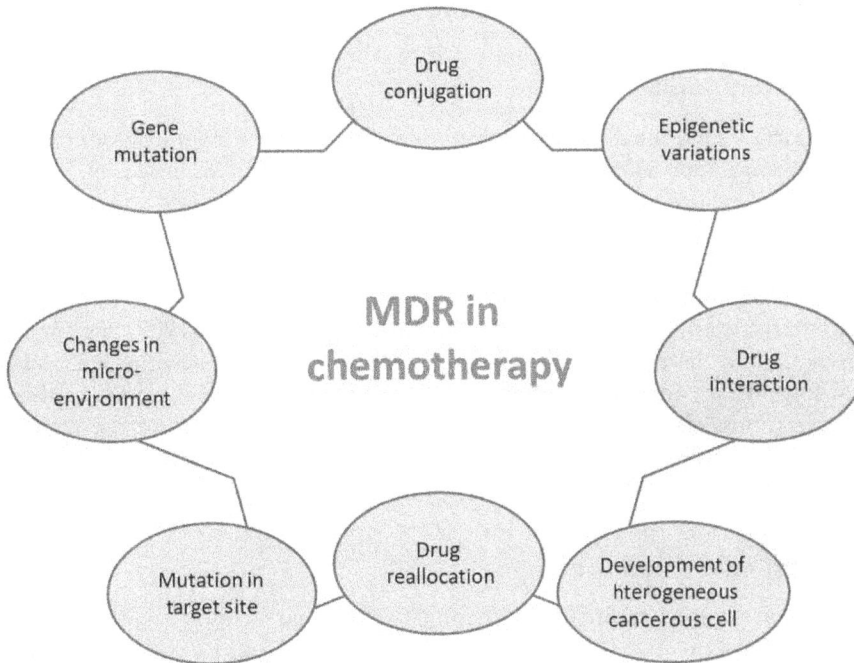

Figure 16.1 Factors leading to MDR in chemotherapy.

statistics on the causes of cellular opposition to anticancer therapies currently utilized in scientific studies. Future research on more efficient cancer treatment methods should be guided by a better knowledge of MDR processes and the targets of innovative chemotherapeutic drugs.

16.2 NANOTHERAPEUTICS

Metal-based or tagged nanocomposites are a family of cutting-edge materials that combine metal nanoparticles and polymer matrices. The spectrum of acceptable features in nanocomposites varies depending on how metal and polymer selection is modulated. Modern materials called metal–polymer nanocomposites combine metal nanoparticles with polymer matrices in useful ways (Bargude, Chopade et al. 2023; Cao, Zhu et al. 2021; Ali, Sharker et al. 2021). Depending on how metal and polymer selection is regulated, there are a variety of desirable properties in nanocomposites. Drugs may be delivered specifically using NPs as delivery mechanisms, and their prolonged, regulated release allows them to reach the therapeutic intracellular level. It is possible to prevent drug degradation before reaching the target and unwanted drug interactions while also optimizing the physicochemical properties of the linked drug (Gao, Guo et al. 2020; Ali, Sharker et al. 2021).

Other benefits of NP delivery methods include improved drug solubility, simultaneous administration of various medicines, and longer systemic circulation, all of which have been supported by numerous studies. This has led to the endorsement of several NP-based therapies, including the use of NPs as vaccine delivery systems, to treat a wide range of illnesses. Several more therapies are now going through the preclinical and clinical testing stages (Wu, Zeng et al. 2020) (Zhang, Li et al. 2019; Curcio, Cirillo et al. 2020).

16.2.1 Metal nanoparticles

Metal nanoparticles are small, submicron-sized objects formed of uncontaminated metals or their composites, such as gold, platinum, silver, titanium, zinc, cerium, iron, and thallium (e.g., oxides, hydroxides, sulfides, phosphates, fluorides, and chlorides) (Xu, Zhang et al. 2022) (Păduraru, Ion et al. 2022). The term metal nanoparticles portray nano-sized metals with aspects (length (l), width (w), or thickness (t)) inside the size range of 1–100 nm. The presence of metallic nanoparticles in the arrangement was first perceived by Faraday in 1857, and a quantitative clarification of their variety was given by Mie in 1908. High biocompatibility, strength, and the potential for enormous scope combination without the utilization of natural solvents are advantages of honorable metal nanoparticles that are significant for clinical applications. This will affect organic frameworks. Silver is the most widely recognized nanomaterial utilized in items, trailed by carbon-based nanomaterials and metal oxides like TiO_2 (Chronopoulou, Scaramuzzo et al. 2020; Khursheed, Dua et al. 2022).

16.2.2 Metal oxide nanoparticles

Because of their higher steadiness when compared to that of metal nanoparticles, metal oxide nanoparticles are the well-known nanomedicines utilized as remedial and indicative specialists for cardiovascular problems (Subhan 2022; Shabade, Sharma et al. 2022). Specific optical characteristics of metal oxide nanoparticles include UV absorption, color absorption in the visible region dichroism, and photoluminescence. Maghemite (Fe_2O_3) and magnetite (Fe_3O_4) are the two most extensively used crystalline iron oxide nanostructures used in a large number of industrial setups, including data storage, color creation, electrochemistry, and more. Several methods are used to create iron oxide nanoparticles. However, it is quiet challenging to attain a monodispersed population of magnetic nanoparticles with colloidal stability. For the most part, magnetic nanoparticles need an organic polymer covering or an inorganic encapsulation to stay in a stable state, avoid aggregates, and endure pH and electrolyte changes. Functionalization also provides a suitable surface design for subsequent functionalization (Shabade, Sharma et al. 2022; Sharma, Kumar et al. 2016).

16.2.3 Metal sulphide nanoparticles

The exceptional qualities and anticipated approaches in electrical, optical, and optoelectronic instruments have brought much attention to metal sulfide nanostructures. Because of their improved characteristics and cutting-edge applications, well-aligned nanostructure arrays on substrates are quite appealing. The directed development of different metal sulfide nanostructure arrays has been demonstrated for use in energy conversion and storage using the general solution approach and thermal evaporation under regulated circumstances. The chapter gives a summary of recent findings and important discoveries that have been mentioned in the published works. It includes works related to the synthesis of such compounds, the properties, and even the applications, with a focus on energy storage and conversion technologies like lithium-ion batteries, fuel cells, solar cells, and piezoelectric nano-generators (Mohammad, Nazli et al. 2022; Shetty, Lang et al. 2023).

16.3 METAL NANOCOMPOSITES

Nanocomposites are environment-friendly materials with a high probability of unique design combined with some unusual properties to create a high-strength material (Pramanik and Das 2019). Nanocomposites are multiphase materials synthesized by mixing two or more materials in diverse phases, in nano-size form, embedded in a matrix made up of a polymer, carbon, metal, or ceramic, in a way to retain the properties of both constituents. The properties of developed nanocomposite depend upon the properties of the constituents and their phases, the proportion in which they are mixed, the geometry of individual constituents, size and shape, spatial arrangement, and distribution. In addition to a resin and a matrix, stuffing and support are also required for the formulation of a nanocomposite. Nanocomposites bear all the properties of nanoparticles in addition to the properties of their individual constituent materials (Pascariu, Airinei et al. 2015; Nayak, Munawar et al. 2021).

Depending upon the nature of the matrix used nanocomposites can be classified into three types: ceramic matrix-based nanocomposites, metal nanocomposites, and polymer nanocomposites. In ceramic nanocomposites, ceramic is an important component of the nanocomposites, that too majorly consisting of borides, oxides, silicides, or nitrides. Additionally, they may contain a metal dispersed evenly in the matrix. Metal matrix-based nanocomposites, sometimes known as reinforced metal matrix nanocomposites, contain a metal-based matrix and can be classified as continuous or noncontinuous metal matrix-based nanocomposites. Polymer matrix-based nanocomposites have improved performance due to the presence of evenly distributed nanoscale polymer matrix.

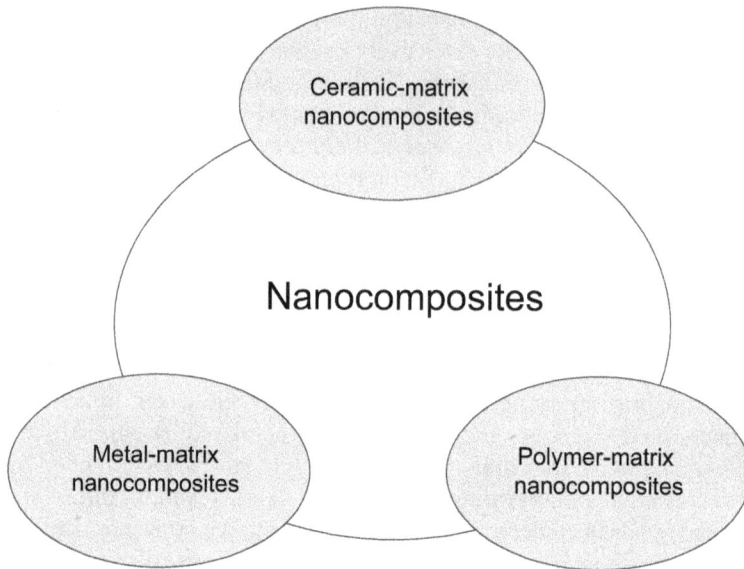

Figure 16.2 Types of nanocomposites based on the nature of the matrix.

Further, nanocomposites are broadly categorized (Mishra, Mishra et al. 2021) into the following.

1. Structural: Can be of two subtypes—laminates and sandwich panels.
2. Fiber-reinforced: Further subcategorized as continuous or discontinuous.
3. Particle-reinforced: Can be classified as large particles or dispersion.
4. Matrix-based: Depending upon the matrix constituent it can be classified as carbon matrix-based, polymer, ceramic, or metal. All these categories can be further subdivided depending upon the nature of the matrix.

16.3.1 Synthesis of metal nanocomposites

Metal nanocomposites can be synthesized using various novel techniques to achieve enhanced properties; some of the methods for synthesis may include (Yadav et al. 2022)

- *Molecular composite formation method* : In this method, two materials are fused to achieve unique properties not present in individual materials.
- *Nanofiller direct dispersion:* Nanofillers are dispersed in a polymer matrix, one of the common methods is the melt blending method.
- *Sol-gel method:* This is a wet chemical method, where metal or metal oxide is liquefied in an aqueous or alcoholic phase and further heated to make a gel.
- *Intercalation:* In this technique, thermoplastic polymer nanocomposites are typically synthesized using a method where the polymer matrix is hardened at high temperatures, then fillers are added and to achieve uniform distribution finally, the composite is kneaded into it.

- *Hydrothermal synthesis:* This is the most usual approach for the synthesis of nanoparticles, based upon solution-reaction technique, providing possibilities for working over large temperature ranges.
- *Ball milling:* This is a grinding methodology, to mix, reshape, grind, and blend and is a successful method for generating a uniform dispersion with size control.
- *Polymerized complex method:* Monomers are polymerized in the presence of the drug molecule or metals ions or metal oxides mechanically to yield metal nanocomposites.
- *Microwave synthesis method:* Usually for lead and sulfur-containing nanocomposite synthesis, with the aid of microwaves.
- *Chemical vapor deposition:* The substrate is exposed to precursors in a gaseous phase, which either reacts with the substrate or decomposes it.
- *Coprecipitation technique:* This is a usual method for metal oxide nanoparticles production, wherein the aqueous solution of slat is coprecipitated with the corresponding hydroxide.
- *Wet chemical synthesis method:* Used for 2D nanoparticles production via hot-injection, solvothermal synthesis, or a similar method.
- *Biological method:* This is the green synthesis method for nanoparticle or nanocomposite synthesis.

Scientists, researchers, and engineers have been interested in these elements, leading to a potential growth in the number of published articles on these materials. Moreover, these NCs are 21st-century materials that provide numerous commercial and technical advances in all spheres of existence. Although these materials have promising features that make them appropriate for several structural and functional applications in various domains, this chapter focuses on metal nanocomposites and metal oxide nanocomposites involved in cancer therapy against MDR (Pascariu, Airinei et al. 2015).

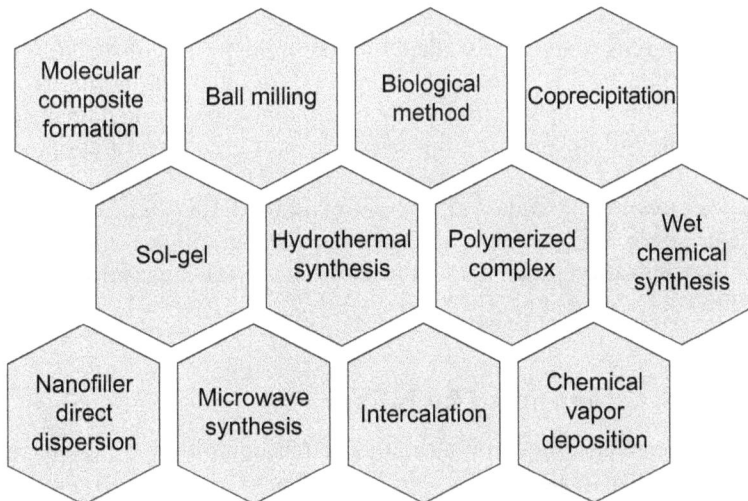

Figure 16.3 Methodologies for the synthesis of nanocomposites.

16.3.2 Silver nanocomposites (Ag NCs)

Among various metallic nanoparticles, developed as nanomedicines, silver nanocomposites or nanoparticles (Ag NCs) are the most vibrant and captivating leading a significant path in cancer imaging and therapy. Acting as an antiviral, antifungal, antibacterial, anti-angiogenic, and anti-inflammatory agent, Ag NCs are used in the treatment of various ailments, in addition to their anticancer activity. Due to the fascinating capabilities of silver, combined with the advantages of nanotechnology, Ag NCs are used in industrial, optical sensors, coating of medical devices, household, and healthcare sectors, including cosmetics, pharmaceuticals, and food industry (Li, Xie et al. 2010; Gurunathan, Park et al. 2015); (Li, Zhang et al. 2014).

MDR malignant growth aggregates decisively lessens the effectiveness of antineoplastic medication frequently prompting the ineffectiveness of chemotherapy. Accordingly, there is a need to design new remedially helpful molecules and propose creative methodologies ready to overcome resistant cancer cells. Further in this direction, Dávid Kovács, Szőke et.al (2016) have suggested the accumulation of drugs by hindering the efflux action of cancerous MDR cells by silver nanoparticles, leading to anti-proliferative consequence and encouraging apoptosis-arbitrated cell destruction, in MDR cases including drug-sensitive cases during cancer treatment (Kovács, Szőke et al. 2016).

Gopisetty, Kovács et.al (2019) developed sliver nanoparticles of 75 nm, capable of obstructing P-glycoprotein activity effectively in cancer cells of the breast, with drug resistance, enhancing the doxorubicin apoptotic capability (Gopisetty, Kovács et al. 2019). Liu, Zhao et.al (2012) synthesized silver nanoparticles of 8 nm average size and studied their anticancer activity on a mice model, with a dose of 1 nmol/kg, related to 4.3 µmol/kg dose of doxorubicin. The silver nanoparticles showed boosted (24 times) anticancer activity for MDR cancerous cells, with suppressed toxicity (Liu, Zhao et al. 2012).

Jianming Liang, Zeng et.al (2015) developed hyaluronic acid (HA)-modified silver nanoparticles, where hyaluronic acid was used as both as a stabilizer and as a reducing agent, which significantly enhanced the CD44-dependent endocytosis, with multiple approaches including autophagy and apoptosis, arrest of cell cycle and process in which mitochondrial membrane potential is reduced (Liang, Zeng et al. 2015).

Jianming Liang and his coworkers studied the coadministration of Ag NCs with the drug albendazole conjugated with bovine serum albumin (BSA). The co-delivery reduced the cellular level of ATP by inhibiting the enzymes involved in glycolysis and damage of mitochondria, inhibiting proliferation, arresting the cell cycle, and induction of apoptosis (Liang, Li et al. 2018).

Ostad with his coworkers, studied tamoxifen-resistant T47D cells, against MDR, using Ag NCs in the presence of silver ions and found the combination of silver nanoparticles or silver ions at a concentration much below the cytotoxic levels, and tamoxifen was found to be cytotoxic (Ostad, Dehnad et al. 2010).

16.3.3 Gold nanocomposites (Au NCs)

Notwithstanding their versatile compound and actual qualities, Au NPs have solid close infrared ingestion, which makes them a brilliant choice for many organic applications. A bio-conjugation surface for subatomic tests, high light dispersing, and the capacity to change their plasmon reverberation range upon collection are only a couple of the vigorous characteristics of colloidal Au NPs. The microbe-prompted NPs collection scattering process is enhanced by empowering the visual recognition of microscopic organisms at a centralization of 100 CFU/mL. Additionally, fluorescent mixtures can be vaguely extinguished by Au NPs. A fluorophore uprooting approach has been utilized for bacterial location because of this peculiarity (Alavi and Rai 2019).

Wu, Liu et al. (2019) developed a drug release system for a gold nanoparticles core with a mesoporous silica shell containing a large dosage of doxorubicin, intelligently controlled via photothermally for the synergistic effect of chemotherapy and photothermal therapy in cases of MDR in breast cancer. This releases doxorubicin with the trigger of IR-radiations, leading to a reduction of MCF-7/ADR cells by about 17%, proving the ability to treat breast cancer in MDR cases.

Yang, Lin et al. (2017) developed and examined the action of Au NPs tagged with doxorubicin-loaded mesoporous silica nanoparticles triggered photothermally for the delivery of the medication at the site of activity. The cellular uptake of the developed nanoparticles was studied with a confocal laser scanning microscope, suggesting their synergistic effects. Yang et al. deliberated a promising and efficient system against the MDR of the HepG2/ADM cells, using photothermally triggered gold nanorods synthesized in situ hollow mesoporous silica, containing doxorubicin, gated with surface-modified-heparin structured-DNA (Yang, Lin et al. 2017).

Jiang and his coworkers developed nanocomposites for the MDR therapy of colorectal cancer; they synthesized mesoporous silica-conjugated gold nano-rods loaded with doxorubicin and coated with poly-histidine to make it pH sensitive; further, to inhibit P-glycoprotein and enhance the drug retention at a cellular level, they coated it with d-α-tocopherol polyethylene glycol 1000 succinate, which resulted in accumulation of the drug with increased anti-tumor activity in MDR SW620/Ad300 cells (Jiang et al., 2020).

Wang, Xiong et al. (2021) developed a co-delivery dosage form containing gold nanorods modified by hyaluronic acid (HA) and layered with mesoporous silica nanoparticles loaded with paclitaxel drug and a micro RNA for the treatment of MDR ovarian cancer, which binds specifically and selectively with CD44 receptor of SKOV3/SKOV3$_{TR}$ cancerous cells, suppressing the cell membrane barrier toward the uptake of the drug and micro RNA, in mice model, reflecting the synergistic effect for the treatment of cancerous cell.

16.3.4 Ruthenium nanoparticles

Ruthenium is a reducible transition metal and its complexes have well-established bioactivity against cancer cells. This is widely used as an alternative therapy in MDR cells, especially resistant to platinum complexes. Its broad-spectrum activities, less toxic nature, and selectivity toward cancer cells make it better for chemotherapy, and participation of ruthenium complexes (such as NAMI-A, KP1019, and TLD1433) in scientific trials signifies its therapeutic potential as a chemotherapeutic agent.

Above that the conversion of ruthenium complexes into nanoparticles has been proven effective in enhancing the chemotherapeutic capabilities, owing to targeted delivery, selectivity, reduced toxicity, and ameliorated lipophilicity. The crystallizing ability of ruthenium, only in a hexagonal structure, further facilitates ruthenium complex assemblies at the nanoscale (Viau, Brayner et al. 2003).

Several researchers in their studies mentioned the synthesis of ruthenium nanoparticles (Ru NPs) by various methods and corroborated their antitumor activities. Srivastava and

Constanti (2012) reported the Ru NPs synthesis from the bacterial extract of *Pseudomonas aeruginosa*, and another group of workers synthesized Ru NPs using *Dictyoma dichotoma* extracts and disclosed cytotoxicity against HeLa, MCF-7, and VERO cell lines. García-Peña, Redon et al. (2015) further reported solventless, green, fast, and reproducible synthesis of Ru NPs. The reports of the transformation of the complexes of Ru into nanoparticles and their anticancer potential are also available in the literature.

Ru complexes can be formulated in nano form by various technologies such as encapsulation in organic or nonorganic nano-carriers and excipient-free nano drugs. Various groups of workers have studied the encapsulation of Ru (II) complexes such as Ru (II)–gold nanomaterials (Zhu, Kuang et al. 2021), Ru(II)–selenium nanoparticles (Huang, Chen et al. 2017), Ru(II)–carbon nanotubes nanomedicine based on Ru(II) polypyridine complexes (Wang et al. 2015), and many more. The other excipient-free Ru NPs offer the main advantage of selectivity and are used in both the diagnosis and treatment of cancer.

In addition, Lu and Zhu et al. (2021) explained dual drug targeting of the Ru complex (ruthenium-complexed carboline acid and four chloride ions, in a nano form, having both antimetastatic and antitumor properties. They demonstrated that it resulted in increased penetrability and retaining (EPR) effect and transferrin/transferrin receptor (TF/TFR) interaction. It also prompted DNA cleavage and encouraged apoptosis. The NPs of the complex were found effective against *in vitro* LLC cell lines and also showed signs of inhibition of lung metastasis (Lu, Zhu et al. 2021).

16.3.5 Copper nanocomposites

Copper is an important element in the physiological processes in humans and plays a profound role in cellular energy processing, hormone maturation, and blood clotting. In excessive amounts, it can be detrimental to cells which can be taken as an advantage against cancer cells, and copper-containing anticancer drugs can be developed (Li, Li et al. 2019; Cobine and Brady 2022; Tsvetkov, Coy et al. 2022).

High concentrations of copper target cancer cells by various mechanisms (Ji, Wang et al. 2023). The first is apoptosis of cancer cells as copper leads to the generation of ROS, which ultimately leads to cytotoxicity in cells (Sies, Belousov et al. 2022; Nakamura and Takada 2021), and this mechanism was supported by the study on copper (II) complex on HeLa cell lines by Guo, Ye et al. (2010). The same kind of results was further confirmed by Liu and coworkers (Liu, Yuan et al. 2022) with nano-copper/catechol-based metal-organic frameworks. They also showed the activity of nano form in preventing the growth of drug-resistant cancer cell lines.

The second mechanism is that it inhibits cell proliferation by stressing the endoplasmic reticulum and causing DNA damage (Oe, Miyagawa et al. 2016). This was investigated by Gul, Khan et al. (2020) with the help of Cu (I) and Cu (II) complexes against A549 cell lines. They also reported the preferential movement of the mixtures on disease cells than ordinary cells. In addition, Shimada and coworkers (Shimada, Reznik et al. 2018) explored that DNA damage owing to copper dysregulation may result in arresting the cell cycle at different phases. The other mechanisms are anti-angiogenesis (Lowndes and Harris 2005) cuproptosis (Wang, Zhang et al. 2022), paraptosis etc.

Cu nanoparticles (Cu NPs), being a targeted agent, constrain cancer growth and its migration to other parts. Xu, Liu et al. (2022) designed and constructed (GOx@[Cu(tz)]) nanomaterial which is a coordination complex of glucose oxidase (GOx) and nonporous copper(I) 1,2,4-triazole ([Cu(tz)]) and demonstrated its significance in cuproptosis. They exploited the hypothesis that for the growth of cancer cells, glucose and intracellular glutathione (GSH) act as copper chelator and promote the growth of cancer cells. So a reduction in the supply of glucose and GSH in cancer cells exerts a synergistic effect in inviting more Cu into the cancer cells and leads to cell death due to cuproptosis (Kennedy, Sandhu et al. 2020). GOx in the NPs helped in catalyzing the oxidation of glucose and diminished the availability of glucose and oxygen in the cancer cells and thus, blocking the energy supply. This reduction enhanced the sensitivity of cancer cells toward GOx@[Cu(tz)]-induced cuproptosis. The positive results of these NPs successfully showed in human bladder cancer cells (Xu, Liu et al. 2022).

In another study by Zhou, Ye et al. (2022) NPs of (Au@MSN-Cu/PEG/DSF) were prepared, and it was found to chelate with Cu^{2+} which in turn increased the concentration of Cu in cancer cells and thus, promoted apoptosis of cancer cells. *In vivo* studies also revealed the photothermal activity of NPs which aided the death of cancer cells.

Chen, Yang et al. (2018) also prepared $Cu(DDC)_2$ NPs using SMILE technology. These Cu-based NPs are stable, having controlled physicochemical properties and high drug-loading capacity. It showed its anticancer potential in MCF-7 cell lines through paraptosis. They explained that Cu (DDC)$_2$ NPs bind to nuclear localization protein 4 (NPL4) and blocked the dilapidation of poly-Ub protein which resulted in its accumulation in the endoplasmic reticulum and caused stress in tumor cells. These events led to the impairment of mitochondria and thus paraptosis in cancer cells.

Apart from these mechanisms, the killing of cancer cells by solely photothermal therapy (PTT) by Cu-containing nanocomposites is also reported in the literature and reflects their ability as a potential PTT candidate (Pan, Svirskis et al. 2021). Cu-containing nanocomposites raise the temperature of cancer cells only instead of laser light treatment which raises the temperature of all cells and causes protein denaturation of normal cells as well.

Li, Wang et al. (2014) reported $Cu_{7.2}S_4$ nanocrystals, coated with amphiphilic polymer. These nanocrystals displayed remarkable photostability and photothermal conversion effectiveness. Owing to these characteristics, $Cu_{7.2}S_4$ nanocrystals were considered good candidates for photothermal therapy (PTT) for cancer cells. Huang, Lai et al. (2015) synthesized copper sulfide nanoparticles (CuS NPs) through a chemical reaction between copper chloride and sodium diethyl dithiocarbamate (SDEDTC) using oleylamine solvent. These NPs were further coated with DSPE-PEG2000 and showed excellent photothermal conversion efficiency when exposed to 808 nm laser light. The authors also acknowledged the preferable killing of cancer cells in comparison to normal healthy cells and no sign of toxicity with good excretion from the body by metabolism.

In addition, chemo-dynamic therapy (CDT) is also used for the treatment of cancer and the competence of Cu-based NPs in this therapy has also been disclosed by many findings. Deng and coworkers (2022) prepared NPs that tend to ambush copper peroxide in the lysosomes and convert it to Cu ions which further catalyze the conversion of H_2O_2 to hydroxyl free radicals (·OH) through the Fenton reaction. These free radicals pass through lysosomes and kill cancer cells.

Liu, Jiang et al. (2022) disclosed copper iron tellurite nanoparticles (A-CFT NPs), having amorphous nature, encapsulated in inositol hexaphosphate (IP6) and bovine serum albumin (BSA). The degradation of these NPs was GSH dependent and their amorphous nature allowed the release of Cu ions in bulk to generate hydroxyl free radicals (·OH) and thus exhibited anticancer activity. In addition to this phenomenon, Cu ions also utilized GSH and prevented hydroxyl free radicals from clearance and thus potentiated chemo-dynamic therapy. On the same principle, Zhou and Li (2022) reported a copper-based nanocomposite (BiOCu$_{2-x}$Te NSS), coated with polyethylene glycol-modified glutathione-degradable Cu$_{2-x}$Te nanosheets. This agent showed its effectiveness as a CDT agent against MCF-7 cancer cells in mice.

16.4 METAL OXIDE NANOCOMPOSITES

16.4.1 Iron oxide nanocomposites (IO NCs)

Iron oxide nanocomposites are widely used in biomedical fields such as photo-responsive therapy, magnetic resonance imaging (MRI), magnetic hyperthermia, drug delivery in cancer, and so on. Their use in the diagnosis and treatment of cancer is the most studied topic of research nowadays. Iron oxide nanoparticles (IONPs) have many advantages like magnetic properties, low toxicity, good imaging, fast clearance by metabolism, and high therapeutic effectiveness. In addition, IONPs have a peroxidase-like activity and also act as a catalyst for generating free hydroxyl radical (•OH) from intracellular hydrogen peroxide (H$_2$O$_2$) (through Fenton reaction) which is responsible for the death of tumor cells (Chandrasekharan, Tay et al. 2020).

These IONPs can be synthesized by various physicals methods (gas phase deposition, electron beam lithography, and ball grinding method), biological methods, and chemical methods (sol-gel method, coprecipitation, hydrothermal solvothermal reaction, thermal decomposition, microemulsion synthesis, sonochemical decomposition, etc.). Out of all the three kind of methods chemical methods of synthesis are preferable due to the many advantages (such as high productivity and easy and economical operation) of it over other methods; in addition, other methods have a few setbacks and pose challenges for the synthesis as physical methods fail to control the size of particles in nano range, and biological methods are tedious, time-consuming with low production rates (Ali, Zafar et al. 2016).

Many researchers have reported the use of IONPs in targeted drug delivery of anticancer agents. Piehler, Dähring et al. (2020) disclosed that the anticancer drug doxorubicin was delivered efficiently in the tumor cells. Doxorubicin prevents cancer cell proliferation through inhibition of topoisomerase2 enzyme and has dangerous side effects on bone marrow cells and heart cells. The authors published that IONPs with doxorubicin resulted in enhanced *in vivo* anticancer effects in breast cancer cells (BT474). Askari, Tajvar et al. further added to the study that PEGylated encapsulation of IONPs along with doxorubicin in liposomes exerted anticancer effects against MCF7 cell lines, at a concentration less than IONPs loaded with doxorubicin (Askari, Tajvar et al. 2020).

Jeon, Lin et al. (2019) prepared NPs of iron oxide with paclitaxel and folic acid, coated with PEG, and displayed good delivery of paclitaxel at the target site with enhanced effectiveness. Park, Park et al. (2019) also reported PTX@PINC complex, i.e., IONPs with PLGA and paclitaxel, and showed anticancer activity along with an appreciable decrease in metastasis. Nosrati, Salehiabar et al. (2018) prepared L-lysine-coated IONPs, conjugated with methotrexate, and found them effective against breast cancer cell lines. In a similar kind of finding (Park, Park et al. 2019), IONPs were coated with arginine which is further bound to methotrexate through carboxylic acid. The authors reported the cytotoxic effects of the NPs against 4T1, MCF7, and HFF-2 cell lines at lower concentrations.

Truffi and coworkers (Truffi, Colombo et al. 2018) constructed IONPs with trastuzumab which resulted in more antitumor activity against HER2+ breast cancer cells. In another study, Badawy and group (Badawy, Abdel-Hamid et al. 2023) also developed chitosan-coated IONPs through gamma radiations, characterized the synthesis of NPS by TEM and XRD, and presented their pronounced anticancer activity. Chitosan in the preparation helped in the opening of intracellular junctions to absorb more IONPs. During anticancer activity evaluation in HCC DEN-induced rats, various signaling pathways such as MAPK and PI3K/Akt/mTOR were studied for identifying their pathway of action. In addition, liver enzymes, caspase-3, and inflammatory markers were also studied. PI3K/Akt/mTOR pathway is crucial for the cancerous growth of cells and drug resistance. The study concluded that IONPs help in the suppression of PI3K/Akt/mTOR and MAPK expressions and thus constitute an effective therapy against drug-resistant cancer cells. The literature is flooded with the data that NPs regulate mTOR expression and cause cycle arrest of tumor cells. In the present study, the liver enzymes such as ALT and AST also decreased significantly.

16.4.2 Copper oxide nanocomposites

Copper is an essential element in the body and has a profound role in cellular metabolism and regulation of physiochemical processes. The use of CuO-based nanocomposites in treating human diseases has been allowed by the United States Environmental Protection Agency (USEPA) since February 2008. Many reports are available for citing their importance in treating cancer (Azizi, Ghourchian et al. 2017). The CuO NPs exert their cytotoxic effects by various mechanisms such as increased formation of ROS, apoptosis, and loss of MMP in cancer cells.

Siddiqui, Alhadlaq et al. (2013) synthesized CuO NPs from copper acetate and sodium hydroxide (NaOH) using the precipitation method. The prepared NPs were examined for antiproliferative activity against human hepatocellular HepG2 cancer cells. The result corroborated their potential as anticancer agents through ROS generation, apoptosis, and MMP alteration.

Sankar, Maheswari et al. (2014) adopted a green approach and prepared CuO NPs from the leaf extract of *Ficus religiosa*. The prepared NPs were characterized by various techniques and were found to be spherical. Their cytotoxic abilities were confirmed against human lung cancer cell lines (A549). In addition, studies also disclosed that mediation of cytotoxicity was mainly due to ROS formation and disturbance of mitochondrial membrane potential in human lung cancer cells.

Sivaraj, Rahman et al. (2014) also reported the synthesis of CuO NPs from aqueous leaf extracts of *Acalypha indica*. The prepared NPs were highly stable and spherical. Its antiproliferative activity was examined and confirmed against breast cancer MCF-7 cell lines by MTT assay. Jeronsia, Raj et al. (2016) disclosed a fast and economical method of preparation of CuO NPs (from cupric acetate, glacial acetic acid, and sodium hydroxide) and isolated the leaf-like crystals. CuO NPs were found to show anticancer activity against human breast cancer (MCF-7) cell lines. In another study by Yugandhar and coworkers (Yugandhar, Vasavi et al. 2017) CuO NPs were prepared from the stem bark of *Syzygium alternifolium*. NPs were found to show antiproliferative activity against human breast cancer lines (MDA-MB-231). The same kind of results was also concluded by Nagaraj and colleagues (Nagaraj, Karuppannan et al. 2019) as the prepared CuO NPs from the leaf extract of *Pterolobium hexapetalum* showed activity against human breast cancer lines (MDA-MB-231). Sundaramurthy and Parthiban (2015) synthesized CuO NPs using an aqueous extract of black bean and characterized for various parameters with the help of a variety of analytical techniques. The NPs were spherical and uniform in shape. They displayed anti-cancer activity against HeLa cancer cell lines through the well-known mechanism of ROS formation and apoptosis.

Xu research group (Xu, Yuan et al. 2019) prepared Zn–CuO NPs and found their anti-tumor activity of it through ROS generation, apoptosis, DNA damage, and autophagy. In another study (Li, Xu et al. 2019), Zn–CuO NPs were prepared from the sonochemical method and were reported to produce an antitumor effect against pancreatic cancer cells by autophagy using AMPK/mTOR pathway. (Yi, Luo et al. 2022) performed another study in which they did micelle encapsulation of Zn–CuO NPs. In comparison to previously prepared zinc-doped CuO NPs, micellar encapsulation resulted in uniform-sized nanoparticles with good stability. They studied and found the anticancer effect of these nanocomposites in six human ovarian cancer cell lines (SNU119, A2780, OVCAR3, OVCAR8, SKOV3, OVSAHO), preferentially in comparison to normal human ovarian cells. Moreover, micelle-encapsulated Zn–CuO NPs reduced the migration of ovarian cancer lines to other parts. It also can repair the homologous recombination (HR) process and reverse drug resistance (Olaparib) in human ovarian cancer cells.

Biswas et al. (2021) presented CuO/GO nanocomposite in which CuO NPs were present over hydrophilic graphene oxide (GO) nanosheets. It was prepared by mixing GO solution in citric acid and copper nitrate solution and confirmed for CuO/GO nanocomposite using various analytical techniques. The CuO/GO nanocomposite showed significant dose-dependent anticancer activity against A-431 cancer cells.

Zughaibi and coworkers (Zughaibi, Mirza et al. 2022) disclosed a green approach for preparing CuO NPs from pumpkin seed extract with an NP size of 20 nm, a size smaller than previously reported NPs. The authors claimed that smaller size led to better penetration into the cells and caused cytotoxicity in MDA-MB-231, a human breast cancer cell line. The main phenomenon of cell cytotoxicity was due to the generation of ROS in dose-dependent manner, confirmed by green fluorescence. The other studied mechanisms were apoptosis at lower concentrations and a dose-dependent decrease in mitochondrial membrane potential (MMP).

Mahmood, Kadhim et al. (2022) prepared CuO NPs from plant extract of *Annona muricata L* (*A. muricata*). They tested NPs for antitumor potential against breast cancer cell lines (AMJ-13 and MCF-7) and normal human breast epithelial cell lines (HBL-100). The results concluded that CuO NPs were selective for inducing cytotoxicity only in cancer cell lines and cause cell death by apoptosis.

Tabrez, Khan et al. (2022) also synthesized CuO NPs from pumpkin seed extract and analyzed antiproliferative activity in the human colorectal cancer cell line (HCT-116). They reported that NPs showed cytotoxicity in cancer cells at lower concentrations, and it was mediated by the dose-dependent generation of ROS, alteration of MMP, and apoptosis in the cancer cells. Sathiyavimal, Vasantharaj et al. (2023) synthesized CuO NPs using *Sida cordifolia* aqueous leaf extract and integrated chitosan into it to make it a nanocomposite. Chitosan provides more stability to the nano-formulation. These nanocomposites were evaluated for cytotoxic potential against human breast cancer MDA-MB-231 and lung cancer A549 cell lines.

16.4.3 Titanium dioxide nanocomposites (TiO₂ NCs)

Titanium dioxide (TiO_2) is widely used in the treatment of cancer as it is readily available, has low toxicity, and is biocompatible. In the lab, titanium dioxide can be prepared by various methods such as hydrothermal methods, microwave methods, sol-gel synthesis, and green methods (Raja, Cao et al. 2020). The reduction of size in the nano range makes it more biocompatible. Various authors have researched the cytotoxic potential of TiO_2 NCs in various types of cancer cell lines and underlined the possible mechanisms/pathways for their activity.

The emergence of drug-resistant tumor cells, particularly those that are multidrug-resistant (MDR), is a significant barrier to chemotherapy's efficacy; as a result, the creation of powerful anti-MDR drugs is crucial to the treatment of tumors. Song et al. (2006) examined the significant impact of TiO_2 NCs and UV illumination on the target cancer cells' drug resistance. The data from fluorescence spectroscopy, microscopy, and electrochemical studies also showed the significant enhancement effect of TiO_2 NCs on drug uptake by drug-resistant leukemia cells. Additionally, it has been found that the combination of TiO_2 NCs, UV irradiation, and the cytotoxic drug daunorubicin can cause significant changes in the target leukemia cells' cell membranes. This finding suggests that TiO_2 NCs did not only increase drug accumulation in target cancer cells but also functioned as a potent anti-MDR agent to inhibit relative drug resistance (Song, Zhang et al. 2006).

Tas et al. (2019) developed a new nanosystem and increased the biocompatibility of TiO_2 NPs by coating them with polyethylene glycol (PEG). The researchers also evaluated the anticancer activity of PEG–TiO_2 NPs loaded with doxorubicin (DOX) on MDA-MB-231 cell lines. The nanostructure system was loaded with DOX after the PEG–TiO_2 NPs and was coated with PEG for this investigation. The MTT technique was used to measure the cytotoxic effect of the fabricated NPs on the MDA-MB-231 cell lines. Several doses of prepared NPs were used against the MDA-MB-231 cells for 24, 48, and 72 hours. Fluorescence microscopy was used to identify apoptosis and necrosis using the Hoechst 33258 (HO)/propidium iodide (PI) double staining. The IC_{50} values for 24, 48, and 72 hours were calculated by comparing the effects of TiO_2, PEG–TiO_2, DOX, and PEG–DOX–TiO_2 NPs on the MDA-MB-231 cell line with the control group. The PEG –DOX–TiO_2 NPs had an inhibitory effect on cancer cell proliferation and an inducer of apoptosis on the MDA-MB-231 cell line when compared to the control group and DOX. PEG coating on TiO_2 NPs improved the efficacy and biocompatibility as an anticancer agent (Tas, Kekliklicioglu Cakmak et al. 2019).

The advanced pulsed laser ablation in liquid (PLAL) technique was utilized to synthesize CuO–TiO_2 nanocomposites to disinfect drug-resistant waterborne bacteria that produce biofilm. The composition of the nanocomposites was varied by incorporating different percentages of copper oxide (5%, 10%, and 20%) with titanium dioxide. This marks the first instance of synthesizing such nanocomposites through this method. Additionally, the cytotoxic activity of both pure TiO_2 and PLAL-synthesized CuO–TiO_2 nanocomposites against normal and healthy cells (HEK-293) and cancerous cells (HCT-116) was evaluated using MTT assay. The results showed no cytotoxic effects on HEK-293 cells, indicating that TiO_2 and PLAL-synthesized CuO–TiO_2 nanocomposites are nontoxic to normal cells (Baig, Ansari et al. 2020).

Jugan et al. (Jugan, Barillet et al. 2012) disclosed the genotoxic ability of TiO_2 NPs against human lung cancer cell lines (A549). The also reported that TiO_2 NPs put stress and cause DNA strand break and damage and thus, display cytotoxic activity. In another research by Biola-Clier, Béal et al. (2017), TiO_2 NPs also reflected cytotoxic and genotoxic nature in BEAS-2B and A549 cancer cell lines by damaging DNA and affecting DNA repair mechanism.

Ahamed, Khan et al. (2017) doped TiO_2 NPs with silver, and characterization studies revealed the even distribution of silver on the surface of TiO_2 NPs. The authors evaluated the antiproliferative activity against human liver cancer cell lines (HepG2), and the results were supported by the study of mechanisms such as ROS formation and exhaustion of antioxidants. In addition, the anticancer potential was also evaluated against MCF-7 and A549 cancer cell lines, and the most probable mechanism was apoptosis. Most importantly, silver-doped TiO_2 NPs showed selective cytotoxicity toward cancer cells in comparison to normal cells.

Kukia, Rasmi et al. (2018) also studied the anticancer potential of TiO_2 NPs against human umbilical vein endothelial cells (HUVEC) and human colorectal cancer cells (HCT116 and HT29) using MTT assay. The results were significant in the case of colorectal cancer cell lines at moderate concentration, whereas the cytotoxic effect in HUVEC was achieved at a very high concentration which is not considerate for human use. In HCT116 cell lines, little proliferation was observed, whereas in HT29 cell lines, reduction in viability of cells was observed in comparison to control cells.

Li, He et al. (2020) studied the antitumor effect of TiO_2 NPs on human liver cancer cells and found a noticeable activity owing to its capacity to produce endoplasmic reticulum stress in cells. The mechanism of activity was investigated by analyzing the expression level of endoplasmic reticulum stress indicators, i.e., PERK and ATF6.

Safaei, Taran et al. (2019) prepared sodium hyaluronate–titanium dioxide bio-nanocomposite in a different ratio by the Taguchi method. All the nanocomposites were evaluated for anticancer activity against breast cancer cell lines (MCF-7). Different ratios showed different cell inhibition in the MTT assay, ranging from 36.19% to 68.89%. The highest anticancer activity was shown when sodium hyaluronate and titanium dioxide were in the ratio 8:2. Elderdery, Alzahrani et al. (2022) synthesized $ZnO-TiO_2$–chitosan–farnesol NCs using the precipitation method and tested them for antitumor activity against leukemia cancer cell lines (MOLT-4). The results confirmed their dose-dependent considerable activity. The investigation of possible pathways revealed an increase in ROS, a decrease in membrane potential, and an increase in caspase 3, 8, and 9 markers, suggesting cell death due to apoptosis.

Apart from its anticancer activity, TiO_2 NPs can be used as a drug delivery means. A study by Mund et al. (Mund, Panda et al. 2014) revealed that TiO_2 NPs loaded with paclitaxel displayed the enhanced anticancer activity of paclitaxel instead of plain paclitaxel.

16.5 CONCLUSION

MDR can be treated with higher chemotherapy dosages, but doing so frequently results in more severe side effects and harm to healthy tissues. To boost the medication concentration inside the tumor and improve treatment efficacy, targeting drug delivery systems may react to many variables (such as pH, enzyme, and light) by releasing the drug. Researchers discovered the drug efflux target and have a preliminary understanding of the MDR process, such as drug efflux. They used nanocarriers to their advantage to avoid P-gp-mediated drug efflux. The P-gp is a significant barrier to treating drug-resistant cancer, and it has been shown that inhibiting it is an effective way to overcome MDR. Two methods might be used to inhibit P-gp activity such as (1) the P-gp may either be directly inhibited by the P-gp inhibitor or have its expression level altered via RNA interference technology. As the P-gp requires ATP for energy, targeting the mitochondria or other methods that disrupt energy metabolism have been routinely used to combat MDR. Drug resistance is a multifaceted and complicated phenomenon with more than one cause (P-gp). Drug efflux is succinctly combined with other non-clear variables (such as gene, metabolize, signal route) into a factor we call pump factors.

Photothermal treatment and photodynamic therapy have been successfully utilized to address non-pump variables, improve the efficacy of chemotherapy, and reverse MDR. One of the recent decade's key areas of study has been autophagy. Although the dual autophagy characteristics are fascinating and endearing, more and more articles have noted that autophagy suppression may help reverse MDR. In the treatment of MDR, MNCs can transport drugs, and even little differences in the formulation of these NCs can have a big impact on the therapeutic outcomes. We anticipate exciting new developments in MDR cancer treatment shortly as mechanistic investigations and clinical evaluations of nanotechnology-based drug delivery approaches in overcoming MDR. The delivery of chemotherapeutic medications for the treatment of cancer with MDR has a lot of potential thanks to the newly developed nanotechnology-based drug delivery systems. This method of drug delivery could lessen the systemic toxicity of medications used to treat MDR cancer and enhance the pharmacokinetic behavior of antitumor medications. It could also deliver chemotherapeutic medications to the intended areas.

REFERENCES

Ahamed, M., M. M. Khan, M. J. Akhtar, H. A. Alhadlaq and A. J. S. r. Alshamsan. 2017. Ag-doping regulates the cytotoxicity of TiO2 nanoparticles via oxidative stress in human cancer cells. *Scientific Reports* 7(1): 17662.

Alavi, M. and M. J. E. r. o. a.-i. t. Rai. 2019. Recent advances in antibacterial applications of metal nanoparticles (MNPs) and metal nanocomposites (MNCs) against multidrug-resistant (MDR) bacteria. *Expert Rev Anti Infect Ther.* 17(6): 419–428.

Alfarouk, K. O., C.-M. Stock, S. Taylor, M. Walsh, A. K. Muddathir, D. Verduzco, A. H. Bashir, O. Y. Mohammed, G. O. Elhassan and S. J. C. c. i. Harguindey. 2015. Resistance to cancer chemotherapy: Failure in drug response from ADME to P-gp. *Cancer Cell International.* 15(1): 1–13.

Ali, A., H. Zafar, M. Zia, I. Ul Haq, A. Phull, J. Ali and A. Hussain. 2016. Synthesis, characterization, applications, and challenges of iron oxide nanoparticles. *Nanotechnology Science and Application.* 9: 49–67.

Ali, E. S., S. M. Sharker, M. T. Islam, I. N. Khan, S. Shaw, M. A. Rahman, S. J. Uddin, M. C. Shill, S. Rehman and N. Das. 2021. Targeting cancer cells with nanotherapeutics and nanodiagnostics: Current status and future perspectives. *Seminars in Cancer Biology*, Elsevier.

Askari, A., S. Tajvar, M. Nikkhah, S. Mohammadi and S. Hosseinkhani. 2020. Synthesis, characterization and in vitro toxicity evaluation of doxorubicin-loaded magnetoliposomes on MCF-7 breast cancer cell line. *Journal of Drug Delivery Science and Technology.* 55: 101447.

Azizi, M., H. Ghourchian, F. Yazdian, F. Dashtestani and H. J. P. o. AlizadehZeinabad. 2017. Cytotoxic effect of albumin coated copper nanoparticle on human breast cancer cells of MDA-MB 231. *PLoS One.* 12(11): e0188639.

Badawy, M. M., G. R. Abdel-Hamid and H. E. Mohamed. 2023. Antitumor activity of Chitosan-Coated Iron Oxide Nanocomposite against Hepatocellular Carcinoma in animal models. *Biological Trace Element Research*. 201(3): 1274–1285.

Baig, U., M. A. Ansari, M. Gondal, S. Akhtar, F. A. Khan, W. J. M. S. Falath and E. C. 2020. Single step production of high-purity copper oxide-titanium dioxide nanocomposites and their effective antibacterial and anti-biofilm activity against drug-resistant bacteria. *Materials Science and Engineering: C*. 113: 110992.

Bargude, S. D., G. L. Chopade, A. V. Pondkule and S. Narayan. 2023. Nanostructured drug delivery systems: An alternative approach to herbal medicine. *World Journal of Biology Pharmacy and Health Sciences*. 13(1): 99–102.

Biola-Clier, M., D. Béal, S. Caillat, S. Libert, L. Armand, N. Herlin-Boime, S. Sauvaigo, T. Douki and M. J. M. Carrière. 2017. Comparison of the DNA damage response in BEAS-2B and A549 cells exposed to titanium dioxide nanoparticles. *Mutagenesis*. 32(1): 161–172.

Biswas, K., Y. K. Mohanta, A. K. Mishra, A. G. Al-Sehemi, M. Pannipara, A. Sett, A. Bratovcic, D. De, B. Prasad Panda and S. J. J. o. M. S. M. i. M. Kumar Avula. 2021. Wet chemical development of CuO/GO nanocomposites: Its augmented antimicrobial, antioxidant, and anticancerous activity. *Journal of Materials Science: Materials in Medicine*. 32: 1–11.

Bukowski, K., M. Kciuk and R. J. I. j. o. m. s. Kontek. 2020. Mechanisms of multidrug resistance in cancer chemotherapy. *Int J Mol Sci*. 21(9): 3233.

Cao, L., Y. Zhu, W. Wang, G. Wang, S. Zhang, H. J. F. i. b. Cheng and biotechnology. 2021. Emerging nano-based strategies against drug resistance in tumor chemotherapy. *Front Bioeng Biotechnol*. 1255.

Chandrasekharan, P., Z. W. Tay, D. Hensley, X. Y. Zhou, B. K. Fung, C. Colson, Y. Lu, B. D. Fellows, Q. Huynh and C. Saayujya. 2020. Using magnetic particle imaging systems to localize and guide magnetic hyperthermia treatment: Tracers, hardware, and future medical applications. *Theranostics*. 10(7): 2965.

Chen, W., W. Yang, P. Chen, Y. Huang and F. Li. 2018. Disulfiram copper nanoparticles prepared with a stabilized metal ion ligand complex method for treating drug-resistant prostate cancers. *ACS Applied Materials & Interfaces*. 10(48): 41118–41128.

Chronopoulou, L., F. A. Scaramuzzo, R. Fioravanti, A. di Nitto, S. Cerra, C. Palocci and I. J. I. j. o. b. m. Fratoddi. 2020. Noble metal nanoparticle-based networks as a new platform for lipase immobilization. *International Journal of Biological Macromolecules*. 146: 790–797.

Cobine, P. A. and D. C. Brady. 2022. Cuproptosis: Cellular and molecular mechanisms underlying copper-induced cell death. *Molecular Cell*. 82(10): 1786–1787.

Curcio, M., G. Cirillo, F. Saletta, F. Michniewicz, F. P. Nicoletta, O. Vittorio, S. Hampel and F. J. C. Iemma. 2020. Carbon nanohorns as effective nanotherapeutics in cancer therapy. *C Journal of Carbon Research*. 7(1): 3.

Dallavalle, S., V. Dobričić, L. Lazzarato, E. Gazzano, M. Machuqueiro, I. Pajeva, I. Tsakovska, N. Zidar and R. J. D. R. U. Fruttero. 2020. Improvement of conventional anti-cancer drugs as new tools against multidrug resistant tumors. *Drug Resistance Updates*. 50: 100682.

Deng, H., Z. Yang, X. Pang, C. Zhao, J. Tian, Z. Wang and X. Chen. 2022. Self-sufficient copper peroxide loaded pKa-tunable nanoparticles for lysosome-mediated chemodynamic therapy. *Nano Today*. 42: 101337.

Elderdery, A. Y., B. Alzahrani, S. Hamza, G. Mostafa-Hedeab, P. L. Mok, S. K. J. B. C. Subbiah and Applications. 2022. Synthesis of Zinc Oxide (ZnO)-Titanium Dioxide (TiO 2)-Chitosan-Farnesol Nanocomposites and assessment of their anticancer potential in human Leukemic MOLT-4 cell line. *Bioinorg Chem Appl*. 2022.

Emran, T. B., A. Shahriar, A. R. Mahmud, T. Rahman, M. H. Abir, M. Faijanur-Rob-Siddiquee, H. Ahmed, N. Rahman, F. Nainu and E. J. F. i. O. Wahyudin. 2022. Multidrug Resistance in Cancer: Understanding molecular mechanisms, immunoprevention, and therapeutic approaches. *Front. Oncol*. 2581.

Gao, D., X. Guo, X. Zhang, S. Chen, Y. Wang, T. Chen, G. Huang, Y. Gao, Z. Tian and Z. J. M. T. B. Yang. 2020. Multifunctional phototheranostic nanomedicine for cancer imaging and treatment. *Materials Today Bio.* 5: 100035.

García-Peña, N. G., R. Redón, A. Herrera-Gomez, A. L. Fernández-Osorio, M. Bravo-Sanchez and G. J. A. S. S. Gomez-Sosa. 2015. Solventless synthesis of ruthenium nanoparticles. *Applied Surface Science.* 340: 25–34.

Gopisetty, M. K., D. Kovács, N. Igaz, A. Rónavári, P. Bélteky, Z. Rázga, V. Venglovecz, B. Csoboz, I. M. Boros, Z. Kónya and M. Kiricsi. 2019. Endoplasmic reticulum stress: Major player in size-dependent inhibition of P-glycoprotein by silver nanoparticles in multidrug-resistant breast cancer cells. *Journal of Nanobiotechnology.* 17(1): 9.

Gul, N. S., T.-M. Khan, M. Chen, K.-B. Huang, C. Hou, M. I. Choudhary, H. Liang and Z.-F. Chen. 2020. New copper complexes inducing bimodal death through apoptosis and autophagy in A549 cancer cells. *Journal of Inorganic Biochemistry.* 213: 111260.

Guo, W.-j., S.-s. Ye, N. Cao, J. Huang, J. Gao and Q.-y. Chen. 2010. ROS-mediated autophagy was involved in cancer cell death induced by novel copper (II) complex. *Experimental and Toxicologic Pathology.* 62(5): 577–582.

Gurunathan, S., J. H. Park, J. W. Han and J. H. Kim. 2015. Comparative assessment of the apoptotic potential of silver nanoparticles synthesized by Bacillus tequilensis and Calocybe indica in MDA-MB-231 human breast cancer cells: Targeting p53 for anticancer therapy. *Int J Nanomedicine.* 10: 4203–4222.

Halder, J., D. Pradhan, B. Kar, G. Ghosh, G. J. N. N. Rath. 2022. Nanotherapeutics approaches to overcome P-glycoprotein-mediated multi-drug resistance in cancer. *Biology and Medicine.* 40: 102494.

Huang, N., X. Chen, X. Zhu, M. Xu and J. J. B. Liu. 2017. Ruthenium complexes/polypeptide self-assembled nanoparticles for identification of bacterial infection and targeted antibacterial research. *Biomaterials.* 141: 296–313.

Huang, Y., Y. Lai, S. Shi, S. Hao, J. Wei and X. Chen. 2015. Copper sulfide nanoparticles with phospholipid-PEG coating for In vivo near-infrared photothermal cancer therapy. *Chemistry—An Asian Journal.* 10(2): 370–376.

Izdebska, M., W. Zielińska, M. Hałas-Wiśniewska and A. J. C. Grzanka. 2019. Involvement of actin in autophagy and autophagy-dependent multidrug resistance in cancer. *Cancers (Basel).* 11(8): 1209.

Jeon, M., G. Lin, Z. R. Stephen, F. L. Kato and M. Zhang. 2019. Paclitaxel-loaded iron oxide nanoparticles for targeted breast cancer therapy. *Advanced Therapeutics.* 2(12): 1900081.

Jeronsia, J. E., D. V. Raj, L. A. Joseph, K. Rubini and S. J. J. J. o. M. S. Das. 2016. In vitro antibacterial and anticancer activity of copper oxide nanostructures in human breast cancer Michigan Cancer Foundation-7 cells. *Journal of Medical Science.* 36(4): 145.

Ji, P., P. Wang, H. Chen, Y. Xu, J. Ge, Z. Tian and Z. Yan. 2023. Potential of copper and copper compounds for anticancer applications. *Pharmaceuticals.* 16(2): 234.

Jiang Y., Z. Guo, J. Fang, B. Wang, Z. Lin, Z. S. Chen, Y. Chen, N. Zhang, X. Yang, W. Gao. 2020. "A multi-functionalized nanocomposite constructed by gold nanorod core with triple-layer coating to combat multidrug resistant colorectal cancer." *Mater Sci Eng C Mater Biol Appl.* 107 (Feb):110224.

Jugan, M.-L., S. Barillet, A. Simon-Deckers, N. Herlin-Boime, S. Sauvaigo, T. Douki and M. J. N. Carriere. 2012. Titanium dioxide nanoparticles exhibit genotoxicity and impair DNA repair activity in A549 cells. *Nanotoxicology.* 6(5): 501–513.

Kennedy, L., J. K. Sandhu, M.-E. Harper and M. Cuperlovic-Culf. 2020. Role of glutathione in cancer: From mechanisms to therapies. *Biomolecules.* 10(10): 1429.

Khursheed, R., K. Dua, S. Vishwas, M. Gulati, N. K. Jha, G. M. Aldhafeeri, F. G. Alanazi, B. H. Goh, G. Gupta, K. R. J. B. Paudel and Pharmacotherapy. 2022. Biomedical applications of metallic nanoparticles in cancer: Current status and future perspectives. *Biomedicine & Pharmacotherapy.* 150: 112951.

Kovács, D., K. Szőke, N. Igaz, G. Spengler, J. Molnár, T. Tóth, D. Madarász, Z. Rázga, Z. Kónya, I. M. Boros and M. Kiricsi. 2016. Silver nanoparticles modulate ABC transporter activity and enhance chemotherapy in multidrug resistant cancer. *Nanomedicine*. 12(3): 601–610.

Kukia, N. R., Y. Rasmi, A. Abbasi, N. Koshoridze, A. Shirpoor, G. Burjanadze and E. J. A. P. j. o. c. p. A. Saboory. 2018. Bio-effects of TiO2 nanoparticles on human colorectal cancer and umbilical vein endothelial cell lines. *Asian Pac J Cancer Prev*. 19(10): 2821.

Li, B., Q. Wang, R. Zou, X. Liu, K. Xu, W. Li and J. Hu. 2014. Cu 7.2 S 4 nanocrystals: A novel photothermal agent with a 56.7% photothermal conversion efficiency for photothermal therapy of cancer cells. *Nanoscale*. 6(6): 3274–3282.

Li, C., Y. Li and C. Ding. 2019. The role of copper homeostasis at the host-pathogen axis: From bacteria to fungi. *International Journal of Molecular Sciences*. 20(1): 175.

Li, C., Y. Zhang, M. Wang, Y. Zhang, G. Chen, L. Li, D. Wu and Q. Wang. 2014. In vivo real-time visualization of tissue blood flow and angiogenesis using Ag2S quantum dots in the NIR-II window. *Biomaterials*. 35(1): 393–400.

Li, W. R., X. B. Xie, Q. S. Shi, H. Y. Zeng, Y. S. Ou-Yang and Y. B. Chen. 2010. Antibacterial activity and mechanism of silver nanoparticles on Escherichia coli. *Appl Microbiol Biotechnol*. 85(4): 1115–1122.

Li, X., H. Xu, C. Li, G. Qiao, A. A. Farooqi, A. Gedanken, X. Liu and X. J. F. i. P. Lin. 2019. Zinc-doped copper oxide nanocomposites inhibit the growth of pancreatic cancer by inducing autophagy through AMPK/mTOR pathway. *Front Pharmacol*. 10: 319.

Li, Z., J. He, B. Li, J. Zhang, K. He, X. Duan, R. Huang, Z. Wu and G. J. J. o. I. M. R. Xiang. 2020. Titanium dioxide nanoparticles induce endoplasmic reticulum stress-mediated apoptotic cell death in liver cancer cells. *J Int Med Res*. 48(4): 0300060520903652.

Liang, J., R. Li, Y. He, C. Ling, Q. Wang, Y. Huang, J. Qin, W. Lu and J. J. N. R. Wang. 2018. A novel tumor-targeting treatment strategy uses energy restriction via co-delivery of albendazole and nanosilver. *Nano Research*. 11: 4507–4523.

Liang, J., F. Zeng, M. Zhang, Z. Pan, Y. Chen, Y. Zeng, Y. Xu, Q. Xu and Y. Huang. 2015. Green synthesis of hyaluronic acid-based silver nanoparticles and their enhanced delivery to CD44+ cancer cells. *RSC Advances*. 5(54): 43733–43740.

Liu, H., R. Jiang, Y. Lu, B. Shan, Y. Wen and M. Li. 2022. Biodegradable amorphous copper iron tellurite promoting the utilization of Fenton-like ions for efficient synergistic cancer theranostics. *ACS Applied Materials & Interfaces*. 14(25): 28537–28547.

Liu, J., Y. Yuan, Y. Cheng, D. Fu, Z. Chen, Y. Wang, L. Zhang, C. Yao, L. Shi and M. Li. 2022. Copper-based metal—organic framework overcomes cancer chemoresistance through systemically disrupting dynamically balanced cellular redox homeostasis. *Journal of the American Chemical Society*. 144(11): 4799–4809.

Liu, J., Y. Zhao, Q. Guo, Z. Wang, H. Wang, Y. Yang and Y. Huang. 2012. TAT-modified nanosilver for combating multidrug-resistant cancer. *Biomaterials*. 33(26): 6155–6161.

Lowndes, S. A. and A. L. Harris. 2005. The role of copper in tumour angiogenesis. *Journal of Mammary Gland Biology and Neoplasia*. 10: 299–310.

Lu, Y., D. Zhu, L. Gui, Y. Li, W. Wang, J. Liu and Y. Wang. 2021. A dual-targeting ruthenium nanodrug that inhibits primary tumor growth and lung metastasis via the PARP/ATM pathway. *Journal of Nanobiotechnology*. 19(1): 115.

Mahmood, R. I., A. A. Kadhim, S. Ibraheem, S. Albukhaty, H. S. Mohammed-Salih, R. H. Abbas, M. S. Jabir, M. K. Mohammed, U. M. Nayef and F. A. J. S. R. AlMalki. 2022. Biosynthesis of copper oxide nanoparticles mediated Annona muricata as cytotoxic and apoptosis inducer factor in breast cancer cell lines. *Scientific Reports*. 12(1): 16165.

Mishra, R., S. Mishra and Y. B. Barot. 2021. Greener synthesis and stabilization of metallic nanoparticles in ionic liquids. *Handbook of Greener Synthesis of Nanomaterials and Compounds*, Elsevier: 245–276.

Mohammad, N. S., R. Nazli, H. Zafar and S. Fatima. 2022. Effects of lipid based multiple micronutrients supplement on the birth outcome of underweight pre-eclamptic women: A randomized clinical trial. *Pak J Med Sci*. 38(1): 219–226.

Mund, R., N. Panda, S. Nimesh and A. J. J. o. n. r. Biswas. 2014. Novel titanium oxide nanoparticles for effective delivery of paclitaxel to human breast cancer cells. *Journal of Nanoparticle Research*. 16: 1–12.

Nagaraj, E., K. Karuppannan, P. Shanmugam and S. J. J. o. C. S. Venugopal. 2019. Exploration of bio-synthesized copper oxide nanoparticles using pterolobium hexapetalum leaf extract by photocatalytic activity and biological evaluations. *Journal of Cluster Science*. 30: 1157–1168.

Nakamura, H. and K. Takada. 2021. Reactive oxygen species in cancer: Current findings and future directions. *Cancer Science*. 112(10): 3945–3952.

Nayak, V., S. M. Munawar, K. B. Sabjan, S. Singh and K. R. Singh. 2021. Nanomaterials' properties, classification, synthesis, and characterization. *Nanomaterials in Bionanotechnology*, CRC Press: 37–68.

Nosrati, H., M. Salehiabar, S. Davaran, H. Danafar and H. K. Manjili. 2018. Methotrexate-conjugated L-lysine coated iron oxide magnetic nanoparticles for inhibition of MCF-7 breast cancer cells. *Drug Development and Industrial Pharmacy*. 44(6): 886–894.

Oe, S., K. Miyagawa, Y. Honma and M. Harada. 2016. Copper induces hepatocyte injury due to the endoplasmic reticulum stress in cultured cells and patients with Wilson disease. *Experimental Cell Research*. 347(1): 192–200.

Ostad, S. N., S. Dehnad, Z. E. Nazari, S. T. Fini, N. Mokhtari, M. Shakibaie and A. R. J. A. j. o. m. b. Shahverdi. 2010. Cytotoxic activities of silver nanoparticles and silver ions in parent and tamoxifen-resistant T47D human breast cancer cells and their combination effects with tamoxifen against resistant cells. *Avicenna J Med Biotechnol*. 2(4): 187.

Păduraru, D. N., D. Ion, A.-G. Niculescu, F. Mușat, O. Andronic, A. M. Grumezescu and A. J. P. Bolocan. 2022. Recent developments in metallic nanomaterials for cancer therapy, diagnosing and imaging applications. *Pharmaceutics*. 14(2): 435.

Pan, P., D. Svirskis, S. W. Rees, D. Barker, G. I. Waterhouse and Z. Wu. 2021. Photosensitive drug delivery systems for cancer therapy: Mechanisms and applications. *Journal of Controlled Release*. 338: 446–461.

Park, J., J. Park, M. A. Castanares, D. S. Collins and Y. Yeo. 2019. Magnetophoretic delivery of a tumor-priming agent for chemotherapy of metastatic murine breast cancer. *Molecular Pharmaceutics*. 16(5): 1864–1873.

Pascariu, P., A. Airinei, M. Grigoras, L. Vacareanu and F. J. A. S. S. Iacomi. 2015. Metal—polymer nanocomposites based on Ni nanoparticles and polythiophene obtained by electrochemical method. *Applied Surface Science*. 352: 95–102.

Piehler, S., H. Dähring, J. Grandke, J. Göring, P. Couleaud, A. Aires, A. L. Cortajarena, J. Courty, A. Latorre and Á. Somoza. 2020. Iron oxide nanoparticles as carriers for DOX and magnetic hyperthermia after intratumoral application into breast cancer in mice: Impact and future perspectives. *Nanomaterials*. 10(6): 1016.

Pramanik, S. and P. Das. 2019. Metal-based nanomaterials and their polymer nanocomposites. *Nanomaterials and Polymer Nanocomposites*, Elsevier: 91–121.

Raja, G., S. Cao, D.-H. Kim and T.-J. Kim 2020. Mechanoregulation of titanium dioxide nanoparticles in cancer therapy. *Materials Science and Engineering: C*. 107: 110303.

Safaei, M., M. Taran, M. M. Imani, H. Moradpoor, A. Golshah, P. J. C. I. i. P. Upadhyay. 2019. Evaluation of anticancer activity of sodium hyaluronate-titanium dioxide bionanocomposite. *Current Issues in Pharmacy and Medical Sciences*. 32(2): 99–103.

Sankar, R., R. Maheswari, S. Karthik, K. S. Shivashangari, V. J. M. S. Ravikumar and E. C. 2014. Anticancer activity of Ficus religiosa engineered copper oxide nanoparticles. *Mater Sci Eng C Mater Biol Appl*. 44: 234–239.

Sathiyavimal, S., S. Vasantharaj, T. Kaliannan, H. A. Garalleh, M. Garaleh, K. Brindhadevi, N. T. L. Chi, A. Sharma and A. J. E. R. Pugazhendhi. 2023. Bio-functionalized copper oxide/chitosan nanocomposite using Sida cordifolia and their efficient properties of antibacterial, anticancer activity against on breast and lung cancer cell lines. *Environmental Research*. 218: 114986.

Shabade, A. B., D. M. Sharma, P. Bajpai, R. G. Gonnade, K. Vanka and B. J. C. S. Punji. 2022. Room temperature chemoselective hydrogenation of C [double bond, length as m-dash] C, C [double bond, length as m-dash] O and C [double bond, length as m-dash] N bonds by using a well-defined mixed donor Mn (i) pincer catalyst. *Chemical Sciences.* 13(46): 13764–13773.

Sharma, H., K. Kumar, C. Choudhary, P. K. Mishra and B. Vaidya. 2016. Development and characterization of metal oxide nanoparticles for the delivery of anticancer drug. *Artif Cells Nanomed Biotechnol.* 44(2): 672–679.

Shetty, A., H. Lang and S. Chandra. 2023. Metal sulfide nanoparticles for imaging and phototherapeutic applications. *Molecules.* 28(6).

Shimada, K., E. Reznik, M. E. Stokes, L. Krishnamoorthy, P. H. Bos, Y. Song, C. E. Quartararo, N. C. Pagano, D. R. Carpizo and A. C. deCarvalho. 2018. Copper-binding small molecule induces oxidative stress and cell-cycle arrest in glioblastoma-patient-derived cells. *Cell Chemical Biology.* 25(5): 585–594. e587.

Siddiqui, M. A., H. A. Alhadlaq, J. Ahmad, A. A. Al-Khedhairy, J. Musarrat and M. J. P. o. Ahamed. 2013. Copper oxide nanoparticles induced mitochondria mediated apoptosis in human hepatocarcinoma cells. *PLoS One.* 8(8): e69534.

Sies, H., V. V. Belousov, N. S. Chandel, M. J. Davies, D. P. Jones, G. E. Mann, M. P. Murphy, M. Yamamoto and C. Winterbourn. 2022. Defining roles of specific reactive oxygen species (ROS) in cell biology and physiology. *Nature Reviews Molecular Cell Biology.* 23(7): 499–515.

Sivaraj, R., P. K. Rahman, P. Rajiv, S. Narendhran, R. J. S. A. P. A. M. Venckatesh and B. Spectroscopy. 2014. Biosynthesis and characterization of Acalypha indica mediated copper oxide nanoparticles and evaluation of its antimicrobial and anticancer activity. *Spectrochim Acta A Mol Biomol Spectrosc.* 129: 255–258.

Song, M., R. Zhang, Y. Dai, F. Gao, H. Chi, G. Lv, B. Chen and X. Wang. 2006. The in vitro inhibition of multidrug resistance by combined nanoparticulate titanium dioxide and UV irradition. *Biomaterials.* 27(23): 4230–4238.

Srivastava, S. K. and M. J. J. o. N. R. Constanti. 2012. Room temperature biogenic synthesis of multiple nanoparticles (Ag, Pd, Fe, Rh, Ni, Ru, Pt, Co, and Li) by Pseudomonas aeruginosa SM1. *Journal of Nanoparticle Research.* 14: 1–10.

Subhan, M. A. J. R. a. 2022. Advances with metal oxide-based nanoparticles as MDR metastatic breast cancer therapeutics and diagnostics. *RSC Advances.* 12(51): 32956–32978.

Sundaramurthy, N. and C. Parthiban. 2015. Biosynthesis of copper oxide nanoparticles using *Pyrus pyrifolia* leaf extract and evolve the catalytic activity. *International Research Journal of Engineering and Technology.* 2(6): 332–338.

Tabrez, S., A. U. Khan, A. A. Mirza, M. Suhail, N. R. Jabir, T. A. Zughaibi and M. J. N. R. Alam. 2022. Biosynthesis of copper oxide nanoparticles and its therapeutic efficacy against colon cancer. *Nanotechnology Reviews.* 11(1): 1322–1331.

Tas, A., N. Keklikcioglu Cakmak, E. Gumus, M. Atabey and Y. Silig. 2019. Chemotherapeutic effects of doxorubicin loaded PEG coated TiO nanocarriers on breast cancer cell lines. *Annals of Medical Research.* 26(5): 821–826.

Truffi, M., M. Colombo, L. Sorrentino, L. Pandolfi, S. Mazzucchelli, F. Pappalardo, C. Pacini, R. Allevi, A. Bonizzi and F. Corsi. 2018. Multivalent exposure of trastuzumab on iron oxide nanoparticles improves antitumor potential and reduces resistance in HER2-positive breast cancer cells. *Scientific Reports.* 8(1): 6563.

Tsvetkov, P., S. Coy, B. Petrova, M. Dreishpoon, A. Verma, M. Abdusamad, J. Rossen, L. Joesch-Cohen, R. Humeidi and R. D. Spangler. 2022. Copper induces cell death by targeting lipoylated TCA cycle proteins. *Science.* 375(6586): 1254–1261.

Viau, G., R. Brayner, L. Poul, N. Chakroune, E. Lacaze, F. Fievet-Vincent and F. J. C. o. M. Fievet. 2003. Ruthenium nanoparticles: Size, shape, and self-assemblies. *Chem. Mater.* 15(2): 486–494.

Wang, J., N. Seebacher, H. Shi, Q. Kan and Z. J. O. Duan. 2017. Novel strategies to prevent the development of multidrug resistance (MDR) in cancer. *Oncotarget.* 8(48): 84559.

Wang, N., Y. Feng, L. Zeng, Z. Zhao, and T. Chen. 2015. "Functionalized multiwalled carbon nanotubes as carriers of ruthenium complexes to antagonize cancer multidrug resistance and radioresistance." *ACS Applied Materials & Interfaces*. 7(27): 14933–14945.

Wang, X., T. Xiong, M. Cui, N. Li, Q. Li, L. Zhu, S. Duan, Y. Wang and Y. Guo. 2021. A novel targeted co-delivery nanosystem for enhanced ovarian cancer treatment via multidrug resistance reversion and mTOR-mediated signaling pathway. *Journal of Nanobiotechnology*. 19(1): 1–18.

Wang, Y., L. Zhang and F. Zhou. 2022. Cuproptosis: A new form of programmed cell death. *Cellular & Molecular Immunology*. 19(8): 867–868.

Wu, Q., Z. Yang, Y. Nie, Y. Shi and D. J. C. l. Fan. 2014. Multi-drug resistance in cancer chemotherapeutics: Mechanisms and lab approaches. *Cancer Lett*. 347(2): 159–166.

Wu, X., J. Liu, L. Yang and F. Wang. 2019. Photothermally controlled drug release system with high dose loading for synergistic chemo-photothermal therapy of multidrug resistance cancer. *Colloids and Surfaces B: Biointerfaces*. 175: 239–247.

Wu, Y., Q. Zeng, Z. Qi, T. Deng and F. J. F. i. C. Liu. 2020. Recent progresses in cancer nanotherapeutics design using artemisinins as free radical precursors. *Frontiers in Chemistry*. 8: 472.

Xu, H., R. Yuan, X. Liu, X. Li, G. Qiao, C. Li, A. Gedanken and X. J. N. Lin. 2019. Zn-doped CuO nanocomposites inhibit tumor growth by NF-κB pathway cross-linked autophagy and apoptosis. *Nanomedicine*. 14(2): 131–149.

Xu, J.-J., W.-C. Zhang, Y.-W. Guo, X.-Y. Chen and Y.-N. J. D. D. Zhang. 2022. Metal nanoparticles as a promising technology in targeted cancer treatment. *Drug Delivery*. 29(1): 664–678.

Xu, Y., S. Y. Liu, L. Zeng, H. Ma, Y. Zhang, H. Yang, Y. Liu, S. Fang, J. Zhao and Y. Xu. 2022. An enzyme-engineered nonporous copper (I) coordination polymer nanoplatform for cuproptosis-based synergistic cancer therapy. *Advanced Materials*. 34(43): 2204733.

Xue, H.-Y., R. Bendayan and H.-L. Wong. 2020. MDR reversal for effective chemotherapy in breast cancer. *Drug Efflux Pumps in Cancer Resistance Pathways: From Molecular Recognition and Characterization to Possible Inhibition Strategies in Chemotherapy*, Elsevier: 121–147.

Yadav, P., S. V. Ambudkar and N. J. J. o. N. Rajendra Prasad. 2022. Emerging nanotechnology-based therapeutics to combat multidrug-resistant cancer. *Journal of Nanobiotechnology*. 20(1): 1–35.

Yang, Y., Y. Lin, D. Di, X. Zhang, D. Wang, Q. Zhao and S. Wang. 2017. Gold nanoparticle-gated mesoporous silica as redox-triggered drug delivery for chemo-photothermal synergistic therapy. *Journal of Colloid and Interface Science*. 508: 323–331.

Yi, J., X. Luo, J. Xing, A. Gedanken, X. Lin, C. Zhang, G. J. B. Qiao. 2022. Micelle encapsulation zinc-doped copper oxide nanocomposites reverse Olaparib resistance in ovarian cancer by disrupting homologous recombination repair. *Bioengineering and Translational Medicine*. e10507.

Yugandhar, P., T. Vasavi, P. Uma and M. Devi. 2017. Bioinspired green synthesis of copper oxide nanoparticles from Syzygium alternifolium (Wt.) Walp: Characterization and evaluation of its synergistic antimicrobial and anticancer activity. *Applied Nanoscience*. 7 (2017): 417–427.

Zhang, Y., M. Li, X. Gao, Y. Chen, T. J. J. o. h. Liu and oncology. 2019. Nanotechnology in cancer diagnosis: Progress, challenges and opportunities. *Journal of Hematology and Oncology*. 12(1): 1–13.

Zhou, G. and M. Li. 2022. Biodegradable copper telluride nanosheets for redox-homeostasis breaking-assisted chemodynamic cancer therapy boosted by mild-photothermal effect. *Chemical Engineering Journal*. 450: 138348.

Zhou, J., Q. Yu, J. Song, S. Li, X.-L. Li, B. Kang, H.-Y. Chen and J.-J. Xu. 2022. *Photothermally Triggered Copper Payload Release for Cuproptosis-Promoted Cancer Synergistic Therapy*. Angewandte Chemie International Edition.

Zhu, L., Z. Kuang, P. Song, W. Li, L. Gui, K. Yang, F. Ge, Y. Tao and W. J. N. Zhang. 2021. Gold nanorod-loaded thermosensitive liposomes facilitate the targeted release of ruthenium (II) polypyridyl complexes with anti-tumor activity. *Nanotechnology*. 32(45): 455103.

Zughaibi, T. A., A. A. Mirza, M. Suhail, N. R. Jabir, S. K. Zaidi, S. Wasi, A. Zawawi and S. J. J. o. N. Tabrez. 2022. Evaluation of anticancer potential of biogenic copper oxide nanoparticles (CuO NPs) against breast cancer. *Journal of Nanomaterials*. 2022: 1–7.

Chapter 17

Nanotherapeutics for Alzheimer's disease using metal nanocomposites

Nitin Verma, Shiwali Sharma, Nikita Thakur, Narinderpal Kaur, and Kamal Dua

LIST OF ABBREVIATIONS

AChE	Acetylcholinesterase
AD	Alzhiemer's disease
ADN's	Adiponectin
Ag	Silver
AgNGSs	Ag nanoprobes shells
Ag NPs	Silver nanoparticles
Akt	Protein kinase B
Al	Aluminum
AlCl$_3$	Aluminum chloride
APP	Amyloid precursor protein
Au	Gold
Aβ	Amyloid-β
Aβ42	Amyloid-beta aggregation
AβOs	Amyloid beta oligomers
BACE1	Amyloid precursor protein cleavage enzyme-1
BBB	Blood–brain barrier
BChE	Butyl cholinesterase
BDNF	Brain-derived neurotrophic factor
CAT	Catalase
CD	Circular dichroism
CdTeQD	*N*-acetyl-l-cysteine-capped cadmium telluride quantum dots
Ch-SeNPs	Selenium nanoparticles containing chitosan
CNS	Central nervous system
CS@Se	Selenium chondroitin sulfate nanoparticles
Cu	Copper
D-FexCuySe NPs	Chiral iron copper selenide nanoparticles
DRP1	Dynamic-related protein-1
EGCG	Epigallocatechin gallate
FDA	Food and Drug Administration
Fe	Iron
GLUT-1	Glucose transporter 1
GNRs	Gold nanorods
GNSs	Gold nanospheres

DOI: 10.1201/9781032621135-17

GQDs	Graphene quantum dots
GSH-Px	Glutathione peroxidase
GSK-3	Glycogen synthesis kinase 3 beta
MDA	Malondialdehyde
MONs	Metal oxidase nanoparticles, or
MPN	Metal-phenolic network
mRNA	Ribonucleic acid
NFkB	Nuclear factor kappa B
NRF-1	Proteins nuclear factor-1
PB	Prussian blue
p-CREB	cAMP-response element binding protein
Pd	Palladium
PEG	Polyethylene glycol
PI3K	Phosphatidylinositol 3-kinase
POD	Peroxidase
Pt	Platinum
PTCN	Prussian blue nano-formulation
QDs	Quantum dots
ROS	Reactive oxygen species
RSV-Se NPs	Resveratrol-selenium nanoparticles
SEF	Surface enhancement fluorescence
Se NPs	Selenium nanoparticles
SERS	Surface-enhanced Raman Scattering
SIRT1	Sirtuin-1
SOD	Superoxide dismutase
SP	Senile plaques
SPIONs	Supraparamagnetic noncarriers
SPR	Surface plasmon resonance
STAT3	Transducers and transcription activation
STIM2	Stromal interaction molecule
TEM	Transmission electron microscopy
TFAM	Mitochondrial transcription factor A
TGN-Res SeNPs	Selenium peptide nanocomposite
Zn	Zinc
ZnO NPs	Zinc oxide nanoparticles

17.1 INTRODUCTION

Alzheimer's disease (AD) is a degenerative condition that is primarily associated with age, in which the brain deteriorates due to the accumulation of plaque in the hippocampus (Rao et al., 2022). A significant number of amyloid beta oligomers (AβOs) are produced, which causes the disease; consequently, detecting the presence of this protein as early as feasible can be beneficial (Murphy and LeVine, 2010). As a result, many efforts have been undertaken, but finding a simple, specific, and cost-effective method for sensitizing AβOs has been problematic (Rostagno et al., 2022). Various symptoms that occur due to neurodegeneration are memory loss, lexical access loss, and judgment impairment (McDade, 2022).

It mostly affects people between the ages of 60 and 85. Obesity, diabetes, cardiovascular disease, and mutations in genes explain the early onset familial type of AD, whereas the late onset sporadic form is caused by the inherited €4 allele of apolipoprotein E, which is located on 19q13 (Tarawneh and Holtzman, 2012). Pollutants in the environment are a key cause of the disease. They are breathed as metals and metal oxides and deposited in the lungs and brain (Siddiqi et al., 2018). There has been a lot of attention paid to dementia because it is a leading cause of death for millions of individuals (Shui et al., 2018). According to the World Health Organization (WHO), AD has resulted in an increased societal burden of USD 1.9 trillion by 2030 (Kim et al., 2020).

Nanomaterials, which are more sensitive and effective in their application, are developed to act as substrates for sensors such as fluorescent aptasensors (Swathi and Sebastian, 2008). AD affected more than 50 million people in 2019 (Khan NH et al., 2021). Amyloid beta (Aβ) is the disease's defining feature. Traditional therapy techniques, such as the use of acetylcholinesterase (AchE) inhibitors, have failed due to numerous limitations, such as solubility concerns, low bioavailability, and difficulties in transporting medications over the blood–brain barrier (Passeri et al., 2022). Memory loss and issues with cognitive conduct are common indicators of AD (Abbas, 2021). Nanotechnology is one of the most advanced scientific approaches for treating and preventing the progression of AD (Fonseca-Santos et al., 2015). Nanoparticles are colloidal and have an active ingredient embedded in them with controlled release or site-specific drug delivery for biological systems (Karthivashan et al., 2018). Neuro-nanotechnology is the use of nanoparticles with properties such as size, shape, increased solubility, biodegradability, surface area, and deep penetration through biological barriers to treating central nervous system disorders in which damaged neurons are treated without affecting healthy neurons (Figure 17.1) (Farheen et al., 2021; Salem,

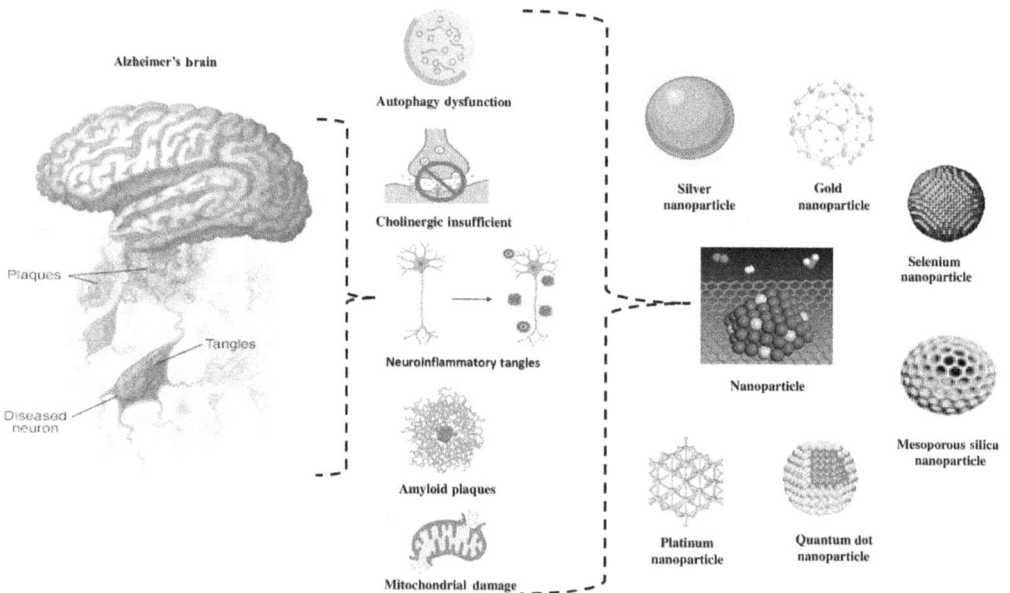

Figure 17.1 Demonstrates the influence of different nanoparticles on different biomarkers in Alzheimer's disease.

2023). Nanotechnology is the engineering of materials or devices with the smallest functional organization on a nanoscale scale (1–100 nm); these materials are composed of many substances such as polymers, lipids, viruses, and organometallic compounds (Gupta et al., 2019; Sofias and Lammers, 2023). Nanotherapeutics and nanodiagnostics have been shown to have an effect on humans and are particular to organs, as well as the circulatory system and brain (Chiang et al., 2020).

Nanoparticles are classified into three kinds based on their size: carbon-based nanoparticles (carbon nanotubes), inorganic nanoparticles (metal oxidase nanoparticles or MONs), and organic nanoparticles (De Matteis and Rinaldi, 2018). Metal nanoparticles are made up of gold, (Au NPs), aluminum, copper oxide, titanium dioxide, zinc oxide, iron oxide, and aluminum oxide (Kotrange et al., 2021). By focusing on protein aggregation, nanotherapeutics pave the path for target-based therapy. Protein-capped metal nanoparticles, such as iron oxide and cadmium sulfide PC-nanoparticles, have been created for this purpose, and they influence tau polymerization efficiency (Sonawane SK et al., 2019). Metal-containing nanoparticles are employed in the diagnosis, monitoring, and therapy of central nervous system pathology, such as AD (Chang et al., 2021). The use of nanoparticles has the advantage of having great penetration qualities into the cell membrane, high strength performance, and low cost. One of the characteristics of these nanoparticles is that they execute identical functions to any biological enzyme found in the human body; thus, they are referred to as nano mimetic enzymes or nanozymes (Gao et al., 2007; Kanaoujiya et al., 2023). Nanostructures and nanomedicine drug delivery systems show effective therapeutic potential in neurodegenerative illnesses; for example, redox nanoparticles have various applications in nanomedicines. It generates nanosize, surface hydrophobicity, biocompatibility, long-term impacts, and low-toxicity actions. They can enter changed cells (for example, cancer cells) and traverse the blood–brain barrier. In addition to these characteristics, redox nanoparticles have antioxidant and anti-inflammatory capabilities (Sadowska-Bartosz and Bartosz, 2018; Magne et al., 2023). When nanotechnology or nanoparticles are utilized to estimate AD, they have a 90% accuracy rate. To detect soluble alpha-synuclein, DNA aptamers based on aptasensors are created; these aptasensors are based on gold nanoparticles with colorimeters, surface plasmon resonance, and electrochemical impedance detection (Negahdary and Heli, 2019).

17.2 ROLE OF METAL NANOCOMPOSITES IN NEURODEGENERATION: FROM INCEPTION TO TREATMENT

The brain is a specialized structure that requires metal ions for various cellular processes to occur. Normally, it has a higher concentration of transitional metals such as iron, zinc, and copper that perform their function to generate neural activity in-between synapses, particularly that of Zn (II), and to maintain a homeostatic balance between different metalloproteins (cytochrome C-oxidase, Cu/Zn superoxide dismutase), so the cell has developed this sophisticated process (Budimir, 2011). AD affects millions of individuals globally, and the number is growing. There is currently no definitive therapy for this illness, but nanomedicine could make a significant contribution by utilizing multifunctionalized nanodevices as drug transporters with added features of brain targeting. Nanomedicine may potentially be a viable option for the pharmaceutical industry, which has transitioned from small molecular weight medicines to bigger biologicals, including antibodies and nucleotides, as the next generation of therapeutics, posing the difficulty of effective drug delivery (Formicola et al., 2019). The development of AD is regulated by reactive oxygen species (ROS), which are significant signaling molecules. Both AD imaging and treatment have benefited from the discovery of numerous useful antioxidant nanomedicines with ROS-scavenging properties (Hasan et al., 2023).

The complicated pathogenic mechanism driving increased AchE activity, Aβ accumulation, and tau hyperphosphorylation in the brain are the major markers of AD disease. There are treatment recommendations in the form of AchE inhibitors, as well as anti-amyloid and anti-tau therapies, which are being researched at a rapid speed (Beata et al., 2023). However, pharmacological restrictions limit the extent of the therapeutic outcome. Biostability and bioavailability of anti-AD medicines are reduced by their susceptibility to peripheral metabolism and fast elimination, as well as their failure to cross the blood–brain barrier (BBB) and reach the target brain region. The nanovesicle technology has developed as a means of preserving the therapeutic efficacy of anti-AD medications while also promoting AD treatment (Roy et al., 2022). The discovery of novel practical therapeutic modalities for the treatment of AD is pursued earnestly by researchers.

There are several reasons why we haven't been able to develop effective treatments for AD, including the complexity of the human brain. Furthermore, it is acknowledged that AD is multisystemic, which presents several difficulties for future treatment of these diseases. The inability to transmit pharmaceuticals to the brain effectively due to the multiple protective barriers surrounding the CNS, such as the BBB, is another important contributing factor to the lack of effective therapies and drug delivery methods for AD (Fazil et al., 2012). Nanoparticles are one of the primary criteria used for medications to cross the BBB, with the transcytosis mechanism passing through endothelial cells via an adsorption pathway that mostly utilizes positively charged nanoparticles (Qu et al., 2014). A receptor-mediated method is also used, which is based on the decorating of nanoparticles with ligands for receptors found on the luminal plasma membrane of BBB endothelial cells (Shao et al., 2015). Many receptors and transporters, such as glucose transporter 1 (GLUT-1) and transferrin receptors, are ligand targets. Crossing the BBB can also be assisted by prolonging the circulation duration of nanoparticles, allowing them to be internalized and quickly operate on the target (Kang et al., 2016). Nanoparticles, which have dimensions measured in billionths of a meter, have paved the door for possible medical applications. Inorganic nanomaterials (metals, metal oxides, quantum dots, and ceramics), organic nanomaterials (liposomes and polymers), and carbon-based nanomaterials are the three categories of nanoparticles (Sim et al., 2020). Nanoparticles with the addition of polyethylene glycol (PEG) offer several advantages, including protection from enzyme degradation, opsonization, and phagocytosis by immune cells (Suk et al., 2016; Liu et al., 2018; de la Torre and Ceña, 2018). Amyloid beta has a high affinity for Cu and Zn, and amyloid precursor proteins (APP) have this affinity as well, becoming connected to the N-terminal metal-binding region (Atwood et al., 2000). In AD, the equilibrium of different metalloproteins such as Cu and Zn is disrupted (Cuajungco et al., 2000). The function of both apolipoprotein E and α2m known risk factors in AD are known to be modified by metals (Liu et al., 2013; White et al., 2006). Increased Zn, Fe, Cu, and Al localization within senile plaques (SP) exacerbates Aβ-mediated oxidative damage and functions as a catalyst for Aβ aggregation in AD. Thus, chelation therapy, which disrupts abnormal metal–peptide interactions, has great potential as a rational therapeutic option against Alzheimer's amyloid etiology (Bandyopadhyay et al., 2010). The presence of conjugated amyloid fibrils before cell exposure enhances the development of new and efficient techniques for transporting metal nanoparticles into dendritic tissues. These kinds of discoveries have been validated on commercial gold nanoparticles, where amyloid fibrils, with colloidal gold as a reference, serve as a helpful control model. A similar pattern of behavior is observed in synthetic gold, silver, and palladium nanoparticles produced by chemical reduction in the presence of amyloid fibrils. Amyloid fibrils function as transport shuttles in diverse cells, opening a novel pathway for tailored drug delivery systems (Bolisetty et al., 2014).

Traditional therapy has encountered several difficulties. Drug transportation can be accomplished using invasive, non-invasive, and other methods, albeit the BBB may impede its efficacy (Upadhyay, 2014). Metals have distinct physical and chemical properties; electrons are moved in our biological system via redox processes that take place in metal catalytic centers of enzymes such as iron, copper, and magnesia. Metals help to keep an electrical gradient across the cell membrane (Anthony et al., 2020). Metal nanoparticles have a variety of biomedical applications. These particle sizes can be regulated from a few nanometers to tens of nanometers, putting them at diameters that are smaller or comparable to the dimensions of a cell (10–100 nm), a virus (20–450 nm), a protein (5–50 nm), or a gene (2 nm wide and 10–100 nm long). Magnetic metal nanoparticles are modulated by an external magnetic field gradient. This inherent penetrability of magnetic NPs into human tissue opens up a wide range of applications involving the transport and/or immobilization (Kogan et al., 2007). Metal NPs can move inside a human biological system through the respiratory system, gastrointestinal tract, circulatory system, and central nervous system (Crisponi et al., 2017; Xuan et al., 2020). Metal NPs or metal nanocomposites (MNCs) are widely used nanomaterials due to their superior conductivity and biocompatibility, which play an essential role in the fabrication of nanocomposites (Díez-Pascual, 2022).

Platinum (Pt), silver (Ag), palladium (Pd), copper (Cu), and gold (Au) are examples of metal nanoparticles that have a unique three-dimensional structure with nanometer size (2–100 nm), minimal polydispersity, and high surface functionality and flexibility (Razzino et al., 2020). With the help of plant phytochemicals such as alkaloids, polysaccharides, amino acids, vitamins, and terpenoids, metal ions or metal oxides are converted into zerovalent metal nanoparticles (Agarwal et al., 2017). Several plants, including *Vitex negundo, Plectranthus amboinicus, and Anisochilus carnosus*, have been studied and discovered to create nanoparticles in a variety of shapes, including rod-shaped, hexagonal, quasispherical, and spherical with agglomeration. The size of NPs decreases as plant content increases. Amides, carbonate, amines, and carboxylic acids are effective capping agents that can be used to investigate a variety of disorders. Antioxidant, anti-Alzheimer, and antimicrobial applications are among the many biomedical applications (Jan et al., 2021). Metal ions have an important role in brain pathology by causing the creation, breakdown, and clearance of amyloid plaques. Zinc, copper, iron, and many other metal compounds have been found in significant concentrations in AD. Protein and peptide aggregation, which has been linked to Aβ-related disorders, is being studied using NPs. Supramagnetic NPs destroy amyloid fibrils (Girigoswami et al., 2019). In target drug delivery systems, metallic nanocarriers such as gold, silver, and iron oxide are employed. Superparamagnetic nanocarriers (SPIONs) with diameters of 10–100 nm, such as hematite, maghemite, and magnetite, are magnetic nanocarriers formed of iron oxide that exhibit super magnetism. Silver and gold-based nanocarriers have good optical properties, are easy to synthesize, and are innocuous and poisonous (Gopalan et al., 2020).

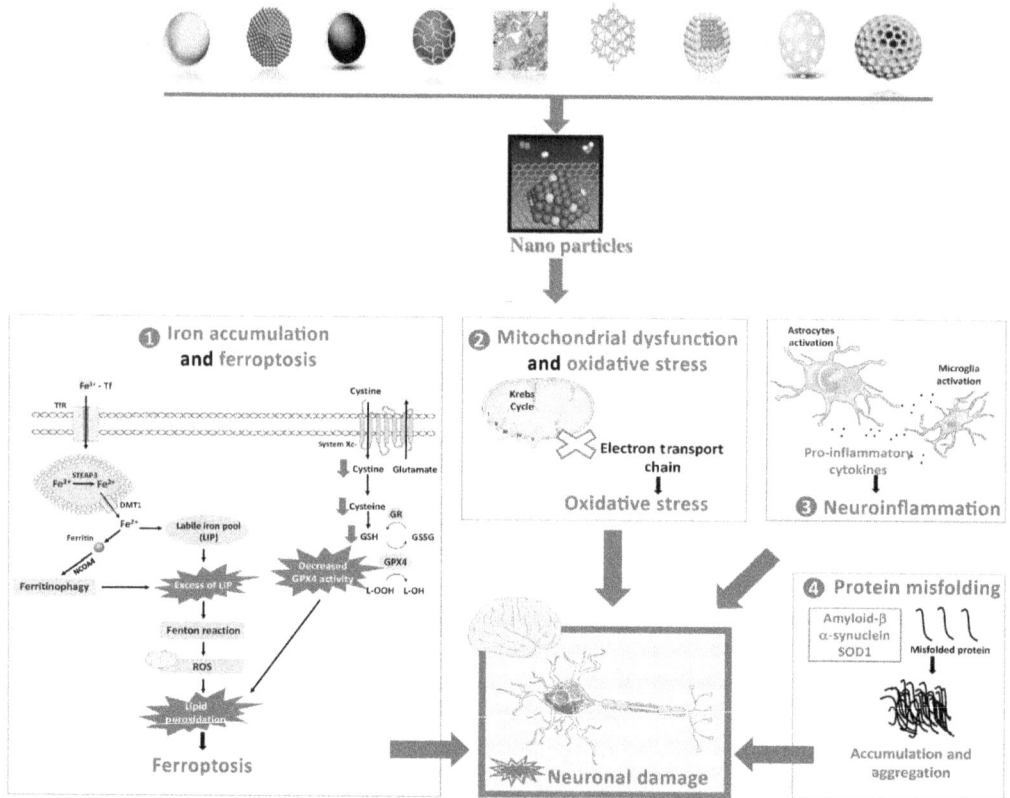

Figure 17.2 Explains the several mechanisms by which metal nanoparticles work to prevent Alzheimer's disease.

17.3 METAL NANOCOMPOSITES UTILIZED IN THE DETECTION AND TREATMENT OF ALZHEIMER'S DISEASE

17.3.1 Prussian blue nanoparticles

Prussian blue (PB) nanoparticles are excellent as that have many advantages such as easy preparation, good stability, biocompatibility, and low cost (Chen et al., 2022). Prussian blue (PB) or ferric ferrocyanide is approved by FDA and acts as an anti-inflammatory, chemotherapeutic, molecular imaging, and biomedical safety field showing its effect by stimulating various enzymes such as SOD, CAT, and POD (Li et al., 2022). PB NPs affect amyloid aggregation by capturing Cu^{2+} ions that preserve Aβ fibrillation and hence prevent the formation of very toxic Cu^{2+}-containing oligomer species, therefore effective in AD (Kowalczyk et al., 2021). PB with CAT-like and or SOD-like activity can remove ROS that is abundant in AD (Estelrich and Busquet, 2021).

PB nano-formulation (PTCN) with dual targeting can target Aβ aggregates by demonstrating its antioxidant qualities and was able to permeate the BBB (Tolar et al., 2020). An adjustable gradient dosing strategy is implemented to extract preventive and therapeutic measures in AD based on its oxidative stress at different stages. It was found that PTCN ameliorated AD by improving cognitive decline and rescuing hippocampus atrophy of APP/PS1 mice in the trials. PTCN has a good combination of biomaterials with good biosafety that has good reach for early prevention (Zhao et al., 2021).

17.3.2 Silver nanoparticles (Ag NPs)

With their antibacterial, anti-inflammatory, and antioxidant capabilities, silver nanoparticles (Ag NPs) have proven to offer broad therapeutic potential. Huang et al. and colleagues investigated the effects of 3–5 nm Ag NPs on gene expression in murine brain ALT astrocytes, microglial BV-2 cells, and neuron N2a cells in a recent study. According to the findings of this study, Ag NPs may have anti-inflammatory and antioxidant properties that could be useful in the treatment of AD. Furthermore, the researchers discovered that Ag NPs were not harmful to brain cells and caused no significant alterations in total cell viability. These findings imply that Ag NPs could provide a novel therapeutic approach for AD (Huang et al., 2015). In comparison to traditional procedures, using plant extract to create Ag NPs has various benefits, including cost and lesser time for synthesis. Ag NPs are a form of noble metal that have applications in the treatment, diagnostics, and fighting of cancer, infectious disease, and neurological disorders (Habeeb et al., 2022).

Youssif et al. (2019) synthesized spherical Ag NPs by using an aqueous extract of *Lampranthus coccineus* and *Malephora lutea* that were further evaluated for antioxidant and cholinesterase inhibitor activity through docking and were found to be linked to human AChE, butyrylcholinesterase (BChE), and glutathione transferase receptor that have a role in neurodegeneration. The study's findings show that these nano silvers have anti-cholinesterase and antioxidant activities that provided neuroprotective effects in AD (Hassan et al., 2022). Ag NPs (50 μg/mL) have microglial phagocytic action and help in delaying the inception and movement of AD (Azeem et al., 2022).

Noninvasive and sensitive Ag nanoprobes shells (AgNGSs) were surface-enhanced Raman Scattering (SERS) colloidal nanoprobes that exhibit selectivity, sensitivity, and multiplex detection for Aβ40 and Aβ42 in the blood. This occurs by conjugating with their specific antibodies, assisting in the identification of AD biomarkers in human serum (Yang et al., 2019). Beta site secretase (BACE1) catalyzes the cleavage of APP, which leads to plaque deposition in the brain of AD patients. The development of inhibitors for the treatment of AD is aided by proper BACE1 monitoring. This is accomplished by using Ag NPs in electrochemical tests (Zhou et al., 2023).

17.3.3 Gold nanoparticles (Au NPs)

Gold nanoparticles have anti-inflammatory and antioxidant activity in the brain via reducing macrophage and microglial activation. It has been shown to influence TNF levels in the hippocampus. The results of an *in vitro* animal model of AD suggests that it reduces oxidative stress in the brain of mice by decreasing the quantity of activated caspase 3 and an inhibitor of nuclear factor kappa B (NF-kB) (Davies et al., 2021). Gold nanoparticles come in a variety of shapes and sizes, including gold nanospheres (GNSs), gold nanorods (GNRs), silica/gold nanoshells, and hollow gold nanospheres (Kulkarni et al., 2022). Au NPs have customizable size, unique optical features, flexible surface modification, good compatibility, and long circulation time (Dheyab et al., 2022).

Gold nanoparticles catalytically supply nucleation sites, and they can be used as a sensing tool due to their optical features. Au NPs-based calorimetry is a simple and inventive way for efficiently monitoring anti-Aβ chemicals. It is a low-cost, rapid-detection tool for AD (Ribeiro et al., 2022). p-CREB, brain-derived neurotrophic factor (BDNF) and amyloid beta fibrillation are all prevented by negatively charged gold nanoparticles (Au NPs), which also break up fibrils and stromal interaction molecule (STIM2) (Sanati et al., 2019). Surface electrons in gold nanoparticles vibrate in response to incident light; this phenomenon is known as surface plasmon resonance (SPR). They absorb light energy, which is converted to heat energy, which causes light scattering and thus increases the fluorescent emission of fluorescent molecules near their surface, resulting in surface enhancement fluorescence (SEF), which is having application in the detection of Aβ aggregation *in vitro*. Gold nanorods with D1 peptide have been developed for early detection of Aβ (Jara-Guajardo et al., 2020). Mimosines are anti-inflammatory, antiapoptotic, and antioxidant. Mimosine–Au NPs were observed to pass the BBB and to be neuroprotective against Aβ-induced toxicity by suppressing Aβ aggregation and activating the disassembly of mature Aβ *in vitro* mouse cortical culture neurons (Anand et al., 2021). The *in vitro* construction of an AD cell model by Wang and his colleagues revealed that there was a decrease in the expression of tau (τ), phosphorylated tau (P-τ), Aβ protein, ROS, and downregulation of apoptosis after treatment with gold nanoparticles (Au NPs). Higher concentrations of the proteins nuclear factor-1 (NRF-1), mitochondrial transcription factor A (TFAM) mRNA, and dynamic-related protein-1 (DRP1) lead to improved mitochondrial activity. By influencing the miR-21–5p/SOCS6 pathway, Au NPs regulate mitochondrial damage (Wang et al., 2023).

17.3.4 Copper nanoparticles (CuO NPs)

Copper nanoparticles are a new technology with several advantages, including low cost, strong electrical and thermal conductivity, long-term stability, and oxidation resistance. Copper (III) oxide is employed for nanoparticle production due to its particular optimum parameter (Ma et al., 2022). Copper nanoparticles (Cu NPs) can be used in their pure form; however, blended compositions provide higher functional properties (Shi et al., 2020). A chiral L/D-FexCuySe nanoparticle was created with specifically characterized peaks at 435, 515, and 780 nm (in the range 400–1,000 nm) on circular dichroism (CD) spectra, and it was discovered that chiral L/D-NPs 14 were able to interfere with amyloid fibrils or monomer plaques by triggering the Aβ42 fibril to change from a dense structure to loosen up. D-FexCuySe NPs have a much higher affinity for amyloid beta 42 fibrils than L-FexCuySe NPs, and chiral D-FexCuySe NPs have a neuroprotective effect against neuron damage in AD as demonstrated in its model through the recovery of 24 cognitive competences, implying their use in neurodegeneration treatment (Zhang et al., 2019). AD is caused by copper dysregulation. Amyloid precursor protein (APP), amyloid peptide (A), amyloid precursor protein cleavage enzyme-1, amyloid precursor protein cleavage enzyme-1 (BACE1), and presenilin secretase are metalloproteins, with APP having a Cu-binding site and acting as a Cu transporter (Vassar et al., 1999). Cu ion imbalance causes oligomerization of BACE1, which leads to neurodegeneration in Alzheimer's disease. Cu (II) complexing NPs play an important role in the loading and release of nanoparticles and have a Cu-chelating moiety on the outer surface of the shell; hence, Cu (II) can be easily encapsulated, resulting in copper release at a somewhat low pH value. Because Cu NPs are less toxic and protect Cu from biological environmental degradation, they can be utilized to enhance Cu levels in the brain (Fehse et al., 2014).

17.3.5 Selenium nanoparticles

Selenium (Se) is an essential micronutrient that is required for normal brain function (Nazıroğlu et al., 2017). Abozaid and colleagues investigated the effect of resveratrol–selenium nanoparticles (RSV–Se NPs) in the etiology of AD in rats using an animal model in which neurotoxicity is caused by 100 mg/kg/day of aluminum chloride ($AlCl_3$) (Abozaid et al., 2022). According to the findings of this study, RSV–Se NPs worked as antioxidants by lowering oxidative stress and mitochondrial dysfunction. Activation of phosphatidylinositol-3-kinase (PI3K) suppressed tau hyperphosphorylation mediated by glycogen synthesis kinase 3 beta (GSK-3). Other pathways for AD prevention included RSV-Se NPs downregulation of transducers and transcription activation (STAT3) expression and interleukin (IL-1) levels. Sirtuin-1 (SIRT1) expression and microRNA-134 reductions that cause neurite outgrowth combine to cause the anti-inflammatory action of adiponectin (AND). Cognitive dysfunction was improved by these mechanisms (Abozaid et al., 2022). AD is a neurodegenerative disease in which a study has discovered that non-occurring enantiomers of amino acids such as D-phenylaniline, D-alanine, and D-glutamate play a key role. As demonstrated in transmission electron microscopy (TEM), D-aspartate and DL-selenomethionine assemble amyloid protein and cause AD.

Selenium nanoparticles containing chitosan (Ch–Se NPs) were synthesized, which were reported to reduce Aβ42 aggregation (Ashraf et al., 2023). A very tiny Res–selenium peptide nanocomposite (TGN–Res–Se NPs) was discovered to alleviate cognitive dysfunction by interacting with Aβ and reducing its aggregation in the hippocampus, demonstrating antioxidant effects. *In vivo* investigations showed that amyloid-induced neuroinflammation is also reduced by altering the NF-κB, mitogen-activated protein kinase, and Akt signal pathways in BV-2 cells (Li et al., 2021). It also modulated gut microorganisms like *Helicobacter*, which are directly or indirectly associated with neurodegeneration, making it a possible new-age weapon against AD (Uberti et al., 2022). Selenium chondroitin sulfate NPs (CS@Se) were created, and their therapeutic impact was observed in an AD animal model *in vitro*. The study discovered that CS@Se tends to reduce Aβ aggregation and protected SH-SY5Y cells against Aβ42-induced neurotoxicity. It affected malondialdehyde (MDA) and ROS. It increased the activity of glutathione peroxidase (GSH-Px). Attenuation of tau hyperphosphorylation (Ser 396/Ser 404) aids in the regulation of GSK-3, increasing the likelihood of effective treatment in AD (Gao et al., 2020). Selenium nanoparticles (Se NPs) are low in toxicity, have a high bioavailability, and have improved bioactivity. Qiao and colleagues created *Lactobacillus casei* ATCC 393, which exhibited antioxidant and anti-inflammatory activities *in vitro* and animal models and aids in the regulation of gut microbiota (Qiao et al., 2022).

17.3.6 Zinc oxide nanoparticles (ZnO NPs)

Zinc oxide nanoparticles (ZnO NPs) are inorganic materials with many applications such as semiconductors, chemical sensors, and catalysts. They are nontoxic, biocompatible, and cheap. ZnO NPs show anti-inflammatory, antimicrobial, antioxidant activity, and bio-imaging activity (Anjum et al., 2021). Zinc-loaded nanoparticles minimize toxicity by inhibiting the expression of zinc-sensitive genes and proteins such as metallothionine and zinc transporters. ZnO NPs function quickly and deliver a substantial amount of zinc to the brain (Chhabra et al., 2015). Zinc (II) binding peptides are loaded to PEG-modified chitosan nanoparticles (NPs) to improve their stability and bioavailability and reduce Zn concentration and amyloid secretion in mouse neuroblastoma N2a cells (Ogunyemi et al., 2019). Zinc oxide nanoparticles derived from methanolic extracts of *S. blackburniana* leaves, fruits, and pollen grains have shown powerful and inhibitory action on metabolomics and hence neuroprotective function. *S. blackburniana* was used in the green synthesis of ZnO NPs with anti-Alzheimer action (El-Hawwary et al., 2021).

Andrikopoulos and colleagues developed a biocompatible metal-phenolic network (MPN) in which the polyphenol epigallocatechin gallate (EGCG) was embedded by physiological zinc (II) adsorbed on gold nanoparticles. This resulted in the development of the MPN@AuNP nanoconstruct, which demonstrated excellent anti-amyloid aggregation and *in vitro* Aβ toxicity. MPN@AuNP demonstrated numerous features, including antioxidant activity, a strong ability to cross the BBB, and anti-amyloidosis (Andrikopoulos et al., 2023). A green synthesis of ZnO NPs was performed in which ibuprofen (IBP) was adsorbed on ZnO NPs, and it was discovered that when used at high concentrations, they showed antioxidant and anti-inflammatory activity by suppressing cyclooxygenase, and prominent anti-Alzheimer's activity was visible as they specifically suppress acetylcholinesterase (AChE) and butyl cholinesterase (BChE) (Alibrahim et al., 2023).

17.3.7 Quantum dots

Quantum dots (QDs) are nanoparticles that can be used to treat diseases like AD. QDs are distinguished by their unique optical characteristics and high sensitivity (Villalva et al., 2021). The application of QDs has progressed from sensing and tagging to serving as a molecular actuator disrupting disease protein pathogenesis. They serve as nanoprobes for AD diagnostics (Matea et al., 2017). To prevent amyloid protein aggregation and buildup, graphene QDs interfere with its hydrophobic interaction (Ning et al., 2022). N-acetyl-l-cysteine-capped cadmium telluride quantum dots (CdTeQDs) demonstrate amyloid plaque degradation (Sikorska et al., 2020). QDs have a three-dimensional structure and operate on the phenomenon of light excitability. QDs conjugated with biological molecules such as antibiotics or biomarkers target specific cells or proteins responsible for disease pathogenesis; they are also used in optogenetics due to their surface chemistry control, stability, sensitivity, biocompatibility, and optical property (Zamaleeva et al., 2015; Rosenthal et al., 2011). A QD has been shown to specifically target microglial cells (Minami et al., 2012). Tang and his colleagues developed a sandwich immunoassay for detecting Aβ42 in human cerebrospinal fluid samples by employing QDs as fluorescent labels, which was reported to be faster and more efficient in detecting Aβ in human cerebrospinal fluid samples (Tang et al., 2018). For green synthesis in the treatment of Alzheimer's disease, graphene quantum dots (GQDs) were synthesized from the flower of *Clitoria ternatea* with one-pot microwave synthesis. CT-GODs were found to inhibit the acetylcholinesterase enzyme, boost glutathione and protein levels, and decrease nitric oxide and lipid peroxide levels. In rodents, combined methods will alleviate the symptoms of AD (Tak et al., 2020; Yang et al., 2010).

Table 17.1 Different types of nanocomposites and their distinct impacts on AD therapy and diagnostics

Type of metal nanoparticle	Example	Target	References
Protein capped	PC-Fe$_3$O$_4$ and PC-CdS	Tau aggregation	Sonawane et al., 2019
Chelators	Fe	Apolipoprotein	Liu et al., 2018
Biogenic metallic nanoparticles	Tet1-Se@EGCH nanosystem Selenium nanoparticles with Epigallocatechin Gallate (EGCG) coated with Tet1 protein	Inhibit fibrillation process of Amyloid beta	Ovais et al., 2018
Plasmonic metal nanoparticles	Gold, silver, platinum (Au NPs, Ag NPs, Pt NPs)	Amyloid beta, apolipoprotein	Oyarzún et al., 2021
Magnetic nanoparticles	Iron oxide magnetic nanoparticles maghemite γ-Fe$_2$O$_3$,	Reactive oxygen species	Shubayev et al., 2009
	Congo red/Rutin-magnetic NPs	Specifically detect amyloid plaque	Hu et al., 2015
Green synthesis metal nanoparticles	Gold and silver nanoparticle synthesis with the help of plants such as *Hypoxis hemerocallidea*	Anti-inflammatory	Aboyewa et al., 2021
	Graphene and monoporous silica hybrid nanomaterial Graphene@mesoporous silica hybrid (GSHs)	Specifically detect mutated apolipoprotein E gene. This has high reproducibility, reliability, and accuracy	Wu et al., 2016
Immunosensor	Gold nanoparticles with functional chitosan aligned carbon nanotubes (CS-ACNT-Au NPs)	Highly efficient for detecting amyloid beta	Ranjan and Khan, 2022

Table 17.2 Different types of metallic nanoparticles and their application on Alzheimer disease

Metallic nanoparticles	Applications in Alzheimer	References
Gold nanoparticles	• Biomedical imaging	Cabuzu et al., 2015
Supra paramagnetic iron oxide (SPIONs)	• Diagnosis and treatment in AD	Amiri et al., 2013
Gold exosomes	• Targeted drug delivery in Alzheimer	Perets et al., 2019
PLGA-functioning quercetin nanoparticles PLGA@NPs	• Increases therapeutic index and reduces side effects	Sun et al., 2016
Silver nanoparticles (Ag NP)	• Investigate specific biomarker amyloid	Soria et al, 2015
γFe$_2$O$_3$@Au core shell magneto-plasmonic nanoparticle	• Improve detection sensitivity of Tau protein	Chen et al., 2022
Bifunctional Au@Pt/Au core @shell nanoparticles	• Detection of altered P^{53} peptide in Alzheimer	Iglesias-Mayor et al., 2020
Transmemberane peptide- Chondroitin sulphate gold nanoparticle (TAT-CS@ Au NPs)	• Inhibit amyloid β accumulation, decreases oxidative stress • Inhibit Tau phosphorylation • Downregulate inflammatory mediators	Feng et al., 2023
Acetylcholinesterase (AChE) as organic component and copper phosphate AChE-Cu$_3$(PO$_4$)$_2$ HNFs nanoparticles	• Direct measurement of acetylcholinesterase	Yang et al., 2022

17.4 CONCLUSION

AD is associated with the abnormal self-assembly of amyloid-β (Aβ) peptides into toxic β-rich aggregates. Experimental studies have shown that hydrophobic nanoparticles retard Aβ fibrillization by slowing the nucleation process. However, the impact of nanoparticles on Aβ oligomeric structure remains elusive. Metal nanoparticles are valuable resource for those researching gold-, silver-, and iron-based drug delivery systems for controlled and targeted delivery of potential drugs and genes for improved clinical efficacy. These fields include biomaterials, nanomedicine, and pharmaceutical sciences. Due to its high solubility, efficiency, relatively low cost of nanoscale products, and ability to improve patient comfort, nanotechnology is widely used in drug delivery. Neurological illnesses will continue to be a therapeutic concern in the future, and biomaterials offer a viable treatment option. Biomaterials, as previously noted, can be made up of both natural and synthetic chemicals. These biomaterials employed in the experiments described earlier have a variety of properties, they are all biocompatible, bifunctional, bioinert, biodegradable, and sterilizable. Nanoparticles have been identified as promising vectors for the delivery of a variety of active and protective compounds for the treatment of neurodegeneration. Super absorbent polymers (SAPs,) nanofibers, and carbon-based nanomaterials, on the other hand, have been reported to be beneficial in the restoration of nerve damage. Both SAPs and nanofibers have a fiber-like structure that allows for proper axon guiding, promoting, and supporting axon development and hence show significant role in defying AD and therefore combined with metal nanoparticles.

REFERENCES

Abbas M. Potential role of nanoparticles in treating the accumulation of Amyloid-Beta Peptide in Alzheimer's patients. *Polymers (Basel)*. 2021 Mar 27;13(7):1051. doi: 10.3390/polym13071051

Aboyewa JA, Sibuyi NRS, Meyer M, Oguntibeju OO. Green synthesis of metallic nanoparticles using some selected medicinal plants from Southern Africa and their biological applications. *Plants (Basel)*. 2021 Sep 16;10(9):1929. doi: 10.3390/plants10091929.

Abozaid OAR, Sallam MW, El-Sonbaty S, Aziza S, Emad B, Ahmed ESA. Resveratrol-Selenium nanoparticles alleviate neuroinflammation and neurotoxicity in a rat model of ADby regulating Sirt1/miRNA-134/GSK3β expression. *Biol Trace Elem Res*. 2022 Dec;200(12):5104–5114. doi: 10.1007/s12011-021-03073-7.

Agarwal, Happy, S. Venkat Kumar, and S. Rajeshkumar. A review on green synthesis of zinc oxide nanoparticles: An eco-friendly approach. *Resource-Efficient Technologies*. 2017;3(4):406–413.

Alibrahim KA. Adsorption of Ibuprofen as a pharmaceutical pollutant from aqueous phase using zinc oxide nanoparticles: Green synthesis, batch adsorption and biological activities. *J Mol Recognit*. 2023 Apr 6:e3015. doi: 10.1002/jmr.3015.

Amiri H, Saeidi K, Borhani P, Manafirad A, Ghavami M, Zerbi V. Alzheimer's disease: Pathophysiology and applications of magnetic nanoparticles as MRI theranostic agents. *ACS Chem Neurosci*. 2013 Nov 20;4(11):1417–1429. doi: 10.1021/cn4001582.

Anand BG, Wu Q, Karthivashan G, Shejale KP, Amidian S, Wille H, Kar S. Mimosine functionalized gold nanoparticles (Mimo-AuNPs) suppress β-amyloid aggregation and neuronal toxicity. *Bioact Mater*. 2021 May 10;6(12):4491–4505. doi: 10.1016/j.bioactmat.2021.04.029.

Andrikopoulos N, Li Y, Nandakumar A, Quinn JF, Davis TP, Ding F, Saikia N, Ke PC. Zinc-Epigallo-catechin-3-gallate Network-coated nanocomposites against the pathogenesis of Amyloid-Beta. *ACS Appl Mater Interfaces*. 2023 Feb 15;15(6):7777–7792. doi: 10.1021/acsami.2c20334.

Anjum S, Hashim M, Malik SA, Khan M, Lorenzo JM, Abbasi BH, Hano C. Recent advances in Zinc Oxide Nanoparticles (ZnO NPs) for cancer diagnosis, target drug delivery, and treatment. *Cancers (Basel)*. 2021 Sep 12;13(18):4570. doi: 10.3390/cancers13184570.

Anthony EJ, Bolitho EM, Bridgewater HE, Carter OWL, Donnelly JM, Imberti C, Lant EC, Lermyte F, Needham RJ, Palau M, Sadler PJ, Shi H, Wang FX, Zhang WY, Zhang Z. Metallodrugs are unique: Opportunities and challenges of discovery and development. *Chem Sci*. 2020 Nov 12;11(48):12888–12917. doi: 10.1039/d0sc04082g.

Ashraf, Hajra, Davide Cossu, Stefano Ruberto, Marta Noli, Seyedesomaye Jasemi, Elena Rita Simula, and Leonardo A. Sechi. Latent potential of multifunctional selenium nanoparticles in neurological diseases and altered gut microbiota. *Materials*. 2023;16(2):699. doi:10.3390/ma16020699.

Atwood CS, Scarpa RC, Huang X, Moir RD, Jones WD, Fairlie DP, Tanzi RE, Bush AI. Characterization of copper interactions with Alzheimer amyloid beta peptides: Identification of an attomolar-affinity copper binding site on amyloid beta1–42. *J Neurochem*. 2000 Sep;75(3):1219–1233. doi: 10.1046/j.1471-4159.2000.0751219.x.

Azeem MNA, Ahmed OM, Shaban M, Elsayed KNM. In vitro antioxidant, anticancer, anti-inflammatory, anti-diabetic and anti-Alzheimer potentials of innovative macroalgae bio-capped silver nanoparticles. *Environ Sci Pollut Res Int*. 2022 Aug;29(39):59930–59947. doi: 10.1007/s11356-022-20039-x. Epub 2022 Apr 9.

Bai C, Tang M. Toxicological study of metal and metal oxide nanoparticles in zebrafish. *J Appl Toxicol*. 2020 Jan;40(1):37–63. doi: 10.1002/jat.3910.

Bandyopadhyay S, Huang X, Lahiri DK, Rogers JT. Novel drug targets based on metallobiology of Alzheimer's disease. *Expert Opin Ther Targets*. 2010 Nov;14(11):1177–1197. doi: 10.1517/14728222.2010.525352.

Beata BK, Wojciech J, Johannes K, Piotr L, Barbara M. Alzheimer's disease-biochemical and psychological background for diagnosis and treatment. *Int J Mol Sci*. 2023 Jan 5;24(2):1059. doi: 10.3390/ijms24021059.

Bolisetty S, Boddupalli CS, Handschin S, Chaitanya K, Adamcik J, Saito Y, Manz MG, Mezzenga R. Amyloid fibrils enhance transport of metal nanoparticles in living cells and induced cytotoxicity. *Biomacromolecules*. 2014 Jul 14;15(7):2793–2799. doi: 10.1021/bm500647.

Budimir A. Metal ions, AD and chelation therapy. *Acta Pharm*. 2011 Mar;61(1):1–14. doi: 10.2478/v10007-011-0006-6.

Cabuzu D, Cirja A, Puiu R, Grumezescu AM. Biomedical applications of gold nanoparticles. *Curr Top Med Chem*. 2015;15(16):1605–1613. doi: 10.2174/1568026615666150414144750.

Chang X, Li J, Niu S, Xue Y, Tang M. Neurotoxicity of metal-containing nanoparticles and implications in glial cells. *J Appl Toxicol*. 2021 Jan;41(1):65–81. doi: 10.1002/jat.4037.

Chen KL, Tsai PH, Lin CW, Chen JM, Lin YJ, Kumar P, Jeng CC, Wu CH, Wang LM, Tsao HM. Sensitivity enhancement of magneto-optical Faraday effect immunoassay method based on biofunctionalized γ-Fe_2O_3@Au core-shell magneto-plasmonic nanoparticles for the blood detection of Alzheimer's disease. *Nanomedicine*. 2022 Nov;46:102601. doi: 10.1016/j.nano.2022.102601.

Chhabra R, Ruozi B, Vilella A, Belletti D, Mangus K, Pfaender S, Sarowar T, Boeckers TM, Zoli M, Forni F, Vandelli MA, Tosi G, Grabrucker AM. Application of polymeric nanoparticles for CNS targeted Zinc delivery in vivo. *CNS Neurol Disord Drug Targets*. 2015;14(8):1041–1053. doi: 10.2174/1871527314666150821111455.

Chiang MC, Nicol CJB, Cheng YC, Yen C, Lin CH, Chen SJ, Huang RN. Nanogold neuroprotection in human neural stem cells against amyloid-beta-induced mitochondrial dysfunction. *Neuroscience*. 2020 May 21;435:44–57. doi: 10.1016/j.neuroscience.2020.03.040.

Crisponi G, Nurchi VM, Lachowicz JI, Peana M, Medici S, Zoroddu MA. Toxicity of nanoparticles: Etiology and mechanisms. In *Antimicrobial Nanoarchitectonics*; Elsevier: Amsterdam, The Netherlands, 2017; pp. 511–546.

Cuajungco MP, Fagét KY, Huang X, Tanzi RE, Bush AI. Metal chelation as a potential therapy for Alzheimer's disease. *Ann N Y Acad Sci.* 2000;920:292–304. doi: 10.1111/j.1749-6632.2000. tb06938.x.

Davies DA, Adlimoghaddam A, Albensi BC. Role of Nrf2 in synaptic plasticity and memory in Alzheimer's disease. *Cells.* 2021 Jul 25;10(8):1884. doi: 10.3390/cells10081884.

de la Torre C, Ceña V. The delivery challenge in neurodegenerative disorders: The nanoparticles role in ADTherapeutics and diagnostics. *Pharmaceutics.* 2018 Oct 17;10(4):190. doi: 10.3390/ pharmaceutics10040190.

De Matteis V, Rinaldi R. Toxicity assessment in the nanoparticle era. *Adv Exp Med Biol.* 2018;1048: 1–19. doi: 10.1007/978-3-319-72041-8_1.

Dheyab MA, Aziz AA, Moradi Khaniabadi P, Jameel MS, Oladzadabbasabadi N, Mohammed SA, Abdullah RS, Mehrdel B. Monodisperse gold nanoparticles: A review on synthesis and their application in modern medicine. *Int J Mol Sci.* 2022 Jul 2;23(13):7400. doi: 10.3390/ijms23137400.

Díez-Pascual AM. PMMA-based nanocomposites for odontology applications: A state-of-the-art. *Int J Mol Sci.* 2022 Sep 7;23(18):10288. doi: 10.3390/ijms231810288.

El-Hawwary SS, Abd Almaksoud HM, Saber FR, Elimam H, Sayed AM, El Raey MA, Abdelmohsen UR. Green-synthesized zinc oxide nanoparticles, anti-Alzheimer potential and the metabolic profiling of *Sabal blackburniana* grown in Egypt supported by molecular modelling. *RSC Adv.* 2021 May 18;11(29):18009–18025. doi: 10.1039/d1ra01725j.

Estelrich J, Busquets MA. Prussian Blue: A nanozyme with versatile catalytic properties. *Int J Mol Sci.* 2021 Jun 1;22(11):5993. doi: 10.3390/ijms22115993. PMID: 34206067; PMCID: PMC8198601.

Farheen, Khan MA, Ashraf GM, Bilgrami AL, Rizvi MMA. New horizons in the treatment of neurological disorders with tailorable gold nanoparticles. *Curr Drug Metab.* 2021;22(12):931–938. doi: 10.2174/1389200222666210525123416.

Fazil M, Shadab, Baboota S, Sahni JK, Ali J. Nanotherapeutics for AD(AD): Past, present and future. *J Drug Target.* 2012 Feb;20(2):97–113. doi: 10.3109/1061186X.2011.607499. Epub 2011 Oct 25. PMID: 22023651.

Fehse S, Nowag S, Quadir M, Kim KS, Haag R, Multhaup G. Copper transport mediated by nanocarrier systems in a blood-brain barrier in vitro model. *Biomacromolecules.* 2014 May 12;15(5):1910–1919. doi: 10.1021/bm500400k.

Feng Y, Li X, Ji D, Tian J, Peng Q, Shen Y, Xiao Y. Functionalised penetrating peptide-chondroitin sulphate-gold nanoparticles: Synthesis, characterization, and applications as an anti-ADdrug. *Int J Biol Macromol.* 2023 Mar 1;230:123125. doi: 10.1016/j.ijbiomac.2022.123125.

Fonseca-Santos B, Gremião MP, Chorilli M. Nanotechnology-based drug delivery systems for the treatment of Alzheimer's disease. *Int J Nanomedicine.* 2015 Aug 4;10:4981–5003. doi: 10.2147/ IJN.S87148.

Formicola B, Cox A, Dal Magro R, Masserini M, Re F. Nanomedicine for the treatment of Alzheimer's disease. *J Biomed Nanotechnol.* 2019 Oct 1;15(10):1997–2024. doi: 10.1166/ jbn.2019.2837.

Gao F, Zhao J, Liu P, Ji D, Zhang L, Zhang M, Li Y, Xiao Y. Preparation and in vitro evaluation of multi-target-directed selenium-chondroitin sulfate nanoparticles in protecting against the Alzheimer's disease. *Int J Biol Macromol.* 2020 Jan 1;142:265–276. doi: 10.1016/j. ijbiomac.2019.09.098.

Gao L, Zhuang J, Nie L, Zhang J, Zhang Y, Gu N, Wang T, Feng J, Yang D, Perrett S, Yan X. Intrinsic peroxidase-like activity of ferromagnetic nanoparticles. *Nat Nanotechnol.* 2007 Sep;2(9):577–583. doi: 10.1038/nnano.2007.260.

Girigoswami A, Ramalakshmi M, Akhtar N, Metkar SK, Girigoswami K. ZnO Nanoflower petals mediated amyloid degradation: An in vitro electrokinetic potential approach. *Mater Sci Eng C Mater Biol Appl.* 2019 Aug;101:169–178. doi: 10.1016/j.msec.2019.03.086.

Gopalan D, Pandey A, Udupa N, Mutalik S. Receptor specific, stimuli responsive and subcellular targeted approaches for effective therapy of Alzheimer: Role of surface engineered nanocarriers. *J Control Release.* 2020 Mar 10;319:183–200. doi: 10.1016/j.jconrel.2019.12.034.

Gupta J, Fatima MT, Islam Z, Khan RH, Uversky VN, Salahuddin P. Nanoparticle formulations in the diagnosis and therapy of Alzheimer's disease. *Int J Biol Macromol.* 2019 Jun 1;130:515–526. doi: 10.1016/j.ijbiomac.2019.02.156.

Habeeb Rahuman HB, Dhandapani R, Narayanan S, Palanivel V, Paramasivam R, Subbarayalu R, Thangavelu S, Muthupandian S. Medicinal plants mediated the green synthesis of silver nanoparticles and their biomedical applications. *IET Nanobiotechnol.* 2022 Jun;16(4):115–144. doi: 10.1049/nbt2.12078.

Hasan I, Guo B, Zhang J, Chang C. Advances in antioxidant nanomedicines for imaging and therapy of Alzheimer's disease. *Antioxid Redox Signal.* 2023 Feb 23. doi: 10.1089/ars.2022.0107.

Hassan NA, Alshamari AK, Hassan AA, Elharrif MG, Alhajri AM, Sattam M, Khattab RR. Advances on therapeutic strategies for Alzheimer's disease: From medicinal plant to nanotechnology. *Molecules.* 2022 Jul 28;27(15):4839. doi: 10.3390/molecules27154839.

Hu B, Dai F, Fan Z, Ma G, Tang Q, Zhang X. Nanotheranostics: Congo Red/Rutin-MNPs with enhanced magnetic resonance imaging and H_2O_2-responsive therapy of AD in APPswe/PS1dE9 transgenic mice. *Adv Mater.* 2015 Oct 7;27(37):5499–5505. doi: 10.1002/adma.201502227.

Huang CL, Hsiao IL, Lin HC, Wang CF, Huang YJ, Chuang CY. Silver nanoparticles affect on gene expression of inflammatory and neurodegenerative responses in mouse brain neural cells. *Environ Res.* 2015 Jan;136:253–263. doi: 10.1016/j.envres.2014.11.006.

Iglesias-Mayor A, Amor-Gutiérrez O, Novelli A, Fernández-Sánchez MT, Costa-García A, de la Escosura-Muñiz A. Bifunctional Au@Pt/Au core@shell nanoparticles as novel electrocatalytic tags in immunosensing: Application for ADBiomarker detection. *Anal Chem.* 2020 May 19;92(10):7209–7217. doi: 10.1021/acs.analchem.0c00760.

Jan H, Shah M, Andleeb A, Faisal S, Khattak A, Rizwan M, Drouet S, Hano C, Abbasi BH. Plant-based synthesis of Zinc Oxide Nanoparticles (ZnO-NPs) using aqueous leaf extract of *Aquilegia pubiflora*: Their antiproliferative activity against HepG2 cells inducing reactive oxygen species and other *In Vitro* properties. *Oxid Med Cell Longev.* 2021 Aug 17;2021:4786227. doi: 10.1155/2021/4786227.

Jara-Guajardo P, Cabrera P, Celis F, Soler M, Berlanga I, Parra-Muñoz N, Acosta G, Albericio F, Guzman F, Campos M, Alvarez A, Morales-Zavala F, Kogan MJ. Gold nanoparticles mediate improved detection of β-amyloid aggregates by fluorescence. *Nanomaterials (Basel).* 2020 Apr 6;10(4):690. doi: 10.3390/nano10040690.

Kanaoujiya R, Saroj SK, Rajput VD, Alimuddin, Srivastava S, Minkina T, Igwegbe CA, Singh M, Kumar A. Emerging application of nanotechnology for mankind. *Emergent Mater.* 2023 Jan 31:1–14. doi: 10.1007/s42247-023-00461-8.

Kang YS, Jung HJ, Oh JS, Song DY. Use of PEGylated immunoliposomes to deliver dopamine across the blood-brain barrier in a rat model of Parkinson's disease. *CNS Neurosci Ther.* 2016 Oct;22(10):817–823. doi: 10.1111/cns.12580.

Karthivashan G, Ganesan P, Park SY, Kim JS, Choi DK. Therapeutic strategies and nano-drug delivery applications in management of ageing Alzheimer's disease. *Drug Deliv.* 2018 Nov;25(1):307–320. doi: 10.1080/10717544.2018.1428243.

Khan NH, Mir M, Ngowi EE, Zafar U, Khakwani MMAK, Khattak S, Zhai YK, Jiang ES, Zheng M, Duan SF, Wei JS, Wu DD, Ji XY. Nanomedicine: A promising way to manage Alzheimer's disease. *Front Bioeng Biotechnol.* 2021 Apr 9;9:630055. doi: 10.3389/fbioe.2021.630055.

Kim K, Lee CH, Park CB. Chemical sensing platforms for detecting trace-level Alzheimer's core biomarkers. *Chem Soc Rev.* 2020 Aug 7;49(15):5446–5472. doi: 10.1039/d0cs00107d.

Kogan MJ, Olmedo I, Hosta L, Guerrero AR, Cruz LJ, Albericio F. Peptides and metallic nanoparticles for biomedical applications. *Nanomedicine (Lond).* 2007 Jun;2(3):287–306. doi: 10.2217/17435889.2.3.287.

Kotrange H, Najda A, Bains A, Gruszecki R, Chawla P, Tosif MM. Metal and Metal Oxide Nanoparticle as a Novel Antibiotic Carrier for the Direct Delivery of Antibiotics. *Int J Mol Sci.* 2021 Sep 4;22(17):9596. doi: 10.3390/ijms22179596.

Kowalczyk J, Grapsi E, Espargaró A, Caballero AB, Juárez-Jiménez J, Busquets MA, Gamez P, Sabate R, Estelrich J. Dual effect of Prussian Blue Nanoparticles on Aβ40 aggregation: β-Sheet Fibril reduction and copper dyshomeostasis regulation. *Biomacromolecules*. 2021 Feb 8;22(2):430–440. doi: 10.1021/acs.biomac.0c01290.

Kulkarni S, Kumar S, Acharya S. Gold Nanoparticles in Cancer Therapeutics and Diagnostics. *Cureus*. 2022 Oct 9;14(10):e30096. doi: 10.7759/cureus.30096.

Li C, Wang N, Zheng G, Yang L. Oral administration of Resveratrol-Selenium-Peptide Nanocomposites Alleviates Alzheimer's Disease-like Pathogenesis by Inhibiting Aβ Aggregation and Regulating Gut Microbiota. *ACS Appl Mater Interfaces*. 2021 Oct 6;13(39):46406–46420. doi: 10.1021/acsami.1c14818.

Li R, Hou X, Li L, Guo J, Jiang W, Shang W. Application of metal-based nanozymes in inflammatory disease: A review. *Front Bioeng Biotechnol*. 2022 Jun 16;10:920213. doi: 10.3389/fbioe.2022.920213

Liu CC, Liu CC, Kanekiyo T, Xu H, Bu G. Apolipoprotein E and AD: Risk, mechanisms and therapy. *Nat Rev Neurol*. 2013 Feb;9(2):106–118. doi: 10.1038/nrneurol.2012.263. Epub 2013 Jan 8. Erratum in: Nat Rev Neurol. 2013. doi: 10.1038/nmeurol.2013.32. Liu, Chia-Chan [corrected to Liu, Chia-Chen].

Liu DZ, Cheng Y, Cai RQ, Wang Bd WW, Cui H, Liu M, Zhang BL, Mei QB, Zhou SY. The enhancement of siPLK1 penetration across BBB and its anti glioblastoma activity in vivo by magnet and transferrin co-modified nanoparticle. *Nanomedicine*. 2018 Apr;14(3):991–1003. doi: 10.1016/j.nano.2018.01.004.

Ma X, Zhou S, Xu X, Du Q. Copper-containing nanoparticles: Mechanism of antimicrobial effect and application in dentistry-a narrative review. *Front Surg*. 2022 Aug 5;9:905892. doi: 10.3389/fsurg.2022.905892.

Magne TM, Alencar LMR, Carneiro SV, Fechine LMUD, Fechine PBA, Souza PFN, Portilho FL, de Barros AODS, Johari SA, Ricci-Junior E, Santos-Oliveira R. Nano-Nutraceuticals for health: Principles and applications. *Rev Bras Farmacogn*. 2023;33(1):73–88. doi: 10.1007/s43450-022-00338-7.

Matea CT, Mocan T, Tabaran F, Pop T, Mosteanu O, Puia C, Iancu C, Mocan L. Quantum dots in imaging, drug delivery and sensor applications. *Int J Nanomedicine*. 2017 Jul 28;12:5421–5431. doi: 10.2147/IJN.S138624.

McDade EM. AD. *Continuum (Minneap Minn)*. 2022 Jun 1;28(3):648–675. doi: 10.1212/CON.0000000000001131.

Minami SS, Sun B, Popat K, Kauppinen T, Pleiss M, Zhou Y, Ward ME, Floreancig P, Mucke L, Desai T, Gan L. Selective targeting of microglia by quantum dots. *J Neuroinflammation*. 2012 Jan 24;9:22. doi: 10.1186/1742-2094-9-22.

Murphy MP, LeVine H 3rd. ADand the amyloid-beta peptide. *J Alzheimers Dis*. 2010;19(1):311–323. doi: 10.3233/JAD-2010-1221.

Nazıroğlu M, Muhamad S, Pecze L. Nanoparticles as potential clinical therapeutic agents in Alzheimer's disease: Focus on selenium nanoparticles. *Expert Rev Clin Pharmacol*. 2017 Jul;10(7):773–782. doi: 10.1080/17512433.2017.1324781.

Negahdary M, Heli H. An ultrasensitive electrochemical aptasensor for early diagnosis of Alzheimer's disease, using a fern leaves-like gold nanostructure. *Talanta*. 2019 Jun 1;198:510–517. doi: 10.1016/j.talanta.2019.01.109.

Ning S, Jorfi M, Patel SR, Kim DY, Tanzi RE. Neurotechnological approaches to the diagnosis and treatment of Alzheimer's disease. *Front Neurosci*. 2022 Mar 24;16:854992. doi: 10.3389/fnins.2022.854992.

Ogunyemi SO, Abdallah Y, Zhang M, Fouad H, Hong X, Ibrahim E, Masum MMI, Hossain A, Mo J, Li B. Green synthesis of zinc oxide nanoparticles using different plant extracts and their antibacterial activity against Xanthomonas oryzae pv. oryzae. *Artif Cells Nanomed Biotechnol*. 2019 Dec;47(1):341–352. doi: 10.1080/21691401.2018.1557671.

Ovais M, Zia N, Ahmad I, Khalil AT, Raza A, Ayaz M, Sadiq A, Ullah F, Shinwari ZK. Phyto-therapeutic and nanomedicinal approaches to cure Alzheimer's disease: Present status and future opportunities. *Front Aging Neurosci*. 2018 Oct 23;10:284. doi: 10.3389/fnagi.2018.00284.

Oyarzún MP, Tapia-Arellano A, Cabrera P, Jara-Guajardo P, Kogan MJ. Plasmonic nanoparticles as optical sensing probes for the detection of Alzheimer's disease. *Sensors (Basel)*. 2021 Mar 16;21(6):2067. doi: 10.3390/s21062067.

Passeri E, Elkhoury K, Morsink M, Broersen K, Linder M, Tamayol A, Malaplate C, Yen FT, Arab-Tehrany E. Alzheimer's disease: Treatment strategies and their limitations. *Int J Mol Sci*. 2022 Nov 12;23(22):13954. doi: 10.3390/ijms232213954.

Perets N, Betzer O, Shapira R, Brenstein S, Angel A, Sadan T, Ashery U, Popovtzer R, Offen D. Golden exosomes selectively target brain pathologies in neurodegenerative and neurodevelopmental disorders. *Nano Lett*. 2019 Jun 12;19(6):3422–3431. doi: 10.1021/acs.nanolett.8b04148.

Qiao L, Chen Y, Song X, Dou X, Xu C. Selenium nanoparticles-enriched *Lactobacillus casei* ATCC 393 prevents cognitive dysfunction in mice through modulating microbiota-gut-brain axis. *Int J Nanomedicine*. 2022 Oct 13;17:4807–4827. doi: 10.2147/IJN.S374024.

Qu B, Li X, Guan M, Li X, Hai L, Wu Y. Design, synthesis and biological evaluation of multivalent glucosides with high affinity as ligands for brain targeting liposomes. *Eur J Med Chem*. 2014 Jan 24;72:110–118. doi: 10.1016/j.ejmech.2013.10.007.

Ranjan P, Khan R. Electrochemical Immunosensor for early detection of β-Amyloid ADBiomarker based on aligned carbon nanotubes gold nanocomposites. *Biosensors (Basel)*. 2022 Nov 21;12(11):1059. doi: 10.3390/bios12111059.

Rao YL, Ganaraja B, Murlimanju BV, Joy T, Krishnamurthy A, Agrawal A. Hippocampus and its involvement in Alzheimer's disease: A review. *3 Biotech*. 2022 Feb;12(2):55. doi: 10.1007/s13205-022-03123-4.

Razzino CA, Serafín V, Gamella M, Pedrero M, Montero-Calle A, Barderas R, Calero M, Lobo AO, Yáñez-Sedeño P, Campuzano S, Pingarrón JM. An electrochemical immunosensor using gold nanoparticles-PAMAM-nanostructured screen-printed carbon electrodes for tau protein determination in plasma and brain tissues from Alzheimer patients. *Biosens Bioelectron*. 2020 Sep 1;163:112238. doi: 10.1016/j.bios.2020.112238.

Ribeiro TC, Sábio RM, Carvalho GC, Fonseca-Santos B, Chorilli M. Exploiting mesoporous silica, silver and gold nanoparticles for neurodegenerative diseases treatment. *Int J Pharm*. 2022 Aug 25;624:121978. doi: 10.1016/j.ijpharm.2022.121978.

Rosenthal SJ, Chang JC, Kovtun O, McBride JR, Tomlinson ID. Biocompatible quantum dots for biological applications. *Chem Biol*. 2011 Jan 28;18(1):10–24. doi: 10.1016/j.chembiol.2010.11.013.

Rostagno AA. Pathogenesis of Alzheimer's disease. *Int J Mol Sci*. 2022 Dec 21;24(1):107. doi: 10.3390/ijms24010107.

Roy R, Bhattacharya P, Borah A. Targeting the pathological hallmarks of AD through nanovesicle aided drug delivery approach. *Curr Drug Metab*. 2022;23(9):693–707. doi: 10.2174/1389200223666220526094802.

Sadowska-Bartosz I, Bartosz G. Redox nanoparticles: Synthesis, properties and perspectives of use for treatment of neurodegenerative diseases. *J Nanobiotechnology*. 2018 Nov 3;16(1):87. doi: 10.1186/s12951-018-0412-8.

Salem SS. A mini review on green nanotechnology and its development in biological effects. *Arch Microbiol*. 2023 Mar 22;205(4):128. doi: 10.1007/s00203-023-03467-2.

Sanati M, Khodagholi F, Aminyavari S, Ghasemi F, Gholami M, Kebriaeezadeh A, Sabzevari O, Hajipour MJ, Imani M, Mahmoudi M, Sharifzadeh M. Impact of gold nanoparticles on amyloid β-induced AD in a rat animal model: Involvement of STIM proteins. *ACS Chem Neurosci*. 2019 May 15;10(5):2299–2309. doi: 10.1021/acschemneuro.8b00622.

Shao K, Zhang Y, Ding N, Huang S, Wu J, Li J, Yang C, Leng Q, Ye L, Lou J, Zhu L, Jiang C. Functionalized nanoscale micelles with brain targeting ability and intercellular microenvironment biosensitivity for anti-intracranial infection applications. *Adv Healthc Mater*. 2015 Jan 28;4(2):291–300. doi: 10.1002/adhm.201400214.

Shi Y, Pilozzi AR, Huang X. Exposure of CuO nanoparticles contributes to cellular apoptosis, redox stress, and Alzheimer's Aβ amyloidosis. *Int J Environ Res Public Health*. 2020 Feb 5;17(3):1005. doi: 10.3390/ijerph17031005.

Shubayev VI, Pisanic II TR, Jin S. Magnetic nanoparticles for theragnostics. *Advanced Drug Delivery Reviews*. 2009;61(6):467–477.

Shui B, Tao D, Florea A, Cheng J, Zhao Q, Gu Y, Li W, Jaffrezic-Renault N, Mei Y, Guo Z. Biosensors for AD biomarker detection: A review. *Biochimie*. 2018 Apr;147:13–24. doi: 10.1016/j.biochi.2017.12.015.

Siddiqi KS, Husen A, Sohrab SS, Yassin MO. Recent status of nanomaterial fabrication and their potential applications in neurological disease management. *Nanoscale Res Lett*. 2018 Aug 10;13(1):231. doi: 10.1186/s11671-018-2638-7.

Sikorska K, Grądzka I, Wasyk I, Brzóska K, Stępkowski TM, Czerwińska M, Kruszewski MK. The impact of Ag nanoparticles and CdTe quantum dots on expression and function of receptors involved in Amyloid-β Uptake by BV-2 microglial cells. *Materials (Basel)*. 2020 Jul 20;13(14):3227. doi: 10.3390/ma13143227

Sim TM, Tarini D, Dheen ST, Bay BH, Srinivasan DK. Nanoparticle-based technology approaches to the management of neurological disorders. *Int J Mol Sci*. 2020 Aug 23;21(17):6070. doi: 10.3390/ijms21176070.

Sofias AM, Lammers T. Multidrug nanomedicine. *Nat Nanotechnol*. 2023 Feb;18(2):104–106. doi: 10.1038/s41565-022-01265-3.

Sonawane SK, Ahmad A, Chinnathambi S. Protein-capped metal nanoparticles inhibit tau aggregation in Alzheimer's disease. *ACS Omega*. 2019 Jul 29;4(7):12833–12840. doi: 10.1021/acsomega.9b01411.

Soria C, Coccini T, De Simone U, Marchese L, Zorzoli I, Giorgetti S, Raimondi S, Mangione PP, Ramat S, Bellotti V, Manzo L, Stoppini M. Enhanced toxicity of silver nanoparticles in transgenic Caenorhabditis elegans expressing amyloidogenic proteins. *Amyloid*. 2015;22(4):221–228. doi: 10.3109/13506129.2015.1077216.

Sun D, Li N, Zhang W, Zhao Z, Mou Z, Huang D, Liu J, Wang W. Design of PLGA-functionalized quercetin nanoparticles for potential use in Alzheimer's disease. *Colloids Surf B Biointerfaces*. 2016 Dec 1;148:116–129. doi: 10.1016/j.colsurfb.2016.08.052.

Swathi RS, Sebastian KL. Resonance energy transfer from a dye molecule to graphene. *J Chem Phys*. 2008 Aug 7;129(5):054703. doi: 10.1063/1.2956498. PMID: 18698917.

Suk JS, Xu Q, Kim N, Hanes J, Ensign LM. PEGylation as a strategy for improving nanoparticle-based drug and gene delivery. *Adv Drug Deliv Rev*. 2016 Apr 1;99(Pt A):28–51. doi: 10.1016/j.addr.2015.09.012.

Tak K, Sharma R, Dave V, Jain S, Sharma S. *Clitoria ternatea* mediated synthesis of graphene quantum dots for the treatment of Alzheimer's disease. *ACS Chem Neurosci*. 2020 Nov 18;11(22):3741–3748. doi: 10.1021/acschemneuro.0c00273.

Tang M, Pi J, Long Y, Huang N, Cheng Y, Zheng H. Quantum dots-based sandwich immunoassay for sensitive detection of Alzheimer's disease-related $Aβ_{1-42}$. *Spectrochim Acta A Mol Biomol Spectrosc*. 2018 Aug 5;201:82–87. doi: 10.1016/j.saa.2018.04.060.

Tarawneh R, Holtzman DM. The clinical problem of symptomatic AD and mild cognitive impairment. *Cold Spring Harb Perspect Med*. 2012 May;2(5):a006148. doi: 10.1101/cshperspect.a006148.

Tolar M, Abushakra S, Hey JA, Porsteinsson A, Sabbagh M. Aducanumab, gantenerumab, BAN2401, and ALZ-801-the first wave of amyloid-targeting drugs for ADwith potential for near term approval. *Alzheimers Res Ther*. 2020 Aug 12;12(1):95. doi: 10.1186/s13195-020-00663-w.

Uberti AF, Callai-Silva N, Grahl MVC, Piovesan AR, Nachtigall EG, Furini CRG, Carlini CR. *Helicobacter pylori* Urease: Potential contributions to Alzheimer's disease. *Int J Mol Sci*. 2022 Mar 13;23(6):3091. doi: 10.3390/ijms23063091.

Upadhyay RK. Drug delivery systems, CNS protection, and the blood brain barrier. *Biomed Res Int*. 2014;2014:869269. doi: 10.1155/2014/869269.

Vassar R, Bennett BD, Babu-Khan S, Kahn S, Mendiaz EA, Denis P, Teplow DB, Ross S, Amarante P, Loeloff R, Luo Y, Fisher S, Fuller J, Edenson S, Lile J, Jarosinski MA, Biere AL, Curran E, Burgess T, Louis JC, Collins F, Treanor J, Rogers G, Citron M. Beta-secretase cleavage of Alzheimer's amyloid precursor protein by the transmembrane aspartic protease BACE. *Science*. 1999 Oct 22;286(5440):735–741. doi: 10.1126/science.286.5440.735.

Villalva MD, Agarwal V, Ulanova M, Sachdev PS, Braidy N. Quantum dots as a theranostic approach in Alzheimer's disease: A systematic review. *Nanomedicine (Lond)*. 2021 Aug;16(18):1595–1611. doi: 10.2217/nnm-2021-0104.

Wang G, Shen X, Song X, Wang N, Wo X, Gao Y. Protective mechanism of gold nanoparticles on human neural stem cells injured by β-amyloid protein through miR-21–5p/SOCS6 pathway. *Neurotoxicology*. 2023 Mar;95:12–22. doi: 10.1016/j.neuro.2022.12.011.

White AR, Barnham KJ, Bush AI. Metal homeostasis in Alzheimer's disease. *Expert Rev Neurother*. 2006 May;6(5):711–722. doi: 10.1586/14737175.6.5.711.

Wu L, Ji H, Sun H, Ding C, Ren J, Qu X. Label-free ratiometric electrochemical detection of the mutated apolipoprotein E gene associated with Alzheimer's disease. *Chem Commun (Camb)*. 2016 Oct 4;52(81):12080–12083. doi: 10.1039/c6cc07099j.

Xuan L, Ju Z, Skonieczna M, Zhou PK, Huang R. Nanoparticles-induced potential toxicity on human health: Applications, toxicity mechanisms, and evaluation models. *MedComm (2020)*. 2023 Jul 14;4(4):e327. doi: 10.1002/mco2.327.

Yang JK, Hwang IJ, Cha MG, Kim HI, Yim D, Jeong DH, Lee YS, Kim JH. Reaction kinetics-mediated control over silver nanogap shells as surface-enhanced raman scattering nanoprobes for detection of AD biomarkers. *Small*. 2019 May;15(19):e1900613. doi: 10.1002/smll.201900613.

Yang L, Zhang X, Li M, Qu L, Liu Z. Acetylcholinesterase-Cu$_3$(PO$_4$)$_2$ hybrid nanoflowers for electrochemical detection of dichlorvos using square-wave voltammetry. *Anal Methods*. 2022 Oct 13;14(39):3911–3920. doi: 10.1039/d2ay01014c.

Yang Z, Zhang Y, Yang Y, Sun L, Han D, Li H, Wang C. Pharmacological and toxicological target organelles and safe use of single-walled carbon nanotubes as drug carriers in treating AD. *Nanomedicine*. 2010 Jun;6(3):427–441. doi: 10.1016/j.nano.2009.11.007.

Youssif KA, Haggag EG, Elshamy AM, Rabeh MA, Gabr NM, Seleem A, Salem MA, Hussein AS, Krischke M, Mueller MJ, Abdelmohsen UR. Anti-Alzheimer potential, metabolomic profiling and molecular docking of green synthesized silver nanoparticles of Lampranthus coccineus and Malephora lutea aqueous extracts. *PLoS One*. 2019 Nov 6;14(11):e0223781. doi: 10.1371/journal.pone.0223781.

Zamaleeva AI, Despras G, Luccardini C, Collot M, de Waard M, Oheim M, Mallet JM, Feltz A. FRET-Based Nanobiosensors for Imaging Intracellular Ca^{2+} and H$^+$ Microdomains. *Sensors (Basel)*. 2015 Sep 23;15(9):24662–24680. doi: 10.3390/s150924662.

Zhang X, Zhong M, Zhao P, Zhang X, Li Y, Wang X, Sun J, Lan W, Sun H, Wang Z, Gao H. Screening a specific Zn(ii)-binding peptide for improving the cognitive decline of ADin APP/PS1 transgenic mice by inhibiting Zn^{2+}-mediated amyloid protein aggregation and neurotoxicity. *Biomater Sci*. 2019 Nov 19;7(12):5197–5210. doi: 10.1039/c9bm00676a.

Zhao D, Tang Y, Suo X, Zhang C, Dou Y, Chang J. A dual-targeted multifunctional nanoformulation for potential prevention and therapy of Alzheimer's disease. *Innovation (Camb)*. 2021 Sep 1;2(4):100160. doi: 10.1016/j.xinn.2021.100160.

Zhou Y, Huang Z, Liu J, Dong H, Zhang Y, Xu M, Bi Y. Tyrosine-conjugation method for evaluating BACE1 activity based on silver nanoparticles/metal-organic framework composites as electrochemical tags. *Chemistry*. 2023 Feb 17:e202300450. doi: 10.1002/chem.202300450.

Chapter 18

Nanotherapeutics for Parkinson's disease using metal nanocomposites

*Rishika Dhapola, Prajjwal Sharma, Sneha Kumari,
Pushank Nagar, Bikash Medhi, and
Dibbanti Harikrishna Reddy*

LIST OF ABBREVIATIONS

6-OHDA	6-hydroxydopamine
ALIDEs	Aluminum interdigitated electrodes
COMT	Catechol O-methyltransferase
CSF	Cerebrospinal fluid
FBXO7	F-box protein 7
f-MWCNT/AgNP	Functionalized multiwalled carbon nanotube/silver nanoparticle
GCE	Glassy carbon electrode
GR-β-CD	Graphene-β-cyclodextrin
IFN-γ	Interferon-gamma
IL	Interleukin
LB	Lewy bodies
LZH	Layered zinc hydroxide
MAO-B	Monoamine oxidase type-B
MOFs	Metal–organic frameworks
MPTP	1-methyl-4-phenyl-1,2,3,6-tetrahydropyridine
NGF	Nerve growth factor
NMDA	N-methyl-d-aspartate
PARK	Parkin
PARK2	Parkinson protein 2
PARK7	Parkinsonism associated deglycase
PD	Parkinson's disease
PET	Positron emission tomography
PINK1	Phosphatase and tensin homolog-induced kinase 1
RPAgNC	Ropinirole–silver nanocomposite
SNc	Substantia nigra pars compacta
SNCA	Synuclein-α
SnS_2	Tin disulfide
TNF-α	Tumor necrosis factor-α
α-Syn	α-Synuclein

DOI: 10.1201/9781032621135-18

18.1 BACKGROUND

In his 1817 publication "Essay on the Shaking Palsy," James Parkinson first described Parkinson's disease (PD) as a progressive neurological condition. It is characterized by rigidity, tremor, bradykinesia, postural deformity, and other motor and non-motor symptoms. It is highly prevalent in old age people between 85 and 89 years of age (Jankovic and Tan 2020). Dopaminergic neuron loss in the substantia nigra pars compacta region of the brain is the disease's primary distinctive characteristic. An abnormal aggregate of α-synuclein proteins is deposited to form a Lewy body which is a major characteristic of PD (Armstrong and Okun 2020) (Figure 18.1).

For clinical diagnosis of PD motor symptoms are of utmost importance. PD significantly affects the nervous system giving rise to non-motor symptoms including olfactory dysfunction and sleep irregularities. Other symptoms also arise late in PD when the effect of the disease spreads to other brain regions apart from basal ganglia, including dyskinesia, pain, urinary incontinence, and so on (Shulman, De Jager, and Feany 2011).

The risk factors for the disease include genetic mutations, environmental factors, and aging (Collier, Kanaan, and Kordower 2011). Other factors which increase the susceptibility toward PD are pesticides, traumatic brain injury, melanoma, intake of dairy products, smoking, consuming alcohol, and use of certain medications like ibuprofen (Ascherio and Schwarzschild 2016; Emamzadeh and Surguchov 2018). Exposure to these risk factors may increase the risk of PD by several folds.

PD diagnosis is based on a history and physical examination. Prodromal symptoms including eye movements, sleep irregularities, constipation, recognizable movement difficulties

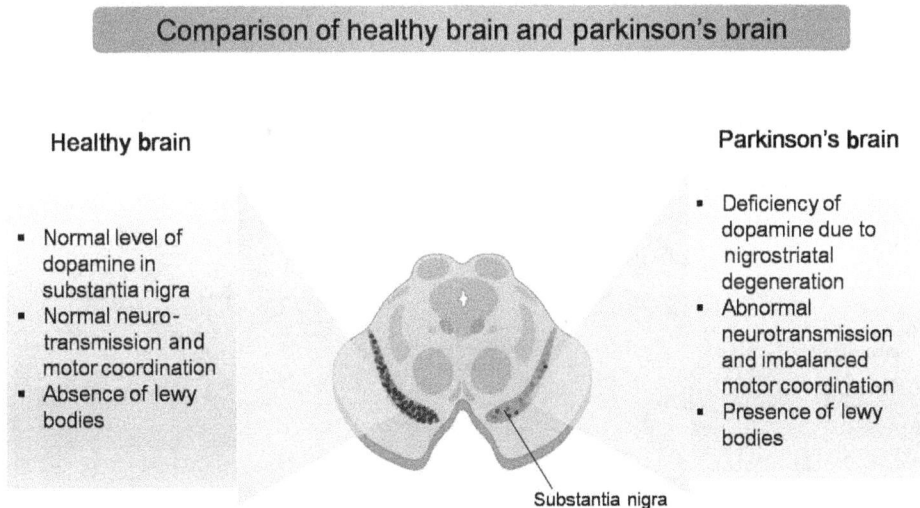

Comparison of healthy brain and parkinson's brain

Healthy brain

- Normal level of dopamine in substantia nigra
- Normal neuro-transmission and motor coordination
- Absence of lewy bodies

Parkinson's brain

- Deficiency of dopamine due to nigrostriatal degeneration
- Abnormal neurotransmission and imbalanced motor coordination
- Presence of lewy bodies

Substantia nigra

Figure 18.1 This figure demonstrates the changes in the PD brain in comparison to the healthy brain. This includes a reduction in the levels of dopamine in specific brain regions, impaired neuronal transmission, dysregulated motor coordination, and formation of Lewy bodies due to the accumulation of α-syn and other proteins.

like tremors, rigidity, sluggishness, and psychological or cognitive issues can all be present in the past. Bradykinesia with tremors, rigidity, or both, is frequently seen during the examination. When the presence of parkinsonism is unclear, dopamine transporter single-photon emission computed tomography can help with the diagnosis (Armstrong and Okun 2020).

18.2 PATHOPHYSIOLOGY OF PD

Abnormal aggregation of α-synuclein, mitochondrial dysfunction, lysosomal and vesicular transport abruption, neuroinflammation, and synaptic dysfunction lead to the progression of PD. These pathways ultimately come up with the degeneration of dopaminergic neurons along with other motor and non-motor pathways (Bloem, Okun, and Klein 2021) (Figure 18.2).

18.2.1 Genetic variability

As most of the cases of PD are inherited, there are chances of a genetic link in the disease development. SNCA is one of the genes that encode α-synuclein the major aggregated protein in PD. Mutations comprising triplication of this gene give rise to autosomal dominant forms of PD (Zafar et al. 2018). Genetics is also involved in the development of Lewy bodies generated from α-synuclein protein. Mitophagy is a process for the removal of aggregated protein or cell organelles, especially damaged and disintegrated mitochondria, and is regulated by various genes including PARK2, PARK7, FBXO7, and PINK1, which are associated

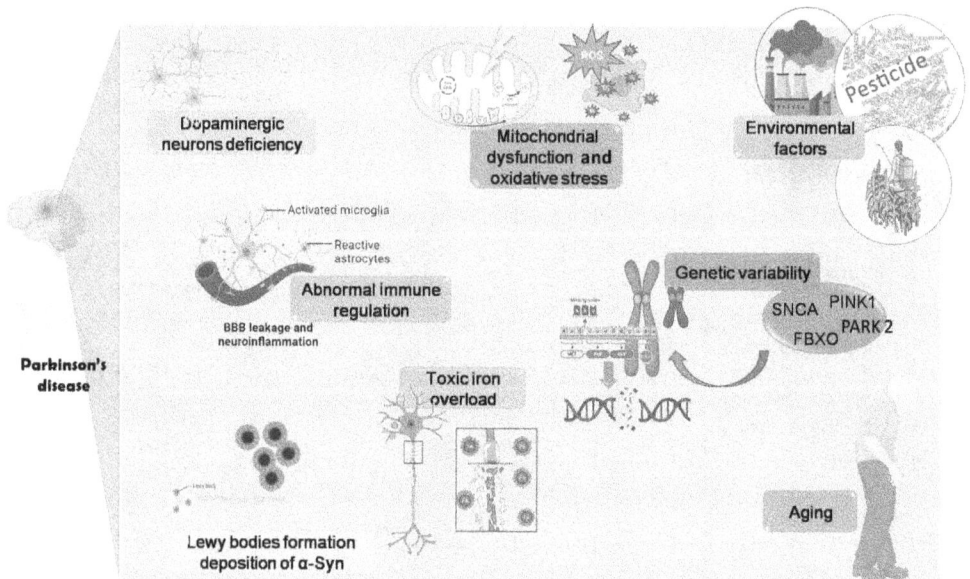

Figure 18.2 This figure shows the pathologies associated with PD, including dopamine insufficiency, Lewy body formation, mutation in various genes associated with PD such as PINK1, SNCA, and PARK2, aging, toxic iron accumulation, abnormal functioning of the immune system, and mitochondrial dysfunction. These pathologies ultimately result in the degeneration of neuronal cells and cause motor and non-motor abnormalities.

with PD (Abeliovich and Gitler 2016). Mutations in these genes lead to the progression of familial PD (Dodson and Guo 2007). PINK1 and parkin are involved in the regulation of mitochondrial function, but mutated forms are unable to do so resulting in mitochondrial dysfunction and cell death in PD. Early onset PD associated with mutations in the PINK1 gene is not associated with the formation of Lewy bodies and accounts for a majority of PD cases (Martin, Dawson, and Dawson 2011).

18.2.2 Mitochondrial defects and oxidative stress

As mitochondria are the major organelles involved in the generation of ATP, they can give rise to a major portion of reactive oxygen species (ROS). ROS generation takes place due to the leaking out of electrons from the electron transport chain. These ROS play a crucial role in the pathogenesis of PD as in other neurodegenerative diseases (Lin and Beal 2006). Along with mitochondrial ROS, mitochondrial DNA also contributes to the disease progression. Mutation in mtDNA-encoded complex I genes increases the risk of PD progression. These mutations result in the dysfunctioning of mitochondria and the generation of oxidative stress as mutations potentially affect the functioning of complex I of the mitochondrial respiratory chain (Yana, Wang, and Zhu 2013).

18.2.3 Faulty immune regulation

It is now known that PD is a multifactorial disorder that affects several systems and is marked by neuroinflammation and immunological dysfunction. PD is characterized by a dopaminergic neuronal decline in the substantia nigra pars compacta (SNc) and intracellular inclusions known as Lewy bodies (LB) in the neurons of afflicted brain areas (Pang et al. 2019). PD suggests changes in immune cell populations and indicators of inflammation that might instigate or aggravate neuroinflammation and exacerbate the neurodegenerative process (Tansey et al. 2022). Immunosenescence is a phenomenon in which there is a decline in the levels of T cells and B cells, which are associated with aging and indirectly to PD. These impairments increase the risk of infection and a kind of aging-related autoimmunity where autoantibodies may manifest through increasing susceptibility to infection. Inflammation can alter the permeability of the blood–brain barrier (BBB) allowing flux of proinflammatory factors from the periphery. Hence, increasing BBB permeability might be a catalytic element in the development of PD. One mechanism contributing to microgliosis and neuroinflammation may be dysfunctional phagocytosis in glial cells brought on by lysosomal abnormalities caused by PD-related mutations (Joshi and Singh 2018). Another mechanism contributing to microgliosis and neuroinflammation may be dysfunctional phagocytosis in glial cells brought on by lysosomal abnormalities caused by PD-related mutations that reduce the clearance of toxic aggregation (Tansey et al. 2022).

Neuronal malfunction external assaults may be the originator of the inflammatory signals. Microglia respond to a variety of signals, including α-syn, viral and bacterial products, antibodies, and cytokines. (Grotemeyer et al. 2022). High levels of α-syn encourage the accumulation of neuronal α-syn aggregates, which is considered to be an important factor in the etiology of PD (PD) by stimulating the immune system. Activation of the innate immune system results in the secretion of several soluble immune mediators such as chemokines, cytokines, and the complement system, which are more prevalent in the post-mortem brain of people with PD. Positron emission tomography (PET) imaging of PD patients demonstrates activated microglia in the midbrain and striatum. As a result of stimulation of the immune system microglia activation results in the release of various pro-inflammatory cytokines and chemokines particularly interleukins (IL-1β, IL-6), tumor necrosis factor-α (TNF-α), and interferon-gamma (IFN-γ) were found to be elevated in the post-mortem of PD brain as well as in the CSF of PD patients (Allen Reish and Standaert 2015).

18.2.4 Abnormal protein processing

Since PD was originally described in 1817, several attempts have been undertaken to determine its etiology. Numerous proteins were found in Lewy bodies, most notably α-syn, which raised the possibility that the degradation of undesired, defective, or altered proteins may be disturbed in PD, causing cellular aggregation and neuronal death. High levels of α-syn encourage the accumulation of neuronal α-syn aggregates, which is considered to be an important factor in the etiology of PD (PD). According to epidemiological research, the majority of PD cases are sporadic, with just 10% having a definite family origin known as familial PD confirming the role of genetic mutation and alteration (Thomas and Beal 2007). Various altered genetic loci are identified for PD which are responsible for genetic variability that leads to disease conditions including PARK2, PINK1, FBX07, alpha-synuclein (SNCA), and so on.

The α-syn is also involved in aberrant protein breakdown to oxidative stress and mitochondrial dysfunction. The parkin (PARK2) gene encodes a 465-amino acid protein with an N-terminal domain that is similar to ubiquitin and functions as an E3 ubiquitin-protein ligase. A significant contributing factor to early onset PD is parkin gene mutations. Current research suggests that parkin activates the NF-κB kinase/nuclear factor-κB signaling pathway, mediating neuroprotective properties, whereas parkin mutants are unable to activate this route. Early onset familial PD has also been linked to phosphatase and tensin (PTEN) homolog-induced putative kinase 1 (PINK1) gene mutations that result in mitochondrial dysfunction. PINK1 is a 581-amino acid protein that resembles the serine/threonine kinases of the Ca^{2+} calmodulin family in its highly conserved protein kinase domain and N-terminal mitochondrial targeting regions (Thomas and Beal 2007). Mutations in PINK1 result in increased lipid peroxidation, abnormalities in the function of mitochondrial complex I, and a compensatory rise in glutathione and superoxide dismutases in the mitochondria (Hoepken et al. 2007). Autosomal PD is also brought on by missense and SNCA triplication mutations (Pang et al. 2019).

18.2.5 Toxic iron accumulation

PD and other related disorders have a long and illustrious history with iron. This component is needed for optimal brain functioning, yet excessive brain iron buildup has been related to extrapyramidal illness for a century (Foley, Hare, and Double 2022). To meet the high energy requirements of the brain, it is necessary to have enough availability of iron since iron is crucial for producing ATP and facilitating electron transport. Moreover, iron is required for a broad range of central nervous system-specific metabolic functions; e.g., as a cofactor in myelin synthesis by oligodendrocytes. Deficits in neurodevelopment are therefore caused by impaired neural network and cell signaling pathway expansion during crucial developmental windows (Foley, Hare, and Double 2022). Iron accumulates gradually in the aging brain especially in the putamen, globus pallidus, red nucleus, and substantia nigra region (Hare and Double 2016). Due to increased iron deposition in the substantia nigra caused by PD, dopaminergic neurons are exposed to a more intense pro-oxidant environment. Iron accumulation in neurons is facilitated by overexpression of the parkinsonian protein α-syn, while iron can also facilitate the aggregation of mutant versions of this protein (Hare et al. 2017). Dysregulation of iron metabolism has been widely hypothesized as one of the key factors contributing to oxidative stress, a known factor in the pathogenesis of PD. At the neuronal level, increasing iron contents in the PD brain may be caused by a rise in iron inflow that is promoted by either ferric citrate diffusion or transferrin receptor-2/ divalent metal transporter-1 endocytosis and impairment of iron efflux. Chelation of iron is the most practical approach to reducing iron burden in PD (Hare and Double 2016).

18.2.6 Environmental factors

The environment and immune system are constantly communicating; immunity tolerates some exposures or antigens while reacting to others, and disease results from dysregulated immunological responses as mentioned earlier (Kannarkat, Boss, and Tansey 2013). Epidemiological studies show that PD disease development involves exposure to pesticides, industrial wastes, germs, viruses, and environmental pollutants (Tansey et al. 2022; Thomas and Beal 2007). Exposure to MPTP, a toxic chemical present in pesticides, is linked with PD progression. Other chemicals which are related to MPTP include Paraquat which is a major constituent of herbicides. Similarly, rotenone, a lipophilic mitochondrial toxin, used in organic pesticides inhibits mitochondrial Complex I increasing ROS and decrease in ATP generation. Thus, these compounds are used as a toxin to develop experimental animal models for PD (Pang et al. 2019).

18.3 CURRENT THERAPEUTICS FOR PD

Current treatment for PD focuses on COMT inhibition, MAO-B inhibition, and dopamine agonism (Finberg 2018). The approved treatments generally contain levodopa and carbidopa. Levodopa works as a precursor for dopamine and the mechanism of carbidopa is the inhibition of dopa-decarboxylase which is involved in the extracellular degradation of levodopa (Yoosefian, Rahmanifar, and Etminan 2018; Colamartino et al. 2015). This prevents the metabolism of levodopa before crossing the BBB thus increasing the bioavailability of levodopa and increasing the level of dopamine (Colamartino et al. 2015). Catechol-o-methyltransferase (COMT) inhibitors are another class of anti-parkinsonian drugs that acts more efficiently when employed along with levodopa/carbidopa. COMT is involved in the degradation of levodopa extra-cerebrally; thus inhibiting this can result in elevating the plasma concentration of levodopa (Fabbri, Ferreira, and Rascol 2022). Another class of drugs called dopamine agonists act by binding the dopamine receptors and can cross the BBB, thus mimicking the effect of dopamine and reducing dyskinesia associated with PD (Pilleri and Antonini 2015).

Carbidopa and COMT inhibitors aid in dose reduction of levodopa; otherwise, a higher dose will be needed to attain the desired effect. Currently, Mirapex, Requip, Apokyn, Kynmobi, and Neupro are the dopamine agonists which are FDA approved for the management of PD. Further, monoamine oxidase Type B (MAO-B) is involved in the breakdown of dopamine inside the brain. Therefore, inhibitors of MAO-B aid in the treatment of PD by increasing the levels of dopamine. Examples of MAO-B inhibitors include Eldepryl, Zelapar, Azilect, and Xadago. Symptomatic treatment of dyskinesia is executed by NMDA receptor antagonists like Symmetrel, Gocovri, and Osmolex (Chopade et al. 2023). Adenosine 2A antagonists have also shown beneficial effects in the prevention of neuronal degradation in PD including Nourianz (Chen and Cunha 2020). Other drugs belonging to the anticholinergics are also potential agents for the treatment of PD. They can mitigate tremors due to a reduction in the imbalance between acetylcholine and dopamine in the brain of PD patients. Examples of anticholinergics include artane and cogentin (Kriebel-Gasparro 2016). However, all the approved drugs provide only symptomatic relief to PD patients (Chopade et al. 2023).

As mentioned earlier, PD treatment is only symptomatic and mainly focuses on the dopamine pathway. The gold standard for the treatment of PD is considered Levodopa, which is the most potent drug for curing motor symptoms associated with PD (Tambasco, Romoli, and Calabresi 2017). In the dopaminergic neurons of the substantia nigra pars compacta region of the brain levodopa is converted into dopamine. Side effects associated with the use of levodopa are nausea, hallucinations, sleepiness, confusion, and hypotension. These side effects are avoided by using carbidopa and benserazide along with levodopa. Complications associated with the use of levodopa are dyskinesia and dystonia. These complications are supposed to arise due to discontinuous stimulation of dopamine receptors, unlike the normal physiological condition in which there is continuous stimulation of dopamine receptors (Borovac 2016).

18.3.1 Dopamine agonists

Dopamine agonists used in the treatment of PD act on the D1–3 dopaminergic receptors. Dopamine agonists are not as effective as levodopa but are quite beneficial in relieving dyskinesia. It is given in the initial stages of PD either alone or in conjugation with levodopa. The major side effects associated with the use of dopamine agonists are similar to levodopa (Abyad and Sami Hammami 2009). Besides those some other side effects are also seen including edema, impulse control disorders, and sleepiness (Abyad and Sami Hammami 2009). Commonly used dopamine agonists are ropinirole and pramipexole. These drugs are given orally. Another drug rotigotine is given as a transdermal patch once daily (Rizos et al. 2016). A drug having a short duration of action is apomorphine, which is generally administered subcutaneously or as a continuous infusion that mitigates motor symptoms linked to PD (Olivola et al. 2019).

18.3.2 Monoamine oxidase B (MAO-B) inhibitors

MAO-B is involved in the metabolism of dopamine, therefore inhibition of MAO-B will increase the duration of dopamine function along with potentiating its stimulation (Finberg 2018). Irreversible inhibitors of MAO-B are rasagiline and selegiline, which are used to provide symptomatic relief in early PD. In later stages, it is used along with levodopa to relieve motor symptoms (Finberg and Rabey 2016). Recent treatment for motor abnormalities includes safinamide, which is a reversible inhibitor of MAO-B. It also modulates excitotoxicity by regulating voltage-dependent sodium channels and calcium channels and hinders the release of glutamate. Apart from these, safinamide has been shown to reduce dyskinesia in animal models of PD (Pisanò et al. 2020).

18.3.3 Catechol-O-methyl transferase (COMT) inhibitors

COMT enzymes are involved in the metabolism of levodopa along with decarboxylase inhibitors. COMT inhibitors are used in conjugation with levodopa for the treatment of PD as these agents are efficient in elevating the half-life of levodopa. Examples of COMT inhibitors include tolcapone, entacapone, and opicapone (Salamon et al. 2022).

Other therapeutic agents including anticholinergics like trihexyphenidyl, benztropine, orphenadrine, and biperiden are efficient in ameliorating motor functions but have the drawback of causing cognitive deficits (Balestrino and Schapira 2020). This is further mitigated by the use of acetylcholinesterase inhibitors including rivastigmine, donepezil, and galantamine. To overcome this, another drug amantadine is employed which is an N-methyl-d-aspartate-type (NMDA) receptor antagonist and also possesses anticholinergic effects along with the reduction in dyskinesia (Müller, Kuhn, and Möhr 2019). Serotonin (5-HT) transmission is another target for the management of PD. The drug buspirone, which is a 5-HT1A and α1-adrenergic receptor agonist, is currently under clinical trial for the treatment of motor dysfunctions and dyskinesia associated with PD. Similarly, Eltoprazine, pimavanserin are 5HT1A, 5HT1B, and 5HT2A inverse agonists which are under different phases of clinical trial due to their anti-dyskinesia properties (Balestrino and Schapira 2020).

18.4 NANOTHERAPEUTICS: A PROMISING APPROACH IN THE TREATMENT OF PD

Due to the drawbacks of current therapeutics including the inability to cross BBB new effective therapeutics are needed. As the inappropriate transfer of drugs inside the CNS leads to lesser bioavailability, targeted delivery of the therapeutic agent is required for treating PD. Nanotechnology giving rise to nanotherapeutics shows a promising effect in overcoming the limitations of current therapeutics (Kiran et al. 2022). In this technology, nanoparticles in the range of 1–1,000 nm are employed for easy penetration into the CNS (Srivastava et al. 2020). Materials like natural polymers, synthetic polymers, and inorganic materials are used to produce the nanoparticles. Emerging from nanotechnology, the concept of nanocarriers which makes it easier to transfer the drugs inside the CNS by crossing BBB and increasing the bioavailability. The favorable characteristics of nanocarriers which make them suitable for use as PD therapeutic include a high capacity for loading the drug, low toxicity, improvised permeability of the drug to the targeted site, and increased stability (Lamptey et al. 2022).

Recently, nanotherapeutics have shown a promising effect in treating various diseases by their size, shape, and morphology (Kim, Ahn, and Kim 2019). Generally, nanoparticles are used to carry the therapeutic agent and various genes to the target site which is however not possible if the therapeutic agent is given alone. The properties of these nanoparticles used in nanotherapeutics can be altered to attain the desired aim. Nanotherapeutics aids in overcoming various physiological hurdles which can be seen in various parts of the body including the BBB. Nanoparticles can be conjugated with polymers and targeting ligands which will increase the specificity and affinity of the therapeutic agent (Dilnawaz and Sahoo 2015). These nanoparticles are combined with noble metals like gold (Au), silver (Ag), platinum (Pt), and palladium (Pd) which further elevates the therapeutic potential of the nanoparticles. For example, gold nanoparticles have been used in the treatment of cancer, AIDS, smallpox, syphilis, and skin ulcers (Honary et al. 2013).

Similarly, silver nanoparticles have been employed as antiviral and antimicrobial used in wound infections (Jagaran and Singh 2021). Another metal palladium is used along with gold to enhance autophagy for the treatment of neurodegenerative disorders like Alzheimer's disease. These nanoparticles were modified by quercetin. Platinum is another metal used as an antioxidant for the reduction of free radicals and is used in anticancer drugs too. Similarly, selenium is used for the treatment of various cardiovascular disorders, neurodegenerative diseases, and diabetes (Jagaran and Singh 2021). Many of the NPs discussed here have demonstrated promise in nanomedicine and could be used to treat neurological conditions including PD.

18.4.1 Nanocomposites: focusing metal nanocomposites

Nanocomposites are made up of two or more materials that have different morphology. It comprises a matrix and a reinforcing phase which may be 1-D, 2-D, or 3-D (Nair et al. 2019). Composites are classified as fibrous, laminated, and particulate composites based on the materials they are manufactured from. If the matrix used to manufacture composites is a polymer the composite is termed a polymer-based composite. Polymers have been used recently in the synthesis of drug-delivery materials for the management of AD/PD. Certain chemical, physical, and mechanical characteristics can be achieved by developing a nanocomposite; in fact, the final product may combine the greatest features of its constituents and exhibit intriguing characteristics that the individual constituents frequently lack.

One, two, or three nanometer-sized phases make up a solid substance known as a nanocomposite, which is composed of several phases. At least one constituent with a nanometric size of 10^{-9} m or smaller is present in the nanocomposite. The major advantages of nanocomposites include better mechanical properties, small filler sizes, and enhanced optical properties. Limitations of nanocomposites include toughness and impact performance and are not cost-effective.

Nanocomposite materials are divided into three categories based on their matrix materials including ceramic matrix nanocomposites (CMNC), polymer matrix nanocomposites (PMNC), and metal matrix nanocomposites (MMNC). Metal matrix nanocomposites (MMNC) are made up of metals as matrix and nanosized material. These nanocomposites are highly ductile, tough, and of high strength. Common metals used in their production are Al, Mg, Pb, Sn, W, and Fe. Metal nanocomposites are prepared by spray pyrolysis, liquid metal infiltration, rapid solidification, vapor techniques, electrodeposition, and chemical methods, which include colloidal and sol-gel processes. The use of polymer–metal nanocomposites, which use metal nanoparticles as nanofillers and polymer as the matrix, in biomedical applications and medical devices is prevalent (Omanović-Miklicanin et al. 2020). Metal nanocomposites have also been used for combating microbial infections. Metal nanocomposites offer an increased area of contact to the bacteria and have the advantage of small size. These nanocomposites destroy the bacterial cell membrane by producing ROS thus further inhibiting microbial growth (Pachaiappan et al. 2021). Electrochemical biosensors have significantly been used in functionalized multiwalled carbon nanotube/silver nanoparticle (f-MWCNT/AgNP). Nanocomposites are used as biosensors for the detection of dopamine levels in the brain (Anshori et al. 2021) (Table 18.1).

Table 18.1 characteristic features of metal nanocomposites

Important characteristics of metal nanocomposites	
Metals employed as matrix in nanocomposites	• Aluminum • Zinc • Gold • Magnesium • Silver
Therapeutic applications	• Biosensors • Nanocarriers • Nanotherapeutics
Targeted pathways	• BBB permeability • α-syn deposition • Dopamine deficiency
Advantages of metal nanocomposites	• Enhanced biocompatibility • Improved electrochemical properties • Large surface area

Metal Nanocomposites Designed to Overcome BBB Permeability

BBB is the main hindrance for the drugs to enter the CNS. Around 98% of the medications are unable to cross the BBB. Therefore, this obstruction offered by BBB paves the way for the development of new therapeutics and drug delivery. When carefully manipulated, nanoparticles or nanocomposites are capable of defeating the BBB. High levels of drug efflux transporter proteins, including P-glycoprotein, multidrug resistance-associated proteins, and breast cancer resistance protein, are present in the endothelial cells of brain capillaries, which limit the intrusion and dissemination of a variety of therapeutic agents. Hence, several transporters expressed by the BBB are responsible for delivering the majority of endogenous substances, including nutrients, into the brain (Teleanu et al. 2019).

Polydopamine (PDA) is highly biocompatible, biodegradable, has adhesive properties, and photothermal, which has various biomedical applications. Recent years have seen many investigations on polydopamine nanocomposites. PDA has been employed to alter a wide range of molecules, like metal nanoparticles, to increase the stability and utility of the nanosystem (Chinchulkar et al. 2022). Taking this into consideration, MgOp@PPLP nanoparticles were found to be capable of penetrating the BBB and are efficiently absorbed by the neuronal cells to exhibit antioxidant effects. These properties along with biocompatibility suggest the use of MgOp@PPLP nanoparticles in neurodegenerative diseases as they show neuroprotective effects (Gao et al. 2022).

Another approach utilizes chitosan/carbon dots (CS/CD) in which CS/CD is used to make the matrix, and dopamine is embedded in it. As mentioned earlier, if we use high doses of levodopa it causes many side effects. Therefore, a proper carrier that minimizes the dose of levodopa is needed. Dopamine@CS/CD nanocomposite offers this advantage when used for the treatment of PD. It when coupled with the chitosan matrix provides an effective drug delivery nanocarrier for neurological illnesses as it promotes better drug distribution and prolonged release of drug (Mathew et al. 2020).

Carbidopa (CD), another effective drug for the treatment of PD, is also modified by incorporating it in layered zinc hydroxide (LZH) using an ion exchange method. As zinc is involved in various physiological processes like immune regulation and synthesis of carbohydrates, proteins, and other enzymes, it is suitable for preparing a nanocomposite which will aid in PD management. Moreover, zinc possesses antimicrobial activities. Studies based on LZH/chitosan intercalated carbidopa disclosed that the developed nanocomposites are suitable for a controlled release of carbidopa employed in PD treatment (Ghamami, Golzani, and Lashgari 2016).

Metal nanocomposites targeting α-synuclein

There is a considerable amount of focus on targeting α-Syn for potential disease alteration because numerous studies link the protein α-synuclein (α-Syn) to the pathogenesis of PD. A significant pathogenic protein called α-Syn is also present in other synucleinopathies, most frequently dementia with Lewy bodies. Immunostaining demonstrated that α-Syn is a key component of Lewy bodies, which is a pathologic hallmark of PD (PD). It has been shown that α-Syn regulates neurotransmitter release, synaptic activity, and plasticity in a range of investigations using cellular and animal models that looked at synaptic homeostasis after enhancing or reducing α-Syn levels (Jasutkar, Oh, and Mouradian 2022).

Early detection of accumulated α-synuclein may aid in the proper management of PD. The traditionally used aluminum interdigitated electrodes (ALIDEs) were modified by using silicon and zinc-oxide nanocomposites by using the spin-coating method. The ZnO nanocomposites offer a potential platform for the incorporation of antibodies that can bind to the antigen, here α-synuclein for early detection of the aggregates. The ZnO nanocomposite layering on the ALIDEs surface has several benefits, including an increased amount of electrochemically active surface area and improved analyte mobility to the electrode surface. This method has immense potential to detect early aggregates of α-synuclein in the serum of PD patients (Adam, Gopinath, and Hashim 2021; Bharti and Sadhu 2022). Gold-metal oxide nanocomposites have also been employed in the area of medicine, agriculture, and industry (Bharti and Sadhu 2022).

In a study on the *Drosophila* PD model, the neuroprotective effects of ropinirole-silver nanocomposite (RPAgNC) were evaluated. Due to the deposition of α-synuclein, the dopaminergic neurons were damaged in the brain of the diseased flies. This dopaminergic neuronal loss was accompanied by irregular muscular coordination, oxidative stress, and cognitive deficits. The neuroprotective role of RPAgNC was found to be much more prominent and significant than ropinirole alone (Naz et al. 2020).

Another study which was performed on a 6-hydroxydopamine (6-OHDA) induced PD model of rat nanocomposites showed a significant neuroprotective effect. In this model, Nano-MgO composites were evaluated for cognitive deficits in diseased rats. In this study, it was seen that the levels of aggregated α-synuclein were increased upon administration of 6-OHDA. This increased α-synuclein level was then mitigated by nano-MgO composites. Further, there was a reduction in dendritic spine density in the CA1, CA3, and dentate gyrus regions of the hippocampus in the PD rats. But upon treatment with nano-MgO composites, the pathological changes in the dendrites were ameliorated. Nano-MgO composites were also found to improve cognitive deficits and promoted long-term potentiation of PD rats (Huang et al. 2022).

Another study utilized a degradable nano-MgO micelle composite (MgO(pDNA)-INS-Plu-mRNA-NGF) with double interference (mediated by RNAi and α-synuclein (α-syn)-targeted mRNA). This nanocomposite was more potent in inhibiting the expression of α-syn when compared to RNAi alone. Further, *in vitro* studies suggested that nano-MgO micelle composite enhanced the cell viability in PD *in vitro* models. Moreover, these nanocomposites can cross BBB and transfer the required gene and mRNA to the neurons using endocytosis. This endocytosis was supported by nerve growth factor (NGF) reducing the α-syn expression in PD animal models (Li et al. 2021).

Metal nanocomposites targeting dopaminergic neuroregeneration

Dopamine is a hormone present in the brain that acts as a neurotransmitter. Dopamine is synthesized in the substantia nigra pars compacta region of the brain along with the ventral tegmental area and hypothalamus. Due to abnormal regulation, functioning, and level of dopamine various neurodegenerative diseases are propagated. There are two types of dopamine receptors D-1 and D-2 family of dopaminergic receptors. Type-1 and type-5 belong to the D-1 family and type-2, -3, and -4 belong to D-2 family receptors. The receptors belonging to similar families possess similar structures (Latif et al. 2021). Dopamine belongs to the monoamine and catecholamine groups. It functions as both an excitatory and inhibitory neurotransmitter. A low level of dopamine is associated with the onset of PD, and high levels of it give rise to other complications including hypertension, cardiotoxicity, and so on. Therefore it is a crucial biomarker for many diseases (Anshori et al. 2022). Similarly, for the treatment of PD, the levels of dopamine need to be monitored properly in the early stages to alter the disease progression. For that, functionalized multiwalled carbon nanotube/silver nanoparticle (f-MWCNT/AgNP) nanocomposites have been synthesized to detect dopamine in PD patients. Carbon-based materials are frequently used due to cost-effectiveness and biocompatibility. The small size of the nanocomposite makes it suitable for amplifying the signals for electrochemical analysis. They have been found to possess high sensitivity toward dopamine and can be used multiple times (Anshori et al. 2021).

Another biosensor used to detect the level of dopamine is tin disulfide (SnS$_2$) nanorods decorated graphene-β-cyclodextrin (SnS$_2$/GR-β-CD) nanocomposites. This dopamine biosensor was manufactured by employing a glassy carbon electrode (GCE) restructured with SnS$_2$/GR-β-CD nanocomposite using sonochemical and hydrothermal methods. In the presence of ascorbic acid and uric acid which act as interfering species, these biosensors become highly selective toward dopamine. This biosensor has been successfully used in determining dopamine levels in rat brains and human serum samples (Balu et al. 2020).

Therefore, these metal nanocomposites have a wide range of therapeutic applications which have been successfully employed for the management of PD. Modifications in this nanotherapeutics by using various nanotechnologies can aid in the development of a potent therapeutic option for PD which will overcome the limitations of current treatment options for PD.

18.5 APPLICATIONS OF METAL NANOCOMPOSITES FOR PD

18.5.1 Biosensors

Metal nanocomposites can be used as biosensors for the detection of aggregated α-syn and decreased dopamine levels in the PD brain. Early detection of aggregated protein and dopamine levels may aid in the potential treatment of PD. Nanocomposites have paved the way to develop better technologies for the detection of PD biomarkers in the early stages of the disease.

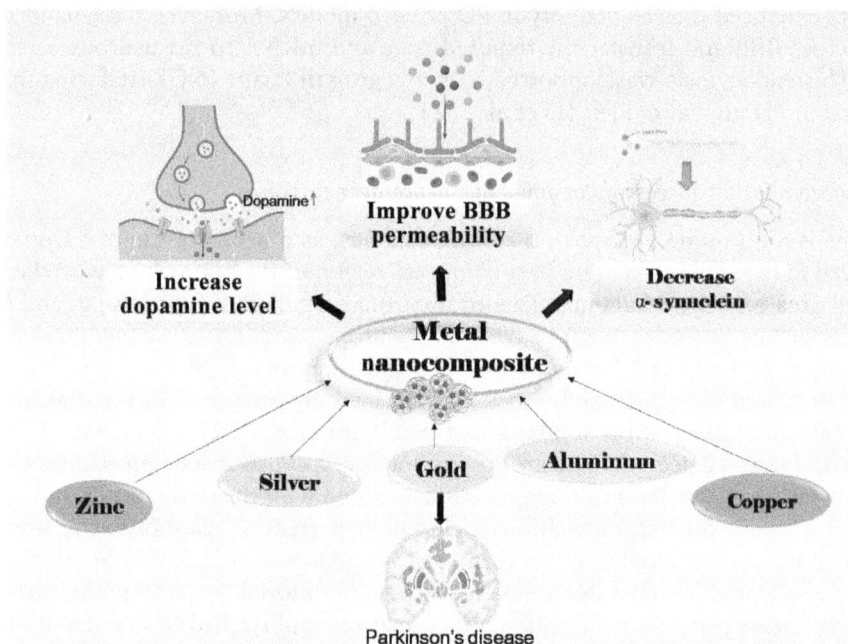

Figure 18.3 This figure shows various metals used in preparing nanocomposites and mechanisms by which metal nanocomposites exhibit their therapeutic effect. Metal nanocomposites act as biosensors for detecting biomarkers of PD, which can be diagnosed early and better treatment can be provided to the PD patients. Further, they can act as carriers to deliver the therapeutic agent across the BBB increasing the bioavailability of the drug and thus contributing indirectly in the treatment of PD.

18.5.2 Nanocarriers

Most of the drugs developed for PD are unable to cross the BBB and reach the targeted site. Therefore, they need to get incorporated with a carrier that can easily overcome this hurdle. Nanocomposites offer the great advantage of delivering the therapeutic agent to the desired location in the brain on account of its properties including nano-size, biocompatibility, and strength.

18.5.3 Nanotherapeutics

Some of the metal nanocomposites themselves act as therapeutic agents to bring down the pathologies associated with PD, including decreased dopamine level, impaired BBB permeability, and Lewy bodies formation, directly or indirectly (Figure 18.3).

18.6 CONCLUSION

PD is a progressive neurodegenerative disease that affects the neurons in specific regions of the brain resulting in neurodegeneration. Restoring the dopaminergic ratio in the striatum and substantia nigra pars compacta is the primary goal of the majority of currently available treatments. There is an urge to develop novel therapies that can overcome the constraints of existing treatments due to the low specificity and bioavailability, lower patient compliance, and many unwanted side effects of current treatments. Moreover, the process of developing drugs is hampered due to the difficulty of overcoming BBB by different substances. Considering these drawbacks nanotherapeutics came into existence which can easily cross the BBB and reach the targeted site increasing the bioavailability of the therapeutic agent.

Recently, a new concept nanocomposite has been developed; it is made up of two or more materials having distinct physical and chemical properties, one being a nanomaterial. Nanocomposite possesses the characteristics of both its components and possesses better therapeutic potential than the components individually. Metal nanocomposite is one of the classes of nanocomposites that comprises metal in their matrix which provide the nanocomposite high strength and better temperature tolerability. The role of metal–organic frameworks (MOF)-based substances has been extensively used in recent years for therapeutic purposes. Ropinirole– silver nanocomposite is one such example that showed better therapeutic potential in increasing the expression of α-syn and dopamine-regulating genes when compared to normal ropinirole. Nanocomposites are also used to deliver therapeutic agents to the specific site. For example, chitosan/carbon dots nanocomposites are used for the delivery of dopamine across the BBB to the specific site which potentiates the sustained release of dopamine in the affected brain area. Another advantage of these metal nanocomposites is that they are more biocompatible. Another target for metal composites for the treatment of PD is BBB permeability. Magnesium oxide-based polydopamine nanocomposites were found to easily cross the BBB and increase neuronal uptake where they show their antioxidant activity. Early detection of aggregated α-syn and reduced dopamine level may promote the prevention of inconsistent accumulation of the proteins and maintain normal dopamine levels. This is exhibited by the use of metal nanocomposites when used as biosensors to detect the biomarkers associated with PD. The aluminum interdigitated electrode is consolidated with zinc oxide as a nanocomposite that is used as a biosensor improving the sensitivity for α-syn antibody detection. Carbidopa, a known drug for PD, is also modified for controlled and extended release and delivery at the specific site. Carbidopa is intercalated with layered zinc hydroxide and chitosan biopolymer. Therefore, these metal nanocomposites are used to improve bioavailability, increasing specificity and potentiating the therapeutic potential of current treatment options for PD.

REFERENCES

Abeliovich, Asa, and Aaron D. Gitler. 2016. "Defects in Trafficking Bridge PD Pathology and Genetics." *Nature* 539 (7628). Nature Publishing Group: 207–216. doi:10.1038/nature20414.

Abyad, Abdulrazak, and Ahmed Sami Hammami. 2009. "An Update on Pathophysiology, Epidemiology, Diagnosis and Management Part 7: Medical Treatment of Early and Advanced PD: Use of Dopamine Agonist." *Middle East Journal of Age and Ageing* 6 (1): 3–6. doi:10.5742/MEJAA.2022.93630.

Adam, Hussaini, Subash C.B. Gopinath, and Uda Hashim. 2021. "Integration of Aluminium Interdigitated Electrodes with Zinc Oxide as Nanocomposite for Selectively Detect Alpha-Synuclein for PD Diagnosis." *Journal of Physics: Conference Series* 2129 (1). IOP Publishing Ltd: 012094. doi:10.1088/1742-6596/2129/1/012094.

Allen Reish, Heather E, and David G Standaert. 2015. "Role of α-Synuclein in Inducing Innate and Adaptive Immunity in Parkinson Disease." *Journal of PD* 5 (1). Netherlands: 1–19. doi:10.3233/JPD-140491.

Anshori, Isa, Komang Arya Attyla Kepakisan, Lavita Nuraviana Rizalputri, Raih Rona Althof, Antonius Eko Nugroho, Rikson Siburian, and Murni Handayani. 2022. "Facile Synthesis of Graphene Oxide/Fe3O4 Nanocomposite for Electrochemical Sensing on Determination of Dopamine." *Nanocomposites* 8 (1). Taylor & Francis: 155–166. doi:10.1080/20550324.2022.2090050.

Anshori, Isa, Lavita Nuraviana Rizalputri, Raih Rona Althof, Steven Sean Surjadi, Suksmandhira Harimurti, Gilang Gumilar, Brian Yuliarto, and Murni Handayani. 2021. "Functionalized Multi-Walled Carbon Nanotube/Silver Nanoparticle (f-MWCNT/AgNP) Nanocomposites as Non-Enzymatic Electrochemical Biosensors for Dopamine Detection." *Nanocomposites* 7 (1). Taylor & Francis: 97–108. doi:10.1080/20550324.2021.1948242.

Armstrong, Melissa J, and Michael S Okun. 2020. "Diagnosis and Treatment of PD A Review." *JAMA* 323 (6): 548–560. doi:10.1001/jama.2019.22360.

Ascherio, Alberto, and Michael A. Schwarzschild. 2016. "The Epidemiology of PD: Risk Factors and Prevention." *The Lancet Neurology* 15 (12). Elsevier: 1257–1272. doi:10.1016/S1474-4422(16)30230-7.

Balestrino, R., and A. H.V. Schapira. 2020. "PD". *European Journal of Neurology* 27 (1). John Wiley & Sons, Ltd: 27–42. doi:10.1111/ENE.14108.

Balu, Sridharan, Selvakumar Palanisamy, Vijaylakshmi Velusamy, Thomas C.K. Yang, and El Said I. El-Shafey. 2020. "Tin Disulfide Nanorod-Graphene-β-Cyclodextrin Nanocomposites for Sensing Dopamine in Rat Brains and Human Blood Serum." *Materials Science and Engineering: C* 108 (March). Elsevier: 110367. doi:10.1016/J.MSEC.2019.110367.

Bharti, Kanika, and Kalyan K. Sadhu. 2022. "Syntheses of Metal Oxide-Gold Nanocomposites for Biological Applications." *Results in Chemistry* 4 (January). Elsevier: 100288. doi:10.1016/J.RECHEM.2022.100288.

Bloem, Bastiaan R., Michael S. Okun, and Christine Klein. 2021. "PD." *Lancet* 397 (10291). Lancet: 2284–2303. doi:10.1016/S0140-6736(21)00218-X.

Borovac, Josip Anđelo. 2016. "Focus: The Aging Brain: Side Effects of a Dopamine Agonist Therapy for PD: A Mini-Review of Clinical Pharmacology." *The Yale Journal of Biology and Medicine* 89 (1). Yale Journal of Biology and Medicine: 37. /pmc/articles/PMC4797835/.

Chen, Jiang Fan, and Rodrigo A. Cunha. 2020. "The Belated US FDA Approval of the Adenosine A2A Receptor Antagonist Istradefylline for Treatment of PD." *Purinergic Signalling* 16 (2). Springer: 167–174. doi:10.1007/S11302-020-09694-2.

Chinchulkar, Shubham Arunrao, Paloma Patra, Dheeraj Dehariya, Aimin Yu, and Aravind Kumar Rengan. 2022. "Polydopamine Nanocomposites and Their Biomedical Applications: A Review." *Polymers for Advanced Technologies* 33 (12). John Wiley & Sons, Ltd: 3935–3956. doi:10.1002/PAT.5863.

Chopade, Puja, Neha Chopade, Zongmin Zhao, Samir Mitragotri, Rick Liao, and Vineeth Chandran Suja. 2023. "Alzheimer's and PD Therapies in the Clinic." *Bioengineering & Translational Medicine* 8 (1). John Wiley & Sons, Ltd: e10367. doi:10.1002/BTM2.10367.

Colamartino, Monica, Massimo Santoro, Guglielmo Duranti, Stefania Sabatini, Roberta Ceci, Antonella Testa, Luca Padua, and Renata Cozzi. 2015. "Evaluation of Levodopa and Carbidopa Antioxidant Activity in Normal Human Lymphocytes In Vitro: Implication for Oxidative Stress in PD." *Neurotoxicity Research* 27 (2). Springer Science and Business Media, LLC: 106–117. doi:10.1007/S12640-014-9495-7.

Collier, Timothy J., Nicholas M. Kanaan, and Jeffrey H. Kordower. 2011. "Ageing as a Primary Risk Factor for PD: Evidence from Studies of Non-Human Primates." *Nature Reviews. Neuroscience* 12 (6). NIH Public Access: 359. doi:10.1038/NRN3039.

Dilnawaz, Fahima, and Sanjeeb Kumar Sahoo. 2015. "Therapeutic Approaches of Magnetic Nanoparticles for the Central Nervous System." *Drug Discovery Today* 20 (10). Elsevier Current Trends: 1256–1264. doi:10.1016/J.DRUDIS.2015.06.008.

Dodson, Mark W., and Ming Guo. 2007. "Pink1, Parkin, DJ-1 and Mitochondrial Dysfunction in PD." *Current Opinion in Neurobiology* 17 (3). Elsevier Current Trends: 331–337. doi:10.1016/J. CONB.2007.04.010.

Emamzadeh, Fatemeh N., and Andrei Surguchov. 2018. "PD: Biomarkers, Treatment, and Risk Factors." *Frontiers in Neuroscience* 12 (AUG). Frontiers Media S.A.: 612. doi:10.3389/ FNINS.2018.00612/BIBTEX.

Fabbri, Margherita, Joaquim J. Ferreira, and Olivier Rascol. 2022. "COMT Inhibitors in the Management of PD." *CNS Drugs* 36 (3). Adis: 261–282. doi:10.1007/S40263-021-00888-9/METRICS.

Finberg, John P.M. 2018. "Inhibitors of MAO-B and COMT: Their Effects on Brain Dopamine Levels and Uses in PD." *Journal of Neural Transmission* 126 (4). Springer: 433–448. doi:10.1007/ S00702-018-1952-7.

Finberg, John P.M., and Jose M. Rabey. 2016. "Inhibitors of MAO-A and MAO-B in Psychiatry and Neurology." *Frontiers in Pharmacology* 7 (OCT). Frontiers Media S.A.: 340. doi:10.3389/ FPHAR.2016.00340/BIBTEX.

Foley, Paul B., Dominic J. Hare, and Kay L. Double. 2022. "A Brief History of Brain Iron Accumulation in PD and Related Disorders." *Journal of Neural Transmission (Vienna, Austria : 1996)* 129 (5–6). Austria: 505–520. doi:10.1007/s00702-022-02505-5.

Gao, Yifei, Yuxue Cheng, Jiapeng Chen, Danmin Lin, Chao Liu, Ling Kun Zhang, Liang Yin, Runcai Yang, and Yan Qing Guan. 2022. "NIR-Assisted MgO-Based Polydopamine Nanoparticles for Targeted Treatment of PD through the Blood—Brain Barrier." *Advanced Healthcare Materials* 11 (23). John Wiley & Sons, Ltd: 2201655. doi:10.1002/ADHM.202201655.

Ghamami, Shahriar, Mojdeh Golzani, and Amir Lashgari. 2016. "New Inorganic-Based Nanohybrids of Layered Zinc Hydroxide/PD Drug and Its Chitosan Biopolymer Nanocarriers with Controlled Release Rate." *Journal of Inclusion Phenomena and Macrocyclic Chemistry* 86 (1–2). Springer Netherlands: 67–78. doi:10.1007/S10847-016-0642-Z.

Grotemeyer, Alexander, Rhonda Leah McFleder, Jingjing Wu, Jörg Wischhusen, and Chi Wang Ip. 2022. "Neuroinflammation in Parkinson's Disease—Putative Pathomechanisms and Targets for Disease-Modification." *Frontiers in Immunology* 13. doi:10.3389/fimmu.2022.878771.

Hare, Dominic J, Bárbara Rita Cardoso, Erika P Raven, Kay L Double, David I Finkelstein, Ewa A. Szymlek-Gay, and Beverley-Ann Biggs. 2017. "Excessive Early-Life Dietary Exposure: A Potential Source of Elevated Brain Iron and a Risk Factor for PD." *Npj PD* 3 (1). Springer US: 1. doi:10.1038/s41531-016-0004-y.

Hare, Dominic J, and Kay L Double. 2016. "Iron and Dopamine: A Toxic Couple." *Brain* 139 (4): 1026–1035. doi:10.1093/brain/aww022.

Hoepken, Hans-Hermann, Suzana Gispert, Blas Morales, Oliver Wingerter, Domenico Del Turco, Alexander Mülsch, Robert L Nussbaum, et al. 2007. "Mitochondrial Dysfunction, Peroxidation Damage and Changes in Glutathione Metabolism in PARK6." *Neurobiology of Disease* 25 (2): 401–411. doi:https://doi.org/10.1016/j.nbd.2006.10.007.

Honary, Soheyla, Soheyla Honary, Eshrat Gharaei-Fathabad, Eshrat Gharaei-Fathabad, Hamed Barabadi, Hamed Barabadi, Farzaneh Naghibi, and Farzaneh Naghibi. 2013. "Fungus-Mediated Synthesis of Gold Nanoparticles: A Novel Biological Approach to Nanoparticle Synthesis." *Journal of Nanoscience and Nanotechnology* 13 (2). American Scientific Publishers: 1427–1430. doi:10.1166/JNN.2013.5989.

Huang, Shu Yi, Zhong Si Wei Dong, Zhao Hui Chen, Zhi Wei Zeng, Wen Qiao Zhao, Yan Qing Guan, and Chu Hua Li. 2022. "Nano-MgO Composites Containing Plasmid DNA to Silence SNCA Gene Displays Neuroprotective Effects in Parkinson's Rats Induced by 6-Hydroxydopamine." *European Journal of Pharmacology* 922 (May). Eur J Pharmacol. doi:10.1016/J.EJPHAR.2022. 174904.

Jagaran, Keelan, and Moganavelli Singh. 2021. "Nanomedicine for Neurodegenerative Disorders: Focus on Alzheimer's and PDs." *International Journal of Molecular Sciences* 22 (16). Multidisciplinary Digital Publishing Institute: 9082. doi:10.3390/IJMS22169082.

Jankovic, J, and E K Tan. 2020. "PD: Etiopathogenesis and Treatment." *J Neurol Neurosurg Psychiatry* 91: 795–808. doi:10.1136/jnnp-2019-322338.

Jasutkar, Hilary Grosso, Stephanie E. Oh, and M. Maral Mouradian. 2022. "Therapeutics in the Pipeline Targeting α-Synuclein for PD." *Pharmacological Reviews* 74 (1). American Society for Pharmacology and Experimental Therapeutics: 207–237. doi:10.1124/PHARMREV.120.000133.

Joshi, Neeraj, and Sarika Singh. 2018. "Updates on Immunity and Inflammation in PD Pathology." *Journal of Neuroscience Research* 96 (3): 379–390. doi:10.1002/jnr.24185.

Kannarkat, George T., Jeremy M. Boss, and Malú G. Tansey. 2013. "The Role of Innate and Adaptive Immunity in Parkinson's Disease." *Journal of Parkinson's Disease* 3 (4): 493–514. doi:10.3233/jpd-130250.

Kim, Jinhwan, Song Ih Ahn, and Yong Tae Kim. 2019. "Nanotherapeutics Engineered to Cross the Blood-Brain Barrier for Advanced Drug Delivery to the Central Nervous System." *Journal of Industrial and Engineering Chemistry* 73 (May). Elsevier: 8–18. doi:10.1016/J.JIEC.2019.01.021.

Kiran, Pallavi, Sujit Kumar Debnath, Suditi Neekhra, Vaishali Pawar, Amreen Khan, Faith Dias, Shubham Pallod, and Rohit Srivastava. 2022. "Designing Nanoformulation for the Nose-to-Brain Delivery in PD: Advancements and Barrier." *Nanomedicine and Nanobiotechnology* 14 (1). John Wiley & Sons, Ltd: e1768. doi:10.1002/WNAN.1768.

Kriebel-Gasparro, Ann. 2016. "PD: Update on Medication Management." *The Journal for Nurse Practitioners* 12 (3). Elsevier: e81–e89. doi:10.1016/J.NURPRA.2015.10.020.

Lamptey, Richard N.L., Bivek Chaulagain, Riddhi Trivedi, Avinash Gothwal, Buddhadev Layek, and Jagdish Singh. 2022. "A Review of the Common Neurodegenerative Disorders: Current Therapeutic Approaches and the Potential Role of Nanotherapeutics." *International Journal of Molecular Sciences* 23 (3). Multidisciplinary Digital Publishing Institute: 1851. doi:10.3390/IJMS23031851.

Latif, Saad, Muhammad Jahangeer, Dure Maknoon Razia, Mehvish Ashiq, Abdul Ghaffar, Muhammad Akram, Aicha El Allam, et al. 2021. "Dopamine in PD." *Clinica Chimica Acta; International Journal of Clinical Chemistry* 522 (November). Clin Chim Acta: 114–126. doi:10.1016/J.CCA.2021.08.009.

Li, Mingchao, Kaikai Hu, Danmin Lin, Zhen Wang, Mingze Xu, Jinpeng Huang, Zhan Chen, et al. 2021. "Synthesis of Double Interfering Biodegradable Nano-MgO Micelle Composites and Their Effect on PD." *ACS Biomaterials Science & Engineering* 7 (3). ACS Biomater Sci Eng: 1216–1229. doi:10.1021/ACSBIOMATERIALS.0C01474.

Lin, Michael T., and M. Flint Beal. 2006. "Mitochondrial Dysfunction and Oxidative Stress in Neurodegenerative Diseases." *Nature* 443 (7113). Nature Publishing Group: 787–795. doi:10.1038/nature05292.

Martin, Ian, Valina L. Dawson, and Ted M. Dawson. 2011. "Recent Advances in the Genetics of PD." *Annual Review of Genomics and Human Genetics* 12 (September). Annual Reviews: 301–325. doi:10.1146/ANNUREV-GENOM-082410-101440.

Mathew, Sheril Ann, P. Praveena, S. Dhanavel, R. Manikandan, S. Senthilkumar, and A. Stephen. 2020. "Luminescent Chitosan/Carbon Dots as an Effective Nano-Drug Carrier for Neurodegenerative Diseases." *RSC Advances* 10 (41). Royal Society of Chemistry: 24386–24396. doi:10.1039/D0RA04599C.

Müller, Thomas, Wilfried Kuhn, and Jan Dominique Möhr. 2019. "Evaluating ADS5102 (Amantadine) for the Treatment of PD Patients with Dyskinesia." *Expert Opinion on Pharmacotherapy* 20 (10). Taylor & Francis: 1181–1187. doi:10.1080/14656566.2019.1612365.

Nair, Arun, Dissertation Director Salvador Barraza-Lopez, Jim Leylek, Paul Millett, and Uche Wejinya. 2019. "Predicting the Mechanical Properties of Nanocomposites Reinforced with 1-D, 2-D and 3-D Nanomaterials Committee Member Committee Member."

Naz, Falaq, Rahul, Mahino Fatima, Swaleha Naseem, Wasi Khan, Amal Chandra Mondal, and Yasir Hasan Siddique. 2020. "Ropinirole Silver Nanocomposite Attenuates Neurodegeneration in the

Transgenic Drosophila Melanogaster Model of PD." *Neuropharmacology* 177 (October). Pergamon: 108216. doi:10.1016/J.NEUROPHARM.2020.108216.

Olivola, Enrica, Alfonso Fasano, Sara Varanese, Francesco Lena, Marco Santilli, Cinzia Femiano, Diego Centonze, and Nicola Modugno. 2019. "Continuous Subcutaneous Apomorphine Infusion in PD: Causes of Discontinuation and Subsequent Treatment Strategies." *Neurological Sciences* 40 (9). Springer-Verlag Italia s.r.l.: 1917–1923. doi:10.1007/S10072-019-03920-5/TABLES/2.

Omanović-Miklicanin, Enisa, Almir Badnjević, Anera Kazlagić, and Muhamed Hajlovac. 2020. "Nanocomposites: A Brief Review." *Health and Technology* 10 (1). Springer: 51–59. doi:10.1007/S12553-019-00380-X.

Pachaiappan, Rekha, Saravanan Rajendran, Pau Loke Show, Kovendhan Manavalan, and Mu Naushad. 2021. "Metal/Metal Oxide Nanocomposites for Bactericidal Effect: A Review." *Chemosphere* 272 (June). Pergamon: 128607. doi:10.1016/J.CHEMOSPHERE.2020.128607.

Pang, Shirley Yin-Yu, Philip Wing-Lok Ho, Hui-Fang Liu, Chi-Ting Leung, Lingfei Li, Eunice Eun Seo Chang, David Boyer Ramsden, and Shu-Leong Ho. 2019. "The Interplay of Aging, Genetics and Environmental Factors in the Pathogenesis of PD." *Translational Neurodegeneration* 8 (1): 23. doi:10.1186/s40035-019-0165-9.

Pilleri, Manuela, and Angelo Antonini. 2015. "Therapeutic Strategies to Prevent and Manage Dyskinesias in PD." *Expert Opinion on Drug Safety* 14 (2). Informa Healthcare: 281–294. doi:10.1517/14740338.2015.988137.

Pisanò, Clarissa A., Alberto Brugnoli, Salvatore Novello, Carla Caccia, Charlotte Keywood, Elsa Melloni, Silvia Vailati, Gloria Padoani, and Michele Morari. 2020. "Safinamide Inhibits in Vivo Glutamate Release in a Rat Model of PD." *Neuropharmacology* 167 (May). Pergamon: 108006. doi:10.1016/J.NEUROPHARM.2020.108006.

Rizos, A., A. Sauerbier, A. Antonini, D. Weintraub, P. Martinez-Martin, B. Kessel, T. Henriksen, et al. 2016. "A European Multicentre Survey of Impulse Control Behaviours in PD Patients Treated with Short- and Long-Acting Dopamine Agonists." *European Journal of Neurology* 23 (8). John Wiley & Sons, Ltd: 1255–1261. doi:10.1111/ENE.13034.

Salamon, András, Dénes Zádori, László Szpisjak, Péter Klivényi, and László Vécsei. 2022. "What Is the Impact of Catechol-O-Methyltransferase (COMT) on PD Treatment?" *Expert Opinion on Pharmacotherapy* 23 (10). Taylor & Francis: 1123–1128. doi:10.1080/14656566.2022.2060738.

Shulman, Joshua M., Philip L. De Jager, and Mel B. Feany. 2011. "PD: Genetics and Pathogenesis." *Annual Review of Pathology: Mechanisms of Disease* 6 (January). Annual Reviews: 193–222. doi:10.1146/ANNUREV-PATHOL-011110-130242.

Srivastava, Rajnish, Pratim Kumar Choudhury, Suresh Kumar Dev, and Vaibhav Rathore. 2020. "Preliminary Physicochemical Screening of Bioactive Based Nanocarrier System." *Journal of Drug Delivery and Therapeutics* 10 (1): 184–198. doi:10.22270/JDDT.V10I1.4378.

Tambasco, Nicola, Michele Romoli, and Paolo Calabresi. 2017. "Levodopa in PD: Current Status and Future Developments." *Current Neuropharmacology* 16 (8). Bentham Science Publishers: 1239–1252. doi:10.2174/1570159X15666170510143821.

Tansey, Malú Gámez, Rebecca L. Wallings, Madelyn C. Houser, Mary K. Herrick, Cody E. Keating, and Valerie Joers. 2022. "Inflammation and Immune Dysfunction in PD." *Nature Reviews Immunology* 22 (11). Springer US: 657–673. doi:10.1038/s41577-022-00684-6.

Teleanu, Daniel Mihai, Irina Negut, Valentina Grumezescu, Alexandru Mihai Grumezescu, and Raluca Ioana Teleanu. 2019. "Nanomaterials for Drug Delivery to the Central Nervous System." *Nanomaterials* 9 (3). Nanomaterials (Basel). doi:10.3390/NANO9030371.

Thomas, Bobby, and M Flint Beal. 2007. "PD." *Human Molecular Genetics* 16 (R2): R183–R194. doi:10.1093/hmg/ddm159.

Yana, Michael H., Xinglong Wang, and Xiongwei Zhu. 2013. "Mitochondrial Defects and Oxidative Stress in Alzheimer Disease and PD." *Free Radical Biology and Medicine* 62 (September). Pergamon: 90–101. doi:10.1016/J.FREERADBIOMED.2012.11.014.

Yoosefian, Mehdi, Elham Rahmanifar, and Nazanin Etminan. 2018. "Nanocarrier for Levodopa Parkinson Therapeutic Drug; Comprehensive Benserazide Analysis." *Artificial Cells, Nanomedicine,
and Biotechnology* 46 (suppl 1). Taylor & Francis: 434–446. doi:10.1080/21691401.2018.14
30583.

Zafar, Faria, Ruksana Azhu Valappil, Sam Kim, Krisztina K. Johansen, Anne Lynn S. Chang, James
W. Tetrud, Peggy S. Eis, et al. 2018. "Genetic Fine-Mapping of the Iowan SNCA Gene Triplication in a Patient with PD." *Npj PD* 4 (1). Nature Publishing Group: 1–7. doi:10.1038/
s41531-018-0054-4.

Index

For Product Safety Concerns and Information please contact our EU
representative GPSR@taylorandfrancis.com
Taylor & Francis Verlag GmbH, Kaufingerstraße 24, 80331 München, Germany

www.ingramcontent.com/pod-product-compliance
Lightning Source LLC
Chambersburg PA
CBHW080137220326
41598CB00032B/5091